Soil Health and Intensification of Agroecosystems

Soil Health and Intensification of Agroecosystems

Edited by

Mahdi M. Al-Kaisi

Iowa State University, Ames, IA, United States

Birl Lowery

University of Wisconsin-Madison, Madison, WI, United States

Academic Press is an imprint of Elsevier
125 London Wall, London EC2Y 5AS, United Kingdom
525 B Street, Suite 1800, San Diego, CA 92101-4495, United States
50 Hampshire Street, 5th Floor, Cambridge, MA 02139, United States
The Boulevard, Langford Lane, Kidlington, Oxford OX5 1GB, United Kingdom

British Library Cataloguing-in-Publication Data
A catalogue record for this book is available from the British Library

Library of Congress Cataloging-in-Publication Data
A catalog record for this book is available from the Library of Congress

ISBN: 978-0-12-805317-1

For Information on all Academic Press publications visit our website at https://www.elsevier.com/books-and-journals

Publisher: Nikki Levy
Acquisition Editor: Nancy Maragioglio
Editorial Project Manager: Billie Jean Fernandez
Production Project Manager: Nicky Carter
Designer: Victoria Pearson

Typeset by MPS Limited, Chennai, India

Contents

List of Contributors

Abdullah Alhameid South Dakota State University, Brookings, SD, United States
Mahdi M. Al-Kaisi Iowa State University, Ames, IA, United States
Francisco J. Arriaga University of Wisconsin-Madison, Madison, WI, United States
Kipling S. Balkcom USDA-ARS National Soil Dynamics Laboratory, Auburn, AL, United States
Marisol Berti North Dakota State University, Fargo, ND, United States
Caroline Colnenne-David INRA Research Center at Versailles, Versailles, France
Christian Dold National Laboratory for Agriculture and the Environment, Ames, IA, United States
Mathew E. Dornbush University of Wisconsin-Green Bay, Green Bay, WI, United States
Yucheng Feng Auburn University, Auburn, AL, United States
Clark J. Gantzer University of Missouri, Columbia, MO, United States
Mohammad H. Golabi University of Guam, Mangilao, Guam
Greta Gramig North Dakota State University, Fargo, ND, United States
Jose Guzman The Ohio State University, Columbus, OH, United States
Jerry L. Hatfield National Laboratory for Agriculture and the Environment, Ames, IA, United States
John Hendrickson USDA-Agricultural Research Service, Mandan, ND, United States
Randall D. Jackson University of Wisconsin-Madison, Madison, WI, United States
Claudia Pozzi Jantalia Embrapa Agrobiology, Rio de Janeiro, Brazil
Virginia L. Jin USDA-ARS, Lincoln, NE, United States
Jane M.-F. Johnson USDA-ARS, Morris, MN, United States
Shibu Jose University of Missouri, Columbia, MO, United States
Robert J. Kremer University of Missouri, Columbia, MO, United States
Sandeep Kumar South Dakota State University, Brookings, SD, United States
Rattan Lal The Ohio State University, Columbus, OH, United States
Yvonne Lawley University of Manitoba, Winnipeg, MB, Canada
Mark Liebig United States Department of Agriculture, Agriculture Research Service, Northern Great Plains Research Laboratory, Mandan, ND, United States
Birl Lowery University of Wisconsin-Madison, Madison, WI, United States
Amadou Maiga South Dakota State University, Brookings, SD, United States; University of Sciences, Technics and Technologies of Bamako, Mali
Kenneth R. Olson University of Illinois, Urbana, IL, United States
Shannon Osborne USDA-ARS, Brookings, SD, United States
Matt Sanderson USDA-Agricultural Research Service, Mandan, ND, United States
Thomas Schumacher South Dakota State University, Brookings, SD, United States
Bobby A. Stewart West Texas A&M University, Canyon, TX, United States

Catherine E. Stewart USDA-ARS, Fort Collins, CO, United States
Colin Tobin South Dakota State University, Brookings, SD, United States
Ranjith P. Udawatta University of Missouri, Columbia, MO, United States
Adam C. von Haden University of Wisconsin-Madison, Madison, WI, United States
Sharon L. Weyers USDA - ARS, North Central Soil Conservation Research Lab, Morris, MN, United States
Abbey Wick North Dakota State University, Fargo, ND, United States
Zhengqin Xiong Nanjing Agricultural University, Nanjing, China

Preface

There is considerable awareness these days about how people worldwide are concerned about their personal health. It is equally important that the world population should have even greater concern about soil health as the soil is the medium where most of our food, and fiber, is derived/produced. Human health is directly related to the food we eat and water we drink. Not only is soil health related to food safety, but it is also directly related to food production and security, and water quality in some settings. Good or healthy soil results in much greater and sustainable food production than degraded soil. Thus, soil health is key to future human health and a sufficient food supply.

The terms soil health and soil quality are closely linked and often used interchangeably as a reference or benchmark for the functionality of soil systems. Attempts to define soil health or quality by scientists all focus on the same fundamental building units of what defines a well-functioning soil ecosystem (good biological, physical, and chemical properties). There have been numerous definitions proposed for soil health by various scientists, but Doran et al. (1996, 1999) have been credited with providing the most widely cited one. Accordingly, one of the most recent definitions associate with Doran is: "Soil quality or health can be broadly defined as the capacity of a living soil to function, within natural or managed ecosystem boundaries, to sustain plant and animal productivity, maintain or enhance water and air quality, and promote plant and animal health" (Doran et al., 1999). The definition by Doran et al. (1999) characterizes the soil as (1) a medium that supports and promotes the growth and development of plants, animals, and humans, while regulating water processes in the ecosystem, (2) an environmental buffer that regulates and degrades hazardous compounds in the ecosystem, and (3) a medium that provides food and fiber services for sustaining animal and human lives. While it would be good if soils were all highly productive, and stagnant with respect to this abundant production, and as such in good health, this is not the case. Soils are very dynamic, ever-changing both chemically, physically, and biologically. Thus, soil health is in a state of constant flux since these characteristics impact soil health. Soil health is of concern for both managed and nonmanaged ecosystems. However, of most concern is the managed agroecosystem. When soils are managed in a manner such that soil erosion and other degrading causes and practices are reduced, or if possible eliminated, good soil health is maintained. In addition to human influence on soil health, the weather, including climate change, plays an important role in affecting soil functions.

Climate change is a threat to Earth as we know it, and to human existence. The reason being that it is one of the greatest threats of the modern era to soil, the most fundamental of all natural resources on Earth. Soil is the beginning and end all for agricultural production and food security. The threat of climate change to soil includes soil health and sustainability, because as the climate warms biological activity will increase and this will result in a reduction in soil organic carbon, which is key to good soil health. The threat of climate change to soil heath can be accelerated by the degree of agriculture intensification. Intensified agriculture evokes more of a process, rather than an explicit method of production. Over millennia, a variety of technological advances in agricultural practices have led to intensification of agriculture. In this sense, intensification of agriculture can be defined as "increasing productivity on a set area of land." This definition distinguishes intensification from extensification, which can be defined as "increasing productivity by increasing land area under production." In both cases, these approaches present detrimental effects to soil health and sustainability. However, with agricultural intensification, genetic and chemical advances in agricultural technologies have led to both stabilization and destabilization of the biological, physical, and chemical nature of soils. The required system inputs and management practices have had a global devastating impact on soil resources, where soil erosion, soil organic matter loss, and decline in soil biodiversity to name a few, are endemic in modern agriculture systems. The link between agroecosystem intensification and soil health is magnified by management practices that led to desertification, deforestation, erosion, and other forms of soil degradation. These dynamics, along with weather variability, such as frequent wet and drought events, are prevalent in different parts of the world and are expected to increase with climate change.

The decline in soil productivity as a result of soil erosion and other forms of degradation is manifested in the deterioration of soil health/quality or functionality, where soil chemical, physical, and biological properties are severely degraded. These soil characteristics are the foundation for a productive soil and its ecosystem services. These soil functions are critical to food and fiber production, including nutrient provision and cycling, protection against pests and pathogens, production of growth factors, water availability, and the formation of stable soil physical structure capable of reducing the potential risks of soil erosion and increasing water processing. These functions are strongly affected by climate variability and extreme weather conditions. Therefore, without stable agricultural conservation systems that encompass practices that mitigate extreme climatic conditions, these soil health functions can be degraded. The adoption of such conservation practices within production fields and on marginal lands can provide solutions to combat natural and anthropogenic management effects on soil resiliency. The task through this book is to identify and present management practices, systems, and alternatives within the confines of intensified agricultural systems to transform such systems into sustainable intensified systems that can provide food, fiber, and animal feed to meet the challenge of increased human population, yet preserving soil resources and ecosystem services.

An attempt has been made in this book to provide an overview of basic or fundamental soil properties and relationships, and climate impacts to complex and wide-ranging and contrasting management systems that will influence soil health under dryland to humid environments. An effort has been made to cover all potential soil and crop management practices for maintaining good soil health under many different environmental conditions including a global prospective where possible. This includes different cropping systems, cover crops, perennial cover crops, livestock integration with cropping, managing intensified agroecosystems, agroforestry, low input systems to intensification, nutrient cycling, and biotechnology use in modern agriculture production role in affecting soil health.

The hope is that this book will contribute to and provide insight to the current dialogue about the importance of soil health and sustainability, by building on the accomplishments and contributions of countless numbers of scientists regarding the concept of soil health/quality during the past few decades.

The editors express their sincere thanks and appreciation to all chapters' authors and the publisher for their excellent cooperation and contributions to this book. As with any scientific endeavor, the extension of knowledge is based on a body of scholarly and discovery work by past and present scientists that we feel indebted to for their contribution.

Mahdi M. Al-Kaisi and
Birl Lowery

References

Doran, J.W., Sarrantonio, M., Liebig, M., 1996. Soil health and sustainability. Adv. Agron. 56, 1–54.
Doran, J.W., Jones, A.J., Arshad, M.A., Gilley, J.E., 1999. Determinants of soil quality and health. In: Lal, R. (Ed.), Soil Quality and Soil Erosion. CRC Press, pp. 17–36.

CHAPTER 1

Fundamentals and Functions of Soil Environment

Mahdi M. Al-Kaisi[1], Rattan Lal[2], Kenneth R. Olson[3] and Birl Lowery[4]

[1]Iowa State University, Ames, IA, United States [2]The Ohio State University, Columbus, OH, United States [3]University of Illinois, Urbana, IL, United States [4]University of Wisconsin-Madison, Madison, WI, United States

1.1 Introduction

The soil system is complex and dynamic. The definition of soil varies widely, as it is dictated by its use and how we perceive soil as a society for providing services, food, habitat, and enjoyment, where these functions are essential to soil health or quality. One well-established definition of soil is a medium that includes minerals, organic matter, countless organisms, liquid, and gases that together support life on earth through many services. Soil is the foundation for early and modern agriculture, and for human civilization. Most people think of farming or gardening when they think of soils (Brevik, 2005). However, the definition of soil depends on the multiple uses of this medium for different purposes such as farming, engineering, and environment. To a farmer, soil is a medium to produce food, which differs from that of a geologist, who considers soil a natural medium and unconsolidated materials above bedrock. An engineer defines soil as a naturally occurring surface layer formed by complex biochemical and physical weathering processes that contains living matter. Soil is considered capable of supporting plant, animal, and human life by agronomists and pedologists (Brevik, 2005). Soil environment and functions are influenced by the parent materials and forming factors that contribute to the physical, chemical, and biological characteristics of soils. The inorganic fractions of mineral soils generally consist of sand, silt, and clay. The proportion of these different fractions determines soil texture, along with its subsequent chemical, physical, and biological properties. Soil formation progresses in steps and stages that are not distinctly separated. These processes are overlapping, and it may not be possible to know when one stage in soil formation stops and another starts (Huggett, 1998). Soil characteristic depends primarily on the parent materials, and secondarily on the vegetation, the topography, and time. These are the five variables known as the *factors of soil formation* (Jenny, 1941). The typical development of a soil and its profile

Soil Health and Intensification of Agroecosystems.
DOI: http://dx.doi.org/10.1016/B978-0-12-805317-1.00001-4

is called *pedogenesis*, which includes physical and chemical processes and disintegration of the exposed rock formation as the soil's parent material (Hillel, 1998). These loosened materials are colonized by living organisms (plant and animal, micro- and macroorganisms). This process leads to accumulation of soil organic matter (SOM) at and below the soil surface resulting in the formation of an A horizon. Important aspects of soil formation and development include two processes of *eluviation* (washing out) and *illuviation* (washing in), where clay particles and other substances, including calcium carbonate, emigrate from the overlay surface, eluvial A horizon, and accumulate in the underlying illuvial B horizon (Jenny, 1941). The formation of the soil profile and its physical, chemical, and biological characteristics through these processes differ from location to location and region to region. In arid regions, for example, salt movement from upper to lower horizons may create physical, chemical, and biological conditions that are different from those in humid areas and the tropical, where there is more of a tendency for leaching of minerals and chemicals through the soil profile because the driving force for this reaction being water, is greater in these environments. Therefore, different soil properties, such as color or SOM concentration, occur in the top soil layer and at subsequent depths of the soil profile (Weil and Brady, 2016). These processes influence soil fertility, water availability, and SOM content, which limit the choice of type of crops and management practices that are essential for sustaining soil health and productivity. Therefore, the level of soil health is different for different soil types.

1.2 Soil Properties and Interrelationships

1.2.1 Soil Physical Environment

The soil physical environment is generally characterized by three distinct phases that include the solid phase that forms the soil matrix, the liquid phase comprised of water in the soil system, called the soil solution, and the gaseous phase or the soil atmosphere. The soil matrix (mineral component) consists of soil particles varying in size, shape, and chemical properties (Fig. 1.1). The formation of the soil matrix through the grouping of different particles with amorphous substances, particularly SOM, when attached to the surface of different mineral particles,

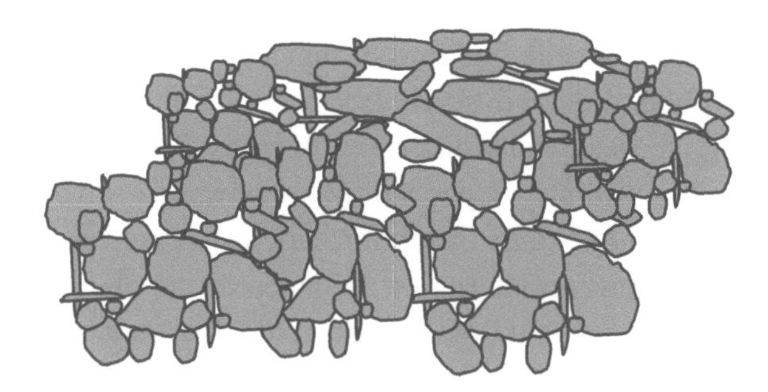

Figure 1.1
Schematic representation of pore spaces between soil aggregates.

form soil aggregates that are essential components of soil health or quality. The formation of soil aggregates determines the soil structure and geometric characteristics of pore spaces in which water and air retention and movement occur (Tisdall and Oades, 1982). The water and air proportions vary in space and time, and the increase in one portion leads to a decrease in another (Fig. 1.2). The relative proportions of the three phases in a soil are not fixed, but are rather dynamic, changing continuously depending on variables such as weather, vegetation, and management by humans. Tillage and cropping systems can significantly impact soil aggregate formation and stability. Generally, soil aggregate formation is highly influenced by plant roots and fungal hyphae as major binding agents for macroaggregates (>0.25 mm), while organic compounds are responsible for the formation of microaggregates (<0.25 mm) (Tisdall and Oades, 1982). Soil structure can influence its environment by providing conditions that impact plant growth such as water availability, nutrient dynamics, and soil tilth (Oades, 1984). One of the quantitative measures to evaluate the soil physical environment is bulk density. This soil property is shaped by soil texture and influenced by management practices through changes in soil structure and porosity. Bulk density is often used as a soil health or quality indicator. Bulk density is defined as mass per volume ($kg\,m^{-3}$ or $Mg\,m^{-3}$) as described by the following equation:

$$\rho_b = \frac{M_s}{V_t} \tag{1.1}$$

where ρ_b is soil bulk density ($kg\,m^{-3}$ or $Mg\,m^{-3}$), M_s is soil solid mass (kg or Mg), and V_t is soil total volume (solids and pores) (m^3). Soil bulk density should be measured for each soil depth separately. It generally depends on soil texture and is affected by management

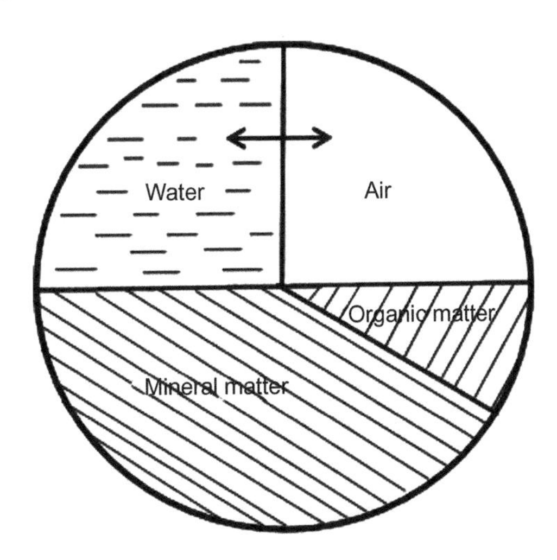

Figure 1.2
Schematic representation by volume of different soil components at optimal condition for plant growth. Total solid matter components (mineral and organic matter) make up 50% and the pore space 50% of the total soil volume, which is divided equally between water and air. The water and air volumes are exchangeable as indicated by the arrows, depending on soil moisture conditions.

practices that include tillage, field equipment and travel pattern, and crop rotation. It should be calculated on oven dry weight for comparison purposes but can be on a moist basis.

1.2.2 Components of the Total Soil–Water Potential

The force that governs water movement in soil is called the soil–water potential, and is of a great fundamental importance to soil and soil health considerations as water availability is key to plant growth. "The water potential concept replaces the arbitrary categorization that prevailed in the early stages of the development of soil physics and that purported to recognize and classify different forms of soil water, such as, gravitational water, capillary water, hygroscopic water" (Hillel, 1998). Soil water is subjected to a number of possible forces that influence water movement, or supply to plants. This force is a result of the interaction between the soil matrix and water, the presence of solutes in the soil solution, action of external gas pressure, and gravity. The sum of these forces forms the total potential, or soil–water potential (Hillel, 1998; Rose, 1966):

$$\Psi_t = \Psi_g + \Psi_p + \Psi_o + \ldots \quad (1.2)$$

Where, Ψ_t is the total potential, Ψ_g the gravitational potential (positive or negative pressure depending on the reference point), and Ψ_p the pressure potential (or matric, as this can be a positive or negative pressure), Ψ_o the osmotic potential (negative pressure). The different components in Eq. (1.2) may not act in the same way, and their separate gradients may not always be equally effective in causing the flow (Hillel, 1998). The pressure at which water is retained in soil is strongly related to soil porosity and pore size distribution and these are keys to soil quality.

1.2.3 Water Movement and Governing Forces

Fluid flow in a complex porous medium such as soil is governed by physical and chemical forces that are incorporated in the concept of total potential. Water flow through soil pores is influenced by the physical formation of their irregular shape, which is far different than if the soil pores are regularly shaped as a bundle of straight tubes (Marshall, 1958; Miller and Low, 1963; Klute et al., 1965). The direction of water movement in soil is dictated by total soil pore water pressure differences, and this difference might be related to antecedent soil moisture conditions. The total potential of water that governs the water flow can be expressed as energy per unit mass or volume. The rate of water flux in the soil is a product of the hydraulic gradient which is the rate of change of the driving force (pressure difference) over a selected distance. This water force was defined by Henry Darcy in 1856:

$$q = -K\,\mathrm{d}H/\mathrm{dx} \quad (1.3)$$

where q is the flux or the volume of water flow through a cross-section area of soil per unit time, K is the hydraulic conductivity of saturated soil, H is the total hydraulic head (positive pressure), and x is the distance of water flow in the direction of flow. A portion of the soil volume is occupied by soil particles, and the water flow is only through the pore spaces (macro and micro sizes). Hydraulic conductivity is a property of soils and rocks that describes the ease with which a fluid (usually water) can move through pore spaces or fractures. It depends on the intrinsic permeability of the material, the degree of saturation, and on the density and viscosity of the fluid. Under saturated soil conditions it is called "hydraulic conductivity in saturated soil, K_{sat}," which describes water movement through saturated porous medium. Soil texture and type of clay (Smiles and Rosenthal, 1968) have significant effects on hydraulic conductivity. In addition, the size of pores (which depends on different portions of sand, silt, and clay, and organic matter content and soil structure) can influence water movement. Sandy soils and soils with good structure tend to have large pore spaces and conduct water easily (Hillel, 1980). In addition, the rate of water movement in soils increases with an increase in the driving force, or potential/hydraulic gradient. However, the hydraulic conductivity of saturated soil does not change with increasing or decreasing this driving force, it is a constant for a given soil. Under saturated conditions, the driving force is the difference in elevation and positive external pore water pressures in the soil (Fig. 1.3). On the other hand, under unsaturated conditions, the dominant driving force for water movement is the attraction of soil matrix to water molecules, which is much greater than that in saturated soils. Also, in unsaturated soil the hydraulic conductivity and water flux are a function of the

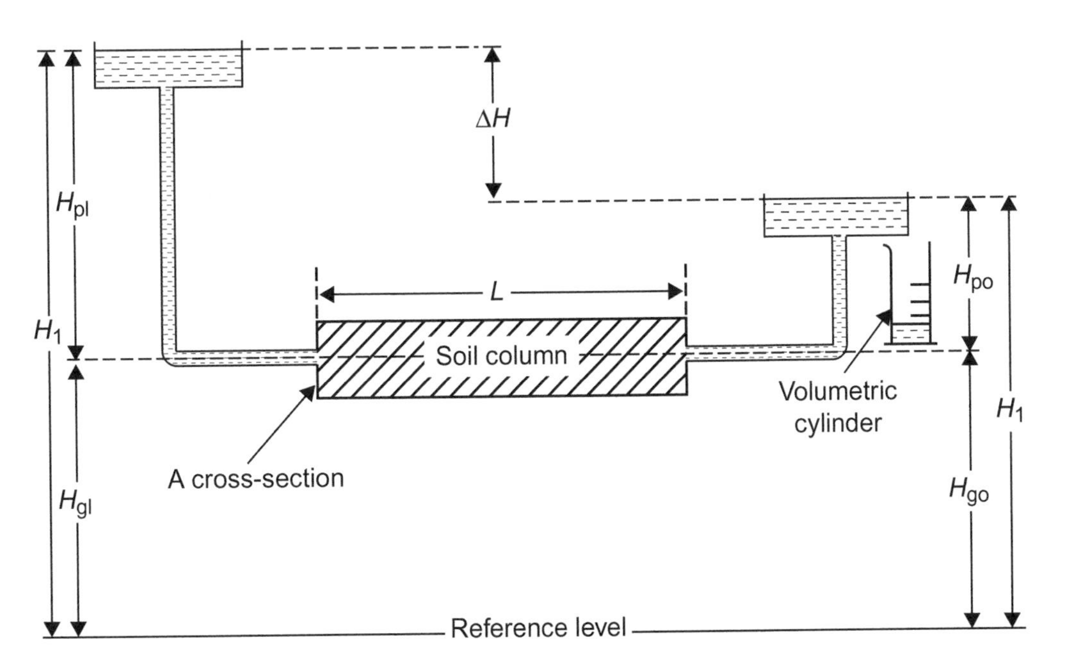

Figure 1.3
Flow in a horizontal saturated column. *After Hillel, D., 1998. Environmental Soil Physics. Academic Press, New York.*

soil water content. That is the flux is greater when the soil water content is close to saturation and it decreases as soil water content decreases. The driving force in both saturated (gravity) and unsaturated (matric potential as affected by adhesion and cohesion forces) flows is influenced by soil properties and management practices (Gardner and Hsieh, 1959; Kutilek and Nielsen, 1994).

1.2.4 Soil Structure and Water Pathways

Water flow through soil is influenced by its structure, where connecting pores are the natural pathway for water and air exchange. Water movement (conductivity) through soil is not only influenced by the total porosity, but also primarily by the size of the pores, and the relative proportion of sand, silt, and clay dictate such property (Gerke and Van Genuchten, 1993; Hillel, 1980). The formation of soil structure is primarily influenced by soil texture (clay content), presence of divalent cations, and SOM (Cambardella, 2002). Also, water movement in soil is affected by pore geometry, where pore size distribution and internal surface area can determine the correlation between permeability and total porosity (Jacob, 1946; Franzini, 1951). Generally, most soil reactions and processes involve interaction of soil and water under unsaturated conditions. These processes may include water and nutrient uptake by plant roots, chemical reactions, and biological activities (Kutilek and Nielsen, 1994). However, water flow, especially under unsaturated conditions, occurs either as a film along the walls of wide pores, or as flow through narrow/small, water-filled pores. These conductive properties in unsaturated soils depend largely on texture and structure (Hamblin, 1985) and forces of adhesion and cohesion. Water movement through different pore sizes as influenced by soil structure (aggregate size) can be classified in three categories: *micropores*, *capillary pores*, and *macropores*.

> Micropores—water moves through pores that are less than a micrometer in diameter and occur typically in clay soils, and the water held in the pores is subject to adsorptive forces which may differ from water present in wider pores.
>
> Capillary pores—are generally found in medium-textured soil, and their width ranges from several micrometers to a few millimeters. Water flow through these pores follows the capillary and Darcy's Law. Water flow through these pores is laminar.
>
> Macropores—these are generally visible to the naked eye and they range in width from several millimeters to centimeters. They appear as cracks or voids, especially in clay soils, or as a result of biological activity such as earth worms or other burrowing animals. They permit fast water flow when filled, but create barriers under dry or unsaturated conditions for capillary water movement.

Conductive soils contain large and continuous pores, which constitute the majority of the overall pore volume under saturated flow conditions. On the other hand, less conductive soils contain a pore volume made up largely of numerous micropores. Therefore, sandy soils, or well-aggregated soils, conduct and discharge water much faster than clay soils (Saxton et al., 1986). The opposite is true in unsaturated conditions, where suction (negative pressure) becomes the dominant force for water transport through capillary movement as greater tension develops with clay soils through water movement compared to sandy soils (Gardner and Hsieh, 1959).

1.2.5 Soil Temperature

Soil temperature is an important property that is essential for many soil processes and reactions that may include, but are not limited to, water and nutrient uptakes, microbial activities, nutrient cycling, root growth, and many other processes (Doran and Smith, 1987). Soil temperature properties change by the radiant, thermal, and latent energy exchange processes that take place at the soil surface. Components of soil thermal properties, such as specific heat capacity, thermal conductivity, and thermal diffusivity, are affected by basic soil properties that include bulk density, texture, and water content (Table 1.1 and Fig. 1.4). There is a strong dependence of thermal conductivity and diffusivity on soil wetness and other soil properties (van Bavel and Hillel, 1976). The flow of water and heat is an interactive process, where temperature gradients affect the moisture potential and both liquid and vapor movement in soil (McInnes, 2002). Heat flow in soil can be described by the following equation:

$$Q = K_t A \frac{dT}{dX} \tag{1.4}$$

where Q is heat flux per unit area, K_t is the soil thermal conductivity of soil ($W\,m^{-3}\,K^{-1}$), T is the soil temperature (K), A is the surface area (m^2), and X is the soil distance (m).

Table 1.1: Average thermal properties of soils and snow

Soil Type	Porosity	Volumetric Wetness ($cm^3\,cm^{-3}$)	Thermal Conductivity ($10^{-3}\,cal\,cm^{-1}\,s^{-1}\,deg^{-1}$)	Volumetric Heat Capacity ($cal\,cm^{-1}\,s^{-1}\,deg^{-1}$)	Damping Depth (Diurnal) (cm)
Sand	0.4	0.0	0.7	0.3	8.0
	0.4	0.2	4.2	0.5	15.2
	0.4	0.4	5.2	0.7	14.3
Clay	0.4	0.0	0.6	0.3	7.4
	0.4	0.2	2.8	0.5	12.4
	0.4	0.4	3.8	0.7	12.2
Peat	0.8	0.0	0.14	0.35	3.3
	0.8	0.4	0.7	0.75	5.1
	0.8	0.8	1.2	1.15	5.4
Snow	0.95	0.05	0.15	0.05	9.1
	0.8	0.2	0.32	0.2	6.6
	0.5	0.5	1.7	0.5	9.7

Source: After van Wijk, W.R., de Vries, D.A., 1963. Periodic temperature variations in homogeneous soil. In: van Wijk, W.R. (Ed.), Physics of Plant Environment. North-Holland, Amsterdam (van Wijk and de Vries, 1963).

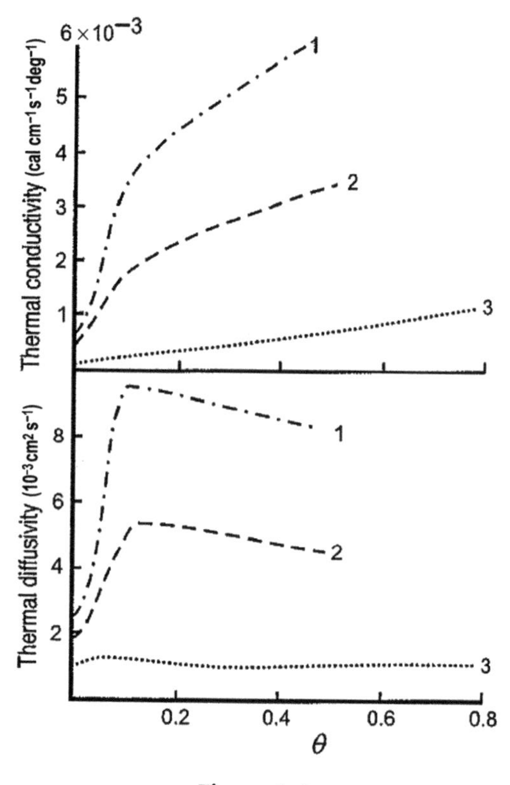

Figure 1.4

Thermal conductivity and diffusivity as functions of volume wetness (volume fraction of water) for (1) sand (bulk density 1460 $kg\,m^{-3}$, volume fractions of solids 0.55); (2) loam (bulk density 1330 $kg\,m^{-3}$, volume fractions of solids 0.50); (3) peat (volume fraction of solids 0.20). *After Hillel, D., 1998. Environmental Soil Physics. Academic Press, New York.*

Soil temperature varies continuously in response to climate and meteorological changes and the interaction of soil and atmosphere. Some of the factors that affect soil temperature include diurnal and annual cycles, and irregular episodic changes in weather (i.e., cloudiness, drought, wet, warm, rainfall, and cold events). Also, landscape formation, regional differences, vegetation, and soil management practices by humans are some other factors. The other dimension of soil temperature variation is within soil depths, where the soil temperature shifts in peaks as it travels deeper in the soil profile (Fig. 1.5). The cause of temperature damping with depth is that a certain amount of heat is absorbed or released along the path when the temperature of the conducting soil materials changes. Those changes in soil temperature influence many soil activities with increases in depth of the soil profile, such as microbial activities (McInnes, 2002), chemical reaction, nutrient cycling, and many other processes. Partitioning of energy takes place at the soil surface, where different energy transformations and pathways are created (de Vries, 1975). Therefore, any changes or modifications at the soil surface through human interference, such as drainage, tillage, and vegetation covers,

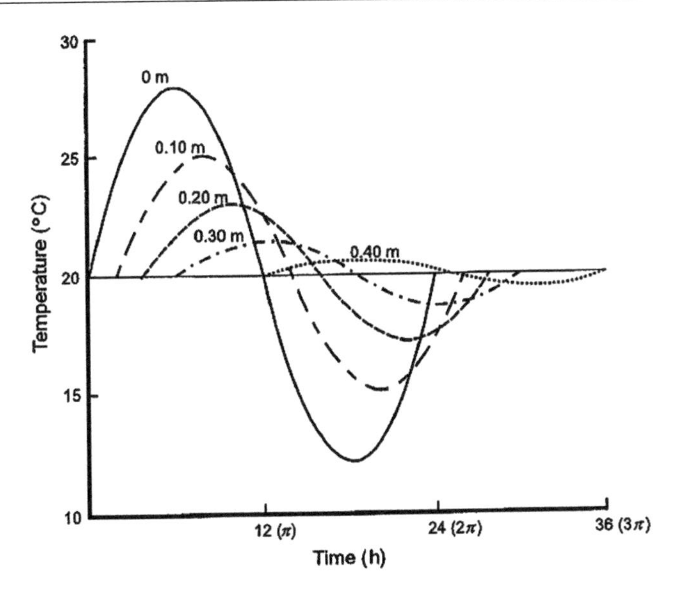

Figure 1.5

Idealized variation of soil temperature with time for various depths. Note that at each succeeding depth the peak temperature is damped and shifted progressively in time. Thus, the peak at a depth of 0.4 m lags about 12 h behind the temperature peak at the surface and is only about 1/16th of the latter. In this hypothetical case, a uniform soil assumed, with a thermal conductivity of $1.68\,J\,m^{-1}\,s^{-1}\,deg^{-1}$ (or $4 \times 10^{-3}\,cal\,cm^{-1}\,s^{-1}\,deg^{-1}$) and a volumetric heat capacity of $2.2 \times 10^{6}\,J\,m^{-3}\,deg^{-1}$ ($0.5\,cal\,cm^{-3}\,deg^{-1}$). *After Hillel, D., 1998. Environmental Soil Physics. Academic Press, New York.*

change the energy balance and have strong effects on soil temperature (Licht and Al-Kaisi, 2005) at the soil surface. Tillage affects soil temperature in two ways, by creating temporary macroporosity and by changing soil thermal properties, especially the albedo at the soil surface (van Duin, 1956).

1.2.6 Soil Aggregate Formation and Structure

In order to understand soil functions and associated ecosystem services provided, it is pertinent to consider the physical orientation or arrangement of different soil particles and binding agents to form a multidimensional framework called *soil matrix*. The arrangement or organization of this soil matrix is called the soil structure (Dexter, 1988). Soil particles are different in shape, size, and orientation, and the mass of such structure becomes very complex and irregular in patterns, making it difficult, or even impossible, to characterize the exact geometric attributes (Tisdall and Oades, 1982; Borie et al., 2008). In addition, the inherent nature of instability of soil structure is influenced by many external and internal processes which add another layer of complication to the inconstancy of time and space.

Soil structure is strongly affected by natural forces, such as climate, biological activities, and soil management practices (Lal, 1993; Troeh et al., 1999; Borie et al., 2008). Given these limitations associated with soil structure, this soil physical property is a qualitative indicator rather than a quantitative attribute. Therefore, the best measure, or close to quantitative aspects of soil structure, is the measure of total porosity and the shape of the pores in the soil and array of their sizes and distribution (Hillel, 1998). The building blocks of soil structure are *soil aggregates*, which involve formation of secondary particles with appreciable contents of clay and SOM, to form structural units (Oades and Waters, 1991). The stability of these units, or aggregates, is affected by management practices, such as tillage system, crop rotation, type of crops, and intensity and duration of vegetative cover (Horne and Sojka, 2002; Lal, 1993). The size of these aggregates varies and ranges in order of micrometers and millimeters. The large aggregates are called macroaggregates or peds, and smaller aggregates are called microaggregates (Duiker, 2002; Tisdall and Oades, 1982; Oades and Waters 1991). The hierarchy theory of aggregate formation states that aggregate formation is highly influenced by roots and fungal hyphae as major binding agents for *macroaggregates* (>0.25 mm), while organic compounds are responsible for the formation of *microaggregates* (<0.25 mm) (Tisdall and Oades, 1982). Macroaggregates form a protection shield by encapsulating microaggregate from microbial activity, but are more susceptible to external forces such as tillage, erosion, rainfall, and other mechanical forces (Duiker, 2002). There is a complex interrelationship of biological, physical, and chemical reactions in the formation and degradation of all sizes of soil aggregates, and these have been investigated by many scientists (Lal, 1993). The root system and the microbial community are the foundation for such processes. Through the physical influence of the root system in building and separating soil aggregates (Cambardella, 2002), and the attachment of mycorrhizae hyphae to the root system for nutrient cycling (Tisdall and Oades, 1982) and production of organic compounds are what is essential to build soil aggregates (Metting, 1993). The type of vegetation plays a significant role in the formation and stability of soil aggregates, where perennials and, in particular, grasses produce more stable and well-formed soil aggregates (Diaz-Zorita et al., 2002) as compared to annual cropping systems (Cambardella, 2002; Bronick and Lal, 2005). However, the influence of cropping systems on soil aggregates is controlled essentially by the root system, where root intensity and morphology and how the system is managed can impact the formation and stabilization of soil aggregate, which is a major driver for enhancing soil health or quality by providing a balanced soil environment through aeration, water, and nutrient supply to the root system (Diaz-Zorita et al., 2002). Other factors that affect soil aggregate formation may include:

1. Climate: which can influence soil structure through a significant effect on aggregate formation or destruction. This may include the intensity and duration of rain, episodic drought conditions, and freezing and thawing cycles. These conditions can impact the soil biological, physical, and chemical environment that is essential to soil aggregate

formation or creating a destruction force leading to the instability of soil aggregates and structure (Diaz-Zorita et al., 2002).

2. Soil management and type of cropping systems: which can accentuate the climate effects on aggregate formation or destruction. The stability of the cropping system such as perennials, no-tillage (NT), and a more diverse cropping system can increase the strength of aggregates by reducing soil erosion or mechanical destruction of aggregates because of intensive tillage (Horne and Sojka, 2002).

In addition to the above two main factors, many other interacting forces influence the formation and stability of aggregates. Soil aggregate is not an absolute attribute, but a function of aggregate bond strength against stresses induced by internal and external forces, such as physical forces produced by wetting and drying cycles, management practices, and plant and weather interactions. The major shapes of aggregates that form soil structure and that can be identified in the field (Fig. 1.6) are summarized as follows (Hillel, 1998):

1. Spherical: soil aggregates have a rounded shape and are not >2 cm in diameter. This type of aggregate is generally found in the topsoil or A horizon in a loose, granular formation and can be characterized as porous crumbs.
2. Blocky: soil aggregates with a cube-like shape or blocks and having up to 10 cm size, and sometimes have angular, with well-defined, planar faces. This structure generally occurs in the upper portion of B horizon.

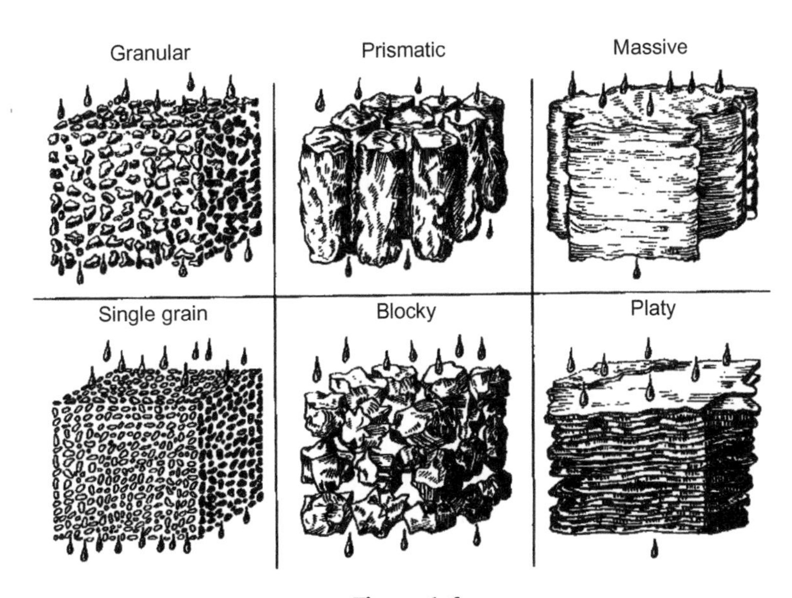

Figure 1.6
Schematic representation of various soil structures. *After Whiting, D., Card, A., Wilson, C., Moravec, C., Reeder, J., 2014. Managing Soil Tilth: Texture, Structure, and Pore Space. GMG GardenNotes #213. Colorado State University Extension, Fort Collins (Whiting et al., 2014), adopted from USDA.*

3. Prismatic or columnar: soil aggregates in this structure have distinct, vertically oriented pillars with six-sided structure up to 15 cm in length. This kind of structure is generally associated with clayey soils in the B horizon in semiarid regions. The flat, vertical aggregates are called *prismatic* and the rounded ones are called *columnar*.
4. Platy: soil aggregates have thin, flat, and horizontal layers. This kind of structure is generally found in recently deposited and compacted clay soils.

The state of aggregate of any soil at any particular time reflects the status of soil structure at that point in time only, because this structure is subject to change over time and with management. The aggregate strength and stability in agriculture soils are highly affected by type of tillage, crop rotation, traffic, and other management inputs.

1.3 Soil Biological Environment

1.3.1 Soil–Plant Relationship

The soil biological environment contains microorganisms such as bacteria, actinomycetes, fungi, and algae; the microflora, and the protozoa, worms, and arthropods; and the microfauna and fauna (Russell, 1973). Organisms in the soil need food for two distinct purposes: to supply energy for their essential and vital processes and to build their body tissues. However, the sources of food for various organisms can be different for these two purposes. Fully autotrophic organisms require much more energy for growth than many heterotrophs, since greater energy is needed to convert the carbon (C) of inorganic sources than that of organic substances such as sugar (Russell, 1973). Also, autotrophic organisms can meet their nitrogen (N) need from ammonium or nitrate salts as preferred sources. However, heterotrophic organisms are classified by their nutritional needs or by the biochemical changes they bring. The majority of heterotrophs can use glucose as their primary source and produce an enzyme, if necessary, to convert a wide range of carbohydrates into sugar (Herman et al., 2006). Plant and soil interaction is characterized by a defined-zone called the "rhizosphere," where the root system influences microbial activity. This zone is distinguished from the rest of the soil mass by the active interaction between soil–plant–microbial communities (Russell, 1982; Herman et al., 2006). The interaction between soil, plant, and microbes is essential to nutrient cycling processes, such as soil organic nitrogen (SON) mineralization and phosphorus (P) transformation for plant nutrient supply (Bregliani et al., 2010). In the root zone within the soil environment the contact between microorganisms and the root system involves a range of activities that contribute to the plant growth. Such soil–microbial interaction plays a significant role in increasing bioavailability and uptake of mineral nutrients by plants (Glick et al., 1999). These biofertilizers such as, plant growth-promoting bacteria or fungi, as Mycorrhizae and *Penicillium bilaii*, can increase nutrient bioavailability (Saleh-Lakha and Glick, 2007). Plant growth-promoting bacteria can contribute to the development of the plant that is more positioned to tolerate adverse growing conditions such as disease, pathogens, and drought stress (Saleh-Lakha and Glick, 2007). Some of the mechanisms that are associated with the promotion of healthy plant growth may include (Glick et al., 1999):

1. Associative N fixation;
2. The lowering of ethylene levels that are otherwise an impediment to plant growth;
3. The sequestration of iron by siderophores;
4. The production of photohormones such as auxin and cytokinins;
5. The introduction of pathogen resistance in the plant;
6. The solubilization of nutrients such P;
7. Promotion of mycorrhizal functioning;
8. Modification of root morphology;
9. Enhancement of legume–rhizobia symbioses; and
10. Decreasing (organic or heavy metal) pollutant toxicity.

1.3.2 Soil–Root Interface and Nutrient Cycling

Root systems are an essential part of the soil biological environment in addition to their basic functions of absorption of water and nutrients, anchorage, storage, and synthesis of diverse organic compounds (Klepper, 1990). The interaction between soil organisms and plants through roots as a symbiotic relationship, and mutual benefits for the microbial community, growth and plant needs for nutrients for production of biomass below and above ground are accomplished (Bardgett and Wardle, 2003). All natural nutrients and water uptake by the plant enter through the root system. It is the rhizosphere where nutrient cycling takes place by the colonized microorganisms of the plant roots. The majority of nutrient cycling mechanisms and processes in soil are performed by macroorganisms. However, the organic compounds provided by the root system, dead plants, and animal materials, along with some inorganic compounds and exudation by plant roots of plant-derived photosynthetic material, provide the basic energy and food sources for microorganisms during the process of nutrient cycling (Prosser, 2007). The interaction between plant roots and diverse microorganisms influences the process of nutrient cycling and the production of organic compounds that are essential for building soil aggregates. The following sections highlight two cycles of the major nutrients in soil that include N and C.

1.3.2.1 Nitrogen cycle

Nitrogen is required by plants and soil organisms, and while it is abundant at the earth's surface, only <2% is available to organisms (Mackenzie, 1998). The majority of N is tied up in different pools that require sources of energy and processing to convert it to available forms for organisms' use. Therefore, the biological and chemical transformation of different sources of N takes place to make it available through the N cycle. The purpose of this discussion is to highlight the main nutrient cycle, such as N as related to soil biological functions.

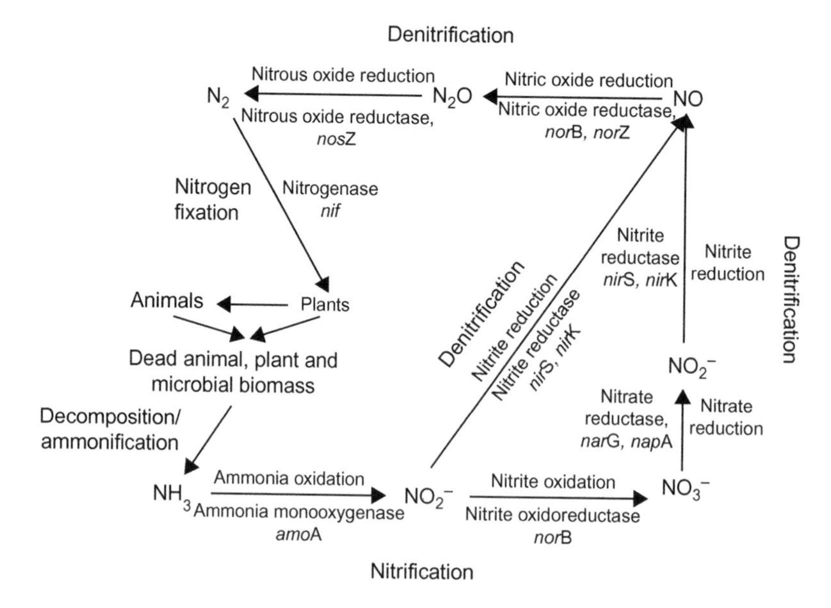

Figure 1.7

The terrestrial nitrogen cycle, including enzymes catalyzing particular transformation and associated functional genes that have been used for analysis. *After Prosser, J.I., 2007. Microorganisms cycling soil nutrients and their diversity. In: van Elsas, J.D., Jansson, J.D., Trevors, J.T. (Eds.), Modern Soil Microbiology, second ed. CRC Press, New York.*

The N cycle is complex, and it is not as it is referred to as a simple two-stage process of nitrification and denitrification. The basis for the N cycle is the decomposition of the organic N pool that is derived from dead animals, plant, and microbial biomass, along with other transformation processes. The complexity of the N cycle, as indicated by Prosser (2007), is within the organic N pool, where little is known of the functional diversity of the organisms decomposing this material. The two main N cycle processes, nitrification and denitrification, are summarized in Fig. 1.7. The nitrification process involves the oxidation of reduced forms of N such as ammonia (NH_3^+) to nitrate (NO_3^-), and occurs as a two-stage process. First NH_3 is oxidized to nitrite (NO_2^-), then NO_2^- is oxidized to NO_3^-, as illustrated in Fig. 1.7, involving different bacteria and enzymes. Denitrification is the reduction of NO_3^- to N gas via NO_2^- and nitric (NO) and nitrous oxides (N_2O). It is an anaerobic process during which NO_3^- acts as an electron acceptor during anaerobic respiration by heterotrophic bacteria.

1.3.2.2 Carbon Cycle

The terrestrial C cycle process mediated by soil microorganisms begins on land with primary production through photosynthetic plants that take up inorganic C as CO_2 and produce organic compounds (Post et al., 1990). The terrestrial C cycle involves two major processes that include (1) fixation of CO_2 into organic materials through autotrophic organisms that

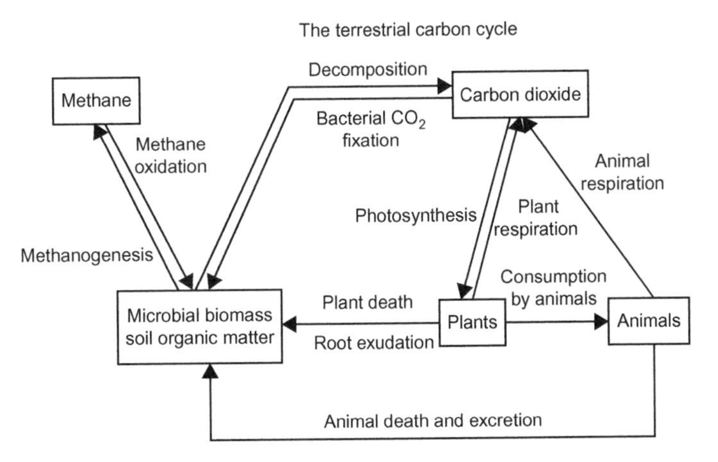

Figure 1.8
The terrestrial carbon cycle, indicating the major processes mediated by soil microorganisms. *After Prosser, J.I., 2007. Microorganisms cycling soil nutrients and their diversity. In: van Elsas, J.D., Jansson, J.D., Trevors, J.T. (Eds.), Modern Soil Microbiology, second ed. CRC Press, New York.*

acquire energy from the photosynthesis process or from the oxidation of reduced inorganic compounds, and (2) decomposition of fixed C in SOM to CO_2 by heterotrophic organisms (Fig. 1.8). In this cycle, also, is the oxidation of methane (CH_4) by anaerobic archaea, which is an important process to control the level of a key greenhouse gas (GHG). The balance between CO_2 assimilation by photosynthesis and the release of C from both living and dead material determines the net exchange of C between the atmosphere and the terrestrial systems (Post et al., 1990). The decomposition process of organic matter (OM) is very complex and is affected by many factors, such as the heterogeneity of SOM, availability of C compounds, the stability of organic compounds within the plant cell, the link to other nutrient cycles, such as N, and the functional diversity of microorganisms, and especially in decomposing organic compounds (Prosser, 2007).

1.3.2.3 Water cycle

The interlink of both C and N cycles to water in the agro-ecosystems and its influence on these two cycles, makes it imperative to shed some light on the water cycle. Water covers nearly 71% of the earth's surface and 97% is held in oceans. The remaining water is in the air as vapor, in soil and groundwater, rivers, frozen in glaciers, humans, and animals. Water moves from the earth's surface to the atmosphere as solar energy absorbed by water on or near the earth's surface stimulating evaporation—the conversion of liquid water into vapor. The vapor water moves up into the atmosphere, eventually forming clouds that can move across regions on the globe. One third of solar energy that reaches the earth's surface is absorbed by water on or near the earth surface (Weil and Brady, 2016). Pressure and temperature differences in the atmosphere cause water vapor to condense into liquid droplets or solid particles which return to the earth as rain or snow. The rain distribution and its intensity influence characteristics of all terrestrial lives including those of agro-ecosystems and their services on earth where soil moisture is essential to the functionality of such systems.

1.3.3 Soil Environment and Microbial Diversity

The soil physical and chemical environment can have a significant impact on microbial habitat through its influence on water and gaseous movement in the soil system, which affects microbial diversity, activities, and functions (Nannipieri et al., 2003). The organization of soil aggregates in different sizes provides different functions for hosting the microbial community, where macroaggregates act as a shield for microaggregates against soil microbial activity as indicated by the hierarchy theory of soil aggregate functions (Tisdall and Oades, 1982). The pores within soil aggregates, which under ideal conditions represent approximately 50% of the total soil volume, provide a natural habitat for microbes, where they occupy the walls of these pores, and water moving through the soil pores may transport significant numbers of freely mobile bacteria (Standing and Killham, 2007). Water movement in soil is the most significant physical function that affects microbial life, where nutrients, gases, and microbes and their precursors move as well. To understand the relationship between soil, water, and biological activity, four major components of the soil environment that interact with microbial activity must be considered. These include: (1) nutrient diffusion and mass flow, (2) mobility, (3) temperature and aeration, and, (4) pH and Eh (Standing and Killham, 2007). The availability of SOM in its different forms (readily decomposable or recalcitrant) provides a source for food and energy to a diverse microbial community (Nannipieri et al., 2003). However, the microbial activity is governed by the above four major factors, where soil temperature, in particular, is a key determinant for both distribution and activity of soil microorganisms. Soil microorganisms' response to changes in soil temperature is not independent of temperature effects on the plant and animals with which they interact (Grayston et al., 1998). Therefore, the quantity and quality of the rhizosphere C sources are essential to the diversity and activity of the rhizosphere microbial community, which is strongly dependent on temperature (Meharg and Killham, 1989) as is root growth and turn over. Soil pH is another determinant factor for soil microbial diversity where different microbial groups have philological preference to low pH (acidophiles), while other groups prefer high pH (alkalophiles) (Staley et al., 2011). Soil microbial habitat, among many other factors, may influence microbial diversity, which is considered an integral part of the soil ecosystem functions and may be used as an indicator of soil quality and fertility (Trosvik and Ovreas, 2007). Whether changes in microbial diversity and composition in local microenvironments influence the overall ecosystem processes in soil remains an open question.

1.4 Soil Chemical Environment

1.4.1 Soil Nutrient Capacity and Supply

The soil physical environment was addressed in Section 1.2.1, where the four major components (mineral, organic matter, water, and air) that constitute the soil system were discussed. These parts are not independent of each other and influence all soil properties including chemical processes and mechanisms. The interaction between mineral and OM in the soil system influences not only the physical aspects and availability of water to the plant, but also the O_2 level, which is essential not only for plant growth, but for various chemical reactions (and biological reactions) and nutrient processes and mechanisms (Skopp et al., 1990). The dissolved nutrients in the soil solution are utilized by plants through different mechanisms of nutrient uptake by the root system, and as the concentration of such nutrients declines, the inorganic and organic parts of the soil system replenish these in the solution. The exchange of cations and anions between soil particles and roots is one of the most important processes to plants and animals, in addition to photosynthesis and respiration (Weil and Brady, 2016). The exchange of cations and anions occurs mainly on the surface of the finer, or colloidal fractions, of inorganic and organic matter (clay and humus). The mechanism of cation and anion exchange is not only for nutrient supply to the plant, but also is important in controlling the movement of some organic and inorganic chemicals to the water systems. The cation exchange capacity (CEC) of a given soil is determined by the relative amounts of different colloids in that soil and by the CEC of each colloid (Fig. 1.9). On the other hand,

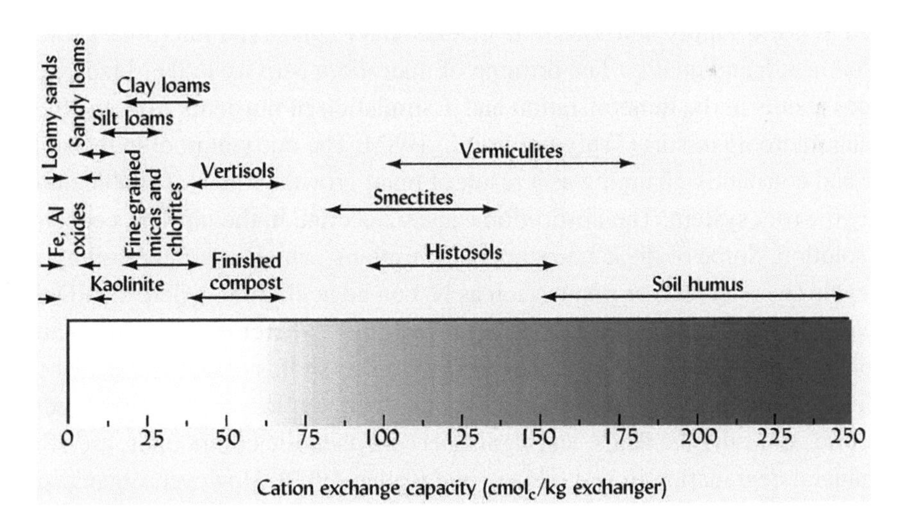

Figure 1.9

Ranges in the cation exchange capacity (at pH 7) that are typical of a variety of soils and soil materials. The high CEC of humus shows why this colloid plays a prominent role in most soils, and especially those high in kaolinite and Fe, Al oxides, clays that have low CECs. *After Weil, R.R., Brady, N.C., 2016. The Nature and Properties of Soils, fifteenth ed. Pearson, New York, 1100 p.*

anions are held by the soil colloids in two ways: (1) by similar adsorption mechanism of cations or (2) by their reaction with surface oxides or hydroxides forming an inner-sphere complex (Haynes, 1990). The anion absorption capacity of most agricultural soils is small compared to CEC. The differences in soil CEC help explain the difference in their nutrient supply abilities. The CEC is expressed in terms of mole of positive charge absorbed per unit mass (Weil and Brady, 2016; Mengel and Kirkby, 1982). The CEC is expressed as a whole number using centimoles of positive charge per kilogram of soil ($+kg^{-1}$). Many factors influence CEC as an essential chemical property in the soil, such as clay type, SOM content, and soil pH (Haynes, 1990). Indeed, different soil compositions and colloids play significant roles in defining the range of CEC (Fig. 1.9). The effect of soil pH is well documented where most of the soils' capacity to hold nutrients increases with an increase in soil pH, where CEC of SOM and clay increases linearly with an increase in pH (Helling et al., 1964). Generally, CEC is low at low pH values. Under these conditions only permanent charges of 2:1 clays and small portions of pH-dependent charges of organic colloids, allophane, and some 1:1 type of clays hold exchangeable ions (Weil and Brady, 2016). The CEC is always dominated by permanent charge for 2:1 minerals, not just under acidic conditions. The effect of pH is much greater for 1:1 minerals or hydrous oxides, and also for organic matter, in which case CEC is entirely due to pH-dependent charge sites.

1.4.2 Nutrient Cycling

Most nutrients are released in the soil environment either from parent materials or decomposed SOM to supply nutrients that are essential to plant and microbial growth, which influence soil health and quality. The priming of microflora activity in the rhizosphere by root exudates results in the mineralization and assimilation of nutrients from the SOM pool by free-living microorganisms (Trofymow et al., 1983). The nutrient pool in the soil system is dynamic and constantly changing as a result of plant growth, where available nutrients are absorbed by the root system. The absorption causes a decline in the nutrient concentration in the soil solution. Some of these nutrient concentrations, which are in the soil by the end of the life cycle (i.e., harvest) of plants such as N, can be negligible, where small amounts can be released by weathering of mineral parent materials. Therefore, other mechanisms, such as biological N fixation (BNF), are needed to replenish this essential nutrient for plant growth (Weil and Brady, 2016). It has been hypothesized that low molar organic acids such as oxalic, citric, and salicylic acids, and lignin-derived phenolic compounds, are effective agents in mineral degradation in soil (Huang and Keller, 1970). However, organic substances constitute approximately 70% of the weight of SOM, but the concentration of low molar mass organic acids is difficult to assess (Guggenberger and Haider, 2002). The plant roots and microorganisms associated with the roots have a role in mineral weathering, where they can contribute to the solubilization and insolubilization processes and promote availability of nutrients such as K, P, Fe, Mg, or adsorbed NH_4^+ in the rhizosphere (Marschner et al.,

1986). This interaction and release of nutrients is governed by abiotic factors that include soil moisture, soil temperature, pH, and oxygen concentration, along with chemical composition of primary and secondary resources (Swift et al., 1979). Another factor that influences the release of nutrients is clay minerology, where they change the microenvironment of microorganisms by affecting soil pH, ionic strength, and availability, as well as enzyme production and activity (Zech et al., 1997). The simultaneous link of nutrient cycling to plant growth and microbial activity in soils of a natural agroecosystem can be disrupted in those of agricultural land, where nutrients mineralized and released into soil during periods without crops are subject to loss by leaching and/or volatilization. Therefore, the gap between nutrient release and uptake can lead to low nutrient use efficiency such as that of N.

1.4.3 Nutrient Pathways and Mechanisms

Many mechanisms govern the chemical movement of nutrients in the soil. These mechanisms include processes such as diffusion/dispersion, absorption, decay, and intra-aggregate diffusion (van Genuchten and Cleary, 1982). The movement of chemicals/ions in the soil to the plant roots may occur by two major mechanisms that include mass flow and diffusion. Mass flow occurs when nutrients are transported to the root system through convective flow of water from soil to the plant roots as influenced by water potential gradients (Mengel and Kirkby, 1982). The rate of nutrient uptake by plant roots depends on the rate of water flow and the rate of water consumption by the plant. The diffusion process, on the other hand, is driven by the nutrient concentration gradient, where nutrient transport occurs from higher to lower concentration by random thermal motion (Brewster and Tinker, 1970). The diffusion process takes place when the concentration at the root surface is either higher or lower than that of the surrounding soil solution. The nutrient is directed towards the root when the concentration is lower at the root surface and away from the root when the concentration is increased at the root surface (Drew et al., 1969). This diffusion mechanism is governed by Fick's Law:

$$F = -D \, \mathrm{d}c/\mathrm{d}x \qquad (1.5)$$

where, F is the diffusion rate (quantity per unit cross section and per unit time), D is diffusion coefficient, c is a concentration, x is distance, and $\mathrm{d}c/\mathrm{d}x$ is the concentration gradient. The diffusion coefficient in Fick's Law is affected by the medium diffusivity characteristic of such air, water, or soil. The homogeneity of the medium plays a role in influencing the diffusion rate as soil is a nonhomogeneous medium, and thus, diffusivity varies within the soil (Nye and Tinker, 1977). The diffusion coefficient is affected by the soil moisture content and nutrient buffer capacity. The rate of ion movement in the soil, whether through mass flow or diffusion, and the impact of soil moisture content have been studied and found to be highly dependent on soil moisture, where a decrease in soil moisture content decreases the diffusion coefficient (Graham-Bryce, 1967). The decrease in soil moisture content can affect the direct pathway contact between the bulk soil and plant roots. The dissolution of mineral materials is a result of three processes that include the access of solvent to the substance, the saturated solution, and the transport of the dissolved materials (Filep, 1999). The dissolution process is much faster than the diffusion, hence the saturated layer forms on the surface layer relatively fast, leading to a concentration gradient between the saturated layer and the rest of the solution, where Fick's Law can be applied because the concentration gradient is constant (Hillel, 1998).

1.5 Conclusions

This chapter covered the basic soil physical, chemical, and biological properties and explored the interrelationships between different soil properties and functions as essential building blocks for a healthy functioning soil system. The soil physical environment, components of the total soil water potential, and water movement, including governing forces, were presented. Soil properties included soil structure and water pathways, soil temperature, and soil aggregation and soil structure formation. The soil biological environment included soil–plant relationships, including plant growth and microorganisms, soil–root interface and nutrient cycling, and soil environment and microbial diversity. The soil chemical environment discussion focused on soil nutrient capacity and supply, nutrient cycling, and nutrient pathways and mechanisms. Soil properties and soil health can be negatively impacted by intensification of the agroecosystem, and it is important to develop agricultural systems to mitigate adverse environmental impacts and restore soil health and productivity. The content of this chapter presents the basic principles for soil physical, chemical, and biological functions that are essential for understanding topics presented in the following chapters. The interrelationship between these functions and the governing factors will determine the services provided by healthy soil within different agro-ecosystems.

References

Bardgett, R.D., Wardle, D.A., 2003. Herbivore mediated linkages between aboveground and belowground communities. Ecology 84, 2258–2268.

Borie, F., Rubio, R., Morales, A., 2008. Arbuscular mycorrhizal fungi and soil aggregation. J. Soil Plant Nutr. 8 (2), 9–18.

Bregliani, M.M., Ros, G.H., Temminghoff, E.J.M., van Riemsdijk, W.H., 2010. Nitrogen mineralization in soils related to initial extractable organic nitrogen: effect of temperature and time. Commun Soil Sci. Plant Anal. 41 (11), 1383–1398. http://dx.doi.org/10.1080/00103621003759387.

Brevik, E.C., 2005. A brief history of soil science. In: Verheye, W. (Ed.), Global Sustainable Development, Theme 1.05: Land Use and Cover. Encyclopedia of Life Support Systems (EOLSS), Developed Under the Auspices of the UNESCO, EOLSS Publishers, Oxford. <http://www.eolss.net>.

Brewster, J.L., Tinker, P.B., 1970. Nutrient cation flows in soil around plant roots. Soil Sci. Soc. Am. Proc. 34, 421–426.

Bronick, C.J., Lal, R., 2005. Soil structure and management: a review. Geoderma 124, 3–22.

Cambardella, C.A., 2002. Aggregation and organic matter. In: Lal, R. (Ed.), Encyclopedia of Soil Science. Marcel Dekker, Inc., New York, pp. 41–44.
de Vries, D.A., 1975. "Thermal Conductivity of Soil." Med., Landbouwhogeschool, Wageningen, Netherlands.
Dexter, A.R., 1988. Advances in characterization of soil structure. Soil Tillage Res. 11, 199–238.
Diaz-Zorita, M., Perfect, E., Grove, J.H., 2002. Soil aggregation, fragmentation, and structural stability measurement. In: Lal, R. (Ed.), Encyclopedia of Soil Science. Marcel Dekker, Inc., New York, pp. 37–40.
Doran, J.W., Smith, M.S., 1987. Organic matter management and utilization of soil and fertilizer nutrients. SSSA. Spec. Pub. No. 19. Paper presented in a Symposium on "Soil Fertility and Organic Matter Components of Production Systems", 3 December 1985, Chicago, IL.
Drew, M.C., Nye, P.H., Vaidyanathan, L.V., 1969. The supply of nutrient ions by diffusion to plant roots in soil. Plant Soil 2, 252–270.
Duiker, S.W., 2002. Aggregation. In: Lal, R. (Ed.), Encyclopedia of Soil Science. Marcel Dekker, Inc., New York, pp. 34–36.
Filep, G., 1999. Soil Chemistry: Process and Constitutes. Akademia Kiado, Budapest.
Franzini, J.B., 1951. Porosity factor for case of laminar flow through granular media. Trans. Am. Geophys. Union 32 (3), 443–446.
Gardner, W.H., Hsieh, J.C., 1959. Water Movement in Soils (Video). Washington State University, Pullman, WA.
Gerke, H.H., Van Genuchten, M.T., 1993. Evaluation of a first-order water transfer term for variably saturated dual-porosity flow model. Water Resour. Res. 29 (4), 1225–1238.
Glick, B.R., Patten, C.L., Holguin, G., Penrose, D.M., 1999. Biochemical and Genetic Mechanisms Used by Plant Growth-Promoting Bacteria. Imperial College Press, London.
Graham-Bryce, I.J., 1967. Adsorption of disulfoton by soil. J. Sci. Food Agric. 18, 72–77.
Grayston, S.J., Wang, S., Campbell, C.D., Edwards, A.C., 1998. Selective influence of plant species on microbial diversity in the rhizosphere. Soil Biol. Biochem. 30 (3), 369–378.
Guggenberger, G., Haider, K.M., 2002. Effect of mineral colloids on biogeochemical cycling of C, N, P, and S in soil. In: Huang, P.M., Bollag, J.M., Senesi, N. (Eds.), Interactions Between Soil Particles and Microorganisms: Impact on the Terrestrial Ecosystem. John Wiley & Sons, Ltd, Chichester.
Hamblin, A.P., 1985. The influence of soil structure on water movement, crop root growth, and water uptake. In: Brady, N.C. (Ed.), Advances in Agronomy, vol. 38, American Society of Agronomy. Academic Press, Inc., New York.
Haynes, R.J., 1990. Active ion uptake and maintenance of cation-anion balance: a critical examination of their role in regulating rhizosphere pH. Plant Soil 126, 247–264.
Helling, C.S., Chesters, G., Corey, R.B., 1964. Contribution of organic matter and caly to soil cation exchange capacity as affected by the pH of the saturated solution. Soil Sci. Soc. Am. Proc. 28, 517–520.
Herman, D.J., Johnson, K.K., Jaeger, C.H., Schwartz, E., Firestone, M.K., 2006. Root influence on nitrogen mineralization and nitrification in *Avena barbata* rhizosphere soil. Soil Sci. Soc. Am. J. 70 (5), 1504–1511. http://dx.doi.org/10.2136/sssaj2005.0113.
Hillel, D., 1980. Fundamentals of Soil Physics. Academic Press, New York.
Hillel, D., 1998. Environmental Soil Physics. Academic Press, New York.
Horne, D.J., Sojka, R.E., 2002. Aeration, tillage effects on. In: Lal, R. (Ed.), Encyclopedia of Soil Science. Marcel Dekker, Inc., New York, pp. 30–33.
Huang, W.H., Keller, W.D., 1970. Dissolution of rock-forming silicate minerals in organic acids: simulated first stage weathering of fresh mineral surface. Am. Mineral. 55, 2076.
Huggett, R.J., 1998. Soil chronosequence, soil development, and soil evolution: a critical review. Catena 32, 155–172.
Jacob, C.E., 1946. Drawdown test to determine effective radius of an artesian well. Am. Soc. Civ. Eng. Proc. 72 (5), 629–646.
Jenny, H.F., 1941. Factors of Soil Formation. McGraw-Hill, New York.
Klepper, B., 1990. Root growth and water uptake. In: Stewart, B.A., Nielsen, D.R. (Eds.), Irrigation of Agricultural Crops. ASA–CSSA–SSSA, Madison, pp. 281–322.

Klute, A., Whisler, F.D., Scott, E.H., 1965. Numerical solution of the flow equation for water in a horizontal finite soil column. Soil Sci. Soc. Am. Proc. 29, 353–358.
Kutilek, M., Nielsen, D.R., 1994. Soil Hydrology: GeoEcology Textbook. Catena Verlag, Cremlingen-Destedt.
Lal, R., 1993. Tillage effects on soil degradation, soil resilience, soil quality, and sustainability. Soil Tillage Res. 27, 1–8.
Licht, M., Al-Kaisi, M.M., 2005. Strip-tillage effect on seedbed soil temperature and other soil physical properties. Soil Tillage Res. J. 80, 233–249.
Mackenzie, F.T., 1998. Our Changing Planet, second ed. Prentice Hall, Upper Saddle River, NJ.486.
Marschner, H., Romheld, V., Horst, W.J., Martin, P., 1986. Root induced changes in the rhizosphere: importance for the mineral nutrition of plants. Plant Nutr. Soil Sci. J. 149, 441–456.
Marshall, T.J., 1958. A relation between permeability and size distribution of pores. J. Soil Sci. 9, 1–8.
McInnes, K., 2002. Temperature measurement. In: Lal, R. (Ed.), Encyclopedia of Soil Science. Marcel Dekker, New York, pp. 1302–1304.
Meharg, A.A., Killham, K., 1989. Distribution of assimilated carbon within the plant and rhizosphere of lolium perenne: influence of temperature. Soil Biol. Biochem. 21, 487–489.
Mengel, K., Kirkby, E.A., 1982. Principles of Plum Nutrition.. International Potash Institute, Worblaufen-Bern.
Metting Jr., F.B., 1993. Soil Microbial Ecology: Applications in Agriculture and Environmental Management. Marcel Dekker, New York.
Miller, R.J., Low, P.F., 1963. Threshold gradient for water-flow in clay systems. Soil Sci. Soc. Am. Proc. 27, 605–609.
Nannipieri, P., Assher, J., Ceccherini, M.T., Land, L., Pietramellara, G., Renella, G., 2003. Microbial diversity and soil functions. Eur. J. Soil Sci. 54, 655–670.
Nye, P.H., Tinker, P.B., 1977. Solute Movement in Soil-Root System. Blackwell, Oxford.
Oades, J.M., 1984. Soil organic matter and structural stability: mechanisms and implications for management. Plant Soil 76, 319–337.
Oades, J.M., Waters, A.G., 1991. Aggregate hierarchy in soils. Aust. J. Soil Res. 29, 815–828.
Post, W.M., Peng, T.H., Emanuel, W.R., King, A.W., Dale, V.H., DeAngelis, D.L., 1990. The global carbon cycle. Am. Sci. 78, 310–326.
Prosser, J.I., 2007. Microorganisms cycling soil nutrients and their diversity. In: van Elsas, J.D., Jansson, J.D., Trevors, J.T. (Eds.), Modern Soil Microbiology, second ed. CRC Press, New York.
Rose, C.W., 1966. Agricultural Physics. Robert Maxwell, Pergamon Press, Paris.
Russell, E.W., 1973. Soil Conditions and Plant Growth. Longman Group Limited, New York.
Russell, R.S., 1982. Plant Root Systems, first ed. McGraw-Hill, New York.214.
Saleh-Lakha, S., Glick, B.R., 2007. Plant growth promoting bacteria. In: van Elsas, J.D., Jansson, J.K., Trevors, J.T. (Eds.), Modern Soil Microbiology, second ed. CRC Press, New York.
Saxton, K.E., Rawls, W.J., Romberger, J.S., Papendick, R.I., 1986. Estimating generalized soil-water characteristics from texture. Soil Sci. Soc. Am. J. 50, 1031–1036.
Skopp, J., Jawson, M., Doran, J.W., 1990. Steady-state aerobic microbial activity as a function of soil water content. Soil Sci. Soc. Am. J. 54, 1619–1625.
Smiles, D.E., Rosenthal, M.J., 1968. The movement of water in swelling materials. Aust. J. Soil Res. 6, 237–248.
Staley, B.F., de los Reyes III, F.L., Barlaz, M.A., 2011. Effect of spatial differences in microbial activity, pH, and substrate levels on methanogenesis initiation in refuse. Appl. Environ. Microbiol. 77 (7), 2381–2391.
Standing, D., Killham, D., 2007. The soil environment. In: van Elsas, J.D., Jansson, J.K., Trevors, J.T. (Eds.), Modern Soil Microbiology, second ed. CRC Press, New York.
Swift, M.J., Heal, O.W., Anderson, J.M., 1979. Decomposition in Terrestrial Ecosystems. Blackwell, Oxford.
Tisdall, J.M., Oades, J.M., 1982. Organic matter and water-stable aggregates. J. Soil Sci. 33, 141–163.
Troeh, F.R., Hobbs, J.A., Donahue, R.L., 1999. Soil and Water Conservation: Productivity and Environmental Protection, third ed. Prentice Hall, Upper Saddle River, NJ.
Trofymow, J.A., Morley, C.R., Coleman, D.C., Anderson, R.V., 1983. Mineralization of cellulose in the presence of chitin and assemblages of microflora and fauna in soil. Oecologia 60, 103–110.

Trosvik, V., Ovreas, L., 2007. Microbial phylogeny and diversity in soil. In: van Elsas, J.D., Jansson, K.J., Trevors, J.T. (Eds.), Modern Soil Microbiology, second ed. CRC Press, New York.

van Bavel, C.H.M., Hillel, D., 1976. Calculating potential and actual evaporation from a bare soil surface by simulation of concurrent flow of water and heat. Agric. Meterol. 17, 453–476.

van Duin, R.H.A., 1956. On the Influence of Tillage on Conduction of Heat, Diffusion of Air, and Infiltration of Water in Soil. Versl. Landbouwk Onderz., Wageningen.62.

van Genuchten, M.Th, Cleary, R.W., 1982. Movement of solutes in soil: computer-simulated and laboratory results. In: Bolt, G.H. (Ed.), Soil Chemistry B: Physico-Chemical Models. Elsevier Scientific, New York.

van Wijk, W.R., de Vries, D.A., 1963. Periodic temperature variations in homogeneous soil. In: van Wijk, W.R. (Ed.), Physics of Plant Environment. North-Holland, Amsterdam.

Weil, R.R., Brady, N.C., 2016. The Nature and Properties of Soils, fifteenth ed. Pearson, New York.1100.

Whiting, D., Card, A., Wilson, C., Moravec, C., Reeder, J., 2014. Managing Soil Tilth: Texture, Structure, and Pore Space. GMG GardenNotes #213. Colorado State University Extension, Fort Collins.

Zech, W., Senesi, N., Guggenberger, G., Kaiser, K., Lehmann, J., Miano, T.M., et al., 1997. Factors controlling humification and mineralization of soil organic matter in the tropics. Geoderma 79, 117–169.

CHAPTER 2

Climate Variability Effects on Agriculture Land Use and Soil Services

Jerry L. Hatfield and Christian Dold

National Laboratory for Agriculture and the Environment, Ames, IA, United States

2.1 Introduction

Climate impacts on agriculture encompass all aspects from the direct impacts on crop and livestock productivity to the indirect impacts on pests, diseases, and weeds. In the recent report from the National Climate Assessment sponsored by the United States Global Change Research Program, the agriculture sector produced several key findings. These key findings developed by Hatfield et al. (2014) were stated as:

1. "Climate disruptions to agricultural production have increased in the past 40 years and are projected to increase over the next 25 years. By mid-century and beyond, these impacts will be increasingly negative on most crops and livestock.
2. Many agricultural regions will experience declines in crop and livestock production from increased stress due to weeds, diseases, insect pests, and other climate change induced stresses.
3. Current loss and degradation of critical agricultural soil and water assets due to increasing extremes in precipitation will continue to challenge both rainfed and irrigated agriculture unless innovative conservation methods are implemented.
4. The rising incidence of weather extremes will have increasingly negative impacts on crop and livestock productivity because critical thresholds are already being exceeded.
5. Agriculture has been able to adapt to recent changes in climate; however, increased innovation will be needed to ensure the rate of adaptation of agriculture and the associated socioeconomic system can keep pace with climate change over the next 25 years.
6. Climate change effects on agriculture will have consequences for food security, both in the U.S. and globally, through changes in crop yields and food prices and effects on food processing, storage, transportation, and retailing. Adaptation measures can help delay and reduce some of these impacts."

Soil Health and Intensification of Agroecosystems.
DOI: http://dx.doi.org/10.1016/B978-0-12-805317-1.00002-6

Impacts of climate on agriculture have been documented in the Intergovernmental Panel on Climate Change (IPCC) and Porter et al. (2014) summarized the available literature to evaluate the impact of a changing climate on agricultural systems. Production changes are not simply a direct function of climate (temperature, precipitation, solar radiation, carbon dioxide) but also from indirect factors (soil organic matter (SOM), soil fertility, erosion, irrigation, fertilizers, biotic stresses), and all these factors directly affect soil health. The continued degradation of soil increases the vulnerability to climate change (Hatfield, 2014). More detailed summaries of climate impacts on agricultural crops have been provided by Hatfield et al. (2011) to document that temperature and precipitation are the two major factors determining crop production. Crops require adequate soil water and temperatures within their range for development and growth and each species has its own specific temperature range; however, all crop plants require adequate soil water for optimum production. Soil health and the relationship to SOM and the ability of soils to provide adequate soil water and nutrients to crops provide a foundation for climate resilience. Integration of climate and soils determines the overall range of crop plants and their economic viability.

A changing climate increases the probability of a change in the frequency, intensity, spatial extent, duration, and timing of extreme weather and climate events. Handmer et al. (2012), in a special report for the IPCC, summarized how extreme events in temperature and precipitation could have a major impact on global ecosystems and the soil resources because of the increased potential for erosion and accelerated loss of soil carbon (C) or organic matter. With climate change, changes will occur in the mean temperature or precipitation; Hansen et al. (2012) showed that the shifts in the distribution of temperature or precipitation will lead to increased occurrence in extreme events. There will be shifts in the climate and development of sustainable intensification will require understanding the linkages between climate variability and intersection between land use and soils. These are often viewed as important but not considered in detail to provide guidance on what is the most profitable avenue to synthesize this information. The continued change in temperature and patterns of precipitation will contribute to increased problems of water in agriculture production, the interaction of water with the physical, chemical, or biological processes within the soil profile (Fuhrer et al., 2006; Calanca, 2007; Torriani et al., 2007) and will increase the competition for land and water resources (Lotze-Campen and Schellnhuber, 2009). Climate change will expand environmental impacts, such as higher erosion rates (Nearing et al., 2004), or faster decomposition of SOM and increased nitrogen (N) leaching (Bindi and Olesen, 2010). As a result, there will be increased pressure to develop strategies to adapt agricultural land management practices to offset climate change. Given the different responses to climate variation, Klein et al. (2013) suggested that adaptation to climate change for agricultural lands would require a multiobjective optimization approach to link climate, management, and soils information together with simulation models to evaluate a range of solutions for management practices. This approach developed an optimum set of solutions that were spatially distributed across the landscape and provides a framework for how we need to be evaluating the linkages among the components of agricultural systems.

2.2 Climate Variables and Temporal Trends

Trends in climate variables have been the subject of several analyses using ensembles for climate models. These trends have been documented in the National Climate Assessment and the IPCC 5th Assessment. Collins et al. (2013) and Kirtman et al. (2013) have provided the details on the projections of climate change in the near term (20–25 years) and the long term (50+ years). Their assessment of the future trends has a large significance for agriculture and will determine both the distribution of cropping systems in arable areas and the variation in production over the years.

2.2.1 Temperature

Temperatures are projected to increase in the near term by a mean global average of 1°C for the period from 2016 to 2035 compared to the 1850–1900 period, but not more than 1.5°C (Kirtman et al., 2013). The projection is that the winter temperatures will increase more than summer temperatures and near-term increases in seasonal mean and annual mean temperatures are expected to be larger in the tropics and subtropics than in mid-latitudes (Kirtman et al., 2013). For the United States, Walsh et al. (2014) evaluated temperature records to show there has been an average increase between 0.5°C and 0.75°C in the 1970–99 period compared to the 1890–1970 period. However, this increase has not been uniform across the United States as shown in Fig. 2.1. Coupled with the regional variation is the seasonal variation with temperature increases greater in winters than summers and a longer frost-free period across the United States. For the Midwest the frost-free period is 9 days longer for the period from 1991 to 2012 compared to 1901 to 1960 with the projection of an additional 30–40 days of frost-free conditions in 2070–2099 compared to the 1970–2000 period (Walsh et al., 2014).

Both Kirtman et al. (2013) and Walsh et al. (2014) project the occurrence of more extreme temperature events to occur during the next few years with climate change. These extreme events can occur at both the lower and upper ends of the temperature range. Collins et al. (2013) summarized that changes in mean temperatures will not be uniform across the globe with changes in land areas being greater than in oceans. They used climate model ensembles to show more of an increase in high-temperature extremes compared to cold temperature extremes. Hansen et al. (2012) have shown that winter warming is exceeding summer warming but the variation in seasonal mean temperatures at the high and mid-latitudes is greater in the winter (2–4°C) compared to summer (1°C). An important component in this change in the temperature signal is the occurrence of extreme outliers which they defined as

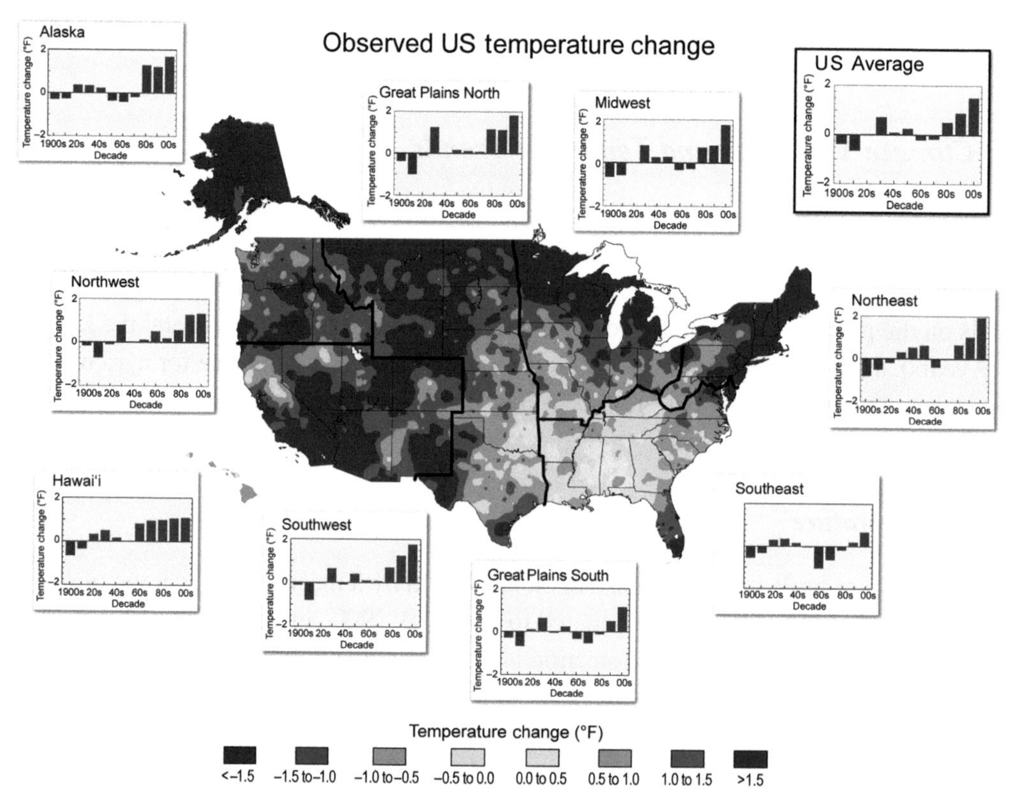

Figure 2.1

The range of colors on the map show the changes in the 22 years from 1991 to 2012 compared to the 1901–1960 average for the continental United States and compared to the 1951–1980 period for Alaska and Hawaii. The bars on graphs show the average temperature change by decade for the 1901–2012 period relative to the 1901–1960 average for each region. The far-right bar in each graph (2000s decade) includes 2011 and 2012. The last decade has been the warmest of any during this comparison period. *Diagram adapted from NOAA NCDC/CICS-NC. Reprinted from Walsh, J., Wuebbles, D., Hayhoe, K., Kossin, J., Kunkel, K., Stephens, G., et al. 2014. Ch. 2: our changing climate. In: Melillo, J.M., Richmond, T.C., Yohe, G.W. (Eds.), Climate Change Impacts in the United States: The Third National Climate Assessment, U.S. Global Change Research Program, pp. 19–67. doi:10.7930/J0KW5CXT.*

temperatures exceeded three standard deviations from the mean. Hansen et al. (2012) found that the frequency of these extremes would typically be 13% and would extend over 0.1–0.2% of the earth; however, during the most recent decade, summer temperatures exceeded this threshold. This large increase in temperature extremes is expected to continue around the globe for the future.

The changing amplitude of the temperature pattern shows an interesting characteristic because the minimum temperatures are increasing while the maximum temperatures are showing less of an increase. This increase is a result of the increasing concentrations of greenhouse gases and water vapor in the atmosphere causing a reduction in the re-radiation of energy at night. There are implications of the change in the diurnal pattern for the physiology of plants and the ability of the plants to efficiently extract C from the atmosphere and increase C storage in the soil as summarized by Hatfield et al. (2011). In the Midwest, there is an increase in the minimum temperatures while the maximum temperatures are showing a decline. Temperature patterns have also revealed that temperatures are increasing more in the winter months than in the summer leading to a longer frost-free period and potentially earlier planting dates and a longer growing season. The changes in temperature will have a big effect on the soil environment which will affect soil temperature and soil water evaporation rates which will in turn affect C and nutrient cycling in soils and the soil biological activity.

2.2.2 Precipitation

Crops and soil biological systems require water to function and the water is ultimately supplied by precipitation. However, projections of changes in precipitation are difficult for scientists to determine the exact amounts, but trends are important to understand what may be occurring relative to agriculture production. Two general trends are present in the precipitation signal for the globe; there is an expected increase in annual precipitation and a shift in the seasonality of precipitation. In the near term, Kirtman et al. (2013) projected an increase in precipitation in the high to mid-latitudes with a concurrent increase in evaporative demand and specific humidity. In the long-term projections of climate, Collins et al. (2013) found that precipitation would increase concurrent with the increasing mean global temperatures at the rate of 1–3% C^{-1} because of the increase in water vapor pressure and increased evaporation to place more water vapor into the atmosphere. They projected an increase in spatial variation in precipitation along with an increasing difference between wet and dry seasons. Trends in evaporation in the long term are expected to continue to show increased amounts because of the increasing temperatures (Collins et al., 2013). The projected increases in the variation between wet and dry seasons would suggest there would be increased variation within the season and among years with the potential for more extreme events. Scaling down to the United States, there has been an increase in flooding and droughts over the past two decades and the expectation is for these extremes to continue with climate change (Walsh et al., 2014). The projections for the United States show the projected changes in seasonality showing more spring precipitation across the higher latitudes with a decreasing summer precipitation (Fig. 2.2). It is expected in the mid-latitudes around the globe that the trends seen in the United States are typical of trends in climate.

Precipitation is exhibiting more variation in both amount and seasonality that will have a large impact on the availability of soil water to crops. As a general relationship, the greater the variation in precipitation, the greater the variation in crop production and with the expected increase in variability in water availability the more we can expect crop production

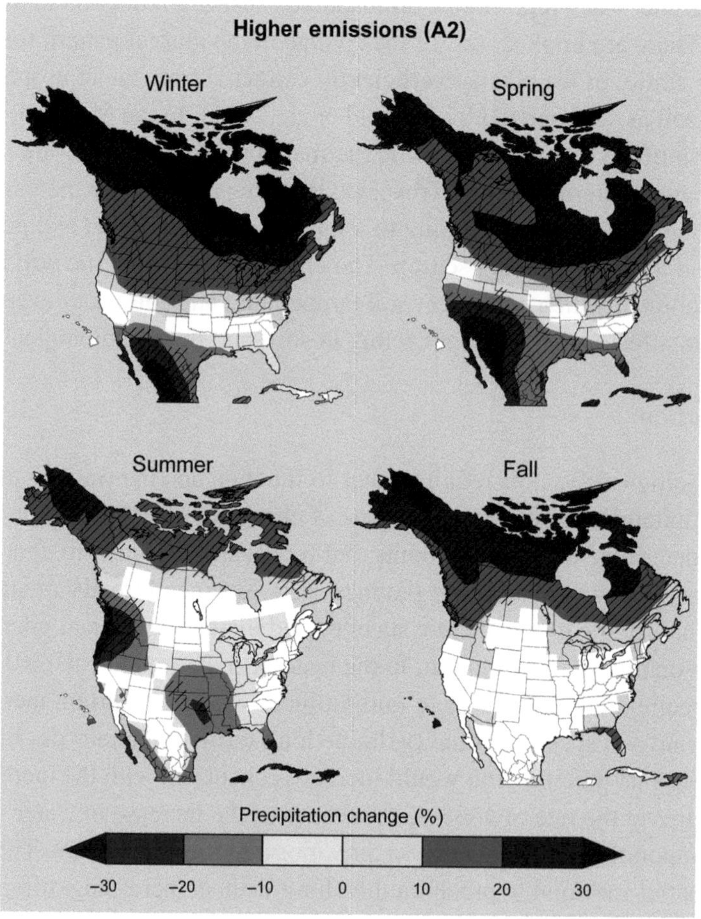

Figure 2.2
Projected change in seasonal precipitation for 2071–99 (compared to 1970–99) under an emission scenario with continued increases in emissions (A2). Hatched areas indicate that the projected changes are significant and consistent among models. White areas indicate the changes are not projected to be larger than expected from natural variability. In general, the northern part of the United States is projected to experience more winter and spring precipitation, while the southeastern United States is projected to experience less precipitation in the spring. *Figure courtesy of NOAA NCDC/CISC-NC. Reprinted from Walsh, J., Wuebbles, D., Hayhoe, K., Kossin, J., Kunkel, K., Stephens, G., et al. 2014. Ch. 2: our changing climate. In: Melillo, J.M., Richmond, T.C., Yohe, G.W. (Eds.), Climate Change Impacts in the United States: The Third National Climate Assessment, U.S. Global Change Research Program, pp. 19–67. doi:10.7930/J0KW5CXT.*

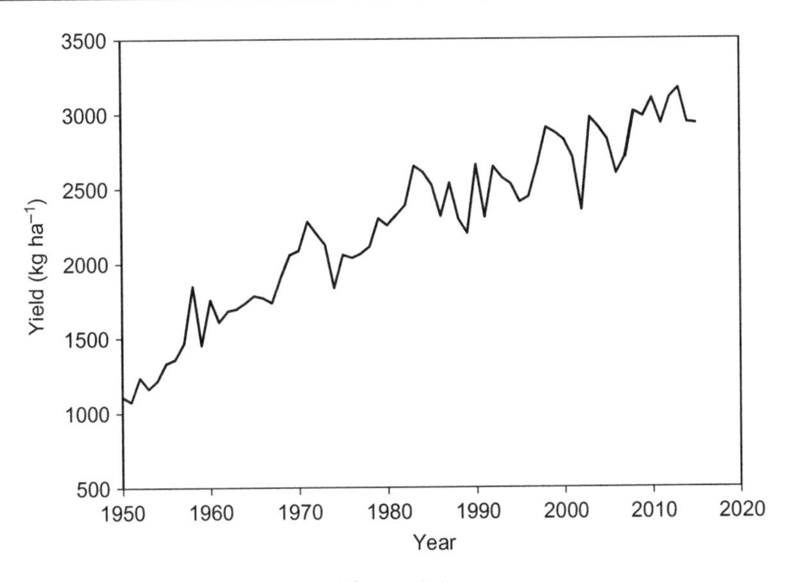

Figure 2.3
Trends of wheat production in the United States since 1950.

to vary. The intersection of soil health and water-holding capacity (WHC) will increase the impact of precipitation variability because soils with low organic matter are dependent upon timely precipitation for optimum production (Hatfield, 2012). The increasing variability in precipitation will directly impact the availability of water in the soil profile to support biological activity. We see this effect in winter wheat yields across the United States as depicted in Fig. 2.3.

2.2.3 Extreme Events in Precipitation

Extreme events in precipitation will continue to increase under climate change. All of the climate projections indicate an increase in extreme precipitation events (Collins et al., 2013; Kirtman et al., 2013; Walsh et al., 2014). Observed data from the United States have already shown there are more intense precipitation events occurring with regional variation in the trends (Fig. 2.4). There have been decadal increases in the events with heavy precipitation, with the largest increases in the Northeast, Midwest, Great Plains, and Southeast (Fig. 2.4). Coupled with more intense events is evidence of a decreasing frequency between precipitation events. These changes will affect water availability for crop production and soil biological activity.

An observation by Hao et al. (2013) on the occurrence of extreme temperature and precipitation events was evaluated with observations from ground-based meteorological stations and a suite of climate models (computer simulations). These evaluations were done for four combinations

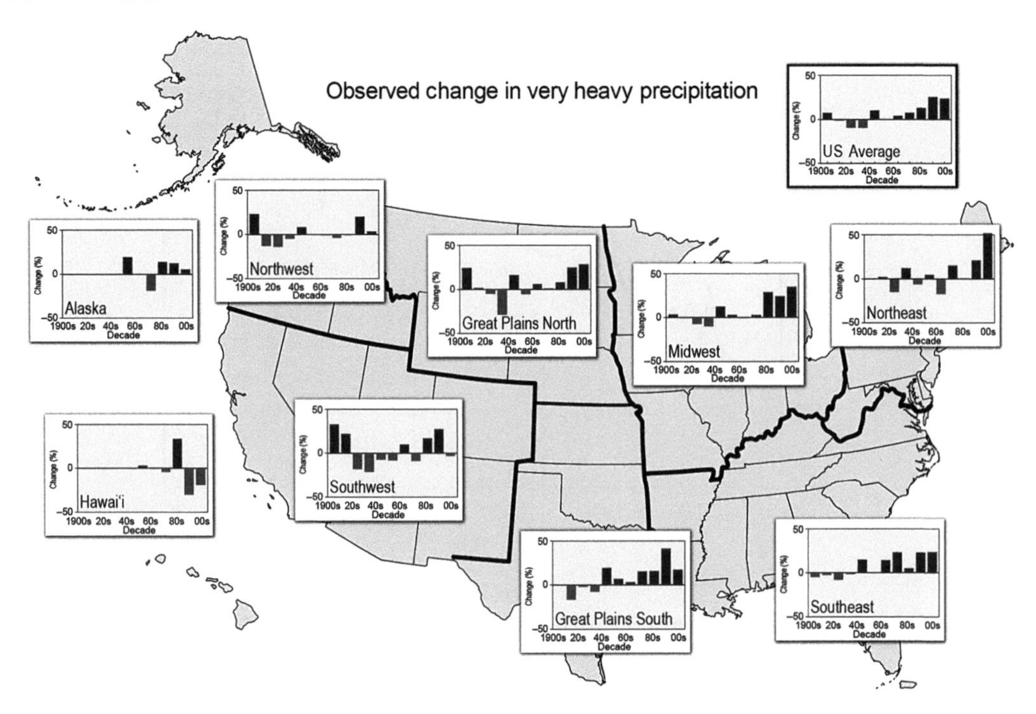

Figure 2.4

Percent changes in the annual amount of precipitation falling in heavy events (defined as the heaviest 1% of all daily events from 1901 to 2012 in each region). *Figure from NOAA NCDC/CSIC-NC. Reprinted from Walsh, J., Wuebbles, D., Hayhoe, K., Kossin, J., Kunkel, K., Stephens, G., et al. 2014. Ch. 2: our changing climate. In: Melillo, J.M., Richmond, T.C., Yohe, G.W. (Eds.), Climate Change Impacts in the United States: The Third National Climate Assessment, U.S. Global Change Research Program, pp. 19–67. doi:10.7930/J0KW5CXT.*

of temperature and precipitation: warm/wet (high temperature/high precipitation), warm/dry (high temperature/low precipitation), cold/wet (cold temperatures/high precipitation), and cold/dry (cold temperatures/low precipitation). They compared the 1978–2004 period with the 1951–77 period on a global scale and found warm/wet and warm/dry extremes to be increased. The warm/wet extremes increased in the high latitudes and tropics while the warm/dry extremes increased in many areas, e.g., central Africa, eastern Australia, northern China, parts of Russia, and the Middle East (Hao et al., 2013). Conversely, the extremes in the cold/wet and cold/dry combinations decreased over most of the earth. The increase in the warm/wet and warm/dry extremes over many of the agriculture areas will have a negative impact on agricultural productivity and change the distribution of viable crop production and more importantly the effect on soil temperature and the functionality of soil biological systems. The extremes observed in the air/atmosphere will carry over to the soil environment and soil biological system will respond to a disruption of a stable soil microclimate in order to efficiently cycle C into organic matter as a major component of soil health.

2.2.4 Meteorological Variables (Solar Radiation, Wind, Humidity)

Changes in climate induce changes in other atmospheric variables affecting plant growth and soil biological activity. With increases in water vapor and concurrent increases in cloud cover there would be a decrease in incoming solar radiation. Observations by Stanhill and Cohen (2001) on global solar radiation showed a 2.7% reduction per decade during the past 50 years, with the current solar radiation totals reduced by $20\,W\,m^{-2}$. Across the United States, Stanhill and Cohen (2005) found that after 1950 the Northeast, West, and Southwest regions in the United States are showing the largest decreases in sunshine duration. Pan et al. (2004) projected solar radiation to continue to decrease because of the increased concentrations of atmospheric greenhouse gases and atmospheric scattering. Medvigy and Beaulieu (2011) examined the variability in solar radiation around the world and concluded there was an increase in solar radiation variability correlated with increases in precipitation variability and convective cloud amounts. The changes in solar radiation will affect growth of plants and there will be local effects because of the increased variability in precipitation. Variation in the growth of plants and biomass production will directly affect the amount of C sequestered in soil and SOM build ups as a key factor for improving soil health.

Projections of climate change by Collins et al. (2013) and Kirtman et al. (2013) suggested that humidity would increase because of the increased evaporation from oceans. Even though there is an increase in humidity, the increase in temperature will lead to increased atmospheric demand. This will lead to more water use by crops and in soils with limited WHC, there will be occurrence of more water stress (Hatfield, 2012).

It is difficult to assess the potential changes in wind speed. Increases in extreme precipitation events are likely to increase the winds during these storms, leading to potential damage of crops or increasing the potential for wind erosion from bare soil. Disturbance of the soil surface by wind or water will have a negative impact on soil health.

2.3 Patterns of Agriculture Land Use

Changing climate will impact the potential areas for crop production and ultimately soil productivity. As a foundation to understand these impacts the development of tools to define where crops can be produced is critical to further understanding the impact of climate variation on crop distribution and soil productivity (van Wart et al., 2013). Crop distribution in arable areas is indicative of the suitability for that species to thrive within a given agroclimatic zone. Zomer et al. (2008) demonstrated how climate zones could be used to evaluate technologies that would enhance the ability of management practices to offset the impacts of climate change and this approach offers a strategy on how to effectively utilize the information gleaned from agroclimatic indices. Development of agroclimatic zones is generally based on a combination of temperature and precipitation and one of the earlier examples to develop agroclimatic normal for a given location was reported by Neild and

Richman (1981) where they used growing degree days (GDD) and seasonal precipitation patterns to evaluate the differences among maize (*Zea mays* L.) hybrids and locations around the world to compare differences in maize phenology. There have continued to be advances in the development of agroclimatic indices to evaluate the suitability of a location for a particular crop. Siddons et al. (1994) found that development of robust agroclimatic indices requires a long time frame for observations and extensive experimentation. They applied this approach to the development of agroclimatic zones for white lupin (*Lupinus albus*) in England and Wales. A recent report by Holzkämper et al. (2013) incorporated six factors into a suitability index for crops. These indicators included average daily minimum temperatures below 0°C for frost impacts, daily mean temperature to determine plant growth, average daily maximum temperature above 35°C for heat stress, average daily soil water availability (precipitation–reference evapotranspiration), and length of the phenological period (days) to account for the effects of changing phenological development on biomass accumulation and crop yield. They were able to relate their suitability index to maize yields for a number of locations around the world with a positive relationship between productivity and the suitability index. This approach is a refinement of the original approach by Neild and Richman (1981) to add more factors into their index to more closely match crop physiological responses.

Agroclimatic zones represent a unique combination of factors affecting plant growth to evaluate the potential for a given crop to produce a grain or forage crop (e.g., Neild and Richman, 1981; Simane and Struik, 1993; Araya et al., 2010; Daccache et al., 2012; Falasca et al., 2012; Moeletsi and Walker, 2012; van Wart et al., 2013). The form of the index depends upon the assumption of the factors limiting growth. For example, Araya et al. (2010) evaluated a combination of factors for the semiarid areas in Ethiopia to determine the suitability of this region for growing barley (*Hordeum vulgare* L.) and teff (*Eragrostis tef*). Because water is a primary limitation in this region and the precipitation pattern during the summer rainfall period is extremely variable, they developed an index using a combination of precipitation and potential evapotranspiration to determine the suitability of being able to plant these crops after June 15. One aspect of their index was to determine the cumulative 5-day total of precipitation relative to the cumulative evapotranspiration over this same period to evaluate whether adequate soil moisture was present in the seed zone to establish the crop. The length of the growing period was determined as the length of time between when adequate soil water was available and the cessation of growth when evapotranspiration began to exceed precipitation for five cumulative days (Araya et al., 2010). Adoption of conservation practices with plant residue on the soil surface will decrease soil water evaporation rate from soils and would alter soil water dynamics in the seed zone (Hatfield et al., 2001). In semiarid regions, where water is the primary limitation, the availability of soil water becomes the dominant factor determining crop suitability. They found this method was superior to the traditional method of using temperature and altitude as the foundation for agroclimatic zones and were able to divide the area into eight regions compared to the traditional five regions and proposed this increase in spatial resolution would help producers better manage drought in this area.

Daccache et al. (2012) incorporated soil variability into their approach to evaluate the potential need for irrigation in the viability of potato (*Solanum tuberosum* L.) production in England and Wales. One of the fundamental components in their index is the potential soil moisture deficit (PSMD) defined as

$$PSMD_i = PSMD_{i-1} + ET_i - P_i \tag{2.1}$$

where $PSMD_i$ is the value in month i and $PSMD_{i-1}$ is the value for the previous month, ET_i is the reference evapotranspiration (ET) for the current month calculated with the Penman–Monteith equation as formulated by Allen et al. (1994), and P_i is the precipitation in the current month. They utilized this approach to demonstrate that under increased variation in precipitation there would be a decrease in potato production in this area on land that is currently suited for production.

Exploration of potential new areas for crop production is one application of agroclimatic analysis and Falasca et al. (2012) showed the utility of this approach to determine potential production areas for castor bean (*Ricinus communis* L.) for the semiarid regions of Argentina. They defined the suitable growing regions relative to the temperature ranges, water requirements, and length of the growing season. Using these characteristics they developed a classification as

Optimal (>750 mm; temperature 24.0–27.0°C; >−8°C; >180 frost-free days);

Very suitable (>750 mm; temperature 21.0–23.9°C; >−8°C; >180 frost-free days);

Suitable with humid regime (>750 mm; temperature 16.0–20.9°C; >−8°C; >180 frost-free days);

Suitable 1 with subhumid regime (450–750 mm; temperature 24.0–27.0°C; >−8°C; >180 frost-free days);

Suitable 2 with subhumid regime (450–750 mm; temperature 21.0–23.9°C; >−8°C; >180 frost-free days);

Suitable 3 with subhumid regime (450–750 mm; temperature 16.0–20.9°C; >−8°C; >180 frost-free days);

Marginal due to humidity (200–450 mm); marginal due to temperature (<16.0°C);

Marginal due to frosts (<180 frost-free days or <−8°C); and

Not suitable areas (combination of two or more of the following variables: <200 mm; <16.0°C; <−8.0°C; <180 frost-free days).

Using this type of approach for different crops one can define the areas suitable for production and begin to develop a system where the impacts of climate change can be assessed through changes in temperature and precipitation. Moeletsi and Walker (2012)

developed a climate risk index for maize in South Africa based on three climatic parameters; onset of rains, frost risk, and drought risk. They combined these parameters into a Poone AgroClimatic Suitability Index (PACSI) and utilized a weighed distribution of these parameters as

$$PASCI = O\,x0.3 + FF\,x0.3x\,WRSI\,x0.4 \tag{2.2}$$

where O is the probability planting conditions are met, FF is the probability of a frost-free growing period, and Water Requirement Satisfaction Index (WRSI) the water requirements satisfaction index. These indices require sufficient data to develop the probability of the different indices including the length of record sufficient to develop robust probability assessments. An aspect of these indices is the drought risk which is directly affected by the soil WHC and any change in the soil affecting water availability to plants will positively affect the WRSI (Eq. (2.2)).

2.3.1 Crop Distribution and Agroclimatology

2.3.1.1 Temperature effects

A foundation of the agroclimatic indices is the GDD or a representation of temperature based on the temperature ranges of each species. Differences among species were summarized in Hatfield et al. (2011) in which each species has a specific lower temperature value or base temperature, an optimum temperature value, and an upper temperature limit. For example, the values for maize are 8–38°C with an optimum of 34°C for vegetative growth and 18–25°C for reproductive growth and for wheat (*Triticum aestivum* L.) are 0–30°C with an optimum of 20–30°C for the vegetative period and 15–25°C for the reproductive period. The temperature ranges for the optimum growth are lower for the reproductive period than the vegetative period for grain crops. Inclusion of the frost period is an important feature for agroclimatic indices for the temperate climate zones because these events define the beginning and end of the growing season.

There has been a large body of work on temperature effects on plant growth and the greater concern today is the impact of the increasing temperatures under climate change. Estimates of the impact of increasing temperatures on crop yield range from less than 5% with temperature increases less than 1°C (Hatfield et al., 2011) to over 50% yield decline in maize and soybean (*Gylcine max* (L.) Merr.) (Schlenker and Roberts, 2009). Evaluations of current yields of different crops have shown a linear decrease in yields with increasing temperatures. In maize, yield decreased 8.3% per 1°C (Lobell and Field, 2007), in wheat 5.3% yield reduction per 1°C (Innes et al., 2015) and 6% decrease per 1°C rise (Asseng et al., 2015). There is a differential response of plants to temperature throughout the growth cycle and the recent results by Laza et al. (2015) for rice (*Oryza sativa* L.) showed high night temperatures had no effect during the vegetative stage; however, high nighttime temperatures during the reproductive stage reduced yields because of the increased dark respiration rate and spikelet degeneration.

Projections of increasing temperatures will alter the growing regions and productivity of crops. Using a statistical analysis of Kansas wheat yields Tack et al. (2015) linked historical yield and combination of freezing impacts and warming impacts in their regression model incorporating meteorological data to evaluate the impact of future temperature conditions. Their analysis projected a 40% reduction in wheat yields with a 4°C temperature increase and found newer varieties were less able to resist heat stress above 34°C than older varieties. They concluded that selection of newer varieties by producers to offset climate impacts may not be effective. As an adaptation strategy, shifts in the growing season by altering planting dates or maturity ratings are often suggested. There have been several approaches to offset the potential exposure to high temperatures and Rezaei et al. (2015) suggested that shifts in phenological development for winter wheat in Germany would prevent exposure to high-temperature events at anthesis; however, it would not offset the detrimental impacts of high temperatures on grain yield because of the shortening of the grain-filling period. Exposure to high temperatures during the grain-filling period was observed to have large impacts on maize yield of over 50% yield reduction when nighttime temperatures during grain-filling were increased by 4°C (Hatfield and Prueger, 2016). The increase in high-temperature events, especially during the pollination period, and the negative impacts on grain yield would suggest that this factor would be as critical in agroclimatic suitability indices as the occurrence of frost.

Temperature impacts from climate change will create an environment in which crops currently suited for a given area will no longer be economically viable because of low productivity or increased variability among years. The projected rise in temperature coupled with the potential for more extreme events will exceed the optimum temperatures for crops and often exceed the upper threshold and as Rezaei et al. (2015) suggest that planting dates could be adjusted, this may not be sufficient to offset the exposure to high temperatures during the grain-filling period. Hansen et al. (2012) have shown warming temperatures increase the potential for extreme temperature events and Walsh et al. (2014) have shown for the United States the occurrence of higher temperatures during the summer is very likely by the mid- to late century. This will cause shifts in the production regions of many crops in the mid- to high latitudes. However, increased temperatures may not always be negative, as suggested by Yang et al. (2015) when they projected a 2.2% increase in maize, rice, and wheat production in China because of the temperature shift into more northern areas of the country. One aspect of increased air temperatures is the effect on soil temperature which will increase the rate of C and nutrient cycling. The ability to improve soil quality may be diminished as soil temperatures increase and even more at the soil surface where more temperature extremes will occur.

2.3.1.2 Precipitation effects

Precipitation effects on crop productivity are determined by the occurrence of the water deficits in the soil profile. Agroclimatic indices for arid and semiarid regions are often based on precipitation amounts adequate to exceed the evapotranspiration at the time of planting (Neild and Richman, 1981; Araya et al., 2010; Daccache et al., 2012; Moeletsi and Walker, 2012; Holzkämper et al., 2013). Daccache et al. (2012) developed the PSMD index as shown in Eq. (2.1). Moeletsi and Walker (2012) provided a framework for the evaluation of water dynamics based on the WRSI in which the ability to meet the water requirements of the crop at any growth stage could be quantified as

$$WR_i = PET_i \, x \, k_{ci} \tag{2.3}$$

where WR_i is the water requirements for a decadal period during the growth of the crop, PET_i is the potential ET during this decadal period, and k_{ci} is the crop coefficient for this corresponding period of growth. They developed a soil–water balance to determine the dynamics during the season using the plant available water (WA) for a given decadal period as

$$WA_i = Prec_i - SW_{i-1} \tag{2.4}$$

where $Prec_i$ is the precipitation in a given decadal period and SW_{i-1} is the soil water in the profile for the previous decadal period. Soil WHC becomes a critical component of this approach because SW is a function of WHC. They computed the WRSI as

$$WRSI_i = WRSI_{i-1} - \frac{WD_i}{\sum_{i=1}^{end} WR} \tag{2.5}$$

with WD_i the water deficit for decadal period i and defined as

$$WD_i = WR_i - Prec_i - SW_{i-1} \, when \, WR_i > Prec_i + SW_{i-1} \tag{2.6}$$

or

$$WD_i = 0 \, when \, WR_i = Prec_i + SW_{i-1} \tag{2.7}$$

In this process soil water in the profile is quantified as

$$SW_i = Prec_i + SW_{i-1} - WR_i \tag{2.8}$$

$$SW_i = WHC \, when \, SW_i = WHC \tag{2.9}$$

$$SW_i = 0 \, when \, SW_i = 0 \tag{2.10}$$

and developed their PACSI (Eq. (2.2)) as a function of different parameters with water requirements being one of the critical factors determining the suitability for maize growth in South Africa (Moeletsi and Walker, 2012). They were able to evaluate the suitability for maize for various planting dates with a correlation of 0.8 between the PACSI and maize yields.

The changing precipitation patterns in intensity and frequency will have a large impact on the water requirement (WR) (Eq. (2.3)) of different crops and will intersect with the WHC (Eq. (2.9)) of the soil to be able to capture and store SW (Eq. (2.8)). Collins et al. (2013) and Walsh et al. (2014) show the changing precipitation patterns under climate change (Fig. 2.2) with decreasing summer precipitation amounts over the United States. If we apply the PACSI (Moeletsi and Walker, 2012) to these changes then maize production could become more variable because of the change in soil water availability. To evaluate this concept, we computed the yield gaps for different counties with maize production in the Midwest following the approach of Egli and Hatfield (2014a) and show the response for maize in Story County, Iowa, since 1950 (Fig. 2.5). To extend this analysis we computed the frequency of the yield gaps over this period of record (Fig. 2.6) and related this to the temperature and precipitation observations for each year. The inability to close the yield gap was due to the lack of soil water to meet the WR of the crop and was due to precipitation events of insufficient amounts to recharge the soil profile during the grain-filling period. The dynamics of this response has been described by Hatfield (2012) to show the largest effect on maize yields in the central United States was the lack of sufficient water availability during the grain-filling period to meet the evaporative demand. The increase in precipitation variability with climate change will increase variation in crop yield as shown in Fig. 2.5. This effect has been shown across ecosystems by Porporato et al. (2004) who found that a soil–water balance model that incorporated the hydroclimatic variability to include frequency and precipitation amounts was positively linked to the soil moisture dynamics in determining vegetative conditions. Soil water becomes the dominant factor affecting vegetative productivity in both

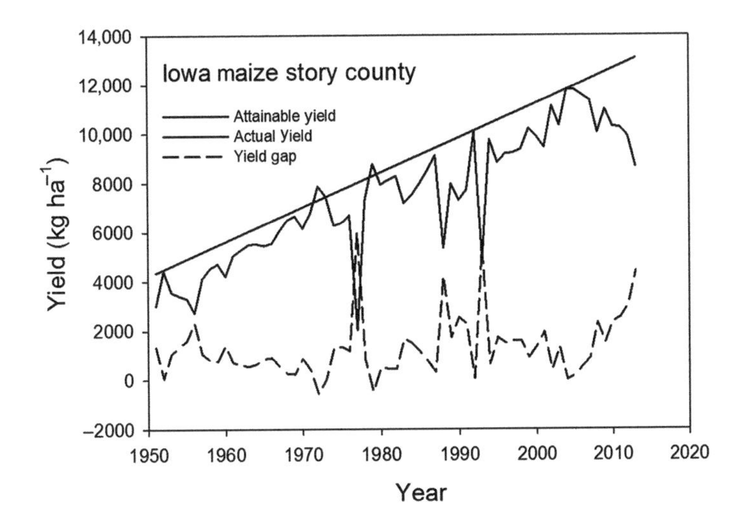

Figure 2.5
Maize yield trends for Story County, Iowa, from 1950 through 2013 with a fitted line to represent attainable yield and yield gaps computed as the difference between attainable and actual yield.

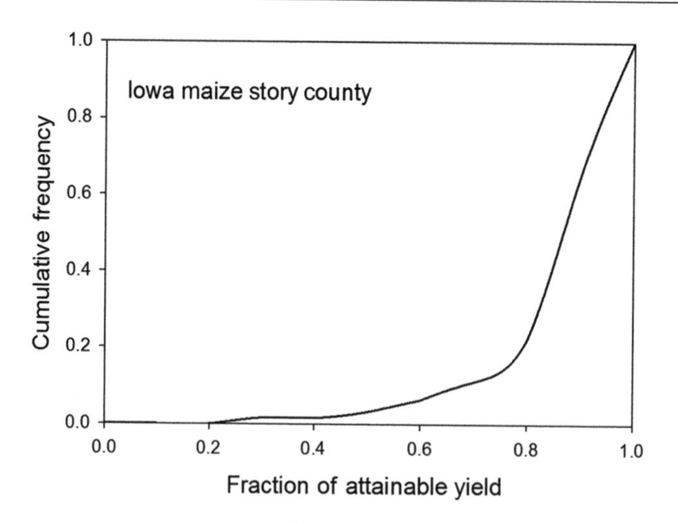

Figure 2.6
Cumulative frequency of attainable yield for the Story County, Iowa, maize yields for the period from 1950 through 2013.

cultivated and natural systems and the ability of the soil to infiltrate and store precipitation will become a critical factor to offset the impact of increasing variability in the precipitation regime. Increases in SOM and the resultant impact on soil WHC will increase the ability of a soil to store water and increase the infiltration rate. Both of these factors will increase the efficiency of a soil to offset variation in precipitation.

2.3.2 Interaction Between Climate Change and Soil Productivity

Climate change and climate variability will impact soil productivity and there have been numerous reports on these impacts. These impacts vary by latitude with the largest reductions in crop yields expected in the mid-latitudes because of the increased temperatures and reduced water availability (Motha and Baier, 2005). Increasing the length of the growing season and the projected increases in CO_2 will enhance crop productivity in the higher latitudes. In Europe, Maracchi et al. (2005) suggested there may be positive impacts because of the introduction of new species and potential expansion into new land areas; however, there would be the potential for increased variation in production because of variations in water availability. Tao et al. (2009) stated that climatic change would provide an opportunity for more diverse cropping systems in China. In the tropics, Sivakumar et al. (2005) found agricultural productivity was sensitive to temperature increases and changes in the nature and characteristics of the monsoon. They projected that changes in agricultural productivity would not only be affected by temperature but also the soil water balance through the combination of the increased temperature and variation in the precipitation patterns. Subash et al. (2014) in India found rice (*Oryza sativa* L.) and wheat yields were sensitive to changes in temperature

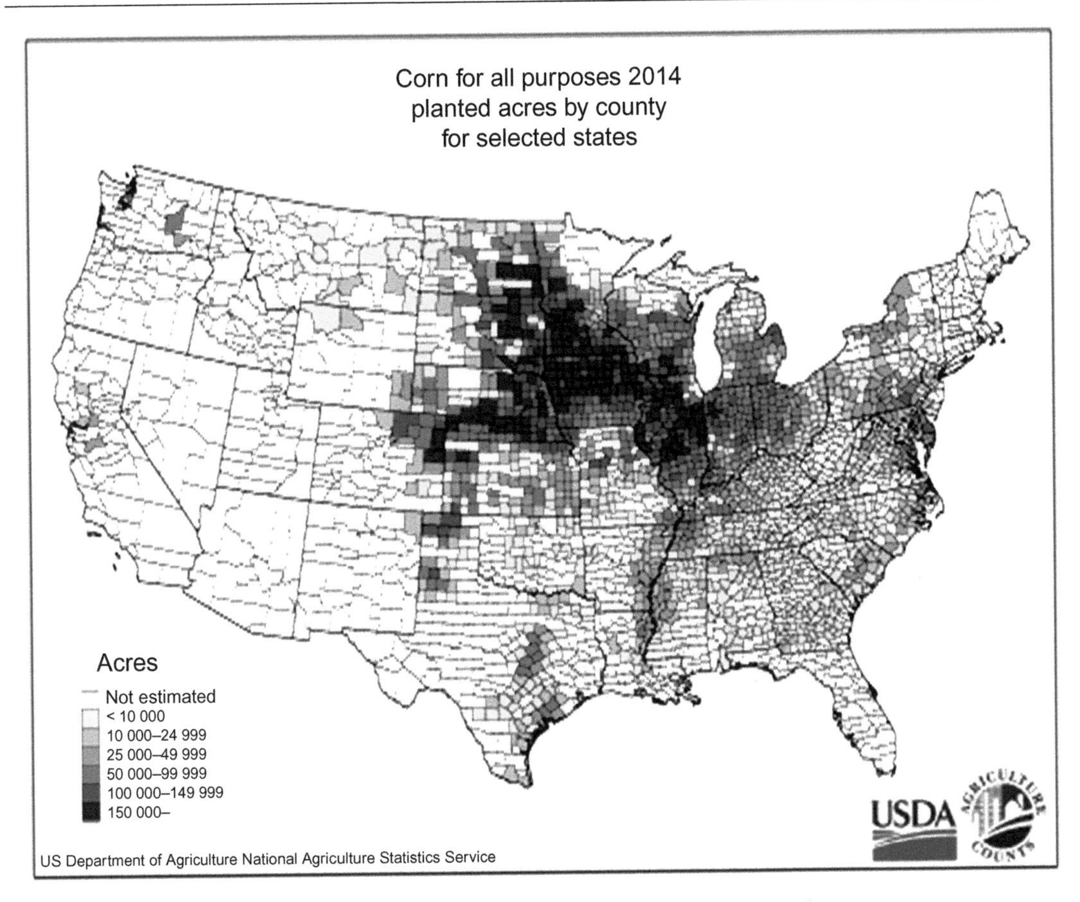

Figure 2.7
Planted area for maize in the United States for 2014. *Data from National Agriculture Statistics Service, www.nass.usda.gov (accessed November 24, 2015).*

and precipitation and management of these crops should include nutrient management as one of the critical input practices. Zabel et al. (2014) using climate projection models found there could be an expansion of suitable cropland by 5.6 million km^2 mainly in the high latitudes of Canada, China, and Russia which would be offset by a decrease in suitability in the tropical regions.

The United States has already experienced a shift in crop distribution as evidenced by the rapid expansion of maize production into the northern Great Plains (Fig. 2.7) with a concurrent decline in wheat, sunflower (*Helianthus annuus* L.), and canola (*Brassica napus* L.) planted areas. Although there is the northern expansion of planted area this doesn't imply that crop yields will be maintained because of the inability of the soil to store sufficient amounts of water. This effect was observed in a recent analysis of county-level maize and soybean yields in the Midwest United States by Egli and Hatfield (2014a,b) where they showed the

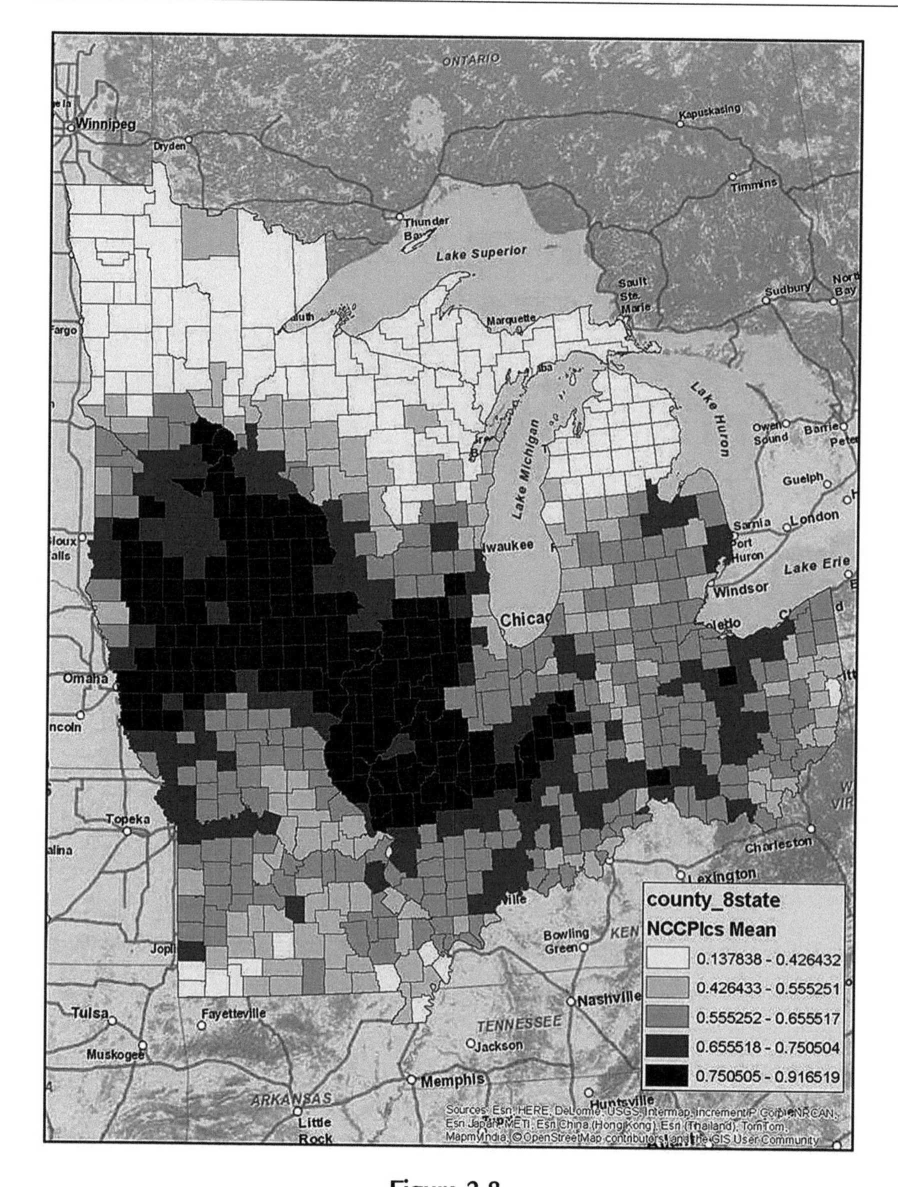

Figure 2.8
Distribution of the National Crop Commodity Productivity Index (NCCPI) for arable agricultural soils in each county of the Midwest United States.

average county yields were positively related to the National Crop Commodity Productivity Index (NCCPI) as developed by NRCS (NRCS, 2012). The distribution of the NCCPI for the Midwestern United States is shown in Fig. 2.8 and shows that as the production areas is expanded north in the United States the productivity of these soils may not achieve the yield levels because of limitations of the soil to provide adequate soil water and would show more variation among years with the increasing variable temperature and precipitation patterns expected in the future. This effect is not unique to the United States and Liu et al. (2013) used a remote sensing index (normalized difference vegetative index) to evaluate land use effect on a global scale and found the differences in productivity changes were more strongly related to precipitation changes than temperature changes. A similar result was observed in Bulgaria on maize by Popova et al. (2014) using the standardized precipitation index to show that precipitation was declining in the past 50 years and the maize yield losses were greater in soils with low WHC. They found that changes in July–August precipitation were the most significant factor related to maize yield variability.

Changing climate throughout the world has prompted a number of studies aimed at evaluating the potential impact of climate change on agricultural productivity. One of the overall summaries was prepared by Walthall et al. (2012) to show the impact on many different aspects of the agricultural system and to document how agricultural systems are part of a sociological framework that determines the decision-making process relative to adaptation to climate change signals. Jackson et al. (2011) documented these responses for producers in Yolo County in California and showed there was a need for decision tools that not only incorporated potential changes in the climate but also economic and sociological factors to help producers understand different adaptation strategies. This would suggest that migration of crops to new areas would not only be driven by changes in agroclimatic factors but also the ability of producers to adapt their management capability for new crops and their ability to capture markets for their production. The use of the NCCPI is similar to the analysis developed by Brown et al. (2011), where they showed land capability classes for production under climate scenarios were dominated by WHC since the ability of the soil to provide adequate soil water would be the primary factor determining crop productivity. Drought risk caused by increasing uncertainty in precipitation requires a linkage with the socioeconomic factors governing the adaptive capacity and knowledge and coping strategies and need to be focused at the local scale (Brown et al., 2011).

Observations from production regions and crops around the world demonstrate the interaction between climate variation and soil productivity (Notaro, 2008). Increasing interannual variation in climate reduces the vegetative cover, especially in semiarid areas, and favors the expansion of grass cover compared to tree cover in these regions (Bounoua et al., 2004; Notaro, 2008). In Niger, Ben Mohamed et al. (2002) observed millet (*Pennisetum glaucum* L.) production was related to sea surface temperature anomalies (temperature variation during the growing season), the amount of rainfall in July, August, and September, the number of rainy days, and a wind erosion factor. The latter parameter was a surrogate for the soil degradation factor. They cautioned that expansion of millet production into marginal areas would create additional risk because increased rainfall variation would increase the potential for crop failure. A similar result was reported by de Vasconcelos et al. (2014) for the Brazilian La Plata Basin with the rapid expansion of maize, soybean, and sugarcane (*Sacharum officinarum* L.)

in this region and the potential susceptibility to climate variation in marginal land areas. In Hungary, Gaál et al. (2014), observed that future productivity of crops would depend upon the soil resource being able to provide adequate water to the crop and that increased variation in precipitation would increase the risk of crop failure. Conversely, winter crops may be less susceptible to climate variation because they are grown in the period of the year where there are cooler temperatures and reduced water demand (Mondal et al., 2014). Observations for wheat and pulses (*Fabaceae* spp.) in India by Mondal et al. (2014) show that climate change, especially increases in mean air temperature and variation in precipitation would be the primary determinants of productivity.

2.4 Role of Soil Services in Mitigating Effects of Climate Variability

Soil water is one of the primary factors affecting the productivity of crops and the increasing variation in precipitation will increase the potential variation in cropping distribution and production (Maracchi et al., 2005; Motha and Baier, 2005; Sivakumar et al., 2005; Tao et al., 2009; Hao et al., 2013; Zabel et al., 2014). Soil water availability is a function of the SOM content (Hudson, 1994) and reduction of SOM by tillage or residue removal will increase the vulnerability to climate variation. The process of soil degradation (chemical, physical, and biological) is extensive throughout the world and especially in the tropics and subtropics (Lal, 1993). Since soil water availability is critical to crop production, changes in soil structure and hydraulic properties are related to cropping systems, i.e., tillage systems, and residue management, led to a degradation of soil structure causing maize yield reductions as large as 50% (Wang et al., 1985). Soil conservation practices which maintained crop residue on the soil surface had a positive impact on water conservation, and translated into increased water availability for the crop in semiarid regions, leading to greater yield potential (Unger et al., 1991). Manipulation of the soil and adoption of conservation practices can have a positive impact on the ability of the soil to supply water for crop demands and practices that enhance soil C provide a means of reducing yield variation under a changing climate (Hatfield et al., 2001). Adoption of agricultural practices directed toward enhancing soil C would provide a benefit to crop productivity and the environment by providing a sink for CO_2 from the atmosphere. Recent studies to evaluate a range of scenarios of climate change relative to the potential change in soil C have shown that combining global change models with ecosystem models produces a consistent projection of soil C increases (Yurova et al., 2010); however, the variation in the amounts of soil C change are a function of temperature, precipitation, evaporation, and the original soil properties (Wang et al., 2011). There remains a large amount of uncertainty in the potential amount of soil C that could be sequestered in agricultural soils (Davidson and Janssens, 2006; Trumbore and Czimczik, 2008; Sun and Mu, 2012).

Increasing soil C will depend upon the cropping system placed on the landscape. For example, in Scotland, Brown et al. (2014) observed that land use changes would have to be linked to climate change scenarios in order to develop a strategy for changing soil C. This is further complicated by the differences among soil types and their response to management systems (Brown and Castellazzi, 2015). Byrd et al. (2015) observed for rangeland systems that land use patterns coupled with precipitation patterns were the dominant factors affecting the change in soil C and land conversion to annual cropping systems coupled with tillage could increase the rate of soil C loss. They suggested that the feedbacks between land use and climate would have to be understood in order to develop effective soil management systems to increase soil C. This will be further complicated by the interaction between land use patterns and soil microbial communities as demonstrated by Lauber et al. (2013), since conversion of organic material to soil C is dependent upon microbial activity. Soil microbial activity will determine the rate of soil respiration and interannual variability was determined by temperature and precipitation patterns (Davidson et al., 1998; Raich et al., 2002; Davidson and Janssens, 2006; Chen et al., 2012).

Land use practices affect the regional climate and modify the temperature and precipitation patterns (Bounoua et al., 2004; Kaharabata et al., 2014). Bounoua et al. (2004) observed for Bolivia that conversion of tropical forest to cropland increased the maximum temperatures by as much as 2°C, with a minor increase in minimum temperatures with an overall increase in the diurnal temperature range. The interactions between the regional climate and plant growth will affect the capacity of the ecosystem to sequester C. Understanding these dynamics would provide a framework for assessing the potential for C sequestration in different ecosystems and the expected change in soil.

2.5 Implications of Soil Health to Offset Climate Variability

Utilization of agroclimatic indices to evaluate the suitability of a given location to produce a crop and the potential productivity of those crops has provided a framework for the assessment of climatic change (Neild and Richman, 1981; Araya et al., 2010; Daccache et al., 2012; Moeletsi and Walker, 2012; Holzkämper et al., 2013, van Wart et al., 2013). These approaches demonstrate the impact of the changing temperature and precipitation regime and the role of adequate soil water storage to offset the variable precipitation regime expected in the coming decades. The approach of Moeletsi and Walker (2012) provides a detailed assessment of the role of soil WHC on the suitability of a given region for a crop. The analyses of future climate on crop productivity demonstrate the sensitivity to temperature and precipitation with precipitation being the dominant factor affecting productivity in the short term (Maracchi et al., 2005; Motha and Baier, 2005; Sivakumar et al., 2005; Tao et al., 2009; Hao et al., 2013; Zabel et al., 2014). Therefore, any soil management practice that can benefit soil water storage and improve soil water availability will have a positive impact on crop productivity and biomass production and subsequent improvement of soil services. These changes will be especially important given the projections that extreme precipitation events, including drought, are expected to increase with climate change (Calanca, 2007; Hansen et al., 2012; Collins et al., 2013; Walsh et al., 2014).

Soil degradation either through erosion (water or wind) or practices that decrease soil C content will reduce WHC and infiltration rates. Degraded soils have been shown to have an increased sensitivity to climate change (Notaro, 2008) and poor-quality soils have shown a reduced yield potential (Egli and Hatfield, 2014a,b). Given that, soil health is often associated with soil C content and therefore enhancing soil WHC, enhancing the ability of the soil to supply more water to the growing crop would increase productivity within a region and also provide the potential to expand and stabilize crop productivity that will occur with changing temperatures. Adequate soil water supplies to the crop can offset the impacts of temperature extremes that are projected to increase during the growing season (Calanca, 2007; Hansen et al., 2012; Collins et al., 2013; Walsh et al., 2014). Sustainable intensification is the goal to provide more output per unit land area; however, the ability to increase production requires developing the linkage between the capacity of the soil to provide water and nutrients and the climate to provide a favorable environment for the crops to thrive.

2.6 Conclusions

Climate change is occurring around the world and impacts the ability to produce agricultural crops because of changing land use patterns and variation in production among years. Temperature and precipitation are the two climatic variables exerting the largest impact on agriculture production because of their direct impact on the rate of growth and the ability to produce harvestable products. Air temperatures are projected to increase with the occurrence of more extremes during the summer growing season while precipitation amounts are projected to increase in the mid-latitudes with a shift in seasonality of precipitation with greater amounts in the spring and less in the summer and with greater occurrences of extreme events. Changes in climate are projected to expand the ranges for some crops and increase the variation in production among years and regions. Utilization of agroclimatic indices to assess the suitability of given regions provides a robust framework for incorporating the climatic variables, and temperature and precipitation represent the dominant factors for climate while soil WHC is the dominant soil factor for crop suitability and crop productivity. Given that, soil WHC is one of the critical factors determining suitability for crop production and production levels, changes in soil water availability will directly affect production. This will subsequently enhance the SOM and soil quality, which offers the potential to offset the impacts of climate variation and increase soil resilience.

References

Allen, R.G., Smith, M., Perrier, A., Pereira, L.S., 1994. An update for the definition of reference evapotranspiration. ICID Bull. 43, 1–34.

Araya, A., Keesstra, S.D., Stroosnijder, L., 2010. A new agro-climatic classification for crop suitability zoning in northern semi-arid Ethiopia. Agric. For. Meteorol. 150, 1057–1064.

Asseng, S., Ewert, F., Martre, P., Rötter, R.P., Lobell, D.B., Cammarao, D., et al., 2015. Rising temperatures reduce global wheat production. Nat. Clim. Change 5, 143–147.

Ben Mohamed, A., Van Duivenbooden, N., Abdoussallam, S., 2002. Impact of climate change on agricultural production in the Sahel- Part 1. Methodological approach and case study for Millet in Niger. Clim. Change 54, 327–348.

Bindi, M., Olesen, J., 2010. The responses of agriculture in Europe to climate change. Reg. Environ. Change 11 (S1), 151–158.

Bounoua, L., DeFries, R.S., Imhoff, M.L., Steininger, M.K., 2004. Land use and local climate: a case study near Santa Cruz. Bolivia. Meteorol. Atmos. Phys. 86, 73–85.

Brown, I., Castellazzi, M., 2015. Changes in climate variability with reference to land quality and agriculture in Scotland. Int. J. Biometeorol. 59, 717–732.

Brown, I., Castellazzi, M., Feliciano, D., 2014. Comparing path dependence and spatial targeting of land use in implementing climate change responses. Land 3, 850–873.

Brown, I., Poggio, L., Gimona, A., Castellazzi, M., 2011. Climate change, drought risk and land capability for agriculture: implications for land use in Scotland. Reg. Environ. Change 11, 503–518.

Byrd, K.B., Flint, L.E., Alvarez, P., Casey, C.F., Sleeter, B.M., Soulard, C.E., et al., 2015. Integrated climate and land use change for California rangeland ecosystem services: wildlife habitat, soil carbon, and water supply. Landscape Ecol. 30, 729–750.

Calanca, P., 2007. Climate change and drought occurrence in the Alpine region: how severe are becoming the extremes? Glob. Planet Change 57, 151–160.

Chen, S., Huang, Y., Zou, J.W., Shi, Y., Lu, Y., Zhang, W., et al., 2012. Interannual variability in soil respiration from terrestrial ecosystems in China and its response to climate change. Sci. China Earth Sci. 55, 2091–2098. http://dx.doi.org/10.1007/s11430-012-4464-6.

Collins, M., Knutti, R., Arblaster, J., Dufresne, J.-L., Fichefet, T., Friedlingstein, P., et al., 2013. Long-term climate change: projections, commitments and irreversibility. In: Stocker, T.F., Qin, D., Plattner, G.-K., Tignor, M., Allen, S.K., Boschung, J., Nauels, A., Xia, Y., Bex, V., Midgley, P.M. (Eds.), Climate Change 2013: The Physical Science Basis. Contribution of Working Group I to the Fifth Assessment Report of the Intergovernmental Panel on Climate Change. Cambridge University Press, Cambridge, and New York. xxxp. 108 p.

Daccache, A., Keay, C., Jones, R.J.A., Weatherhead, E.K., Stalham, M.A., Knox, J.W., 2012. Climate change and land suitability for potato production in England and Wales: impacts and adaptation. J. Agric. Sci. 150, 161–177.

Davidson, E.A., Janssens, I.A., 2006. Temperature sensitivity of soil carbon decomposition and feedbacks to climate change. Nature 440, 165–173.

Davidson, E.A., Belk, E., Boone, R.D., 1998. Soil water content and temperature as independent or confounded factors controlling soil respiration in a temperate mixed hardwood forest. Glob. Change Biol. 4, 217–227.

de Vasconcelos, A.C.F., Schlindwein, S.L., Lana, M.A., Fantini, A.C., Bonatti, M., D'Agostini, L.R., et al., 2014. Land use dynamics in Brazilian La Plata Basin and anthropogenic climate change. Clim. Change 127, 73–81.

Egli, D.B., Hatfield, J.L., 2014a. Yield gaps and yield relationships in central U.S. soybean production systems. Agron. J. 106, 560–566.

Egli, D.B., Hatfield, J.L., 2014b. Yield gaps and yield relationships in central U.S. maize production systems. Agron. J. 106, 2248–2256.

Falasca, S.L., Ulberich, A.C., Ulberich, E., 2012. Developing an agro-climatic zoning model to determine potential production areas for castor bean (*Ricinus communis* L.). Ind. Crops Prod. 40, 185–191.

Fuhrer, J., Beniston, M., Fischlin, A., Frei, C., Goyette, S., Jasper, K., et al., 2006. Climate risks and their impact on agriculture and forests in Switzerland. Clim. Change 79, 79–102.

Gaál, M., Quiroga, S., Fernandez-Haddad, Z., 2014. Potential impacts of climate change on agricultural land use suitability of the Hungarian counties. Reg. Environ. Change 14, 597–610.

Handmer, J., Honda, Y., Kundzewicz, Z.W., Arnell, N., Benito, G., Hatfield, J., et al., 2012. Changes in impacts of climate extremes: human systems and ecosystems. In: Field, C.B., Barros, V., Stocker, T.F., Qin, D., Dokken, D.J., Ebi, K.L., Mastrandrea, M.D., Mach, K.J., Plattner, G.K., Allen, S.K., Tignor, M., Midgley, P.M. (Eds.), Managing the Risks of Extreme Events and Disasters to Advance Climate Change Adaptation. A Special Report of Working Groups I and II of the Intergovernmental Panel on Climate Change (IPCC). Cambridge University Press, Cambridge, UK, and New York, NY, USA, pp. 231–290.

Hansen, J., Sato, M., Ruedy, R., 2012. Perception of climate change. PNAS 109, 14726–14727.
Hao, Z., AghaKouchak, A., Phillips, T.J., 2013. Changes in concurrent monthly precipitation and temperature extremes. Environ. Res. Lett. 8 (034014), 7. http://dx.doi.org/10.1088/1748-9326/8/3/034014.
Hatfield, J.L., 2012. Spatial patterns of water and nitrogen response within corn production fields. In: Aflakpui, G. (Ed.),. Agricultural Science Intech Publishers, pp. 73–96. ISBN 978-953-51-0567-1.
Hatfield, J.L., 2014. Soil degradation, land use, and sustainability. In: Songstad, D.D., Hatfield, J.L., Tomes, D.T. (Eds.), Convergence of Food Security, Energy Security, and Sustainable Agriculture. Biotechnology in Agriculture and Forestry 67. Springer, Berlin Heidelberg, pp. 61–74. doi 10.1007/978-3-642-55262-4_4.
Hatfield, J.L., Prueger, J.H., 2016. Temperature extremes: effects on plant growth and development. Weather Clim. Extremes 10, 4–10.
Hatfield, J.L., Sauer, T.J., Prueger, J.H., 2001. Managing soils for greater water use efficiency: a review. Agron. J. 93, 271–280.
Hatfield, J.L., Boote, K.J., Kimball, B.A., Ziska, L.H., Izaurralde, R.C., Ort, D., et al., 2011. Climate impacts on agriculture: implications for crop production. Agron. J. 103, 351–370.
Hatfield, J., Takle, G., Grotjahn, R., Holden, P., Izaurralde, R.C., Mader, T., et al., 2014. Ch. 6: agriculture. In: Melillo, J.M., Richmond, T.C., Yohe, G.W. (Eds.), *Climate Change Impacts in the United States: The Third National Climate Assessment*, U.S. Global Change Research Program, pp. 150–174. doi:10.7930/J02Z13FR. http://nca2014.globalchange.gov/report/sectors/agriculture.
Holzkämper, A., Calanca, P., Fuhrer, J., 2013. Identifying climatic limitations to grain maize yield potentials using a suitability evaluation approach. Agric. For. Meteorol. 168, 149–159.
Hudson, B.D., 1994. Soil organic matter and available water capacity. J. Soil Water Conserv. 49, 189–194.
Innes, P.J., Tan, D.K.Y., Van Ogtrop, F., Amthor, J.S., 2015. Effects of high-temperature episodes on wheat yields in New South Wales, Australia. Agric. For. Meteorol. 208, 95–107.
Jackson, L.E., Wheeler, S.M., Hollander, A.D., O'Geen, A.T., Orlove, B.S., Six, J., et al., 2011. Case study on potential agricultural responses to climate change in a California landscape. Clim. Change 109 (Suppl. 1), S407–S427.
Kaharabata, S.K., Banerjee, S.M., Kieser, M., Desardins, R.L., Worth, D., 2014. Determining the influence of agricultural land use on climate variables for the Canadian Prairies. Int. J. Climatol. 34, 3849–3862.
Kirtman, B., Power, S.B., Adedoyin, J.A., Boer, G.J., Bojariu, R., Camilloni, I., et al., 2013. Near-term climate change: projections and predictability. In: Stocker, T.F., Qin, D., Plattner, G.K., Tignor, M., Allen, S.K., Boschung, J., Nauels, A., Xia, Y., Bex, V., Midgley, P.M. (Eds.), Climate Change 2013: The Physical Science Basis. Contribution of Working Group I to the Fifth Assessment Report of the Intergovernmental Panel on Climate Change. Cambridge University Press, Cambridge, United Kingdom and New York, NY, USA, pp. 75.
Klein, T., Holzkämper, A., Calanca, P., Seppelt, R., Fuhrer, J., 2013. Adapting agricultural land management to climate change: a regional multi-objective optimization approach. Landscape Ecol. 28, 2029–2047.
Lal, R., 1993. Tillage effects on soil degradation, soil resilience, soil quality, and sustainability. Soil Tillage Res. 27, 1–8.
Lauber, C.L., Ramirez, K.S., Aanderud, Z., Lennon, J., Fierer, N., 2013. Temporal variability in soil microbial communities across land-use types. ISME J. 7, 1641–1650.
Laza, M.R.C., Sakai, H., Cheng, W., Tokida, T., Peng, S., Hasegawa, T., 2015. Differential response of rice plants to high night temperatures imposed at varying developmental phases. Agric. For. Meteorol. 209–210, 69–77.
Liu, G., Liu, H., Yin, Y., 2013. Global patterns of NDVI-indicated vegetation extremes and their sensitivity to climate extremes. Environ. Res. Lett. 8 http://dx.doi.org/10.1088/1748-9326/8/2/025009.
Lobell, D.B., Field, C.B., 2007. Global scale climate-crop yield relationships and the impacts of recent warming. Environ. Res. Lett. 2, 1–7.
Lotze-Campen, H., Schellnhuber, H.-J., 2009. Climate impacts and adaptation options in agriculture: what we know and what we don't know. J. Verbrauch. Lebensm. 4 (2), 145–150.
Maracchi, G., Sirotenko, O., Bindi, M., 2005. Impacts of present and future climate variability on agriculture and forestry in the temperate regions: Europe. Clim. Change 70, 117–135.
Medvigy, D., Beaulieu, C., 2011. Trends in daily solar radiation and precipitation coefficients of variation since 1984. J. Climate 25, 1330–1339.

Moeletsi, M.E., Walker, S., 2012. Simple agroclimatic index to delineate suitable growing areas for rainfed maize production in the Free State Province of South Africa. Agric. For. Meteorol. 162–163, 63–70.
Mondal, P., Jain, M., Robertson, A.W., Galford, G.L., Small, C., DeFries, R.S., 2014. Winter crop sensitivity to inter-annual climate variability in central India. Clim. Change 126, 61–76.
Motha, R.P., Baier, W., 2005. Impacts of present and future climate change and climate variability on agriculture in the temperate regions: North America. Clim. Change 70, 137–164.
Natural Resources Conservation Service (NRCS), 2012. User Guide for the National Crop Commodity Productivity Index (NCCPI) Version 2.0. USDA-NRCS, Washington, DC89., (http://www.nrcs.usda.gov/Internet/FSE_DOCUMENTS/nrcs142p2_050734.pdf).
Nearing, M., Pruski, F., O'Neal, M., 2004. Expected climate change impacts on soil erosion rates: a review. J. Soil Water Conserv. 59, 43–50.
Neild, R.E., Richman, N.H., 1981. Agroclimatic normals for maize. Agric. Meteorol. 24, 83–95.
Notaro, M., 2008. Response of the mean global vegetation distribution to interannual climate variability. Clim. Dyn. 30, 845–854.
Pan, Z., Segal, M., Arritt, R.W., Takle, E.S., 2004. On the potential change in solar radiation over the U.S. due to increases of atmospheric greenhouse gases. Int. J. Renewa. Energy 29, 1923–1928.
Popova, Z., Ivanova, M., Martins, D., Pereira, L.S., Doneva, K., Alexandrov, V., et al., 2014. Vulnerability of Bulgarian agriculture to drought and climate variability with focus on rainfed maizeSystems. Nat. Hazards 74, 865–886.
Porporato, A., Daly, E., Rodriguez-Iturbe, I., 2004. Soil water balance and ecosystem response to climate change. Am. Nat. 164, 625–632.
Porter, J.R., Xie, L., Challinor, A.J., Cochrane, K., Howden, S.M., Iqbal, M.M., et al., 2014. Food security and food production systems. In: Field, C.B., Barros, V.R., Dokken, D.J., Mach, K.J., Mastrandrea, M.D., Bilir, T.E., Chatterjee, M., Ebi, K.L., Estrada, Y.O., Genova, R.C., Girma, B., Kissel, E.S., Levy, A.N., MacCracken, S., Mastrandrea, P.R., White, L.L. (Eds.), Climate Change 2014: Impacts, Adaptation, and Vulnerability. Part A: Global and Sectoral Aspects. Contribution of Working Group II to the Fifth Assessment Report of the Intergovernmental Panel on Climate Change. Cambridge University Press, Cambridge, United Kingdom and New York, NY, USA, pp. 485–533.
Raich, J.W., Potter, C.S., Bhagawati, D., 2002. Interannual variability in global soil respiration, 1980–94. Glob. Change Biol. 8, 800–812.
Rezaei, E.E., Siebert, S., Ewert, F., 2015. Intensity of heat stress in winter wheat-phenology compensates for the adverse effect of global warming. Environ. Res. Lett., 10. http://dx.doi.org/10.1088/1748-9326/10/2/024012.
Schlenker, W., Roberts, M.J., 2009. Nonlinear temperature effects indicate severe damages to U.S. crop yields under climate change. Proc. Natl. Acad. Sci. USA. 106, 15594–15598.
Siddons, P.A., Jones, R.J.A., Hollis, J.M., Hallett, S.M., Huyghe, C., Day, J.M., et al., 1994. The use of a land suitability model to predict where autumn-sown determinate genotypes of the white lupin (*Lupinus albus*) might be grown in England and Wales. J. Agric. Sci. Cambridge 123, 199–205.
Simane, B., Struik, P.C., 1993. Agroclimatic analysis: a tool for planning sustainable durum wheat (*Triticum turgidum* var. durum) production in Ethiopia. Agric. Ecosys. Environ. 47, 31–46.
Sivakumar, M.V.K., Das, H.P., Brunini, O., 2005. Impacts of present and future climate variability and change on agriculture and forestry in the arid and semi-arid tropics. Clim. Change 70, 31–72.
Stanhill, G., Cohen, S., 2001. Global dimming: a review of the evidence for a widespread and significant reduction in global radiation with discussion of its probable causes and possible agricultural consequences. Agric. For. Meteor. 107, 255–278.
Stanhill, G., Cohen, S., 2005. Solar radiation changes in the United States during the twentieth century: evidence from sunshine duration measurements. J. Climate 18, 1503–1512.
Subash, N., Gangwar, B., Singh, S., Koshal, A.K., Kumar, V., 2014. Long-term yield variability and detection of site-specific climate-smart nutrient management practices for rice–wheat systems: an empirical approach. J. Agric. Sci. 152, 575–601.
Sun, G., Mu, M., 2012. Responses of soil carbon variation to climate variability in China using the LPJ model. Theor. Appl. Climatol. 110, 143–153.

Tack, J., Barkley, A., Nalley, N.N., 2015. Effect of warming temperatures on US wheat yields. Proc. Natl. Acad. Sci. USA. 112, 6931–6936.

Tao, F., Yokozawa, M., Liu, J., Zhang, Z., 2009. Climate change, land use change, and China's food security in the twenty-first century: an integrated perspective. Clim. Change 93, 433–445.

Torriani, D., Calanca, P., Schmid, S., Beniston, M., Fuhrer, J., 2007. Potential effects of changes in mean climate and climate variability on the yield of winter and spring crops in Switzerland. Clim. Res. 34, 59–69.

Trumbore, S.E., Czimczik, C.I., 2008. An uncertain future for soil carbon. Science 321, 1455–1456.

Unger, P.W., Stewart, B.A., Parr, J.F., Singh, R.P., 1991. Crop residue management and tillage methods for conserving soil and water in semi-arid regions. Soil Tillage Res. 20, 219–240.

van Wart, J., van Bussel, L.G.J., Wolf, J., Licker, R., Grassini, P., Nelson, A., et al., 2013. Use of agro-climatic zones to upscale simulated crop yield potential. Field Crops Res. 143, 44–55.

Walsh, J., Wuebbles, D., Hayhoe, K., Kossin, J., Kunkel, K., Stephens, G., et al., 2014. Ch. 2: our changing climate. In: Melillo, J.M., Richmond, T.C., Yohe, G.W. (Eds.), *Climate Change Impacts in the United States: The Third National Climate Assessment*, U.S. Global Change Research Program, pp. 19–67. http://dx.doi.org/10.7930/J0KW5CXT.

Walthall, C.L., Hatfield, J., Backlund, P., Lengnick, L., Marshall, E., Walsh, M., et al., 2012. Climate Change and Agriculture in the United States: Effects and Adaptation. USDA Technical Bulletin 1935, Washington, DC186., pages.

Wang, C., McKeague, J.A., Switzer-Howse, K.D., 1985. Saturated hydraulic conductivity as an indicator of structural degradation in clayey soils of Ottawa area, Canada. Soil Tillage Res. 5, 19–31.

Wang, S., Wilkes, A., Zhang, Z., Chang, X., Lang, R., Wang, Y., et al., 2011. Management and land use change effects on soil carbon in northern China's grasslands: a synthesis. Agric. Ecosyst. Environ. 142, 329–340.

Yang, X., Chen, F., Lin, X., Liu, Z., Zhang, H., Zhao, J., et al., 2015. Potential benefits of climate change for crop productivity in China. Agric. For. Meteorol. 208, 76–84.

Yurova, A.Y., Volodin, E.M., Ågren, G.I., Chertov, O.G., Komarov, A.S., 2010. Effects of variations in simulated changes in soil carbon contents and dynamics on future climate projections. Glob. Chang. Biol. 16, 823–835.

Zabel, F., Putzenlechner, B., Mauser, W., 2014. Global agricultural land resources – a high resolution suitability evaluation and its perspectives until 2100 under climate change conditions. PLOS ONE 9, e107522. www.plosone.org.

Zomer, R.J., Trabucco, A., Bossio, D.A., Verchot, L.V., 2008. Climate change mitigation: a spatial analysis of global land suitability for clean development mechanism afforestation and reforestation. Agric. Ecosyst. Environ. 126, 67–80.

CHAPTER 3

Soil Health Concerns Facing Dryland Agroecosystems

Bobby A. Stewart
West Texas A&M University, Canyon, TX, United States

3.1 Introduction

Soil is one of the three major natural resources, alongside air and water. Unlike water and air, however, soil is a living and life-giving resource that is at the heart of all ecosystems. Soils are highly variable. Some are sandy, some are silty or have high amounts of clay particles, and some are highly productive, while others are not even suitable for growing crops. There are many soil characteristics that can be measured and used to describe soils and indicate how well they can be used for particular functions. Since soils are used for many different functions, a particular soil can be highly suitable for one function while being unsuitable for another. Soil directly and indirectly affects agricultural productivity, water quality, and the global climate through its function as a medium for plant growth, a regulator and partitioner of water flow, and an environmental buffer (National Research Council, 1993). The terms "soil quality" and "health" were not widely used historically by soil scientists because of the lack of understanding what the terms meant, and there were no well-defined methods for quantitative measurement. However, following the publishing of the report entitled "Soil and Water Quality: An Agenda for Agriculture" (National Research Council, 1993), there was increased interest in soil quality and also soil health which are generally considered analogous. Notable publications about soil quality and soil health include those by Doran et al. (1994), Karlen et al. (1997), and Acton and Gregorich (1995). The interest in soil quality and soil health, however, was seriously challenged by a number of acclaimed soil scientists (Letey et al., 2003). They criticized the soil quality paradigm as a misleading "fad" lacking in scientific rigor, fraught with societal value intrusion, and conceptually incompatible with air and water quality. Although defining and measuring soil quality and soil health have proven difficult, the terms and particularly soil health resonate with producers and the public at large. On October 11, 2012, the US Department of Agriculture's Natural Resources Conservation Service (USDA-NRCS) launched a new soil health awareness and education effort (USDA-NRCS, 2015a).

Soil Health and Intensification of Agroecosystems.
DOI: http://dx.doi.org/10.1016/B978-0-12-805317-1.00003-8

The initiative is intended to focus on the benefits of improving and maintaining America's soil. Therefore, even though it is well recognized that there are shortcomings associated with the term soil health, it is becoming increasingly important to describe the condition and productivity of soils. Doran (2002) broadly defined soil health, or quality, as the capacity of a living soil to function, within natural or managed ecosystem boundaries, to sustain plant and animal productivity, maintain or enhance water and air quality, and promote plant and animal health. He also noted that soil organic matter (SOM) serves as a primary indicator of soil health. Lal (2014) emphasized the importance of SOM by stating that it is the basis of all physical, chemical, biological, and ecological transformations and reactions within a soil. SOM increases water-holding capacity, improves infiltration, reduces erosion, improves soil structure, enhances soil fertility, increases biological activity, reduces air pollution, and improves overall soil health. Therefore, even though there are exceptions for most rules, the "rule" for discussing soil health in dryland regions in this chapter is that soil health declines when SOM decreases, and improves when SOM levels increase. SOM is generally assumed to contain 58% carbon (C) (Lal, 2014). An extensive list of soil health indicators and methods of measurement have been compiled by USDA-NRCS (USDA-NRCS, 2015b).

This chapter will focus largely on dryland cropping systems and their effects on soil health in the United States. However, there are other large dryland farming regions, particularly in Australia, India, China, and Mediterranean countries. Noteworthy books describing dryland cropping systems in these regions include those by Li Shengxiu (2007), Koohafkan and Stewart (2008), Singh (2005), and Squires and Tow (1991).

3.2 Soil Health in Dryland Agroecosystems

3.2.1 Difficulty of Enhancing Soil Health in Areas with Increasing Temperatures and Decreasing Precipitation

An agroecosystem is a system where agriculture production, either crops or animals or a combination of both, has been introduced into an ecosystem. Whenever a natural ecosystem is transformed, particularly by removing the native vegetation and tilling it for growing crops, several soil-degradative processes are set into motion. Hornick and Parr (1987) reported that most agroecosystems have degradation processes and conservation practices occurring simultaneously. The relation of soil health to soil degradation processes and soil conservation practices is illustrated in Fig. 3.1. As soil degradation processes proceed and intensify, there is a concomitant decrease in soil health. Conversely, soil conservation practices tend to increase soil health. Therefore, the overall soil health of an agroecosystem at any time is a result of the interaction of degradation processes and conservation/reclamation practices, some of which are shown in Fig. 3.1. Maintaining soil health is achieved through the efficient but delicate balance between all necessary inputs and outputs. Failure of agroecosystems to maintain this balance will ultimately lead to declining soil health and unsustainability.

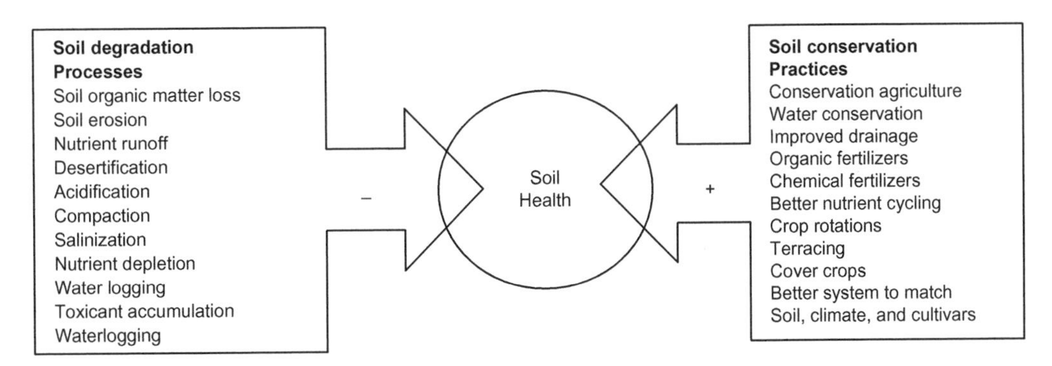

Figure 3.1
Relation of soil health to soil degradation processes and soil conservation practices for an agroecosystem. *Modified from Hornick, S.B., Parr, J.F., 1987. Restoring the productivity of marginal soils with organic amendments. Am. J. Alt. Agric. 2, 64–68.*

The relation shown in Fig. 3.1 applies to all agroecosystems, although the importance of specific processes will vary substantially between agroecosystems. Also, the processes included in Fig. 3.1 are not all-inclusive and since this chapter is focusing on drylands, several of the listed processes are not major factors. Also, as stated earlier for examples used in this chapter, the change in soil organic carbon (SOC) will be used as the sole indicator of whether soil health is deteriorating or being enhanced. Although it is recognized that there are exceptions, it is true in most cases and seems particularly true in dryland regions.

The concepts illustrated in Fig. 3.1 apply to all climates and all levels of inputs. In every case, a truly sustainable ecosystem results when the effects of the conservation practices equal or exceed the effects of the soil degradation processes. Climate is usually the most critical factor determining the sustainability of an agroecosystem. Stewart et al. (1991) presented a generalized view, shown in Fig. 3.2, of the effect of varying temperature and moisture regimes in semiarid regions on the difficulty of achieving sustainability in an agroecosystem. As temperatures increase and the amounts of precipitation decrease or remain the same, the difficulty of developing a sustainable cropping system increases dramatically. The most serious soil degradation process in dryland regions is the loss of SOC. As temperatures rise, SOC decline, particularly in frequently tilled soils, is greatly accelerated. This makes the soil more vulnerable to wind and water erosion and a serious decline of soil health. Soil degradation processes shown in Fig. 3.2 tend to accelerate under hot and arid conditions while conservation practices, while extremely important, become more difficult to implement and benefits are often limited. For example, the most important practices for increasing SOC involve crop residues or cover crops, but the availability of crop residues decreases sharply in semiarid regions and producers often resist growing cover crops because this might reduce water available for cash crops.

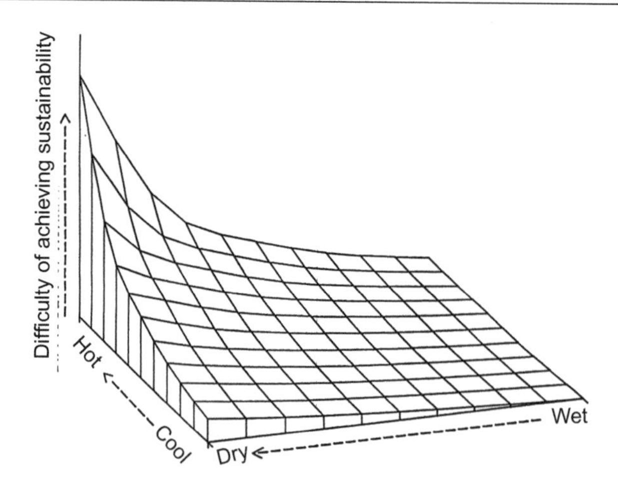

Figure 3.2
Generalized representation of the effects of temperature and precipitation on the difficulty of developing sustainable agricultural systems (Stewart et al., 1991).

Stewart et al. (1991) also applied the concept presented in Fig. 3.2 to some worldwide locations and the results are shown in Fig. 3.3. The points for the various locations were obtained by plotting the annual mean temperature, annual mean precipitation, and the temperature/precipitation value. The authors stressed that the index values were developed only for a general comparison and should not be considered in any way as representing a specific quantitative difference. The authors were also focusing on semiarid regions and the index would likely not apply to humid regions where other factors would perhaps be more important than precipitation and temperature.

The US Great Plains is one of the largest dryland farming regions in the world but was once mapped as "The Great American Desert." Annual precipitation is highly variable with annual amounts periodically ranging from more than two times average to less than one-half average. Because of the low and highly variable amounts of precipitation, crop production has always been challenging. Bennett et al. (1938) stated that the significant point about the precipitation of the Great Plains is that it varies widely about a critical point in crop production. In some parts, the average annual precipitation is sufficient to produce good yields of drought-resistant crops like wheat (*Triticum aestivum*), sorghum (*Sorghum bicolor*), and cotton (*Gossypium hirsutum*). In abundant precipitation years, abundant crops are common. But even in these areas, crops may fail in years of subnormal or poorly distributed rainfall events. In other portions of the region, profitable crops can be produced only in years of abnormally high rainfall and even average rainfall years produce crop failures. Bennett et al. (1938) further stated that while the soils of the Great Plains are generally fertile, they exhibit wide diversity in their texture, depth, and water-holding capacity. Almost any of them will produce fairly

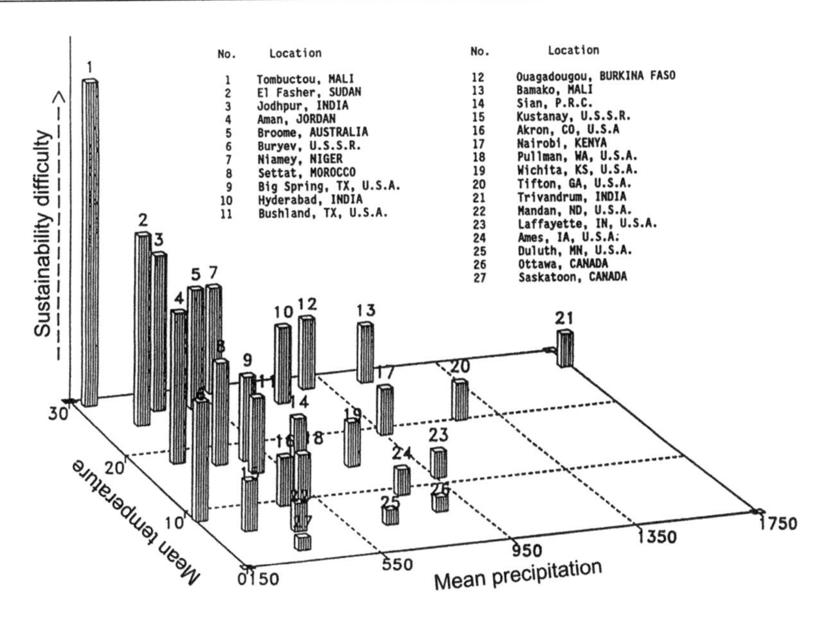

Figure 3.3

Mean annual temperature and precipitation for various locations and temperature/precipitation as an "index" indicating difficulty of developing a sustainable agriculture system (Stewart et al., 1991).

good crops in wet years, but in dry years only those that absorb and hold large quantities of water and resist wind erosion can be farmed successfully. Today, more than 75 years later, the same statements are generally true. The Great Plains has a distinct north–south gradient in average temperature patterns, with a hotter south and colder north. For precipitation, the gradient runs east–west, with a wetter east and much drier west (USGCRP, 2014). Therefore, the climatic conditions for crop production are considerably more challenging in the southern parts of the Great Plains than in the central and northern parts. This can be seen in Fig. 3.3 by comparing Mandan, ND (number 22, Fig. 3.3); Akron, CO (16); Bushland, TX (11); and Big Spring, TX (9). Saskatoon, Canada (27), is in the Canadian Great Plains that has many of the same characteristics as the US Great Plains. Dryland farming began in the Great Plains in the late 1800s and developed rapidly in the early 1900s. Precipitation amounts were generally favorable during the early 1900s and wheat yields were good but soil degradation, particularly SOM decline, was very pronounced. This, along with multiple years of low precipitation, led to the infamous Dust Bowl considered by many as the worst manmade ecological disaster ever to occur in the United States (Egan, 2006). SOM loss was rapid and severe because of intensive tillage and the widespread use of cropping systems using long fallow periods. Wheat-fallow was a common cropping system that resulted in only one crop of wheat every 2 years meaning that the land was fallowed for about 16 months between winter wheat crops and as long as 20 months in the northern plains where spring wheat was the dominant crop. Several tillage operations occurred during the fallow periods to control weeds and this accelerated organic matter decomposition. Although only about 15% of the precipitation occurring during the fallow period was stored in the soil profile for use by the subsequent wheat crop, this water supplemented the precipitation that occurred during the wheat growing season and the production from half the land that had been fallowed was almost equal to that from farming all the land each year (Johnson et al., 1974). Even more important, risk was greatly reduced because there was seldom a complete crop failure on fallowed land. The loss of SOM, however, made the soils much more vulnerable to erosion, particularly wind erosion, and was a major factor contributing to the Dust Bowl. SOC was inherently low in Great Plains soils, particularly in the southern Great Plains, because of the natural low precipitation amounts and high temperatures, so a severe decrease left insufficient binding of soil particles in many cases. Table 3.1 lists the amounts of nitrogen (N) and organic C in the 0–15 cm depth of virgin and cropped soils on selected soils in the US Great Plains (Haas et al., 1957). Three of the sites—Mandan, ND (22); Akron, CO (16); and Big Spring, TX (9)—are included in Fig. 3.3 as mentioned above. Although some of the soils in the northern Great Plains were relatively high in organic C at the time they were converted to cropland, those in the central and particularly those in the southern regions tended to be low. In less than 40 years, the organic C and N concentrations decreased about 40%. Many of the cropped soils, particularly those with high sand percentages, contained less than 0.50% organic C. Although the Dust Bowl encompassed the entire US Great Plains and even much of the Canadian Great Plains, the most severe damage was in the southern areas where the soils tended to be sandier and contained less SOC.

3.2.2 Examples of Summer Fallow Effects on Decline of Soil Organic Carbon

Summer fallowing intensifies SOC loss and when it is combined with intensive tillage, the loss can be dramatic. At Moscow, ID, in the Pacific Northwest, the loss of SOC was extensive with a wheat-oat (*Avena sativa*)-fallow cropping system (Fig. 3.4). Even when manure was applied, organic C declined. When wheat was grown continuously, the decline was much less but still considerable unless manure additions were part of the cropping system which actually increased SOC levels. Manure additions of 33 $t\,ha^{-1}$ were made every third year. Rasmussen and Albrecht (1998) stated that the detrimental effect of fallowing on SOC is twofold: (1) fallow systems produce no above- or belowground residue since soils are kept weed- and crop-free, and (2) biological oxidation of existing SOC increases substantially because of more-favorable soil moisture during the fallow year than when cropped annually. Summer fallow in the Pacific Northwest consists of a 14-month period from mid-August to mid-October of the following year, and the soil is normally kept weed- and crop-free by tillage and herbicide application (Rasmussen and Albrecht, 1998). They also reported that there was extensive loss of SOC in the 30–60-cm soil zone in a wheat-fallow cropping system at Pendleton, OR (Fig. 3.5). This loss was not related to residue treatment, suggesting that

Table 3.1: Nitrogen and organic carbon contents in 0–15 cm depth of virgin and cropped soils for selected soils in the US great plains

	Mean Temperature (C°)	Mean Precipitation (mm)	Texture	Virgin Organic-C (%)	Cropped Organic-C (%)	Virgin Soil-N (%)	Cropped Soil-N (%)	Cropped (years)
Mandan, North Dakota	5.0	401	Fine sandy loam	2.11	1.45	0.160	0.116	30
Dickinson, North Dakota	4.6	397	Loam	3.64	1.51	0.293	0.149	40
Havre, Montana	5.6	292	Clay loam	1.75	0.83	0.150	0.090	31
Moccasin, Montana	5.7	384	Clay loam	3.24	2.19	0.300	0.205	39
(mean)	5.2	369		2.69	1.52	0.226	0.140	35
Akron, Colorado	9.3	442	Silt loam	1.42	0.77	0.134	0.086	39
Colby, Kansas	10.6	469	Silt loam	1.83	1.01	0.165	0.105	41
Hays, Kansas	12.2	587	Silt clay loam	2.47	1.21	0.220	0.122	43
Garden City, Kansas	12.1	458	Fine sandy loam	1.13	0.69	0.120	0.077	39
(mean)	11.1	489		1.71	0.92	0.160	0.098	41
Woodward, Oklahoma	14.5	604	Fine sandy loam	–	–	0.080	0.032	33
Lawton, Oklahoma	16.3	729	Silt loam	–	–	0.154	0.074	31
Dalhart, Texas	12.4	460	Fine Sandy Loam	0.72	0.44	0.067	0.042	39
Big Spring, Texas	17.2	479	Fine Sandy Loam	–	–	0.060	0.041	41
(mean)	15.1	569		0.72	0.44	0.090	0.047	36

Source: Data compiled from Haas, H.J., Evans, C.E., Miles, E.F., 1957. Nitrogen and carbon changes in Great Plains soils as influenced by cropping and soil treatments. Technical Bulletin 1164, Washington, DC. Available online: <https://books.google.com/books?id=iLYXAAAAYAAJ&pg=PP3&lpg=PP3&dq=Haas,+Evans,+Miles+1957&source=bl&ots=_0ATa02RNE&sig=9onkDpv1mIbgHGVZh4E-TjYQlTY&hl=en&sa=X&ved=0CC0Q6AEwBGoVChMI3fa14MizyAIVAxc-Ch3_hAVq#v=onepage&q=Haas%2C%20Evans%2C%20Miles%201957&f=false>.

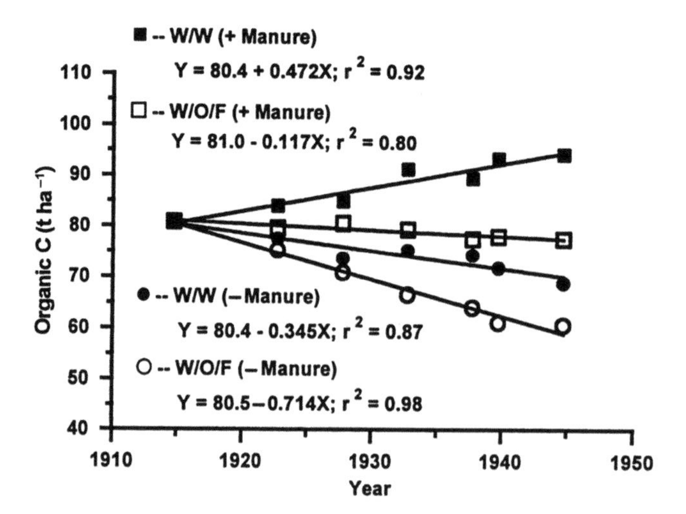

Figure 3.4
Effect of crop rotation (W/O/F, wheat-oat-fallow; W/W, continuous wheat) and manure addition on organic C in the 0 to 30 cm soil layer at Moscow, ID, 1915 to 1945 (Rasmussen and Albrecht, 1998).

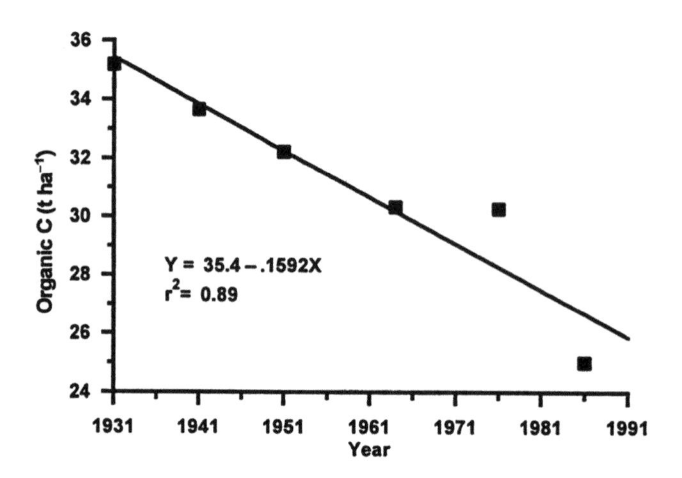

Figure 3.5
Loss of organic C with time in to 30–60 cm soil layer at Pendleton, OR (Rasmussen and Albrecht, 1998).

the decline was due to conversion from grassland to cultivation, where annual biological oxidation loss of C exceeded the C input from roots. SOC losses below 30 cm have not been extensively measured but may be a substantial source of CO_2 loss and it is also difficult to sequester C at these depths.

Stubble-mulching played an important role in the southern Great Plains for controlling wind erosion following the Dust Bowl of the 1930s. The Johnson et al. (1974) study from 1942 to 1970 compared stubble-mulching (pulling a 75-cm wide v-shaped blade sweep about 10 cm below the soil surface) to a one-way disk plow, and continuous wheat compared to wheat-fallow. The one-way disc plow was the most commonly used tillage implement in the plains from the 1930s through the 1960s. Instead of completely burying the plant residues as previous plows had done, the one-way incorporated residues into the upper layer of the topsoil to serve as a mulch and help conserve moisture and reduce wind and water erosion. However, continued use of one-way tillage resulted in accelerated loss of SOC and soil structure and often resulted in plowpans that reduced water movement in the soil profile. As a general guideline, a single tillage operation with a one-way plow buries about 60% of surface residue compared with only about 15% for a sweep plow. The summer-fallow period in the wheat-fallow system was about 15 months long, extending from wheat harvest near the end of June until wheat was seeded at the end of September of the following year. Although variable depending on the precipitation amount and distribution, there were usually about four tillage operations during the 14-month fallow period. For continuous wheat, two tillage operations were typical. There was also a delayed stubble-mulch treatment where tillage was delayed from immediately following wheat harvest until the spring of the year—some 10 months after wheat harvest. During this period, weeds and volunteer wheat were allowed to grow.

Although stubble-mulch tillage was primarily developed to control wind erosion, it soon became apparent that it decreased soil water evaporation so that more water was stored in the soil profile during the fallow period. The 30-year study of Johnson et al. (1974) showed the average amounts of plant-available soil water in the 1.5-m soil profile at time of seeding were 103 mm for stubble-mulch tillage compared to 91 mm for one-way tillage. The increased soil water also increased grain yield (GY) from 600 kg ha^{-1} to 700 kg ha^{-1}. For the wheat-fallow system, plant-available soil water at seeding was 154 mm for stubble-mulch treatment compared to 128 mm for one-way tillage, and the respective GYs were 1070 and 950 kg ha^{-1}. Somewhat surprising was that the delayed tillage treatment with weeds and volunteer wheat growing for several months following harvest stored 144 mm plant-available water during the shortened fallow period and yielded 1040 kg ha^{-1} which were similar to the values for traditional stubble-mulch treatment. However, the delayed tillage system was not considered acceptable because of weed seed and uncontrolled vegetative growth.

The most important finding of the Johnson et al. (1974) study was the extensive loss of SOC in the southern Great Plains with intensive tillage and summer fallowing (Fig. 3.6). The values shown are for SOM so SOC values would be about 58% of the values shown. The loss of SOC was much greater for the wheat-fallow system than for continuous wheat, and it was greater for one-way disk tillage treatment than for stubble-mulch treatment that was tilled with subsurface sweeps. Loss of SOC was least for the delayed tillage treatment because of the large amounts of vegetation produced during the approximate 10-month period between harvesting the wheat and the first tillage the following spring. While the delayed tillage treatment was considered unacceptable, it had a marked beneficial influence on SOC

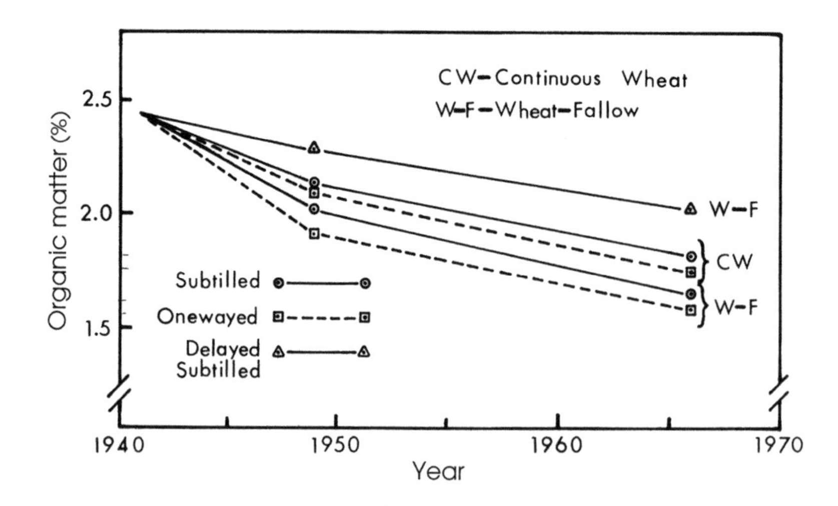

Figure 3.6
Soil organic matter, 0–15 cm, of stubble-mulch plots on a clay loam soil at Bushland, TX, as related to cropping system and tillage method (Johnson et al., 1974).

without causing a significant reduction in water storage during the fallow period. This raises the question as to whether a cover crop could be successfully grown. The main constraints are likely whether or not it would be economically feasible and if producers would harvest or graze the cover crop rather than leave it on the soil surface for sequestering SOC.

The severe decline of SOM recorded by Johnson et al. (1974) was further verified by the presence of large amounts of nitrate-N in the soil profile. At the end of the study, 210 kg ha^{-1} of nitrate-N were in the 180 cm profile of the one-way tilled plots, and 131 kg ha^{-1} in the stubble-mulched plots. More than 60% of these amounts were in the lower half of the profiles, indicating that the mineralization of organic N resulting from tillage exceeded the N needs of the crops produced. In contrast, only 35 kg ha^{-1} of nitrate-N were present in the delayed tillage plots, and more than 50% was in the upper half of the profile. Therefore, the weeds and volunteer wheat that were allowed to grow during the early part of the fallow period used much of the mineralized N and it became part of the SOM.

3.2.3 Cropping Systems to Reduce the Length of the Summer Fallow Period

As the severe soil degradation effects of long fallow periods became documented and it was concluded that one of the unintended consequences was that this practice was a major factor causing the Dust Bowl, efforts were made to eliminate or shorten fallow periods. Fig. 3.7 illustrates three cropping systems at Bushland, TX, in the southern Great Plains with fallow periods between crops of 3 months, 11 months, or 15 months and harvesting a wheat crop every year, a wheat crop and a sorghum crop every 3 years, or a wheat crop every 2 years. Stewart and Robinson (1997) summarized some long-term studies conducted at Bushland,

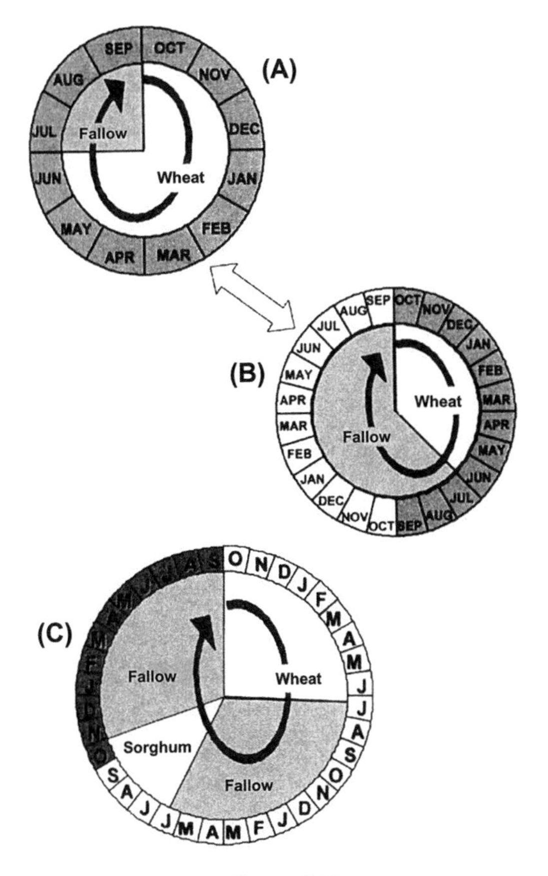

Figure 3.7
Illustrations of (A) continuous wheat, (B) wheat fallow (one crop every 2 years), and (C) wheat-sorghum-fallow (two crops every 3 years) in the southern Great Plains (Unger et al., 2010).

TX, and results are presented in Table 3.2. For the continuous wheat cropping system, the average amount of precipitation received between crops was 202 mm but only 37 mm was stored in the soil at time of seeding. There was 256 mm precipitation during the growing season and with the 37 mm soil water change, there was a total of 293 mm, 64% of the 458 mm annual precipitation, used as evapotranspiration (ET), and the remaining 165 mm was lost as evaporation during the fallow period or as runoff. In contrast, when one wheat crop was grown every 2 years, and the fallow period was increased from 3 months to 15 months, there was an average of 660 mm precipitation during the fallow period but only 98 mm was stored in the soil profile at the time of seeding the wheat. The 98 mm along with the 256 mm average precipitation during the growing season totaled 354 mm that was used for ET, an increase of only 61 mm over 293 mm for the continuous wheat. The 354 mm used for ET was only 39% of the total precipitation, meaning that 61% was lost as evaporation and runoff during the fallow period. Although the results clearly show the inefficiency of the

Table 3.2: Water balance for various cropping systems summarized from long-term studies at bushland, TX, for cropping systems with varying fallow periods illustrated in Fig. 3.7

Continuous Wheat (One Crop Each Year)			
	Wheat	Fallow	Total
Precipitation[a] (mm)	256	202	458
Evapotranspiration[b] (mm)	293		293
Soil water change[c] (mm)	−37	37	
Evaporation and runoff[d] (mm)		165	165

Wheat-Sorghum-Fallow (two Crops In 3 Years)					
	Wheat	Fallow	Sorghum	Fallow	Total
Precipitation[a] (mm)	256	462	241	241	1375
Evapotranspiration[b] (mm)	329		286	286	615
Soil water change[c] (mm)	-86	86	−72	72	
Evaporation and runoff[d] (mm)		389		371	760

Wheat-Fallow (One Crop in 2 Years)			
	Wheat	Fallow	Total
Precipitation[a] (mm)	256	660	916
Evapotranspiration[b] (mm)	354		354
Soil water change[c] (mm)	−98	98	
Evaporation and runoff[d] (mm)		562	562

Source: Adapted from Stewart, B.A., Robinson, C.A., 1997. Are agroecosystems sustainable in semiarid regions. Adv. Agron. 60, 191–228.
[a]Precipitation during the crop growing period, fallow period, and total for the cropping system.
[b]Precipitation plus soil water change between time of seeding and harvest.
[c]Change in stored soil water during growing season or fallow period.
[d]Evaporation of water from soil surface during the fallow period and runoff.

wheat-fallow cropping system, it is still widely used in dryland areas and the reason is illustrated in Fig. 3.8. The relationship between GY of wheat and ET based on many years of dryland and irrigated studies at Bushland, TX, shows there is a threshold of 207 mm ET required before any grain is produced, but 12.2 kg of grain is produced for every additional mm of ET. Using this relationship, the 293 mm of ET shown in Table 3.2 for continuous wheat would produce a yield of 1.05 $Mg\,ha^{-1}$. The wheat-fallow system, while increasing the ET only from 293 mm to 354 mm, would result in an average yield of 1.80 $Mg\,ha^{-1}$ and only half the land would need to be seeded and harvested each year. Even more, the risk of a crop failure is greatly reduced. The values shown in Table 3.2 are long-term averages and a year when the growing season precipitation was only 20% below the average of 256 mm, the GY would be essentially zero. Therefore, most farmers have historically considered fallow an essential part of a successful cropping system even though it was inefficient from a water-use standpoint and led to a decline of soil health. Many farmers, however, have adopted cropping systems that have a shorter fallow period. In the southern Great Plains, the dominant dryland

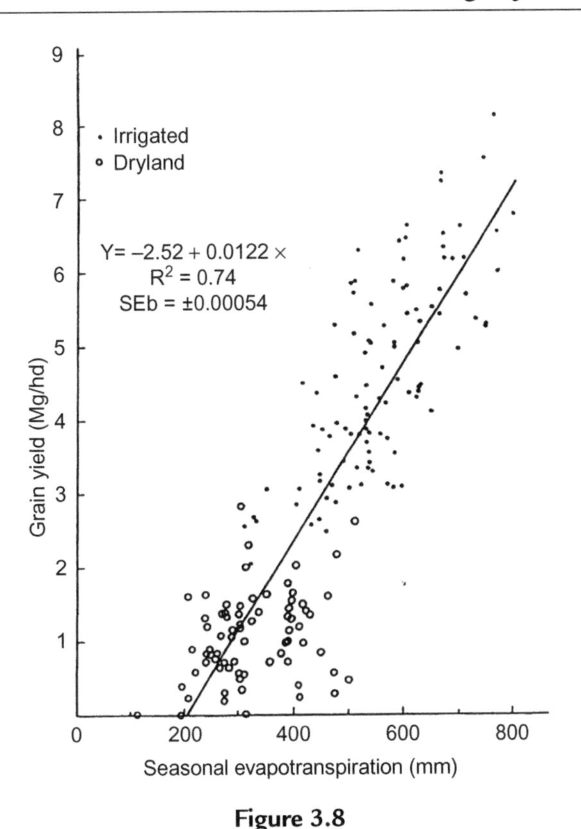

Figure 3.8

Relation of wheat grain yield to seasonal evapotranspiration at Bushland, TX. *From Musick, J.T., Jones, O.R., Stewart, B.A., Dusek, D.A., 1994. Water-yield relationships for irrigated and dryland wheat in the U.S. Southern Plains. Agron. J. 86, 980–986, Musick et al. (1994).*

cropping system is wheat-sorghum-fallow illustrated in Fig. 3.7 that has 11 months of fallow between the wheat and grain sorghum crops. Long-term studies showed that almost as much water was stored in the soil profile during the 11-month fallow periods as for a 15-month period (Johnson et al., 1974). The long-term averages shown in Table 3.2 shows that the 11-month fallow period between the harvesting of grain sorghum and the seeding of wheat increased the stored soil water by 72 mm compared to 98 mm for the 15-month period, and the 11-month period between harvesting wheat and seeding grain sorghum the soil water stored in the profile increased by 86 mm. Overall, during the 3-year period of the wheat-sorghum-fallow system, 615 mm, 45% of the total precipitation, were used as ET which is more efficient than for the wheat-fallow system. While the percentage of time the land is in fallow is about 60% for both the wheat-fallow and the wheat-sorghum-fallow systems, the cropping intensity differs from 50% to 67%, so there is more crop production and an added benefit is that the grain sorghum is a C_4 crop that produces more biomass per unit of water than wheat, which is a C_3 crop so there tends to be more crop residues to better control wind and water erosion, and maintain the level of SOC. Cotton (*Gossypium hirsutum*) and corn (*Zea mays*) are sometimes used in place of grain sorghum in cropping systems (Peterson and Westfall, 2004; Baumhardt and Anderson, 2006).

3.2.4 Loss of SOC Significantly Reduces Water Available for Transpiration (T) by Plants in Dryland Areas

Obviously, lack of water is the primary constraint for growing crops in dryland areas, but a decline in soil health through the loss of SOC greatly exacerbates the situation. The amount of water held by a soil between the field capacity value and the permanent wilting point is the water that is available for use by plants. The field capacity value is the amount of water that is held by a soil after a soil has been wetted and drained by gravitational forces. The permanent wilting point value is the amount of water held by soil that is held so tightly by the soil that plants cannot extract the water so it is not available for T. This water, however, as well as the plant-available water, is subject to evaporation and the closer the water is to the soil surface, the greater its availability for evaporation. When a soil is air-dry, there is very little water in the soil and in dryland areas where the humidity is low, the water present when the soil is air-dry is generally only about 1–2% by volume. Hudson (1994) reported the volumetric water contents at permanent wilting point and at field capacity for a silt loam soil containing varying amounts of SOM (Fig. 3.9). The field capacity value is two times greater when the soil has 5% SOM than the same soil with 1% SOM. In contrast, the permanent wilting point values are only minimally affected by the SOM content. The reason is SOM has a major effect on soil structure, which affects the size of aggregates and more importantly the size of pores within the aggregates. Thus, the field capacity is largely controlled by soil structure while the permanent wilting point is primarily controlled by soil texture. Soils with good structure contain more pores that retain water against gravitational forces. The soil illustrated in Fig. 3.9 shows that when only 1% SOM is contained in the soil, the soil will contain about 22% water by volume after it has been wetted and allowed to drain by gravity. Therefore if the soil was wetted to field capacity to a depth of 10 cm, there would be 2.2 cm of water in the soil. However, since the permanent wilting point value is 11% by volume, 1.1 cm will not be available to plants and the 1.1 cm that is available to plants is also subject to evaporation so the plants have access to only a limited portion of the water. When the soil contains 5% SOM, the soil wetted to field capacity to 10 cm would hold 4.4 cm of water and only about 1.3 cm would not be extractable by plants so the plants would have access to almost three times more water for T. Even more discerning for many dryland areas is how often the surface soil becomes air-dry during a growing season. In humid areas, the soil seldom dries below the permanent wilting point value so plants have access to all added water. In contrast, the upper few cm of soil in dryland areas frequently become air-dry because of high temperatures, high radiation, and high winds. Therefore, most, if not all, of the water that plants cannot extract is lost by evaporation. Then, this water must be replaced by the next rainfall event before

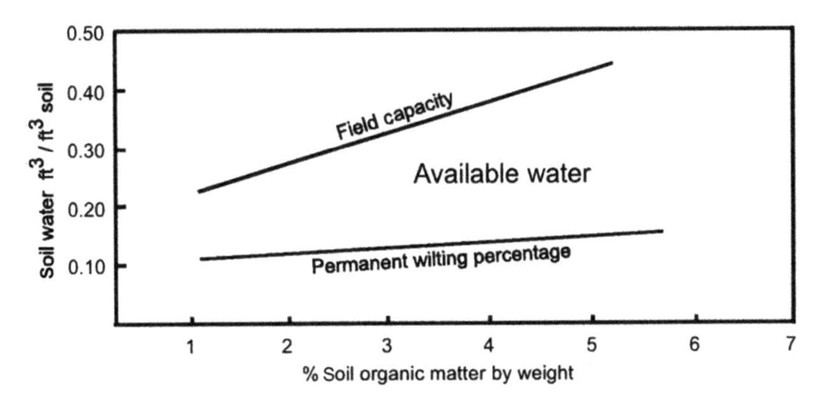

Figure 3.9

Effect of increasing soil organic matter on plant-available water of a silt loam soil. *Adapted from data of Hudson, B.D., 1994. Soil organic matter and available water capacity. J. Soil Water Conserv. 49 (2), 189–194.*

Table 3.3: Averages values for number of events (24-hour total), amount of events, and percentage of total annual precipitation for seven sizes of precipitation events for 22 N-S locations of the US great plains with records ranging from 62 to 104 years

	0.25–2.5	2.5–5	5–12	12–25	25–50	50–75	75–100	Total
	(mm)							
Average number[a]	22	11	13	8	3	1	0	58
Average amount[a]	24	38	106	135	108	31	10	452
Average % total[a]	5	9	23	30	24	7	2	100

Source: Data from Chebrola, K.K., 2006. Precipitation Patterns along a N-S Transect in the Ogallala Aquifer Region. Master of Science Thesis, West Texas A&M University, Canyon, TX, Chebrola (2006).
[a]Rounded to nearest whole number.

there is any water that plants can even compete with evaporative forces to obtain water for T. In addition, much of the precipitation in semiarid regions occurs in small amounts. Using long-term data from 22 locations in the US Great Plains as an example, precipitation occurred on 58 days annually to supply a total of 452 mm (Table 3.3). However, 44 precipitation days had daily amounts less than 12 mm but accounted for 37% of the annual precipitation. Daily events between 12 and 25 mm accounted for 30% and those between 25 and 50 accounted for 24% but occurred only three times a year on average. Therefore, much of the precipitation in semiarid regions occurs in such small quantities that only the top few cm of the soil are wetted to field capacity, much of which quickly evaporates during hot and windy days that are common. Fig. 3.10 illustrates a theoretical example of a 12-mm rainfall event on dry soil

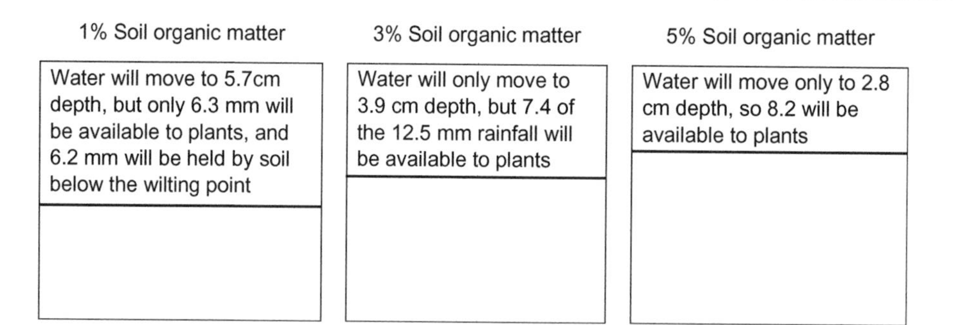

1% Soil organic matter
Water will move to 5.7cm depth, but only 6.3 mm will be available to plants, and 6.2 mm will be held by soil below the wilting point
3% Soil organic matter
Water will only move to 3.9 cm depth, but 7.4 of the 12.5 mm rainfall will be available to plants
5% Soil organic matter
Water will move only to 2.8 cm depth, so 8.2 will be available to plants

Figure 3.10
Theoretical distribution and availability 12 mm of precipitation falling on 10 cm depth of dry silt loam soil of varying soil organic matter content based on properties shown in Fig. 3.9.

of varying SOM contents. As much as half the water will never be available for plant use because it will be held so tightly by the soil particles that it cannot be extracted by plant roots. Therefore only half of the water in the low SOM soil is potentially available for use by plants, and the water that is potentially available for plant use is also subject to evaporation and much of it will be evaporated, particularly early in the growing season when there is not a plant canopy to shade the soil. Surface mulch can greatly reduce this loss and allow more time for the plants to use a greater portion of the plant-available water for T which is directly related to biomass production. However, in semiarid regions the top few cm of the soil can cycle between field capacity air-dry several times during the growing season so the portion of ET that is used as transpiration is much lower than in humid areas, and the differences become even greater with a decline of SOM.

3.2.5 Large Decline in Area of Cultivated Summer Fallow in Dryland Areas of the United States

Cultivated summer fallow refers to cropland in semiarid and subhumid regions of the western United States that are cultivated for one or more seasons to control weeds and accumulate moisture before grain crops are planted (USDA-ERS, 2015). Although the practice is optional in some areas, it is a requirement for successful crop production in the drier cropland areas. As already discussed, however, it is well documented that cultivated summer fallow accelerates soil degradation processes, particularly the loss of SOM. Fig. 3.11 shows the numbers of cropland ha in cultivated summer fallow between 1910 and 2014 as well as the hectares of failed crops. Although summer fallow was used in some of the drier regions almost from the beginning of cropping, there was a large expansion of the adoption of summer fallowing in response to the huge rise in failed crops during the drought years of the 1930s and the Dust Bowl era. The expansion of cultivated summer fallowing peaked in 1969 at more than 16 M ha, about 10% of all US cropland. Without question, this large amount of

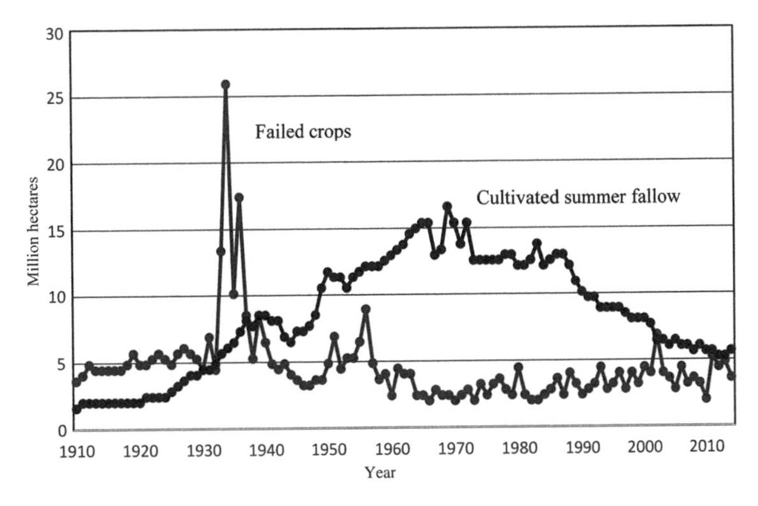

Figure 3.11

Hectares of US cropland that had failed crops and cultivated summer fallow from 1910 to 2014. *Data from USDA-ERS, 2015. Major Land Uses. United States Department Agriculture, Economic Research Service, Washington, DC. Available online:* <http://www.ers.usda.gov/data-products/major-land-uses/.aspx>.

cultivated summer fallow had a negative impact on soil health, water quality, air quality, and other environmental factors. Summer fallowing began to decrease in 1970 and had declined by about 65% by 2014 (Fig. 3.11). The decrease resulted from several factors but among the most important were the rapid increase in oil prices imposed by the Organization of Petroleum Exporting Countries in 1973 and the beginning of the environmental movement in the United States as evidenced by the formation of the Environmental Protection Agency in 1970. These developments resulted in an increased interest in the use of chemicals to replace tillage that led to a reduction of summer fallow. Then, in 1985 the USDA Food Security Act established the Conservation Reserve Program (CRP) that is a cost-share and rental payment program that encouraged farmers to convert highly erodible cropland or other environmentally sensitive acreage to vegetative cover. By 1990, some 13 M ha were enrolled in the program and much of this was in the dryland areas of the Great Plains where cultivated summer fallow was widely used (USDA-ERS, 2015). There would no doubt have been an even larger enrollment of marginal dryland cropland into the program except for the provision that limited enrollment to 25% of the cropland for any county.

3.3 Can SOC Depleted from Dryland Soils be Restored?

3.3.1 Restoring Organic Carbon Requires Restoring Plant Nutrients

The reduction in use of cultivated summer fallow on cropland resulted in a significant reduction in wind and water erosion that had important positive impacts on water quality, air quality, soil health, and the environment as a whole. Much of the reduced erosion was due to the CRP program that took millions of hectares of cropland out of production for varying periods of time, but generally for at least 10 years, and much of the land remains in the program. Some of the land has returned to cropping while some that is no longer in the program remains in some form of grass cover and will likely never return to growing grain crops. The decline of soil health and loss of SOC on millions of hectares of cropland was arrested by the CRP but it is less clear how much soil health was improved.

Sequestration of C in SOC in land enrolled in the CRP program and in cropland where no-tillage is practiced has been widely discussed and many scientists have concluded that there is a high potential for sequestering large amounts of C in SOC that would otherwise be emitted into the atmosphere as CO_2. Lal et al. (1998) estimated that the conversion of natural ecosystems to agriculture in the United States had resulted in a loss of 4000–6000 M Mg of C from cropland. However, they estimated that much of this could be restored by identifying relevant policies, developing appropriate programs, and successfully implementing those policies and programs. They stated the practical upper-limit potential for sequestering C in US cropland between 2000 and 2050 was 5000 M Mg. At the same time, they stated that the technology and practices needed to realize this potential are often difficult to implement, and that it may be easier to prevent the loss of SOC than to restore it.

In reality, sequestering organic C in soil is a complex process that requires simultaneous sequestration of all the other essential elements required for plant growth. C is only one of many constituents that make up SOM and all are necessary to accumulate SOC. Himes (1997) stated that the sequestration of C in SOM is more like building a village than building a house or store. In a village, some buildings are built to last a few years and others are built to last centuries. Similarly, Jenkinson (1988) stated that SOM is composed of a large variety of compounds; some are decomposed within a few years and others last centuries. The important point is that SOM is sequestered and while one of the constituents is C, plant nutrients are sequestered simultaneously with C. Himes (1997), based on a review of literature, reported that decomposed SOM, commonly called humus, contains approximately 58% C (w/w) and the C/N, C/P, and C/S ratios (w/w) are about 12/1, 50/1, and 70/1, respectively. Thus, humus contains about 58% C, 4.8% N, 1.2% P, and 0.8% S as well as amounts of all the other essential plant nutrients. Therefore, the restoration of C lost from soil will require the simultaneous restoration of plant nutrients. For example, to reach the C sequestration potential of 5000 M Mg C in 50 years calculated by Lal et al. (1998) would require about 417 M Mg N and 100 M Mg of P along with large amounts of other plant nutrients. The United States uses about 11 M Mg and 1.8 M Mg of P annually as chemical fertilizers so if the C sequestration potential proposed by Lal et al. (1998) was reached, it would require almost all of the added N and P be added as chemical fertilizer. Therefore, while it has been well established that the decline of SOC in cropland can be stopped, and some restoration can occur, the extent of restoration may be limited, particularly in dryland areas. Conservation agriculture (CA) is being applied and promoted worldwide as the most promising technology for achieving good crop yields while increasing SOC and improving soil health at the same time. As defined by FAO (2015), CA is an approach to managing agroecosystems for improved and sustained productivity, increased profits and food security, while preserving and enhancing the resource base and the environment. CA is characterized by three linked principles, namely: (1) continuous minimum mechanical soil disturbance; (2) permanent organic soil cover; and (3) diversification of crop species grown in sequences and/or associations. Numerous publications have reported the benefits of CA and it is being adopted worldwide at an increasing rate. Derpsch (2001) reported that in 1999 there were 45 M ha worldwide in no-tillage, synonymous with zero-tillage. By 2013, CA was practiced on about 157 M ha (FAO, 2015). Although the adoption has increased dramatically, CA is still being practiced on little more than 10% of the 1400 M ha of cropland worldwide.

The CA principles are closely linked to soil health because they not only tend to increase SOC, the growing of multiple species tend to greatly increase biological activity. These benefits have been clearly shown in favorable environments, but benefits have been fewer and more difficult to achieve in dryland regions. The reasons are clearly evident in Figs. 3.1–3.3 that were presented and discussed earlier. The primary emphasis for increasing yields in dryland areas since about 1970 has been to reduce, or eliminate, tillage between crops by the use of herbicides to control weeds. This has reduced erosion, increased water use efficiency, increased yields, and enhanced soil health. However, the increases in SOC have been less than many projected, and in many cases there have been no measureable increases. The primary reason is the limited amount of carbonaceous material produced because of low and highly variable amounts of precipitation and the small portion of the precipitation used by the plants as T which is the only water that results in increased biomass production. For example, it was shown earlier in Table 3.2 that long-term studies at Bushland, TX, showed that the amount of annual precipitation that was used as ET varied from a high of 64% for continuous wheat to 45% for wheat-sorghum-fallow to a low of 39% for wheat-fallow. Furthermore, Stewart and Peterson (2015) estimated that only about 55% of ET was used for T by dryland grain sorghum grown at Bushland, TX, so the amount of annual precipitation that is actually used for T by the crop can be in the range of 25%, and it is only the water transpired that results in biomass production. Sinclair and Weiss (2010) stated that C_4 species plants like grain sorghum and corn growing in a somewhat average 2 kPa T environment have a T rate of approximately 220 g water for each g of plant growth. For an arid region with a T environment of 2.5 kPa, they stated that the T ratio is about 280, but only about 160 in a humid environment of 1.5 kPa. For a C_3 crop like wheat, they stated that the T rate is about 1.5 times greater than for C_4 crops.

Stewart and Peterson (2015) stated that the yield of a grain crop can be expressed by the following equation:

$$GY = ET \times T/ET \times 1/TR \times HI$$

where GY is $kg\,ha^{-1}$ of dry GY, ET is $kg\,ha^{-1}$ of evapotranspiration (water use by evaporation from soil surface and transpiration by the crop between seeding and harvest), T/ET is the fraction of ET transpired by the crop, TR is the transpiration ratio (number of kilograms water transpired to produce 1 kg of aboveground biomass), and HI is the harvest index (kg dry grain/kg aboveground dry biomass). While the equation applies to all situations where grain crops are produced, the ranges of values for each of the components become considerably greater and more variable in dryland farming regions. Stewart and Peterson (2015) used the equation along with data from the literature reporting values for ET, HI, and GY to estimate TR and T/ET values. Since the equation is linear, increasing any value by 5% will only increase GY by 5%. However, they showed that when ET was increased that generally every one of the other values was favorably affected so much larger increases in GY occurred. In a hypothetical example, they showed that doubling the ET increased the GY fourfold. Although it is unrealistic to think that ET can be doubled in dryland regions by improved management, Unger and Baumhardt (1999) reported that soil water storage during an 11-month fallow period at Bushland, TX, averaged about 100 mm from 1954 to 1970 when clean or stubble mulch tillage was used compared to about 170 from 1971 to 1997 when a major shift to no-tillage crop production occurred. Retaining crop residues on the soil surface were largely responsible for the increased water conservation, and on average, grain sorghum GYs were increased 22.1 $kg\,ha^{-1}$ for each additional mm of plant-available water stored in the soil profile at the time of seeding. Stone and Schlegel (2006) obtained similar results at Tribune, KS, where they reported an increased yield of 21 $kg\,ha^{-1}$ for each additional mm of stored soil water at seeding. Therefore, while results from many studies have clearly shown that improved management can increase precipitation utilization and crop yields in dryland regions, the potential is limited, particularly where annual precipitation is below about 500 mm. Even if 40% of the average annual precipitation of 500 mm could be used for transpiration, and a C_4 crop was grown, only about 8000 $kg\,ha^{-1}$ of aboveground biomass would be produced and assuming a HI of 0.4, 3200 $kg\,ha^{-1}$ would be removed as grain, leaving 4800 $kg\,ha^{-1}$ as crop residue. This biomass could potentially be converted to SOM. Himes (1997) estimated that plant residues contained 45% C, and that about 35% of the C in plant residue would be retained in the decomposed residue considered as humus. However, Rasmussen and Albrecht (1998) summarized long-term dryland field experiments at Lind, WA, and Pendleton, OR, and concluded that 17–18% of C added as plant residues is incorporated into SOC. Assuming 25%, somewhat of an average of these two references, the 4800 $kg\,ha^{-1}$ crop residue in the example above, there would be 1680 $kg\,ha^{-1}$ of C and 420 $kg\,ha^{-1}$ sequestered as SOC. However, based on the C/N and C/P ratios of 12/1 and 50/1, respectively, in SOC documented by Himes (1997), there would have been 35 kg N and 8.4 kg P simultaneously sequestered with the C. Although variable depending on yield and soil fertility status, N and P concentrations in grain sorghum residues average about 0.83% and 0.13%, respectively (Schomberg et al., 1994). Therefore, 4800 $kg\,ha^{-1}$ crop residue at time of harvest would contain 40 kg N and 6.2 kg P, roughly the amounts sequestered along with the C in SOC. The 3200 kg ha^{-1} removed would contain approximately 50 kg N and 10 kg P, so these amounts would have to come from mineralization of existing SOM or from added nutrients.

A somewhat typical dryland soil in the southern US Great Plains could contain 1% SOC in the top 15 cm layer. Assuming a dry bulk density of 1.5 Mg m^{-3}, the layer would weigh 2,250,000 kg containing 22,500 kg C, 1875 kg N, and 450 kg P. Therefore, the 420 kg of C that might be sequestered by the grain sorghum crop discussed above would be only about 2% of the amount already present, and increase the SOC only from 1.0% by weight to 1.02%, and this assumes that there is no decomposition of the existing SOM. Many studies have shown that about 1.5–3% of organic matter in soils mineralizes annually (Brady and Weil, 2008). With no-tillage, mineralization might be considerably less, but if there is no mineralization, the N and P requirements for growing crops would depend almost entirely on added nutrients. Even if 1% is mineralized, only about 20 kg N and 5 kg P would become available annually for crop production and this is less than one-half of the amount that will likely be removed with a grain crop. The significant point of this discussion and example is that even with no-tillage and relatively good yields, C sequestration will be limited at best from growing grain crops in semiarid regions. In many cases, the challenge will be to maintain the present level. It might be feasible to increase SOC levels if cover crops, including leguminous species, are grown and most or all the biomass is left on the soil or used for grazing by animals and the manure remains on the land.

Even though the discussion above suggests that SOC restoration is extremely limited in dryland regions, it is not saying that soil health cannot be improved. With CA practices, newly sequestered SOC will accumulate almost entirely in the top 1–3 cm of the soil, and this will enhance water infiltration and retention because of improved soil physical properties, greatly stimulate biological activity, and improve chemical properties. Therefore, CA principles should be practiced to the fullest extent feasible in dryland regions to maintain and possibly enhance SOC.

3.3.2 Strategies for Increasing Water Storage in Dryland Areas to Increase Productivity

In Section 3.3.1 it was stated that only about 25% of the annual precipitation in many semiarid regions is used as T and this is the only water that produces crop biomass. For a given transpiration environment, there is a direct straight line relationship between the amount of water transpired and the amount of biomass produced. The slope of the line, however, depends on the average vapor pressure deficit (VPD) and is significantly steeper in humid climates than in arid climates showing that considerably more units of water are required as T to produce a unit of biomass. The biggest challenge for growing grain crops in dryland regions where water is limited, but also highly variable both in amount and distribution, is having water available during the grain-filling growth stage. In many dryland areas, it is essential to have a significant amount of plant-available soil water in the soil profile at the time of seeding to supplement the precipitation received during the growing season. However, it is critical that all of the stored plant-available water is not utilized during the vegetative growth stages because unless precipitation is plentiful during the grain-filling growth stage, the harvest index will be greatly reduced, resulting in a lot of biomass but little grain. Dryland producers use several strategies to address this challenge, but often with limited success. The first strategy is almost always to reduce plant population. However, using the equation, $GY = ET \times T/ET \times 1/TR \times HI$, presented above by Stewart and Peterson (2015), the possible effects of various strategies on GY can be assessed. For example, reducing plant population is considered a necessity in dryland regions, but reducing plant population too much can result in not utilizing all the stored soil water, and fewer plants also results in less canopy so the portion of ET used as T (T/ET) is reduced, and less canopy also tends to increase VPD so the TR is increased. Therefore, reducing the plant population tends to make all the factors in the equation less favorable except for HI. While this strategy is essential for successful grain production in water-deficient areas, it results in low water use efficiency values.

Another strategy sometimes used in limited water areas is using wider rows or skip-rows. The hypothesis is that soil water between the rows where the plants are growing will not be available to the plants during early vegetative growth periods but will be available during the grain-filling growth stage. Again, the only factor in the equation that is favorably affected is the HI because all of the water available for ET may not be utilized because of the wide spaces between rows, and less of the soil surface will be shaded by the plant canopy so a smaller portion of the ET will be used for T and there will be more evaporation from the soil surface. It is conceivable that the microenvironment of the plants in the skip rows could be slightly improved with skip rows because if the plant population remains the same, the plants within the planted rows would be closer together than those where every row was planted. However, the primary benefit is increasing the HI so this strategy is basically a risk-aversion strategy rather than a yield-increasing strategy in that while it greatly diminishes a crop failure, it limits taking advantage of the favorable precipitation years.

The most successful strategy in recent years for increasing GYs in dryland areas has been the use of mulch. Mulch can significantly decrease evaporation of water from the soil surface which increases soil water storage between crops, and also during the crop-growing season. This increases the water used for T. In most cases, mulch results from using limited or no-tillage to leave crop residues on the surface. This also increases water storage because soil almost always dries to the air-dry water content to the depth of tillage. When several tillage operations are performed, the amount of water lost can be very significant as discussed earlier in Section 3.2.4 that greatly reduces the amount of water available for T. The difficulty with this strategy in dryland areas is that there is often insufficient mulch to significantly reduce

Figure 3.12
Use of plastic film in China dryland areas to reduce evaporation from soil surface to increase the amount of water available for transpiration.

evaporation, and in many cases, particularly in developing countries, crop residues are highly valued for animal feed and fuel and removed from the land. Mulch, however, tends to improve or at least not decrease every factor in the equation so GYs are usually significantly increased when mulch is used. Without question, this has been the most successful strategy for increasing GYs in dryland areas and was largely the result of effective herbicides that allowed weed control during fallow periods without tillage. This increased soil water storage to supplement growing season precipitation to increase water for ET, and the mulch also significantly increased the portion of ET used for T (T/ET) so that GYs increased markedly where there were sufficient crop residues to form a beneficial mulch. Although not practical for large farms, smallholder farmers in China are using plastic film on millions of hectares of cropland in dryland areas to drastically reduce evaporation from the soil surface so that more water is available for transpiration (Fig. 3.12). In 2011, 1.2 M tonne of plastic film were applied to 19.8 M ha of cropland, equivalent to 12.2% of the arable land in China (Liu et al., 2014). This has increased water use efficiency for maize in some cases to 3 kg m^{-3} compared to values of 1 kg m^{-3} or less commonly obtained for dryland crops in the Texas High Plains (Stewart and Liang Wei-li, 2015). Although use of plastic film for dryland crop production is mostly confined to China, the technology might be applicable to smallholder farmers in other countries. However, there are growing environmental concerns in China about possible soil pollution from the plastic film (Liu et al., 2014).

A rather unique strategy being tried for dryland agriculture is growing grain sorghum and maize in clumps rather than equally spaced rows (Bandaru et al., 2006; Kapanigowda et al., 2010;

Krishnareddy et al., 2010). The hypothesis is that clumped plants, particularly grain sorghum, will produce fewer tillers that utilize water and nutrients but little or no grain when water is limited, improve microclimate to lower VPD, and fully utilize plant-available soil water during the growing season. A possible disadvantage is that reduced shading of the soil might reduce T/ET so it is recommended that clump-planting be combined with mulching. While the potential increase in yield from this strategy is not great, it might be beneficial because no increased inputs are required. It can also reduce labor in developing countries where much of the maize and grain sorghum are planted by hand. The use of clumps as well as the use of skip rows is based on the principle of nonuniformity (Loomis, 1983; Connor et al., 2011). They stated a useful generalization is that when a soil resource such as water is limiting, nonuniform treatment of the land or crop can be an advantage. Where soil resources are nonlimiting, however, they stated that uniform cropping will provide the greatest efficiency in light interception and photosynthesis.

In some cases, the best strategy for crop production in dryland areas is to grow forage rather than grain crops. For forage production, HI is not a factor because only biomass is the product, so higher plant populations can be used and closer row spacing to insure all the plant-available soil water is extracted, and a faster canopy develops so that there is less evaporation from the soil surface, leaving more water for transpiration. The downside is that if the forage is removed from the field, there is little or no crop residue left on the land that leads to soil degradation. A crop–livestock system where the forage is grazed by livestock and the manure stays on the land is a good system for utilizing limited water efficiently and also sustaining the soil resource.

3.4 Conclusions

Agriculture, and human life itself, depends on soil as a limited source. Of greater concern in dryland regions is that soils have been seriously degraded by erosion, excessive tillage, compaction, and other processes, which are accelerated by lack of CA systems use that can mitigate the adverse effects of climate variability. As a result, soil health has declined to critical levels on many of our croplands and rangelands in dryland areas. Although soil health can decline as a result of many factors, the loss of SOC, most commonly by tillage, is considered the most serious factor because it simultaneously lowers the chemical, physical, and biological properties resulting in a greatly degraded medium for growing plants. While declining SOC is a concern for all soils, it is particularly critical for soils in dryland areas because loss of SOC results in dry areas becoming even drier. While it is exceedingly difficult to restore SOC and soil health in dryland areas, it is feasible to do so, or at a minimum to maintain them. In this chapter, the attempt is to highlight management practices and systems such as CA that are critical to the maintenance or improvement of soil health and productivity in such areas. These are minimum soil disturbance, permanent organic soil cover with crops or plant residues, and diversification of crop species. For maximum effectiveness, these practices must be linked and benefits will be slow in developing, particularly in semiarid regions under limited water conditions, so patience can be considered a fourth principle for dryland areas.

References

Acton, D.F., Gregorich, L.J. (Eds.), 1995. The Health of Our Soils. Agriculture and Agri-Food Canada, Publication 1906/E, Ottawa, Canada.

Bandaru, V., Stewart, B.A., Baumhardt, R.L., Ambati, S., Robinson, C.A., Schlegel, A., 2006. Growing dryland grain sorghum in clumps to reduce vegetative growth and increase yield. Agron. J. 98, 1109–1120.

Baumhardt, R.L., Anderson, R.L., 2006. Crop choices and rotation principles. In: Peterson, G.A., Unger, P.W., Payne, W.A. (Eds.), Dryland Agriculture 2nd Edition, Agronomy Monograph no. 23. American Society of Agronomy, Inc., Crop Science Society of America, Inc., Soil Science Society of America, Inc., Madison, Wisconsin WI, pp. 113–139.

Bennett, J.B., Kenney, F.R., Chapline, W.R., 1938. The problem: subhumid areas Soils and Men, Yearbook of Agriculture. United States Department Agriculture, Washington, DC, pp., 68–76.

Brady, N.C., Weil, R.R., 2008. The Nature and Properties of Soils, fourteenth ed. Pearson Prentice Hall, Upper Saddle River, New Jersey NJ.

Chebrola, K.K., 2006. Precipitation Patterns along a N-S Transect in the Ogallala Aquifer Region. Master of Science Thesis, West Texas A&M University, Canyon, TX.

Connor, D.J., Loomis, R.S., Cassman, K.G., 2011. Crop Ecology: Productivity and Management in Agricultural Systems, second ed. Cambridge University Press, Cambridge, UK.

Doran, J.W., 2002. Soil health and global sustainability: translating science into practice. Agric. Ecosyst. Environ. 88, 119–127.

Doran, J.W., Coleman, D.C., Bezdicek, D.E., Stewart, B.A. (Eds.), 1994. Defining Soil Quality for a Sustainable Environment. Soil Science Society of America Special Publication 35, Madison, WI.

Derpsch, R., 2001. Frontiers in conservation tillage and advances in conservation practice. In: Stott, D.E., Mohtar, R.H. (Eds.), Sustaining the Global Farm. Selected papers, 10th International Soil Conservation Organization Meeting, May 24–29, 1999, Purdue University and USDA-ARS National Soil Erosion Research Laboratory, Lafayette, Indiana, pp. 248–254.

Egan, T., 2006. The Worst Hard Time: The Untold Story of Those Who Survived the Great American Dust Bowl. Mariner Books, New York, NY.

FAO, 2015. Conservation Agriculture. Food and Agriculture Organization of the United Nations, Rome, Italy, (Available online: http://www.fao.org/ag/ca/1a.html).

Haas, H.J., Evans, C.E., Miles, E.F., 1957. Nitrogen and Carbon Changes in Great Plains Soils as Influenced by Cropping and Soil Treatments. Technical Bulletin 1164, Washington, DC (Available online: https://books.google.com/books?id=iLYXAAAAYAAJ&pg=PP3&lpg=PP3&dq=Haas,+Evans,+Miles+1957&source=bl&ots=_0ATa02RNE&sig=9onkDpv1mIbgHGVZh4E-TjYQlTY&hl=en&sa=X&ved=0CC0Q6AEwBGoVChMI3fa14MizyAIVAxc-Ch3_hAVq#v=onepage&q=Haas%2C%20Evans%2C%20Miles%20 1957&f=false).

Himes, F.L., 1997. Nitrogen, sulfur, and phosphorus and the sequestering of carbon. In: Lal, R., Kimble, J.M., Follett, R.F., Stewart, B.A. (Eds.), Soil Processes and the Carbon Cycle. Advances in Soil Science. CRC Press, Boca Raton, Florida, pp. 315–319.

Hornick, S.B., Parr, J.F., 1987. Restoring the productivity of marginal soils with organic amendments. Am. J. Alt. Agric. 2, 64–68.

Hudson, B.D., 1994. Soil organic matter and available water capacity. J. Soil Water Conserv. 49 (2), 189–194.

Jenkinson, D.S., 1988. Soil organic matter and its dynamics. In: Wild, A. (Ed.), Russell's Soil Conditions and Plant Growth. Longman Scientific & Technical, Essex, U.K., pp. 564–607.

Johnson, W.C., Van Doren, C.E., Burnett, E., 1974. Summer fallow in the southern Great Plains. In: Summer Fallow in the Western United States. Conservation Research Report. 17, USDA-ARS, Washington, DC. pp. 86–109.
Koohafkan, P., Stewart, B.A., 2008. Water and Cereals in Drylands. Food and Agriculture Organization of the United Nations, Rome and Earthscan, London.
Kapanigowda, M., Stewart, B.A., Howell, T.A., Kadasrivenkata, H., Baumhardt, R.L., 2010. Growing maize in clumps as a strategy for marginal climatic conditions. Field Crops Res. 118, 115–125.
Karlen, D.L., Maubach, M.J., Doran, J.W., Cline, R.G., Harris, R.E., Schuman, G.E., 1997. Soil quality: a concept, definition, and framework for evaluation. Soil Sci. Soc. Am. J. 61, 4–10.
Krishnareddy, S., Stewart, B.A., Payne, W.A., Robinson, C.A., 2010. Grain sorghum tiller production in clump and uniform planting geometries. J. Crop Improv. 24, 1–11.
Lal, R., 2014. Societal value of soil carbon. J. Soil Water Conserv. 69, 186A–192A.
Lal, R., Kimble, J.M., Follett, R.F., Cole, C.V., 1998. The Potential of U.S. Cropland to Sequester Carbon and Mitigate the Greenhouse Effect. Sleeping Bear Press, Inc., Chelsea, Michigan.
Letey, J., Sojka, R.E., Upchurch, D.R., Cassel, D.K., Olson, K.R., Payne, W.A., et al., 2003. Deficiencies in the soil quality concept and its application. J. Soil Water Conserv. 58, 180–187.
Li Shengxiu, 2007. Dryland Agriculture in China. Science Press, Beijing, China.
Liu, E.K., He, W.Q., Yan, C.R., 2014. "White revolution" to "white pollution" – agricultural plastic film mulch in China. Environ. Res. Lett. 9 (9) http://dx.doi.org/10.1088/1748-9326/9/9/091001.
Loomis, R.S., 1983. Crop manipulations for efficient water use. In: Taylor, H.M., Jordan, W.R., Sinclair, T.R. (Eds.), Limitations in Efficient Water Use in Crop Production. American Society of Agronomy, Crop Science Society of America, Soil Science Society of America, Madison, Wisconsin, pp. 345–360.
Musick, J.T., Jones, O.R., Stewart, B.A., Dusek, D.A., 1994. Water-yield relationships for irrigated and dryland wheat in the U.S. Southern Plains. Agron. J. 86, 980–986.
National Research Council, 1993. Soil and Water Quality: An Agenda for Agriculture. National Academy Press, Washington, D.C.
Peterson, G.A., Westfall, D.G., 2004. Managing precipitation use in sustainable dryland agroecosystems. Ann. Appl. Biol. 144, 127–138.
Rasmussen, P.E., Albrecht, S.L., 1998. Crop management effects on organic carbon in semi-arid Pacific Northwest soils. In: Lal, R., Kimble, J.M., Follett, R.F., Stewart, B.A. (Eds.), Management of Carbon Sequestration in Soil. Advances in Soil Science. CRC Press, Boca Raton, Florida, pp. 209–219.
Schomberg, H.H., Ford, P.B., Hargrove, W.L., 1994. Influence of crop residues on nutrient cycling and soil chemical properties. In: Unger, P.W. (Ed.), Managing Crop Residues. Lewis Publishers, Boca Raton, FL.
Sinclair, T.R., Weiss, A., 2010. Principles of Ecology in Plant Production, second ed. CAB Int., Cambridge, Massachusetts.
Singh, R.P., 2005. Sustainable Development of Dryland Agriculture in India. Scientific Publishers, Jodhpur, India.
Squires, V., Tow, P. (Eds.), 1991. Dryland Farming: A Systems Approach – An Analysis of Dryland Agriculture in Australia. Sydney University Press, Sydney.
Stewart, B.A., Liang Wei-li, 2015. Strategies for increasing the capture, storage, and utilization of precipitation in semiarid regions. J. Integr. Agric. 14 (8), 1500–1510.
Stewart, B.A., Peterson, G.A., 2015. Managing green water in dryland agriculture. Agron. J. 107, 1544–1553.
Stewart, B.A., Robinson, C.A., 1997. Are agroecosystems sustainable in semiarid regions. Adv. Agron. 60, 191–228.
Stewart, B.A., Lal, R., El-Swaify, S.A., 1991. Sustaining the resource base of an expanding world agriculture. In: Lal, R., Pierce, F.J. (Eds.), Soil Management for Sustainability. Soil and Water Conservation Society, Ankeny, Iowa, pp. 125–144.
Stone, L.R., Schlegel, A.J., 2006. Yield-water supply relationships of grain sorghum and winter wheat. Agron. J. 98, 1359–1366.
Unger, P.W., Baumhardt, R.L., 1999. Factors related to dryland grain sorghum yield increases: 1939 through 1997. Agron. J. 91, 870–875.

Unger, P.W., Kirkham, M.B., Nielsen, D.C., 2010. Water conservation for agriculture. In: Zobeck, T.M., Schillinger, W.F. (Eds.), Soil and Water Conservation Advances in the United States. Soil Science Society of America Special Publication 60, Madison, Wisconsin, pp. 1–45.

USDA-ERS, 2015. Major Land Uses. United States Department Agriculture, Economic Research Service, Washington, DC, Available online: http://www.ers.usda.gov/data-products/major-land-uses/.aspx.

USDA-NRCS, 2015a. USDA Launches National Soil Health Initiative at Central Ohio Farm. Available online: <http://nemwuppermiss.blogspot.com/2012/10/usda-launches-new-soil-health.html>.

USDA-NRCS, 2015b. Soil Quality Indicator Sheets. Available online: <http://www.nrcs.usda.gov/wps/portal/nrcs/detail/soils/health/assessment/?cid=stelprdb1237387>.

USGCRP, 2014. Great Plains Regional Report. United States Global Change Research Program. Cambridge University Press, New York, NY.

CHAPTER 4

Conservation Agriculture Systems to Mitigate Climate Variability Effects on Soil Health

Mahdi M. Al-Kaisi[1] and Rattan Lal[2]

[1]Iowa State University, Ames, IA, United States [2]The Ohio State University, Columbus, OH, United States

4.1 Introduction

Climate change is the greatest challenge of the modern era, and to the use of our fundamental resource of soil and water for food production and security. This threat to soil health can be intensified by management practices that lead to soil desertification, deforestation, erosion, and other forms of degradation. These dynamics, along with weather variability, such as frequent wet and drought events, are prevalent in different parts of the world. The negative impact of climate change on soil resources has been reported to increase soil erosion, which potentially could reduce agricultural productivity by 10–20% (Delgado et al., 2011). The decline in soil productivity, as a result of soil erosion and degradation, is manifested in the deterioration of soil health and quality or functionality, where soil chemical, physical, and biological properties are severely degraded. These soil characteristics are the foundation for a productive soil and its ecosystem services. However, to understand the level of impact of climate change on soil resources, it is important to clearly define the concept of soil health and quality. The terms soil health and soil quality are closely linked, and these two terms are often used interchangeably, as a reference or a benchmark for the functionality of soil systems (Lehman et al., 2015). Attempts to define soil health or quality all focus on the same fundamental building units of what defines a well-functioning soil ecosystem (biological, physical, and chemical). In a more direct and simplistic way, soil quality is the capacity of the soil to function (Karlen et al., 1997), and soil health is "the continued capacity of soil to function as a vital living system, within ecosystem and land use boundaries, to sustain biological productivity, maintain the quality of air and water environments, and promote plant, animal, and human health" (Doran et al., 1996). This definition characterizes the soil as (1) a medium that supports and promotes the growth and development of plants, animals,

Soil Health and Intensification of Agroecosystems.
DOI: http://dx.doi.org/10.1016/B978-0-12-805317-1.00004-X

and humans, while regulating water processes in the ecosystem, (2) an environmental buffer that regulates and degrades hazardous compounds in the ecosystem, and (3) a medium that provides food and fiber services for sustaining animal and human lives (Weil and Brady, 2016). These soil functions are critical to food and fiber production that include nutrient provision and cycling, protection against pests and pathogens, production of growth factors, water availability, and formation of stable soil physical structure "aggregate" to reduce potential erosion risks and increase water processing (Lehman et al., 2015). These functions are strongly affected by climate variability and extreme weather conditions. As a consequence, in the absence of stable agriculture systems such as conservation practices that mitigate such climate extremes, these soil health functions can be jeopardized (Lal et al., 2011). The adoption of such practices within production fields and on marginal land can provide solutions for climate mitigation and adaptation (Lal, 2015). Such management practices can contribute to the reduction of soil erosion, increase carbon (C) sequestration (Guzman and Al-Kaisi, 2010) to improve soil functions, soil health (Guzman and Al-Kaisi, 2011), and contribute to the resilience of soils and cropping systems that will be needed to respond to climate change and related challenges such as food security (Lal et al., 2011). The integration of soil conservation principles, such as no-tillage (NT) and conservation tillage systems (CTS) (i.e., strip-tillage, minimum tillage, etc.), within the conservation agriculture (CA) system as defined by the Food and Agriculture Organization of the United Nation (FAO, 2009), include: minimal soil disturbance, while providing continuous plant residue cover along with the use of extended crop rotations and cover crops to enhance soil health indicators such as soil C sequestration and help mitigate and adapt to climate change (Hobbs et al., 2008; Kassam et al., 2009; Lal et al., 2011).

4.2 Agriculture Conservation Practices in Row Cropping Systems

4.2.1 No-Tillage, Conservation-Tillage, and Soil Biophysical Health

The increase in human population after World War II led many experts to predict widespread starvation (Weil and Brady, 2016). This prediction has led to the increase of agriculture production on already cultivated land as a primary means to meet such a challenge. This intensification has led to a significant increase in grain production worldwide from 1950 to 90 (Table 4.1). While the introduction of intensified farming systems increased food production, such intensified agriculture systems of land cultivation have direct and indirect effects on soil health. The documentation of soil health under a wide range of soil and crop management systems has been investigated by many scientists across the world to quantify the positive and negative effects on soil functions such as erosion, and aggregate stability (Six et al., 2000), soil C stocks (Oades, 1984; Al-Kaisi and Yin, 2005; Lal, 2015), water infiltration (Guzman and Al-Kaisi, 2011), and many other soil health indicators. The adoption of CA systems should include the integration of soil conservation practices in a system approach

Table 4.1: Percent of increase in food production in different regions between 1961 and 63 and 1989 and 90 attributable to increases in area cropped and to increase in yields per hectare

	Increase Attribute to	
Region	**Increased Area (%)**	**Increased Yield[a] (%)**
Low-income countries		
Sub-Saharan Africa	47	52
Latin America	30	71
Middle East/North Africa	23	77
South Asia	14	86
East Asia	6	94
High-income countries	2	98
World	8	92

Source: Data from the Food and Agriculture Organization (FAO).
[a]Includes both increasing the number of crops per year and increased yields per hectare.

within diverse cropping systems (Fig. 4.1). The introduction of NT as a component of CA systems is in response to the negative effects of conventional tillage (CT) practices that caused soil degradation in many parts of the world, and was examined for its advantages and limitations within different regions of various soil and climate conditions (Table 4.2). The design of agriculture systems within a wide range of continuum that accommodates the production challenges dictated by climate, soil, and hydrological constraints can include five basic principles: (1) residue management, (2) minimum soil disturbance, (3) inclusion of cover crop, (4) crop rotation, and (5) integrated nutrient management (Lal, 2015). The adoption of CA systems has been successful in many parts of the world, especially with highly commercialized production systems in North and South America and Australia (Derpsch, 2002; Bolliger et al., 2006), but it is lagging behind on smallholder farms (Ngwira et al., 2014). The challenge of climate change to soil health, particularly on smallholders' farms, can be mitigated by the adoption of CA that potentially can increase smallholders' farms' resilience to weather variability, reduction of soil degradation, and an increase in food production and profitability (Hobbs et al., 2008; Rockström et al., 2009). The key property that influences soil health and its biophysical and chemical functions is soil organic matter (SOM), which is greatly influenced by management and cropping systems and potential climate change. Maintaining SOM is well-linked to the type of tillage and crop rotations or cropping systems that have been practiced. NT and CTS increase soil aggregate stability and resiliency over time when subjected to prolonged wet conditions (Fig. 4.2), soil C stock, and water infiltration (Guzman and Al-Kaisi, 2011, 2014). These soil health indicators are strongly affected by climate parameters; specifically, the change in SOM, where a significant decline in SOM has been observed in the US Great Plains with different cropping systems (Fig. 4.3). In semidry regions, as found across all land uses, a warmer climate decreased soil organic carbon (SOC) and soil organic nitrogen (SON), but the decreases were mediated by

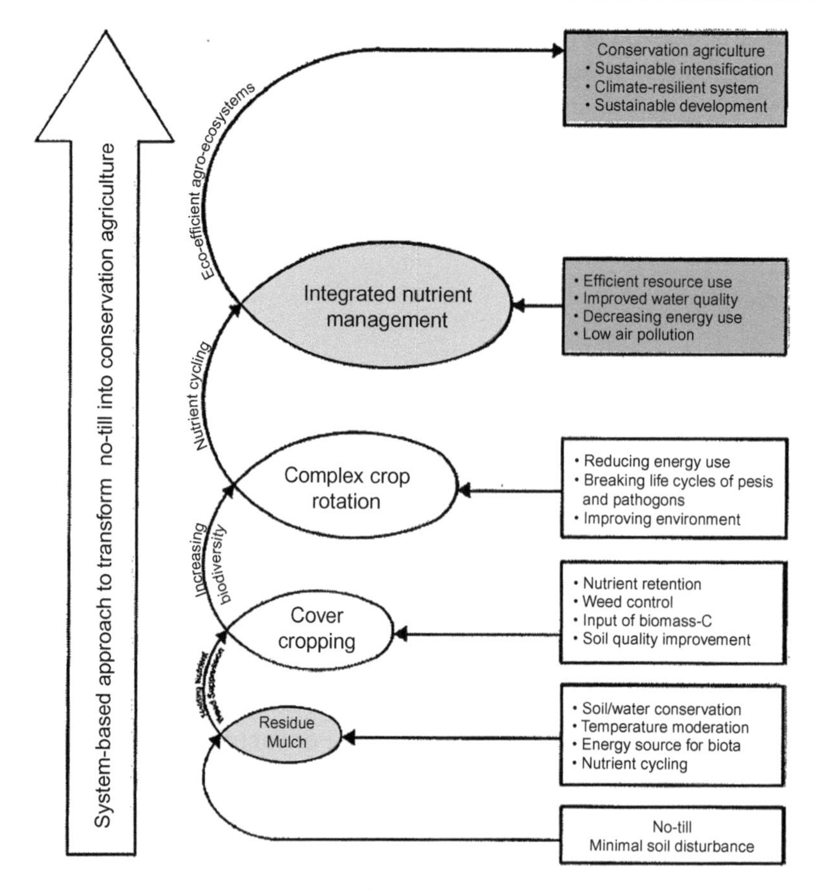

Figure 4.1

Integrating four basic components for transforming no-tillage into conservation agriculture. *After Lal, R., 2015. A system approach to conservation agriculture. J. Soil Water Conserv. 70 (4), 82–88.*

water availability. These results suggest that in dry regions, similar to the US Great Plains, where the trend toward more frequent drought events continues, continual losses of SOC from the system can be expected (Follett et al., 2012). This continual decline in SOC as an essential soil health component can be mitigated by the implementation of CA systems (Fig. 4.3). The improvement of soil biophysical and chemical functions with NT and CTS can influence nutrient and water retention (Oades, 1984; Hillel, 2004; Al-Kaisi et al., 2014). A well-structured soil as an outcome of CA systems and the formation of soil aggregates can provide a balanced soil environment of water storage, air, and nutrients for plant growth.

4.2.2 Conservation Systems and Soil Structure Dynamics

A stable and well-defined soil structure is one of the fundamental functions of healthy soils that can mitigate and adapt to climate variability. A stable soil structure value in sustaining

Table 4.2: Limitations and uncertainties of no-till (NT) farming, which must be addressed through a system-based approach to enhance merits of conservation agriculture to advance climate resilience and promote sustainable intensification

Merits	Limitations	Issues and Uncertainties (For Small Landholders)
1. Erosion control and reduced sedimentation	1. High incidence of weeds, especially perennials	1. Land tenure and economic factors
2. Water conservation and high water use efficiency	2. Greater use of pesticides, including herbicides	2. Access to market and credit
3. Savings in time and labor	3. Need for new seed drill and other farm machinery	3. Availability of inputs
4. Low energy use	4. More insects, pests, and pathogens	4. Changing and uncertain climates (extreme events)
5. Less equipment used	5. Challenges with residue management	5. Nutrient management (N, P, and Ca) and fertilizer placement
6. Low wear and tear of machinery	6. High level of management skills	6. Soil acidification
7. Less nonpoint source pollution	7. Potential emission of N_2O	7. Lack of proper tools and equipment
8. Soil quality improvement and better structure	8. Poor quality of seed placement, low crop stand	8. Shift in weed spectrum
9. Soil carbon sequestration	9. Risks of yield reduction (5–10%)	9. Time required for NT to become fully functional
10. Better environment	10. Suboptimal soil temperatures in spring	10. Slow seed germination in poorly drained soils
11. Climate-resilient system	11. Potential wet conditions in clayey soils	11. Harvesting residues for cellulose ethanol and other uses
12. Sustainable intensification	12. Competing uses of crop residues	12. Nutrient and water interaction effect on crop yield
13. Low production cost and high net profit	13. Potential increase in fertilizer immobilization and low uptake	13. Ammonia volatilization
14. Enhanced fungal hyphae network and increased glomalin	14. Sulfur (S) deficiency at seeding stage	14. Low efficacy of pesticide/herbicide use with mulch
15. High activity and diversity of soil biota including microbial biomass carbon	15. Buildup of soil P in the surface and enhanced risks of eutrophication	15. Changes in soil fertilizer recommendation over time

Source: After Lal, R., 2015. A system approach to conservation agriculture. J. Soil Water Conserv. 70 (4), 82–88.

soil health is rooted in its function to resist water erosion effects through (1) accommodating water infiltration, (2) facilitating water transfer and retention, (3) resisting degradation, and (4) sustaining plant growth (Weil and Brady, 2016). The relative weight of each of these soil functions to resist soil erosion is summarized in Table 4.3, where accommodating water

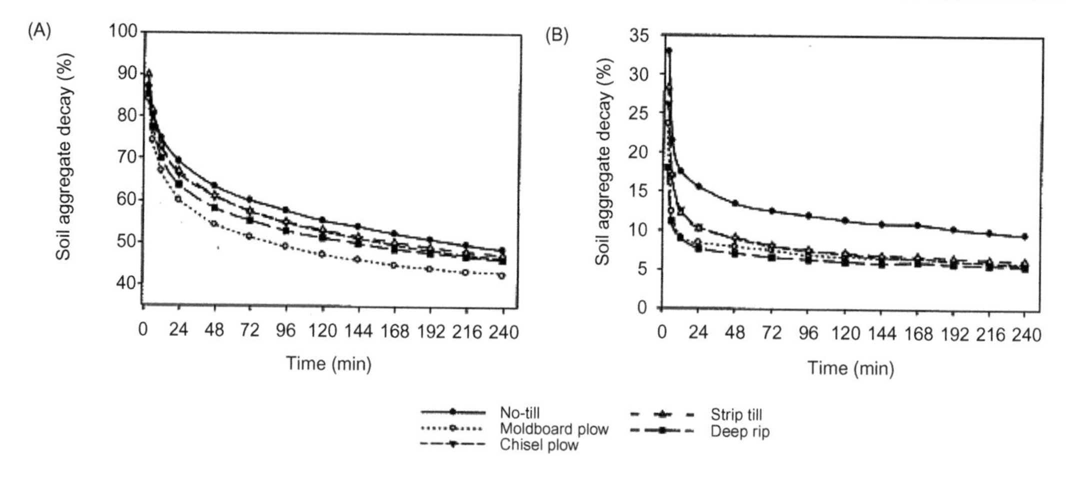

Figure 4.2

Kinetics of wet soil (A) microaggregate and (B) macroaggregate stability decay over time at the top 15 cm of five tillage systems of a long-term tillage and crop rotation study. *After Al-Kaisi, M.M., Archontoulis, S.V., Kwaw-Mensah, D., Miguez, F., 2015. Tillage and crop rotation effects on corn agronomic response and economic return at seven Iowa locations. Agron. J. 107, 1411–1424. doi:10.2134/agronj14.0470.*

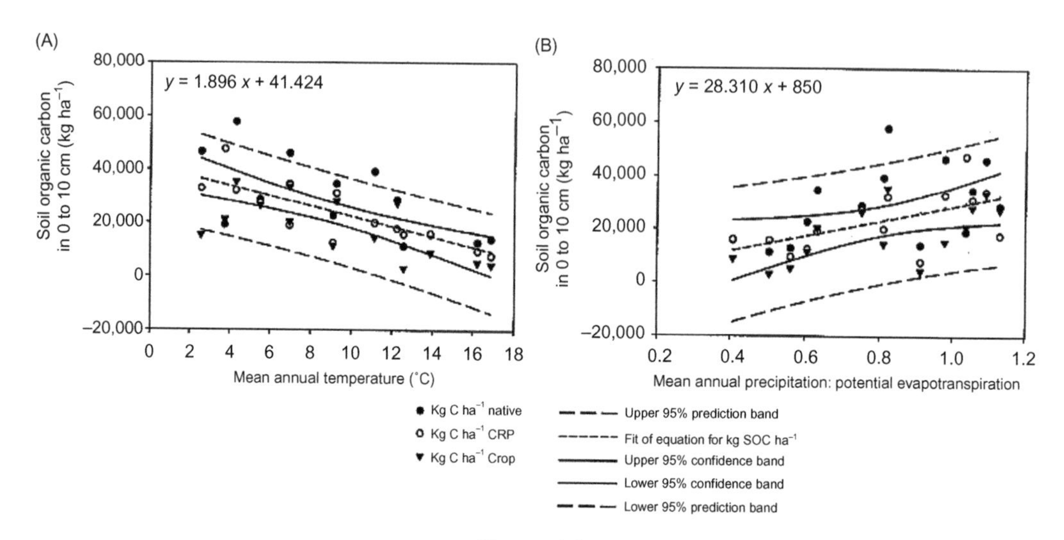

Figure 4.3

Soil organic carbon (kg SOC ha^{-1}) as a function of (A) mean annual temperature and (B) the ratio of mean annual precipitation to potential evapotranspiration for native, Conservation Reserve Program (CRP), and cropped sites across the US Great Plains. *After Follett, R.F., Stewart, C.E., Pruessner, E.G., Kimble, J.M., 2012. Effects of climate change on soil carbon and nitrogen storage in the US Great Plains. J. Soil Water Conserv. 67, 331–342.*

Table 4.3: Four possible soil-quality functions and their relative weights in determining the resistance of soil erosion, along with measurable indicators for each function and their weights

Soil Quality Function	Function Weight	Measurable Indicator	Indicator Weight
1. Accommodate water entry	50	Infiltration rate	50
2. Resist degradation	35	Aggregate stability	27
		Shear strength	4
		Soil texture	2
		Heat transfer capacity	2
3. Facilitate water transfer and absorption	10	Hydraulic conductivity	5
		Porosity	2
		Macropores	3
4. Sustain plant growth	5	Rooting depth	1
		Water relations	2
		Nutrient relations	1
		Chemical barriers	1

Source: After Karlen, D.L., Stott, D.E., 1994. A frame work for evaluating physical and chemical indicators of soil quality. In: Doran, et al. (Eds.), Defining Soil Quality for a Sustainable Environment, SSSA Special Publication no. 35, (Madison, WI, Soil Sci. Soc. Amer.).
Note that with the exception of soil texture; most of the indicators are properties that can be significantly influenced by soil-management practices. Note that accommodating water entry, measurable by infiltration rate, is thought to provide about half (50%) of this function. Resistance degradation, measured primarily by aggregate stability, is of second importance.

infiltration counts for 50%, as compared to the other soil functions. The conceptual model for aggregate formation and the hierarchy theory describes the process by which soil particles are bound together by bacterial, fungal, and organic compounds into microaggregates. These microaggregates are bound together into macroaggregates by transient agents such as microbial and plant-derived polysaccharides and temporary binding agents such as roots and fungal hyphae (Tisdall and Oades, 1982). The hierarchal formation of aggregates can exist in certain soils, as expressed in Alfisols and Mollisols, because organic materials are major binding agents for aggregate formation and stabilization, but not in Oxisols, where oxides rather than organic compounds are the dominant stabilizing agents (Elliot, 1986). Intensified agroecosystems can have a positive effect on soil aggregate and structure dynamics through the increase of biomass production where crop residue is left on the soil surface. In addition, the inclusion of cover crops within CA systems that include NT and CTS has contributed to the improvement of soil physical functions such as water infiltration (Fig. 4.4). The soil structure dynamics and formation are defined by building soil aggregates that are formed through complex interactions between physical, biological, and chemical processes (Tisdall and Oades, 1982). The factors that affect soil aggregate formation and stability can be grouped into biotic (SOM, plant root activities, soil fauna, and microorganisms), abiotic (clay minerals, sesquioxide, exchangeable cations), and environmental (soil temperature and moisture) (Chen et al., 2007). The stability of the soil aggregate depends on the forces

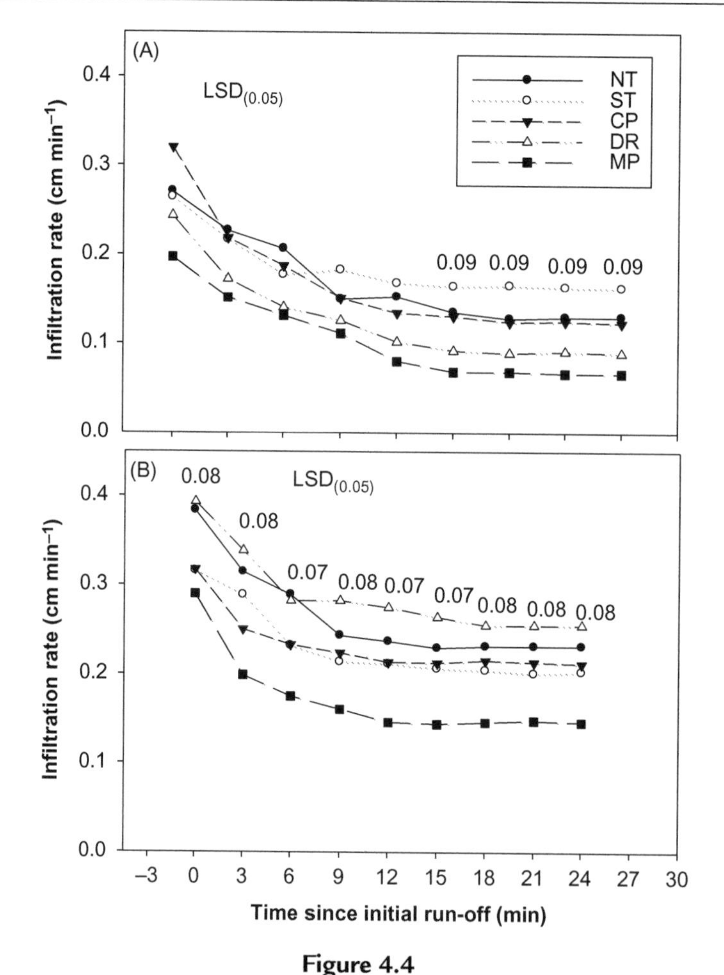

Figure 4.4
Infiltration rate of five tillage systems over time under (A) no-cover crop treatment and (B) cover crop treatment with five tillage systems. Differences between tillage treatments' infiltration rates within each time period greater than the LSD value are significantly different at $p = 0.05$. *Al-Kaisi (unpublished data).*

that bind particles together and on the nature and type of soil disturbance that an aggregate is subject to with different soils and cropping systems practices (Beare and Bruce, 1993). In general, soil cultivation reduces SOM and, consequently, changes the distribution and stability of soil aggregates (Six et al., 2000). While investigating soil aggregate formations and stability in three agroecosystems that include native vegetation, NT system, and CT, it was found that macroaggregates accounted for 85% of the dry weight, and were similar across all systems (Six et al., 2000; Guzman and Al-Kaisi, 2011). However, aggregate distribution from slaked soils increasingly shifts toward more microaggregate and fewer macroaggregates, with increasing soil disturbance or cultivation intensity (Fig. 4.5). There

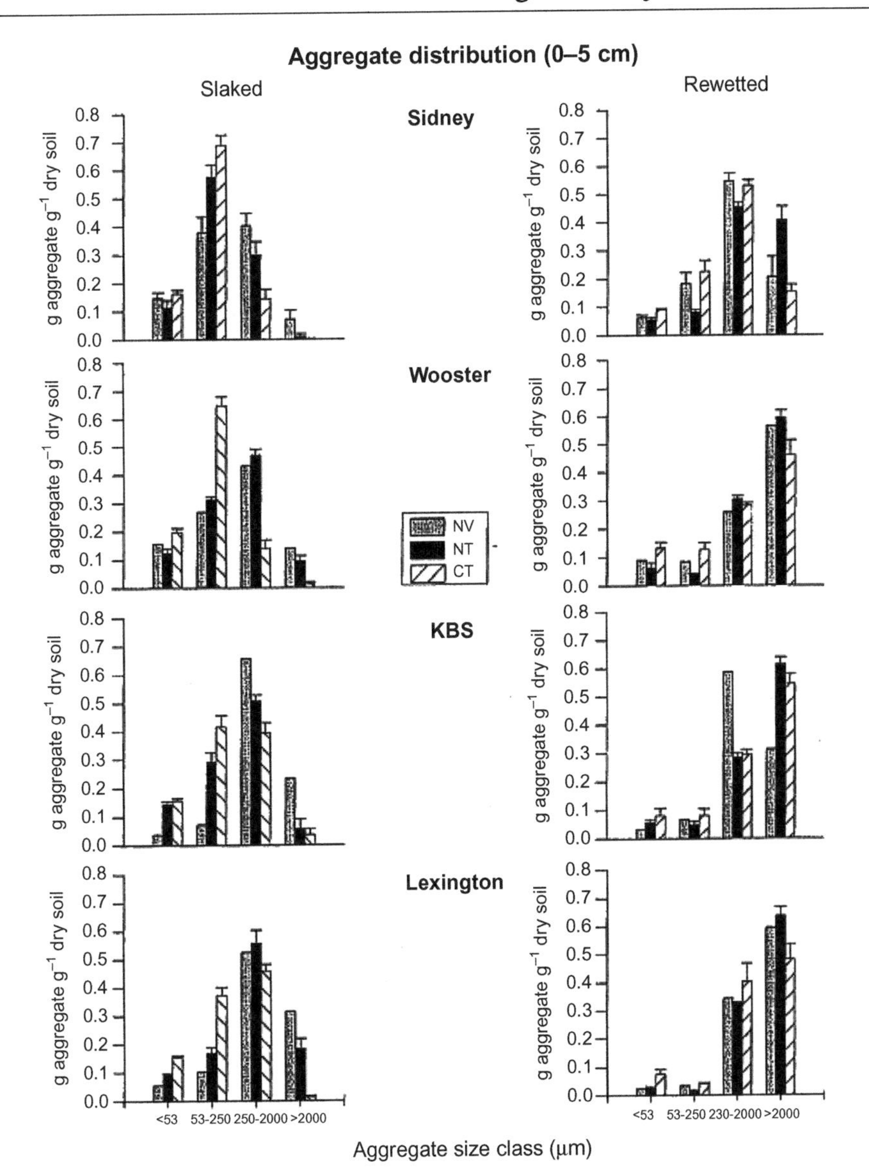

Figure 4.5

Slaked and rewetted aggregate-size distributions in the surface layer (0–5 cm) of four long-term agricultural experiment sites (SID 5 Sidney, NE; WO 5 Wooster, OH; KBS 5 Kellogg Biological Station, MI; LX 5 Lexington, KY) with three management treatments (NV 5 native vegetation; NT 5 no-tillage; CT 5 conventional tillage). Bars are standard deviations. *After Six, J., Paustian, K., Elliott, E.T., Combrink, C., 2000. Soil structure and organic matter: I. Distribution of aggregate-size classes and aggregate-associated carbon. Soil Sci. Soc. Am. J. 64, 681–689.*

is a successful trend in the adoption of CA systems in many countries, and particularly in South America, with 64 million hectare (M ha), or 60%, of the arable land under CA in 2014 (Kassam et al., 2014). The incorporation of cover crops within cropping systems and the particular system approach adopted has led to an increase in soil C sequestration and associated soil structure. In a 22-year experiment on an Oxisol in Parana State, southern Brazil, Sá et al. (2001), reported the SOC sequestration rate under CA of 806 kg C ha^{-1} y^{-1} for 0–20 cm depth and 994 kg C ha^{-1} y^{-1} for 0–40 cm depth. The increase in SOC is one of the essential mechanisms in aggregate formation and stability. The CA systems are characterized by a diversity of cropping systems and minimal soil disturbance, which enhances the process of soil structure formation and stability.

4.2.3 Conservation Agriculture Systems and Water Processing

Water processing in the soil system is governed by its physical properties, as explained in Chapter 1, Fundamental of Soil Environment and Functions. The type of cropping systems, intensity of crop rotations, and tillage systems can influence water movement within the soil profile, soil water storage, and surface water runoff. The CA systems present an integrated approach for building soil biological, physical, and chemical functions that can be affected by extreme weather events (Lal, 2015). In order to sustain soil health, as the demand for food and fiber production increases in a highly intensified agriculture system, adoption of an integrated conservation system is essential to increase water and nutrient use efficiencies and reduce surface runoff (Licht and Al-Kaisi, 2005b). The CA system, as defined by three principles, is keeping crop residues on the soil surface, no soil disturbance (i.e., NT), and diversified crop rotations (Kassam et al., 2009), which have significant advantages for soil water dynamics (Guzman and Al-Kaisi, 2011). The improvement in soil physical properties, such as bulk density, infiltration rate, and an increase in soil porosity (Table 4.4), is well-correlated with an increase in soil aggregate formations as a contribution of root biomass due to root exudates acting as binding agents for soil particles (Tisdall and Oades, 1982; Oades and Waters, 1991). The increase in SOC and the formation of soil aggregates can increase water storage in a CA system, as compared to other CT systems (Fig. 4.6). The CA systems, as diverse production systems, may present well-positioned systems that can mitigate weather variability by conserving water and lessening drought effects on crop production, along with providing good protection to soil function by reducing soil erosion and increasing water recharge (Guzman and Al-Kaisi, 2011; Weil and Brady, 2016). This trend will lead to greater water use efficiency (WUE) in both dryland and irrigated agriculture by reducing the amount of water used for irrigation (Al-Kaisi and Yin, 2003), and will reduce the energy input by using low-input irrigation systems, such as drip and sprinkler irrigation systems (Hillel, 1968). Water dynamics in a CA system, where crop residue is an integral part, is much different from tilled or bare surface where there is a potential for soil to lose as much as 50% of its seasonal storage of moisture to evaporation (Hillel, 1998). In principle, water flux from the soil surface

Table 4.4: Correlations between soil organic carbon (SOC) concentration, root biomass (RB), bulk density (ρ_b), mean weight diameter (MWD), infiltration rate (I_t), infiltration rate the first 3 minutes (3-min I_t), and years since establishment (Years)

	SOC		RB		ρ_b		MWD		I_t		3-min I_t	
	r	*p*-value	*r*	*P*-value	*r*	*P*-value	*r*	*P*-value	*r*	*P*-value	*r*	*P*-value
Years[a]	0.80	<0.0001	0.03	0.8548	−0.66	<0.0001	0.17	0.3258	0.53	0.0008	0.60	0.0001
SOC	–	–	0.13	0.4497	−0.69	<0.0001	0.08	0.6341	0.58	0.0002	0.63	<0.0001
RB	0.13	0.4497	–	–	−0.25	0.1461	0.24	0.1642	0.09	0.5830	0.20	0.2500
ρ_b	−0.69	<0.0001	−0.25	0.1467	–	–	−0.29	0.0895	−0.53	0.0009	−0.70	<0.0001
MWD	0.08	0.6341	0.24	0.1642	−0.29	0.0895	–	–	−0.02	0.8876	0.30	0.0764

Source: After Guzman, J.G., Al-Kaisi, M.M., 2011. Landscape position effect on selected soil physical properties of reconstructed prairies in southcentral Iowa. J. Soil Water Conserv. 66,183–191.

[a] For the P-Remnant site, 150 years was assumed for establishment year in model. Significance level $p \leq 0.05$.

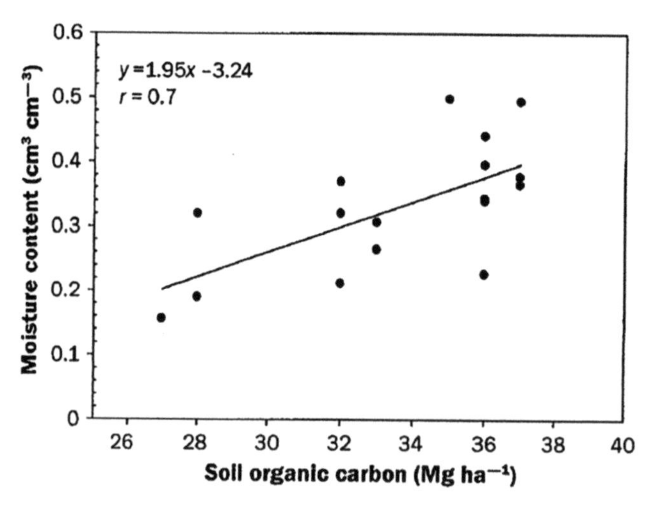

Figure 4.6

Correlation between average soil moisture content and soil organic carbon content at the top 15 cm of a 10-year long-term five tillage systems in corn-soybean rotation study. *After Al-Kaisi, M.M., Duelle, A., Kwaw-Mensah, D., 2014. Soil microaggregate and macroaggregate decay over time and soil carbon change as influenced by different tillage systems. J. Soil Water Conserv. 69, 574–580. http://dx.doi.org/10.2489/jswc.69.6.574.*

can be controlled by three processes: (1) controlling the energy supply to the soil surface such as level crop residues (changing surface albedo), where residue has a lighter color, leading to the reflection of sun radiation and reducing soil temperature at the evaporative surface, (2) reducing potential gradient that drives water toward the soil surface (i.e., lowering of the water table or warming the soil surface), or (3) decreasing the conductivity or diffusivity of the soil profile by changing the soil conditions (i.e., tilling the surface, incorporating organic materials at the soil surface, or soil conditioning). The most natural and sustainable option for building soil, yet conserving soil moisture and sustaining productivity, is what the CA system offers, particularly in dry regions where water conservation and preserving soil health are intertwined (Lal, 2015).

4.3 Crop Rotation Effects on Soil Health

4.3.1 Soil Aggregate Formation Process

Crop rotation is an essential component of CA systems, where a sequence of crops is rotated every season with grasses or legumes to achieve a diverse system to enhance soil biodiversity and physical structure. Over and above the effects of reducing weeds, disease, and insect problems, crop rotation provides additional benefits for soil health functions such as aggregation. The diversity of the root system that promotes a diverse microbial community is reflected in the development of soil aggregates (Tisdall and Oades, 1982). Crop rotation coupled with an NT system can contribute significantly to soil aggregation and water conservation, especially in arid and semiarid climate regions within intensified agriculture cropping systems (Lal, 2015). Root systems of winter cover crops and the lack of soil disturbance during 19 years under NT, contributed more to soil aggregation (>2.00 mm) than under a fallow or a CT system (Calegari et al., 2010). Soil aggregate formation, governed by biotic and abiotic processes, plays a significant role in nutrient cycling, SOC storage, and providing a growth environment for the root system (Oades and Waters, 1991). These functions are influenced by crop rotation as a driver for facilitating the biochemical and physical conditions that are responsible for forming and binding soil particles. Soil structure is defined by aggregates and their stability which most often are governed by the type of vegetation system and how stable and diverse this system is (Six et al., 2000). The mechanisms and processes of soil aggregates of different sizes that involve biological, chemical, and physical interrelationships are well-studied and addressed by many scientists (Tisdall and Oades, 1982; Oades and Waters, 1991; Six et al., 1999). The interaction of root systems with soil particles and production of organic compounds through biochemical processes is essential in the rearrangement, flocculation, and cementation of soil particles, where SOC, poly-cations, clay, minerals, and microbial community play a key role (Chenu et al., 2000; Schulten and Leinweber, 2000; Borie et al., 2008). Many organic compounds, as a product of biotic activities that include substances derived from decomposed SOM, which are recalcitrant, and other organic compounds such as polysaccharides, carbohydrates, lignin, and lipids within the soil environment as a product of root and microbial interactions with inorganic binding agents, are essential in forming soil aggregates. The effect of crop rotation on soil aggregation processes is highly influenced by the diversity of the plant community and the stability of the system (Chan and Heenan, 1996; Bronick and Lal, 2005). The effect of crop rotation is due to its rhizospheric effects, where roots release exudates, play a significant role in the rearrangement of soil particles and changes in the physical, chemical, and biological characteristics leading to soil aggregation (Rilling et al., 2002; Bronick and Lal, 2005). Soil aggregate formation and stability are highly increased as a result of an increased rhizodeposition, root density, root turnover, hyphae growth, and microbial biomass (Caravaca et al., 2002). Size and distribution of soil aggregates can significantly differ among different crop rotations in the top 0–10 cm of soil depth (Holeplass et al., 2004). In Norway, the meadow-based rotation plus grain rotation increased the percentage of aggregates distribution of the 0.6–2.0 mm and the < 0.6 mm fractions, but may decrease at the 6–20 mm and > 20 mm fractions. The aggregate stability for meadow-based rotation can be greater than that for grain rotation (Table 4.5), because of an increase in SOC concentration due to an increase in root biomass (Holeplass et al., 2004).

4.3.2 Crop Rotation and Soil Biology

Soil biology, size and diversity, is a reflection of the agroecosystem plant diversity, vegetation structure, and management practices. Soil environment, whether in temperate or tropical

Table 4.5: Effect of crop rotation on size distribution of water-stable aggregates (WSA)

Crop Rotation	2–6 mm	1–2 mm	0.25–1 mm	<0.25 mm	WSA (%)	MWD (mm)
	0–10 cm depth					
6 year grain	9.6b	15.7	30.8ab	44.0a	56.0b	0.85b
3 year+ 3 year row crops	11.7b	14.7	33.4a	40.2a	58.8b	0.90b
2 year grain+ 4 year meadow	31.7a	25.3	27.0b	16.0b	84.0a	1.8a
$LSD_{0.05}$	11.3	ns	4.9	6.6	6.6	0.3
	10–25 cm depth					
6 year grain	10.5b	14.7b	30.7	44.1a	55.9b	0.9b
3 year grain+ 3 year row crops	11.6b	11.9b	31.2	45.3a	54.7b	0.9b
2 year grain+ 4 year meadow	29.9a	22.6a	29.6	17.9b	82.1a	1.7a
$LSD_{0.05}$	5.5	5.1	ns	7.2	7.2	0.2

Source: After Holeplass, H., Singh, B.R., Lal, R., 2004. Carbon sequestration in soil aggregates under different crop rotations and nitrogen fertilization in an inceptisol in southeastern Norway. Nutr. Cycl. Agroecosyst. 70, 167–177.

Mean values in the same column (crop rotation) followed by the same letter or letters are not significantly different at $p = 0.05$ level.

regions, cropland or grassland, is highly favorable to the microbial community. This trend may be due to better quality and greater abundance of organic matter sources, favorable water conditions, and temperature (Weil and Brady, 2016). Soil invertebrate communities generally are good indicators of soil functionality and soil management effects in response to changes in land use and management practices that include tillage, crop diversity, and a change in vegetation such as the conversion of forestland to row crops (Bongers, 1990). Change in land use differs widely in effects on different microbial communities. However, annual crop production has caused severe depletion of microinvertebrate communities in certain areas, leading to a significant decline in earthworms, such as under rice (*Oryza sativa* L.), corn (*Zea mays* L.), beans (*Glycine max* L.), for a wide range of temperate and tropical environments (Lee, 1985). Yet, there must be an increased emphasis on soil health and agriculture sustainability under annual cropping systems. Thus, it is imperative to determine the effect of row cropping systems, and particularly, the role of crop rotation in maintaining soil function and services. The key property that influences soil health is SOM, which is highly affected by crop rotation (Six et al., 2000; Guzman and Al-Kaisi, 2014; Lal, 2015). Many scientists have reported the positive effect of crop rotations; particularly those which include small grains and legumes, along with CA systems in maintaining or improving SOM (Odell et al., 1984; Campbell et al., 1991). In a row cropping system, the effect of crop rotation on SOC and N fractions and microbial biomass C (MBC) is related to the amount of crop residues returned to the soil (Omay et al., 1997) and root system (Guzman and Al-Kaisi, 2011). Indeed, MBC and microbial biomass N (MBN) can be more under continuous wheat than under wheat-fallow rotations and lower than that under grass pasture (Collins et al., 1992). Further, retention of crop residues impacts MBC and N, where burning residues reduce MBC compared with soil receiving barnyard manure. The decline in seasonal MBC is highly correlated with the decrease in SOC in annual cropping systems (Collins et al., 1992). The data from a 16-year Canadian study, where four wheat rotations were assessed for certain biological properties, showed the lowest bacterial numbers under fallow-wheat, intermediate in fallow-wheat-wheat (*Triticum aestivum* L.), higher in continuous wheat receiving N and P, and the highest in continuous wheat receiving only P (Biederbeck et al., 1984). The other benefits of crop rotation or cropping sequence are in reducing the incidence of soil-borne diseases, as compared to mono cropping, which increases such incidents due to the availability of plant hosts for certain diseases (Schjonning et al., 2004). Continuous cropping can lead to a decrease in the incidence of some specific diseases over a number of years, which is called "take-all decline" (TAD) in cereals. Probable reasons for TAD include an increase in populations of antagonistic microorganisms. For example, the increase in the populations of antagonistic fluorescent pseudomonads in the rhizosphere has been associated with the TAD phenomenon (Weller, 1983). Although crop rotation provides a soil environment that is conducive for disease control through natural processes, the available evidence is not conclusive (Gardner et al., 1998). However, since the control of pathogens in soil is frequently associated with the effects of soil organisms and the stimulation of microbial activity by growing rotational crops, this makes crop rotation more effective than continuous monocrop or fallow systems. The effect of crop rotation in building soil biota and disease control is due to microbial diversity, where, e.g., an increase in bacterial diversity in the rhizosphere of wheat following legumes was higher than in wheat following wheat or wheat following fallow (Lupwayi et al., 1998).

4.3.3 Crop Rotation and Soil Organic Carbon Stocks

The SOC is an important component of soil sustainability, soil fertility, soil health, and productivity (Bauer and Black, 1994). The inclusion of crop rotation and type of tillage system in the agricultural system can change (increase or decrease) the SOC stocks. The degree by which the adoption of such practices influences change in the SOC stocks is well-linked to the rate of atmospheric CO_2, which is sequestered by the plant and soil, where the types of crop rotation and tillage system play a significant role (Al-Kaisi and Yin, 2005). The main objective of implementing CA systems (e.g., crop rotation and NT) is to increase input of biomass-C, decrease the rate of SOM decomposition or oxidation, or both (Follett, 2001). The shift in management practices to less intensive soil cultivation and more diverse and extended crop rotations can reduce SOC loss or reverse it. In a 17-year tillage experiment in Europe, Smith et al. (1998), found an average increase in SOC, with change from CT to NT, at $0.73 \pm 0.39\%$ year^{-1}. However, changing the crop rotation paradigm from monocrop to crop rotation, crop-fallow to monoculture or rotation cropping, or increasing the number of crops in the rotation, caused an average increase in SOC sequestration of $15 \pm 11\,\mathrm{g\,C\,m^{-2}}$ year^{-1}, but not as much as changing the tillage system CT to NT (West and Post, 2002). Generally, continuous corn produces more crop residue and C input than a corn-soybean rotation system (Al-Kaisi and Yin, 2005). The low input of biomass-C may reduce the SOC addition in soil (Duiker and Lal, 1999; Clapp et al., 2000). The enhancement of rotation complexity with NT may not always lead to a significant increase in SOC. It is possible that SOC under NT is close to a maximum steady-state level than that under CT and, therefore, stands to gain less SOC under rotation enhancement (Table 4.6) (West and Post, 2002). The effect of crop rotation on SOC as an essential soil parameter for improving soil health has been well-documented in many long-term studies from around the world, especially those that contain a wide range of cropping sequences. In south Australia, e.g., a long-term study of cropping systems, established in 1925 (that include continuous pasture, continuous wheat, wheat-fallow rotation, and different phases of a 2-year wheat-4-year pasture rotation), shows differential effects on soil aggregate stability of a Calcic Rhodoxeralf (Kay et al., 1994). Further, Kay and colleagues observed a decrease in SOC with a decrease in frequency of pasture in the rotation, and the least SOC was observed when a fallow period was included in the rotation. The Morrow plots in Illinois, USA, established in 1876, are the oldest research plots in the United States (Odell et al., 1982). These plots were established to evaluate the continuous cropping on prairie-based soil in the Midwest (a silt loam, Aquic Argiudoll) for

Table 4.6: Soil organic carbon sequestered in response to enhancing crop rotation[b]

Enhanced Crop Rotation	Number of Paired Treatments	Average Soil Depth (cm)	Average Duration of Experiments (yr)	Mean Increase in SOC		Linear Regression Between Initial System and Enhanced Rotation§	Mean C Sequestration Rate[d]	
				$g\,m^{-2}$	$g\,kg\,C^{-1}$[c]		$g\,m^{-1}\,yr^{-1}$	$g\,kg\,C^{-1}\,yr^{-1}$[c]
All rotations	97	22	25	218 ± 118[a]	56 ± 24	y = 0.98x + 320.12	15 ± 11	4 ± 2
All rotations (no c to c-s) [e]	85	21	26	293 ± 118[a]	70 ± 25	y = 1.00x + 286.29	20 ± 12	5 ± 2
All CT rotations (no c to c-s) [e]	48	21	28	280 ± 167[a]	75 ± 40	y = 0.95x + 527.84	16 ± 14	4 ± 3
All NT rotations (no c to c-s)	14	25	15	171 ± 377	47 ± 56	y = 0.93x + 524.14	26 ± 56	6 ± 8
All rotations with grass, hay, or pastor	18	33	20	538 ± 243[a]	108 ± 64	y = 1.02x + 382.69	19 ± 8	5 ± 4
All corn rotations	35	23	30	163 ± 212	58 ± 424	y = 0.84x + 1021.40	6 ± 11	2 ± 2
All corn rotations (no c to c-s)	23	22	34	412 ± 209[a]	97 ± 53	y = 0.89x + 966.94	19 ± 11	4 ± 2
Continuous corn to corn-soybeans	12	24	21	−311 ± 367	−46 ± 51	y = 0.77x + 904.24	−19 ± 19	−3 ± 4
All wheat rotations	32	19	24	271 ± 139[a]	64 ± 28	y = 1.10x − 142.5	27 ± 22	6 ± 4
All wheat rotations (no w-f to cont. w)[f]	15	20	20	446 ± 274[a]	97 ± 50	y = 1.09x + 13.83	51 ± 47	11 ± 8
Wheat-fallow to continuous wheat	11	17	25	104 ± 100[a]	33 ± 38	y = 1.02x + 54.73	6 ± 8	2 ± 3
All soybean rotation	13	25	11	253 ± 473	57 ± 82	y = 0.85x + 1081.52	26 ± 46	6 ± 8

Source: After West, T.O., Post, W.M., 2002. Soil organic carbon sequestration rates by tillage and crop rotation: a global data analysis. Soil Sci. Soc. Am. J. 66, 1930–1946.

[a]Indicates significant differences between soil organic carbon (SOC) under baseline condition and rotation enhancement at the $p = 0.05$ level.

[b]Consists of changing from monoculture to continuous rotation cropping, crop-fallow to continuous monoculture or rotation cropping, and increasing the number of crops in a rotation system.

[c]Represents an increase in SOC per kg SOC, as opposed to SOC per kg soil.

[d]Sequestration rate was calculated as an average of sequestration rates from each experiment, not by division of the mean increase in SOC with the average duration of experiments.

[e]A change from continuous corn (c) to corn-soybean (c-s) rotation was not shown to sequester a significant amount of C, and was therefore excluded from some analyses.

[f]A change from wheat-fallow to continuous wheat (w) was not shown to sequester a significant amount of C, and was therefore excluded from some analyses. Treatments with a change from wheat-fallow to nonfallow, wheat rotations were included.

SOC stocks (Mitchell et al., 1991). The cropping systems include continuous corn, corn-oat (1876–1966) rotation, corn-soybean (1966–present) rotation, a corn-oat-legume (1901–53) rotation, and alfalfa (1954–present) rotation. All rotations received adequate fertilization with a plow tillage (PT) system. The corn yields have been consistently the lowest with monoculture and the highest with a corn-oat-legume rotation. However, SOC decreased under all cropping systems and fertility treatments, and the decline was greater with continuous corn and the least with corn-oat-legume rotation and was greatest where no fertilizer or manure were added (Odell et al., 1984). Another example of crop rotation effects on SOC was established at the Agriculture Canada Research Station at Lethbridge, Alberta, Canada (Bremer et al., 1994; Johnson et al., 1995). Crop rotations included fallow-wheat combination (1 or 2 years) and continuous wheat with or without 80 kg N $ha^{-1}year^{-1}$. The rotations also included fallow-2-year-wheat with annual application of livestock manure, a fallow-2-year-wheat-3-year-alfalfa (*Medicago sativa* L.)-crested wheat grass (*Agropyron cristatum* L.) mixed ley, and a native grass system. Averaged across rotation phases, total and labile SOC fractions were increased by manure addition, ley period, native grass sod, and reductions in frequency of fallow. Native grass and continuous wheat rotations resulted in the greatest concentration of total and labile SOC in the 0–7.5 cm depth, while with hay production from mixed species, the ley increased SOC in the 15–30 cm depth (Bremer et al., 1994). The role of crop rotation and diversity in improving SOC (Kay et al., 1994) as a central soil property to improving soil health is essential in mitigating weather variability effects on soil quality. The improvement in soil physical and biological properties with crop rotation can improve soil water-holding capacity (Hudson, 1994) that is essential for improving productivity, especially in dry regions, where climate challenges are great.

4.3.4 Soil Water and Nutrient Movement

The impact of crop rotation on soil structure, as illustrated in Section 4.3.1, through the formation of soil aggregates, also influences water and nutrient dynamics, storage, movement, and supply to the plant. Improvement in SOM has a positive effect on soil physical, chemical, and biological functions. Over five decades (1965–2015), there was inconsistent evidence that SOM has little or no effect on soil available-water capacity (AWC) (Bouyoucos, 1939; Hudson, 1994). However, current published research has demonstrated significant positive correlation between SOM and AWC. It was documented that in sand, silt loam and silty clay loam soils, an increase in SOM was significantly correlated with the increase in AWC (Hudson, 1994; Al-Kaisi et al., 2014). It was also found that soil water retention was greater with sandy and silty soils than fine-textured soil with increasing levels of SOM (Rawls et al., 2003). The plant–water–nutrient relationship is highly influenced by agriculture system diversity and water and nutrients availability. Even though this relationship between microbial diversity and the functioning of the agriculture ecosystem (Kennedy and Smith, 1995; Pankhurst et al., 1996) is not well-defined, the available evidence shows that diversity of soil biota is an important factor of agroecosystem physical functions (Swift and Anderson, 1993; Bear et al., 1995), which influence water movement in the soil system as porosity changes, with a change in soil aggregate formation. The management of agriculture systems that include continuous crop rotations and reduced tillage can result in an increase in input of biomass-C, improvement in soil aggregation, and strengthening of nutrient cycling (Odell et al., 1984; Pankhurst et al., 1996). The effect of crop rotation on nutrient dynamics and availability in the soil with different crops within a rotation, such as deep-rooted legumes (i.e., alfalfa), can scavenge deep residual N and increase availability to subsequent shallow-rooted crops (Mathers et al., 1975). Similarly, the diversity of crop root structures and depth in the soil profile can influence water availability (as water requirements differ among crops) for different crops, particularly in dryland or rain-fed agriculture (Nielsen et al., 2011). In addition to the effect of crop rotation on water and nutrient movement in the soil profile, the use of crop rotation improves (as discussed in the previous section) soil structure and SOM, thereby increasing WUE by improving water harvesting and reducing surface water runoff (Varvel, 1994; Tanaka et al., 2005). The dynamics of water and nutrient movement under diverse cropping systems has resulted in increased nutrient use efficiency (Karlen and Stott, 1994). The improvement in soil biological and physical functions as a result of continuous crop rotation can increase yield compared to that under a monocrop system (Al-Kaisi et al., 2015), and to a greater extent, increase concentration of N, P, and K levels in soil (Bolton et al., 1976; Copeland and Crookston, 1992) as crop residue decomposes over time. The climate variability and the episodic drought or wet events in different regions of the world require cropping systems of extended crop rotations that can mitigate the adverse effects of such events on food production. The increase in soil biological and physical functions' resiliency is well-linked to the type of agroecosystem and plant diversity, where water availability influences nutrient supply to plants.

4.4 Conservation Systems and Soil Environment Dynamics

4.4.1 Soil Organic Matter Mineralization

The SOC is a major part (1600×10^{15} g C) of the total terrestrial C pools (2200×10^{15} g). It is significantly greater than the amount of C stored in living vegetation (approximately 600×10^{15} g C) (Bouwman, 1990). Mineralization of SOM is a microbial-mediated transformation of organically bound elements such as C, N, S, and P into inorganic compounds such as, CO_2, CH_4, NH_4^+ NO_3^-, SO_4^{-2}, H_2S, and HPO_4^{-2}. The mineralization process affects different SOM pools, above and below ground as primary resources, organic compounds, and dissolved organic C. The rate of mineralization and the amount of nutrients released differ significantly among different fractions of SOM (Fig. 4.7). Also, the process of mineralization is influenced by biotic and abiotic factors that are governed by the agroecosystems. Therefore, the characteristics and functions of CA systems contribute significantly to nutrient dynamics

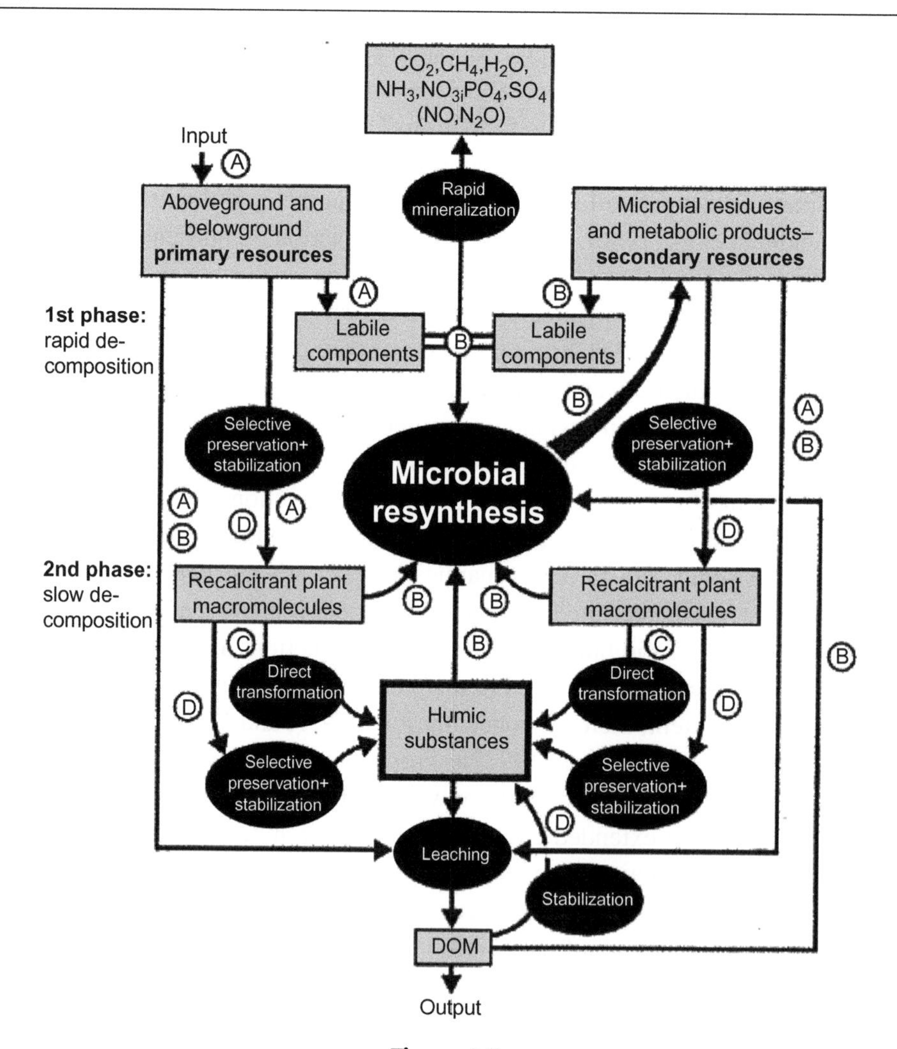

Figure 4.7
SOM pools and SOM dynamics with special reference to litter decomposition and humification (including microbial resynthesis, selective preservation, and direct transformation), mineralization, leaching of dissolved organic matter (DOM), and stabilization by interactions with inorganic compounds. These processes are controlled by different factors: (A) control of the quality and quantity of the primary resources; (B) control of the microbial activities; (C) possible catalytic effects on direct transformation; and (D) control by stabilization with inorganic compounds. *After Zech, W., Senesi, N., Guggenberger, G., Kaiser, K., Lehmann, J., Miano, T.M., et al., 1997. Factors controlling humification and mineralization of soil organic matter in the tropics. Geoderma 79, 117–169.*

and availability, where soil microfauna contributions to rhizosphere functions and potential mechanisms play essential roles (Bouwman, 1990). One of the many ecological functions that are provided by the rhizosphere microbial system is the mineralization and release of plant-available nutrients (Lu and Zhang, 2006). The interaction between soil organisms and plants (through plant roots), is essential as a symbiotic relationship and mutually beneficial for the microbial community growth and plant needs for nutrients for production of biomass below and above ground (Bardgett and Wardle, 2003). The decomposition of different SOM fractions is strongly related to the decomposition of SOM or mineralization, and SOM stabilization, where nutrients release proportionally, are affected by physical protection, chemical and structure composition of different fractions. Rapid mineralization of labile-C components is the dominant process during the first phase of decomposition. While the mineralization continues in the second phase of decomposition, it is slowed down because of the accumulation of refractory molecules (Zech et al., 1997). However, it has been argued that contemporary analytical approaches suggest that the molecular structure is not necessarily a good measure for C stabilization and tracking the fate of labeled organic compound (Dungait et al., 2012). Rather, SOC pool is an ecological property and is stored in biologically nonpreferred soil space where it is physically protected from microbial activity regardless of the initial chemical structure (Six et al., 2000). Change in management practices and cropping systems that include tillage and crop rotation can cause significant losses of both total and mineralizable SOM (Schimel et al., 1985). The effects of cultivation on C and N mineralization cause a reduction in mineralizable C and N, which is proportional to losses of C and N in the surface horizons, where cultivated soils have a higher mineralization ratio of N:biomass than rangeland soils. These changes suggest that substrates are less decomposable, and possibly that more reimmobilization of N into biomass occurs in rangeland than in cultivated soils (Schimel et al., 1985). Conversion of grassland to cropland in North America has resulted in a significant decline in the SOM content, with a loss of approximately 40% of the total soil C during a period of 60 years (Jenny, 1941; Haas et al., 1957; Tiessen and Stewart, 1983; Lal, 2004). Conversion of CT to NT can increase SOC, especially in the top soil layer. Further, the mineral-associated C content in the bare-fallow treatment (Cambardella and Elliott, 1992) can be significantly less than that under native grass, as is also the case under CT than that under native grass (Guzman and Al-Kaisi, 2010). The intensity of total C mineralization (mineralized C:total C) differs among tillage systems, where it can be constant under plow tillage for different soil depths, but is much greater in the soil surface layer, and decreases with increase in depth under NT and chisel plow tillage (Alvarez et al., 1995). In general, however, C input from crops may not differ among tillage systems. Therefore, the greater C input to soil under NT may be due to lower mineralization intensity of SOM, leading to accumulation of SOC over time (Alvarez et al., 1995). Therefore, mulched soils, such as under NT, are cooler than plowed soils by 2.4 and 5°C (Licht and Al-Kaisi, 2005a). Under such soil conditions, labile organic forms accumulate under NT in the top soil layers because of residue cover, leading to lower soil temperature and greater moisture content as drivers for low mineralization rates compared to those under plow tillage.

4.4.2 Soil Environment and Root Growth

Among the notable changes caused by tillage and cropping systems to soil environment include soil temperature (Licht and Al-Kaisi, 2005a), soil aggregation (Six et al., 2000), and soil hydraulic properties (Cox et al., 1990). Such tillage-induced changes in soil properties can influence plant root growth and its depth distribution. Soil temperature plays a significant role in root growth. Barber and Kovar (1991) documented that an increase in soil temperature from 11°C to 27°C increased the corn root growth rate. However, Ma et al. (1994) observed that a reduction in water and nutrient uptake at low temperature caused a reduction in root growth. The distribution of water within the soil profile also has a significant effect on plant root distribution, which is influenced by the tillage system. In general, the soil water distribution pattern and root density peak midway between the rows under ridge tillage (RT). In CT, however, soil water and root density peak in the row (Barber and Kovar, 1991). Accordingly, rooting depth is often shallower under NT than that under CT because of greater soil moisture content associated with NT in the upper 30–50 cm, which may cause reduction in root exploration of deeper soil depths (Raczkowski, 1988; Dwyer et al., 1995). Tillage systems can affect soil water storage during the rainy season, but few studies show that tillage practices can modulate the use efficiency of stored water for winter crops such as wheat in the dry season. The role of the root system is a bridge between the soil environment as influenced by management practices and changes in the above-ground biomass yield (Klepper, 1990). Conversion to NT may cause greater and deeper water accumulation in the soil profile and greater root growth than under CT (Lampurlanés et al., 2001). Merrill et al. (1996) observed that spring wheat roots penetrate to deeper soil depths under NT than under spring disking, with larger root length density because of the cooler soil and superior soil water conservation in the near-surface zone. However, NT can gradually increase surface soil bulk density, limiting the distribution of roots in the upper soil profile and reducing the downward progression of roots (Mosaddeghi et al., 2009). Field experiments show that root surface density of winter wheat under rain-fed agriculture in north central China with PT and RT was greater than that under NT across 110 cm soil depth (Fig. 4.8) at the tillering and flowering stages. However, the difference in soil water content (SWC) in the 0–80 cm soil depth was significant, where NT had the greatest amount of SWC compared to that under PT and RT in the top 0–20 cm soil depth. One of the main effects of CA systems, especially in arid and semiarid regions, is water conservation. In arid environments, root growth and distribution in utilizing soil moisture under different tillage systems can have a significant effect on yield and WUE. In Aragon (NE Spain), where precipitation is limited, a fallow system for cereal production is practiced for water conservation, along with NT and RT with some inconsistent results for improving yield and WUE (Lopez and Arrue, 1997).

4.4.3 Soil Environment and Nutrient Availability

The CA systems are characterized by minimum or no soil disturbance. Those conditions are the opposite of those under CT systems and accelerate oxidation of SOM mediated by microorganisms through changes in soil moisture, aeration, temperature, aggregation, and nutrient availability (Doran and Smith, 1987). In CA systems, residue cover on the soil surface plays a role in the distribution of nutrients within the soil profile because of lack

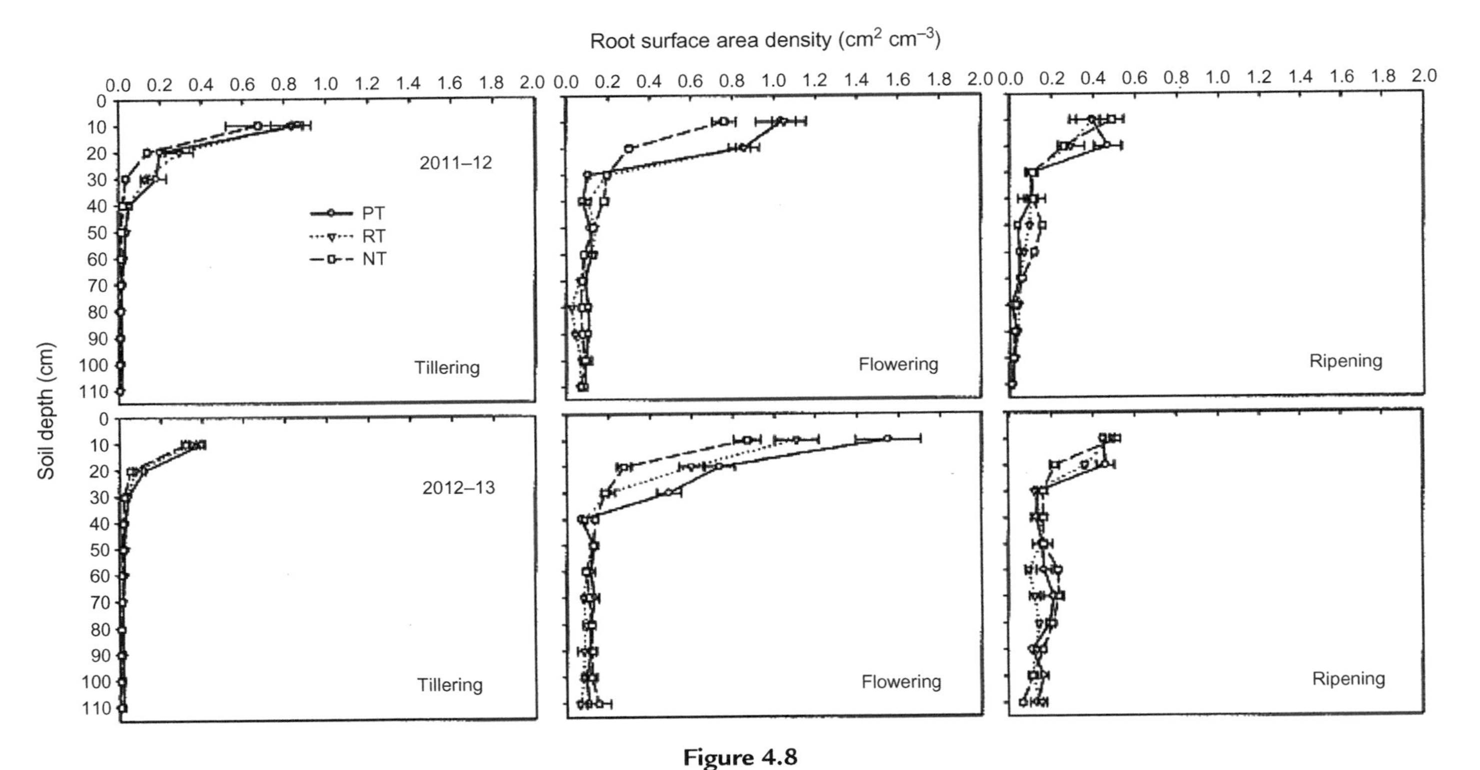

Figure 4.8

Effect of different tillage practices on root surface area density at tillering stage (157 DAS in 2011–12 and 144 DAS in 2012–13), flowering stage (212 DAS and 213 DAS) and ripening stage (249 DAS and 245 DAS) in two winter wheat growing seasons. Horizontal bars are standard errors. *After Guan, D., Zhang, Y., Al-Kaisi, M.M., Wang, Q., Zhang, M., Li, Z., 2015. Tillage practices effect on root distribution and water use efficiency of winter wheat under rain-fed condition in the North China Plain. Soil Till. Res. 146, 286–295, Guan et al. (2015).*

of soil disturbance compared with that under CT systems (Hargrove, 1985; Unger, 1991). Therefore, nutrient distribution within the soil profile under CA systems is characterized by stratification, which may affect crop production (Bruce et al., 1995). The effect of the previous year crop residue on nutrient availability to the next year's crops has been addressed by adopting diverse and continuous crop production rotation systems in many parts of the world to enhance nutrient cycling and biodiversity as essential mechanisms for providing and distributing nutrients within the soil profile. Crop residues are significant sources for C, N, P, S, and other macro- and micronutrients. The concentration and type of nutrients released by crop residue into the soil system depend on crop residue type. Nutrients in crop residues are ultimately recycled into the soil system during the decomposition process and added to the nutrient pool available for plant uptake. In a study in Iowa, evaluating the effects of long-term P and K application with three different tillage systems, pronounced vertical stratification of P and K occurred in the RT and NT systems (Robbins and Voss, 1991). Further, significant surface accumulations of P and K in the upper 5 cm of soil was also observed, which was on average 3.5 times greater than those for the 5–15-cm soil layer. This stratification can be attributed to the lack of incorporation of P, K, and C, over the time associated with NT or minimum tillage (Franzluebbers, 2002; Wright et al., 2007). In contrast, more uniform P and K distributions in the upper 15 cm of soil have been reported for CT systems (Triplett and Van Doren, 1969; Wright et al., 2007). The challenges with CA systems, as far as nutrient availability influenced by low soil temperatures, may include slowing of the mineralization process, especially in cold regions, limited nutrient supply during the early period of plant growth, and potential stratification of certain nutrients, such as P and K, within the soil profile for lack of soil mixing (Hargrove, 1985; Unger, 1991; Al-Kaisi and Kwaw-Mensah, 2007). These challenges are real, but the current development of new seed varieties and management technology can overcome such limitations by adopting certain practices, such as strip-tillage (Licht and Al-Kaisi, 2005a), where minimum soil disturbance and nutrients can be applied at the same time in close proximity to the root system.

4.5 Conclusions

This chapter focuses on the role of CA systems in affecting soil biological, chemical, and physical functions to sustain soil health in order to minimize weather variability on soil and reducing the negative impacts on productivity. The interaction among different tillage and cropping systems effects on soil structure dynamics and the controlling factors that accelerate the enhancement of soil aggregation have been discussed to understand the principal impacts of improving soil structure, taking into consideration soil temperature and moisture regimes, and nutrient availability. The discussion explored the impacts of CA system in comparison with other agriculture systems (i.e., conventional system) in developing a soil biological environment and its role in enhancing soil health through physical and biological improvement. Over the long term, continuous use of CA will lead to improvement of soil structure that will contribute to resilient soil systems that can withstand climate variability. Finally, the discussion explored the fundamental characteristics of CA systems and their effects on soil environment and processes for sustaining soil productivity. Further, the limitations associated with the CA systems in affecting SOM pools and stability, nutrient supply through mineralization process, and the dynamics of nutrients availability for food and fiber production are also discussed.

References

Al-Kaisi, M.M., Kwaw-Mensah, D., 2007. Effect of tillage and nitrogen rate on corn yield and nitrogen and phosphorus uptake in a corn-soybean rotation. Agron. J. 99, 1548–1558.

Al-Kaisi, M.M., Yin, X., 2003. Effects of nitrogen rate, irrigation rate, and plant population on corn yield and water use efficiency. Agron. J. 95, 1475–1482.

Al-Kaisi, M.M., Yin, X., 2005. Tillage and crop residue effects on soil carbon and carbon dioxide emission in corn-soybean rotations. J. Environ. Qual. 34, 437–445.

Al-Kaisi, M.M., Duelle, A., Kwaw-Mensah, D., 2014. Soil microaggregate and macroaggregate decay over time and soil carbon change as influenced by different tillage systems. J. Soil Water Conserv. 69, 574–580. http://dx.doi.org/10.2489/jswc.69.6.574.

Al-Kaisi, M.M., Archontoulis, S.V., Kwaw-Mensah, D., Miguez, F., 2015. Tillage and crop rotation effects on corn agronomic response and economic return at seven Iowa locations. Agron. J. 107, 1411–1424. http://dx.doi.org/10.2134/agronj14.0470.

Alvarez, R., Diaz, R.A., Barbero, N., Santanatoglia, O.J., Blotta, L., 1995. Soil organic carbon, microbial biomass and CO_2-C production from three tillage systems. Soil Tillage Res. 33, 17–28.

Barber, S.A., Kovar, J.L., 1991. Effect of tillage practice on maize (*Zea mays* L.) root distribution In: McMichael, B.L. Parsson, H. (Eds.), Developments in Agricultural and Managed-Forest Ecology, vol. 24. Elsevier Press, New York, NY, pp. 402–409.

Bardgett, R.D., Wardle, D.A., 2003. Herbivore mediated linkages between aboveground and belowground communities. Ecology 84, 2258–2268.

Bauer, A., Black, A.L., 1994. Quantification of the effect of soil organic matter content on soil productivity. Soil Sci. Soc. Am. J. 58, 185–193.

Beare, M.H., Bruce, R., 1993. A comparison of methods for measuring water-stable aggregates: implications for determining environmental effects on soil structure. Geoderma 56, 87–104.

Beare, M.H., Coleman, D.C., Crossley Jr., D.A., Henrix, P.F., Odum, E.P., 1995. A hierarchical approach to evaluating the significance of soil biodiversity to biological cycling. Plant Soil 170, 5–22.

Bolton, E.F., Dirks, V.A., Aylesworth, J.W., 1976. Some effects of alfalfa, fertilizer and lime in corn yield on rotations on clay soil during a range of seasonal moisture conditions. Can. J. Soil Sci. 56, 21–25.

Bouwman, A.F., 1990. Soils and the Greenhouse Effect. John Wiley, New York, NY.

Bremer, E., Janzen, H.H., Johnston, A.M., 1994. Sensitivity of total light fraction and mineralizable organic matter to management practices in a Lethbridge soil. Can. J. Soil Sci. 74, 131–138.

Biederbeck, V.O., Campbell, C.A., Zentner, R.P., 1984. Effect of crop rotation and fertilization on some biological properties of a loam in southwestern Saskatchewan. Can. J. Soil Sci. 64, 355–367.

Bolliger, A., Magid, J., Amadon, T.C., Neto, F.S., Ribeiro, M.D.D., Calegari, A., et al., 2006. Taking stock of the Brazilian "zero-till revolution": a review of landmark research and farmers' practice. Adv. Agron. 91, 47–100.

Bongers, T., 1990. The maturity index: an ecological measure of environmental disturbance based on nematode species composition. Oecologia 83, 9–14.

Borie, F., Rubio, R., Morales, A., 2008. Arbuscular mycorhizal fungi and soil aggregation. J. Soil Plant Nutr. 8 (2), 9–18.

Bouyoucos, G.J., 1939. Effects of organic matter on the water holding capacity and the wilting points of mineral soils. Soil Sci. 47, 377–383.

Bronick, C.J., Lal, R., 2005. Soil structure and management: a review. Geoderma 124, 3–22.

Bruce, R.R., Langdale, G.W., West, L.T., Miller, W.P., 1995. Surface soil degradation and soil productivity, restoration, and maintenance. Soil Sci. Am. J. 59, 654–660.

Calegari, A., dos Santos Rheinheimer, D., de Tourdonnet, S., Tessier, D., Hargrove, W.L., et al., 2010. Effect of soil management and crop rotation on physical properties in a long term experiment in Southern Brazil. 19th World Congress of Soil Science, Soil Solutions for a Changing World, 1-6 August 2010, Brisbane, Australia.

Cambardella, C.A., Elliott, E.T., 1992. Participate soil organic-matter changes across a grassland cultivation sequence. Soil Sci. Soc. Am. J. 56, 777–783.

Campbell, C.A., Leyshon, A.J., Zentner, R.P., LaFond, G.P., Janzen, H.H., 1991. Effect of cropping practices on the initial potential rate of N mineralization in a thin Black Chernozem. Soil Sci 7 (1), 43–53.

Caravaca, F., Garcia, C., Hernández, M.T., Roldán, A., 2002. Aggregate stability changes after organic amendment and mycorrhizal inoculation in the afforestation of a semiarid site with *Pinus halepensis*. Appl. Soil Ecol. 19, 199–208.

Chan, K.Y., Heenan, D.P., 1996. The influence of crop rotation on soil structure and soil physical properties under conventional tillage. Soil Till. Res. 37, 113–125.

Chenu, C., Le Bissonnais, Y., Arrouays, D., 2000. Organic matter influence on clay wettability and soil aggregate stability. Soil Sci. Soc. Am. J. 64, 1479–1486.

Chen, X., Liu, M., Hu, F., Mao, X., Li, H., 2007. Contributions of soil micro-fauna (protozoa and nematodes) to rhizosphere ecological functions. Acta Ecol. Sin. 27 (8), 3132–3143.

Clapp, C.E., Allmaras, R.R., Layese, M.F., Lindena, D.R., Dowdy, R.H., 2000. Soil organic carbon and ^{13}C abundance as related to tillage, crop residue, and nitrogen fertilization under continuous corn management in Minnesota. Soil Till. Res. 55, 127–142.

Collins, H.P., Rasmussen, P.E., Douglas, C.L., 1992. Crop rotation and residue management effects on soil carbon and microbial dynamics. Soil Sci. Soc. Am. J. 56, 783–789.

Copeland, P.J., Crookston, R.K., 1992. Crop sequence affects nutrient composition of corn and soybean grown under high fertility. Agron. J. 84, 503–509.

Cox, W.J., Zobel, R.W., van Es, H.M., Otis, D.J., 1990. Tillage effects on some soil physical and corn physiological characteristics. Agron. J. 82, 806–812.

Delgado, J.A., Groffman, P.M., Nearing, M.A., Goddard, T., Reicosky, D., Lal, R., et al., 2011. Conservation practices to mitigate and adapt to climate change. J. Soil Water Conserv. 66, 118–129.

Derpsch, R., 2002. Making conservation tillage conventional, building a future on 25 years of research: research and extension perspective. In: Santen, E.V. (Ed.), Proceedings of 25th Annual Southern Conservation Tillage Conference for Sustainable Agriculture. Alabama Agricultural Experiment Station, Auburn University, Auburn, AL, USA.

Doran, J.W., Smith, M.S., 1987. Organic matter management and utilization of soil and fertilizer nutrients. Soil Sci. Soc. Am. and Am. Soc. of Agron. In: Soil fertility and organic matter components of production systems. SSSA. Spec. Pub. No. 19.

Doran, J.W., Sarrantonio, M., Liebig, M., 1996. Soil health and sustainability. Adv. Agron. 56, 1–54.

Duiker, S.W., Lal, R., 1999. Crop residue and tillage effects on carbon sequestration in a Luvisol in central Ohio. Soil Till. Res. 52, 73–81.

Dungait, J.A.J., Hopkins, D.W., Gregory, A.S., Whitmore, A.P., 2012. Soil organic matter turnover is governed by accessibility not recalcitrance. Global Change Biol. 18 (6), 1781–1796.

Dwyer, L.M., Ma, B.L., Stewart, D.W., Hayhoe, H.N., Balchin, D., Culley, J.L.B., et al., 1995. Root mass distribution under conventional and conservation tillage. Can. J. Soil Sci. 95, 23–28.

Elliott, E.T., 1986. Aggregate structure and carbon, nitrogen, and phosphorus in native and cultivated soils. Soil Sci. Soc. Am. J. 50, 627–633.

Follett, R.F., Stewart, C.E., Pruessner, E.G., Kimble, J.M., 2012. Effects of climate change on soil carbon and nitrogen storage in the US Great Plains. J. Soil Water Conserv. 67, 331–342.

Follett, R.T., 2001. Soil management concepts and carbon sequestration in cropland soils. Soil Till. Res. 61, 77–92.

Food and Agriculture Organization of the United Nations (FAO), 2009. Conservation Agriculture. FAO Agriculture and Consumer Protection Department. <http://www.fao.org/ag/ca/>.

Franzluebbers, A.J., 2002. Water infiltration and soil structure related to organic matter and its stratification with depth. Soil Till. Res. 66, 197–205.

Gardner, P.A., Angus, J.F., Pitson, G.D., Wong, P.T.W., 1998. A comparison of six methods to control take-all in wheat. Aust. J. Agric. Res. 49, 1225–1241.

Guan, D., Zhang, Y., Al-Kaisi, M.M., Wang, Q., Zhang, M., Li, Z., 2015. Tillage practices effect on root distribution and water use efficiency of winter wheat under rain-fed condition in the North China Plain. Soil Till. Res. 146, 286–295.

Guzman, J.G., Al-Kaisi, M.M., 2010. Soil carbon dynamics and soil carbon budget of newly reconstructed tall-grass prairies in south central Iowa. J. Environ. Qual. 39, 136–146.

Guzman, J.G., Al-Kaisi, M.M., 2011. Landscape position effect on selected soil physical properties of reconstructed prairies in southcentral Iowa. J. Soil Water Conserv. 66, 183–191.

Guzman, J.G., Al-Kaisi, M.M., 2014. Residue removal and management practices effects on soil environment and carbon budget. Soil Sci. Soc. Am. J. 78, 609–623.

Haas, H.J., Evans, C.E., Miles, E.F., 1957. Nitrogen and carbon changes in Great Plains soils as influenced by cropping and soil treatments. USDA Tech. Bul. 1164. U.S. Gov. Print Office, Washington, DC.

Hargrove, W.L., 1985. Influence of tillage on nutrient uptake and yield of corn. Agron. J. 77, 763–767.

Hillel, D., 1968. Soil water evaporation and means of minimizing it, Report to U.S. Dept. Agr., Hebrew Uni. of Jerusalem, Israel.

Hillel, D., 1998. Environmental Soil Physics. Academic Press, San Diego.

Hillel, D., 2004. Introduction to Environmental Soil Physics. Elsevier Science, Oxford.

Hobbs, P.R., Sayre, K., Gupta, R., 2008. The role of conservation agriculture in sustainable agriculture. Philos. Trans. Royal Soc. B 363, 543–555.

Holeplass, H., Singh, B.R., Lal, R., 2004. Carbon sequestration in soil aggregates under different crop rotations and nitrogen fertilization in an inceptisol in southeastern Norway. Nutr. Cycl. Agroecosyst. 70, 167–177.

Hudson, B.D., 1994. Soil organic matter and available water capacity. J. Soil Water Conserv. 49 (2), 189–194.

Jenny, H., 1941. Factors of Soil Formation. McGraw-Hill, New York, NY.

Johnston, A.M., Janzen, H.H., Smith, E.G., 1995. Long-term spring wheat response to summer fallow frequency and organic amendment in southern Alberta. Can. J. Plant Sci. 75, 347–354.

Karlen, D., Mausbach, M., Doran, J., Cline, R., Harris, R., Schuman, G., 1997. Soil quality: a concept, definition, and framework for evaluation (a guest editorial). Soil Sci. Soc. Am. J. 61 (1), 4–10.

Karlen, D.L., Stott, D.E., 1994. A frame work for evaluating physical and chemical indicators of soil quality. In: Doran, Defining Soil Quality for a Sustainable Environment. SSSA Special Publication no. 35, Madison, WI. Soil Sci. Soc. Amer.

Karlen, D.L., Wollenhaupt, N.C., Erbach, D.C., Berry, E.C., Swan, J.B., Eash, N.S., et al., 1994. Long-term tillage effects on soil quality. Soil Till. Res. 32, 313–327.

Kassam, A., Friedrich, T., Shaxson, F., Pretty, J., 2009. The spread of conservation agriculture: justification, sustainability and uptake. Int. J. Agric. Sust. 7 (4), 292–320.

Kassam, A., Fredrich, T., Derpsch, R., Kienzle, J., 2014. Worldwide adopting of conservation agriculture. 6th World Congress of Conservation Agriculture, 22-27 June, 2014, Winnipeg, Canada.

Kay, B.D., Dexter, A.R., Rasiah, V., Grant, C.D., 1994. Weather, cropping practices and sampling depth effects on tensile strength and aggregate stability. Soil Till. Res. 32, 135–148.

Kennedy, A.C., Smith, K.L., 1995. Soil microbial diversity and the sustainability of agricultural soils. Plant Soil Sci. 170, 75–86.

Klepper, B., 1990. Root growth and water uptake. In: Stewart, B.A., Nielsen, D.R. (Eds.), Irrigation of Agricultural Crops. ASA–CSSA–SSSA, Madison, pp. 281–322.

Lal, R., 2004. Soil carbon sequestration impacts on global climate change and food security. Science 304, 1623–1627.

Lal, R., 2015. A system approach to conservation agriculture. J. Soil Water Conserv. 70 (4), 82–88.

Lal, R., Delgado, J.A., Groffman, P.M., Millar, N., Dell, C., Rotz, A., 2011. Management to mitigate and adapt to climate change. J. Soil Water Conserv. 66, 276–285.

Lampurlanés, J., Angás, P., Cantero-Martínez, C., 2001. Root growth: soil water content and yield of barley under different tillage systems on two soils in semiarid conditions. Field Crops Res. 69, 27–40.

Lee, K.E., 1985. Earthworms: Their Ecology and Relationships With Soils and Land Use. Academic Press, Sydney.

Lehman, R.M., Acosta-Martinez, V., Buyer, J.S., Cambardella, C.A., Collins, H.P., Ducey, T.F., et al., 2015. Soil biology for resilient, healthy soil. J. Soil Water Conserv. 70 (1), 12–18.

Licht, M., Al-Kaisi, M.M., 2005a. Strip-tillage effect on seedbed soil temperature and other soil physical properties. Soil Till. Res. 80, 233–249.

Licht, M.A., Al-Kaisi, M.M., 2005b. Corn response, nitrogen uptake, and water use in strip-tillage compared with no-tillage and chisel plow. Agron. J. 97, 705–710.

Lopez, M.V., Arrue, J.L., 1997. Growth, yield and water use efficiency of winter barley in response to conservation tillage in a semi-arid region of Spain. Soil Till. Res. 11, 35–54.

Lu, Y.H., Zhang, F.S., 2006. The advances in rhizosphere microbiology. Soils 38 (2), 113–121.

Lupwayi, N.Z., Rice, W.A., Clayton, G.W., 1998. Soil microbial diversity and community structure under wheat as influenced by tillage and crop rotation. Soil Biol. Biochem. 30, 1733–1741.

Ma, B.L., Dwyer, L.M., Hayhoe, H.N., Culley, J.L.B., McAndrew, D., Evenson, L., et al., 1994. Soil temperature profiles under conventional and no-tillage from latitude 45°N to latitude 55°N. Abstract of the 28th Canadian Metereorological and Oceanographic Society Congress, Ottawa, ON, Canada. pp. 133.

Mathers, A.C., Stewart, B.A., Thomas, J.D., 1975. Residual and annual effects of manure on grain sorghum yields Managing Livestock Wastes (Proc., Int. Symp.). ASAE, St. Joseph, Michigan.

Merrill, S.D., Black, A.L., Bauer, A., 1996. Conservation tillage affects root growth of dryland spring wheat under drought. Soil Sci. Soc. Am. J. 60, 575–583.

Mitchell, C.C., Westerman, R.L., Brown, J.R., Peck, T.R., 1991. Overview of long-term agronomic research. J. Agron. 83, 24–29.

Mosaddeghi, M.R., Mahboubi, A.A., Safadoust, A., 2009. Short-term effects of tillage and manure on some soil physical properties and wheat root growth in a sandy loam soil in western Iran. Soil Till. Res. 104, 173–179.

Ngwira, A., Johnsen, F.H., Aune, J.B., Mekuria, M., Thierfelder, C., 2014. Adoption and extent of conservation agriculture practices among smallholder farmers in Malawi. J. Soil Water Conserv. 69 (2), 107–119.

Nielsen, D.C., Vigil, M.F., Benjamin, J.G., 2011. Evaluating decision rules for dryland rotation crop selection. Field Crops Res. 120, 254–261.

Oades, J.M., 1984. Soil organic matter and structural stability: mechanisms and implications for management. Plant Soil 76, 319–337.

Oades, J.M., Waters, A.G., 1991. Aggregate hierarchy in soils. Aust. J. Soil Res. 29, 815–828.

Odell, R.T., Walker, W.M., Boone, L.V., Oldham, M.G., 1982. The Morrow Plots: A Century of Learning. Agricultural Experiment Station, College of Agriculture. Univ. of Illinois, Bul. 775, Urbana-Champaign, IL.22.

Odell, R.T., Melsted, S.W., Walker, W.M., 1984. Changes in organic carbon and nitrogen of Morrow Plot soils under different treatments, 1904-1973. Soil Sci. 137, 160–171.

Omay, A.B., Rice, C.W., Maddux, L.C., Gordon, W.B., 1997. Changes in soil microbial and chemical properties under long-term crop rotation and fertilization. Soil Sci. Soc. Am. J. 61, 1672–1678.

Pankhurst, C.E., Ophel-Keller, K., Doube, B.M., Gupta, V.V.S.R., 1996. Biodiversity of soil microbial communities in agricultural systems. Biodivers. Conserv. 5, 197–209.

Raczkowski, C.W., 1988. Effects of four tillage systems on corn (*Zea mays* L.) root distribution in the North Carolina Piedmont. Ph.D. dissertation DA8909228. North Carolina State University, Raleigh, NC.

Rawls, W.J., Pachepskyb, Y.A., Ritchiea, J.C., Sobecki, T.M., Bloodworth, H., 2003. Effect of soil organic carbon on soil water retention. Geoderma 116, 61–76.

Rillig, M.C., Wright, S.F., Eviner, V.T., 2002. The role of arbuscular mycorrhizal fungi and glomalin in soil aggregation: comparing effects of five plant species. Plant Soil 238, 325–333.

Robbins, S.G., Voss, R., 1991. Phosphorus and potassium stratification in conservation tillage systems. J. Soil Water Conserv. 46 (4), 298–300.
Rockström, J., Kaumbutho, P., Mwalley, J., Nzabi, A.W., Temesgen, M., Mawenya, L., et al., 2009. Conservation farming strategies in East and Southern Africa: yields and rain water productivity from on-farm action research. Soil Till. Res. 103 (1), 23–32.
Sá, J.C.D., Cerri, C.C., Dick, W.A., Lal, R., Venske, S.P., Piccolo, M.C., et al., 2001. Organic matter dynamics and carbon sequestration rates for a tillage chronosequence in a Brazilian Oxisol. Soil Sci. Soc. Am. J. 65 (5), 1486–1499.
Schimel, D.S., Coleman, D.C., Horton, K.A., 1985. Soil organic matter dynamics in paired rangeland and cropland toposequences in North Dakota. Geoderma 36, 201–214.
Schjonning, P., Elmholt, S., Christensen, B.T. (Eds.), 2004. Managing Soil Quality: Challenges in Modern Agriculture. CABI Publishing, Wallingford, Oxon.
Schulten, H.R., Leinweber, P., 2000. New insights into organic-mineral particles: composition, properties and models of molecular structure. Biol. Fertil. Soils 30, 399–432.
Six, J., Elliott, E.T., Paustian, K., 1999. Aggregate and soil organic matter dynamics under conventional and no-tillage systems. Soil Sci. Soc. Am. J. 63, 1350–1358.
Six, J., Paustian, K., Elliott, E.T., Combrink, C., 2000. Soil structure and organic matter: I. Distribution of aggregate-size classes and aggregate-associated carbon. Soil Sci. Soc. Am. J. 64, 681–689.
Smith, P., Powlson, D.S., Glendining, M.J., Smith, J.U., 1998. Preliminary estimates of the potential for carbon mitigation in European soils through no-till farming. Global Change Biol. 4, 679–685.
Swift, M.J., Anderson, J.M., 1993. Biodiversity and ecosystem function in agricultural systems. In: Schultz, E.D., Mooney, H. (Eds.), Biodiversity and Ecosystem Function. Springer, Berlin, pp. 15–42.
Tanaka, D.L., Anderson, R.L., Rao, S.C., 2005. Crop sequencing to improve use of precipitation and synergize crop growth. J. Agron. 97, 385–390.
Tiessen, H., Stewart, J.W.B., 1983. Particle-size fractions and their use in studies of soil organic matter. II. Cultivation effects on organic matter composition in size fractions. Soil Sci. Soc. Am. J. 47, 509–514.
Tisdall, J.M., Oades, J.M., 1982. Organic matter and water-stable aggregates. J. Soil Sci. 33, 141–163.
Triplett Jr., G.B., Van Doren Jr., D.M., 1969. Nitrogen, phosphorus, and potassium fertilization of non-tilled maize. J. Agron. 61, 637–639.
Varvel, G.E., 1994. Rotation and nitrogen fertilization effects on changes in soil carbon and nitrogen. Agron. J. 86, 319–325.
Uger, P.W., 1991. Organic matter, nutrient, and pH distribution in no- and conventional-tillage semiarid soils. Agron. J. 83, 186–189.
Weil, R.R., Brady, N.C., 2016. The Nature and Properties of Soils, fifteenth ed. Pearson, New York, NY.1100.
Weller, C.E., 1983. Colonization of wheat roots by a fluorescent pseudomonad suppressive to take-all. Phytopathology 73, 1548–1553.
West, T.O., Post, W.M., 2002. Soil organic carbon sequestration rates by tillage and crop rotation: a global data analysis. Soil Sci. Soc. Am. J. 66, 1930–1946.
Wright, A.L., Hons, F.M., Lemon, R.G., McFarland, M.L., Nichols, R.L., 2007. Stratification of nutrients in soil for different tillage regimes and cotton rotations. Soil Till. Res. 96, 19–27.
Zech, W., Senesi, N., Guggenberger, G., Kaiser, K., Lehmann, J., Miano, T.M., et al., 1997. Factors controlling humification and mineralization of soil organic matter in the tropics. Geoderma 79, 117–169.

CHAPTER 5

Conventional Agricultural Production Systems and Soil Functions

Francisco J. Arriaga[1], Jose Guzman[2] and Birl Lowery[1]

[1]*University of Wisconsin-Madison, Madison, WI, United States* [2]*The Ohio State University, Columbus, OH, United States*

5.1 Introduction

Conventional agriculture production systems (CAPS) can be defined as those cropping systems that are most commonly used for a given area that utilize synthetic agricultural chemicals (fertilizers and pesticides). For the purpose of this discussion, CAPS will be limited to cropping systems for the production of human and animal food/feed and fiber. Management of these systems is often "optimized" to increase output with little regard to other factors, such as the environment and long-term soil productivity. This pursuit to increase outputs is often called intensification. Intensification will be regarded in this context as complete and sole use of synthetic chemicals in all aspects of plant or animal production with a goal of maximum output.

In addition to having an understanding of the definition of CAPS as it relates to inputs, it is important to understand the role of different tillage systems in CAPS. Tillage is a major component of agricultural production systems, and as such, there should be a focus on tillage systems and their impacts on soil health. The type of tillage used in CAPS can set the stage for the kind of agriculture management practices that will be used. Therefore, before a discussion of CAPS can be undertaken, the type of tillage systems must be identified and understood (Lal et al., 1994b).

It should be noted that from a global perspective, what is considered as conventional agriculture can vary substantially from one location to another. Thus, the CAPS description provided in this chapter is an attempt to make this clear. Conventional agricultural production systems are not only a spatial variant, but they also vary with time. A prime example is the agriculture systems before the industrial revolution all around the world would be considered organic in today's terms since agriculture production at that time used only organic or nonsynthetic fertilizers, and weeds were controlled by tillage, while diseases and weeds

Soil Health and Intensification of Agroecosystems.
DOI: http://dx.doi.org/10.1016/B978-0-12-805317-1.00005-1

were abated by crop rotations and cover crops. As for the tillage used in these systems, for many years moldboard plowing was considered the conventional tillage system for most of the world, with the exception of developing countries. While moldboard plowing was used in much of the developed countries, many developing countries were using various forms of minimum tillage or no-tillage. It is ironic that, with time, the so-called developed countries would come to realize that very aggressive tillage associated with moldboard plowing was not only contributing to excessive soil erosion, but reducing soil health. Now with the advent of synthetic chemicals, zero or minimum tillage is considered a sustainable farming practice. The impact of CAPS on soil health as related to soil erosion will be covered later in this chapter in greater detail in the section on soil erosion.

Minimum tillage was the norm in places like Africa and by other native people, including Native Americans (Landon, 2008), who planted crops in mounds using simple tools to plant seeds before the introduction of the steel plow. One of the many Native American cropping systems consisted of planting three sister crops (maize [*Zea mays*], beans [*Phaseolus vulgaris* L.], and squash [*Cucurbita* spp.)] in mounds (Fig. 5.1). The bean plants supplied nitrogen (N) and squash provided protection from small predatory animals. These types of systems were very sustainable with respect to reducing soil erosion and maintaining soil health.

Perhaps the key point to make about CAPS is that it is considered to be food production systems that utilize chemical fertilizers, pesticides, antibiotics, and hormones to produce large quantities of food and fiber at a low cost but with limited attention to sustainability or environmental impact.

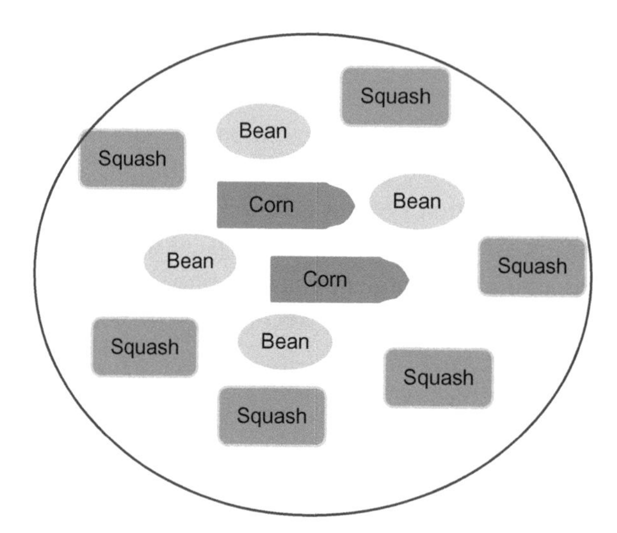

Figure 5.1

Schematic of Three Sister bed planting of corn, bean, and squash in Mesoamerica. As diagramed here Native Americans planted corn in the center of the bed surrounded by bean then in the final ring was squash, forming the sustainable Three Sister crop.

5.2 Conventional Agricultural Crop Production Systems

A conventional agricultural crop production system can be defined as the cropping system with a similar set of crop rotations, tillage, and other management factors, that is most prevalent in a given region. Therefore, such systems will vary across the globe and can include continuous monoculture to extensive crop rotations with tillage ranging from full-width high-disturbance systems (e.g., moldboard and chisel plowing), to no-tillage and various other forms of minimum tillage. In the United States, for example, there will be significant variation in cropping systems from state to state and county to county, and even within a given farm, depending on the predominant crop being grown.

After Europeans came to the United States the initial agricultural practice consisted of extensive crop rotation with moldboard plowing as the primary tillage operation followed by a secondary tillage operation such as disk and/or harrowing. This type of aggressive tillage resulted in a soil surface void of plant residue, but during this period of time this was considered as the conventional tillage and cropping system most widely used by family farmers over much of the developed world. While not the focus here, it should be noted that animal agriculture was an integral part of farming for early American farmers. There was not much monoculture other than forestry, and even on forest land the plantation was mixed vegetation. However, this is no longer the case as there is now more large-scale monoculture agriculture with many forms of minimum tillage in use as the most common system, with the exception of some developing countries where crop diversification, including animal production, is still common.

It has been proven that minimum or no-tillage leads to better soil health, especially in combination with crop rotation and cover crops (Triplett and Dick, 2008). Yet, it remains that extensive crop rotation, planting of cover crops, and conservation tillage, such as no-tillage, in general is not the most commonly used as primary tillage globally.

5.2.1 Conventional Agricultural Production Systems' Impact on Soil Chemistry

Soil chemistry can be considered as the natural chemical composition of a given soil. This natural chemical composition of a soil is a function of that soil's parent material. In many areas of the world soil is formed in place and derived directly from the weathering and degradation of rocks. When soil is derived from rocks, its soil chemistry is a direct reflection of the rocks' chemistry, including the minerals found in the rocks. There are different forms of rocks, including the major classifications of sedimentary, metamorphic, and igneous (Fig. 5.2). Within each of these major separations there are different categories of rocks. Thus, rocks are a collection of different minerals. Soils derived from these different rocks will have different chemical complexes. For example, soils derived from sandstone will have a limited array of different chemicals, whereas soils derived from igneous rock might have a

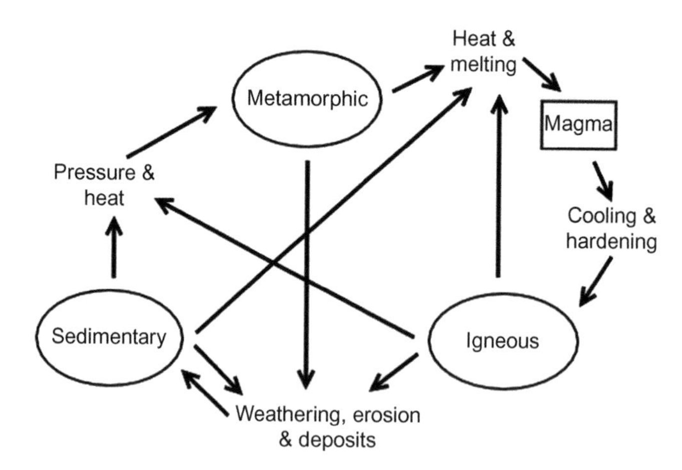

Figure 5.2
Simplified rock cycle and major rock material available to form soil. Note that rocks vary in their mineral and chemical composition, which in turn affects the type of soil derived from these rocks.

greater diversity of minerals and chemicals. However, not all soils are formed in situ from the rocks that exist in place.

Soil parent materials range from glacial drift (glacial till), to water deposited (fluvial) material, and mass wasting of primary rock or other deposit caused by gravitational forces (colluvium), among others. Soil that is derived from glacial drift or any of these other sources will likely have a very complex chemical composition reflecting the many rock types and other materials entrained in the parent material. Thus, a natural soil's chemical composition is a reflection of its source materials. While soil will inherit a given mineral chemical composition, the organic composition is derived from living organisms (e.g., animals, plants, fungi). It is this organic matter that is often considered the key to good soil health.

Natural soil chemistry can be changed by various natural forces such as leaching of chemical elements by water moving through the soil, chemical reactions, and biological activity. However, soil chemistry can also be altered by human impact from various land uses, including farming. It is a given that soil chemistry can, and will, be changed by natural forces, and one might consider these changes negative or positive impacts depending on the intended future use of the soil. Similar to natural changes, management of soils by humans can result in positive or negative changes to soil chemistry, such as increases in soil organic matter or soil erosion because of aggressive tillage practices. For example, fertilizer and manure applications to a silt loam soil increased the percentage of water-stable aggregates of a silt loam in Romania because of increases in soil organic matter (Fig. 5.3). Other management practices, such as inversion tillage, can increase the erodibility of soil because of soil aggregate disturbance (Fig. 5.4).

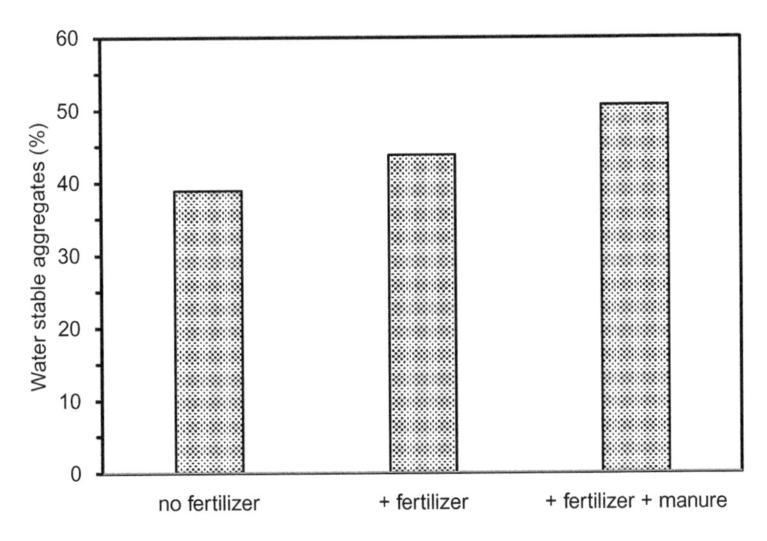

Figure 5.3
Impact of fertilizer and manure application to soybean-wheat-corn-barley rotation on water-stable aggregates of a silt loam from northeastern Romania. *Adapted from Jitareanu, G., Ailincai, C., Bucur, D., 2006. Influence of tillage systems on soil physical and chemical characteristics and yield in a soybean and maize grown in the Moldavian Plain (North-Eastern Romania). In: Horn, R., Fleige, H., Peth, S., Peng, X. (Eds.), Soil Management for Sustainability. Catena Verlag GmbH, Reiskirchen, Germany (Jitareanu et al., 2006).*

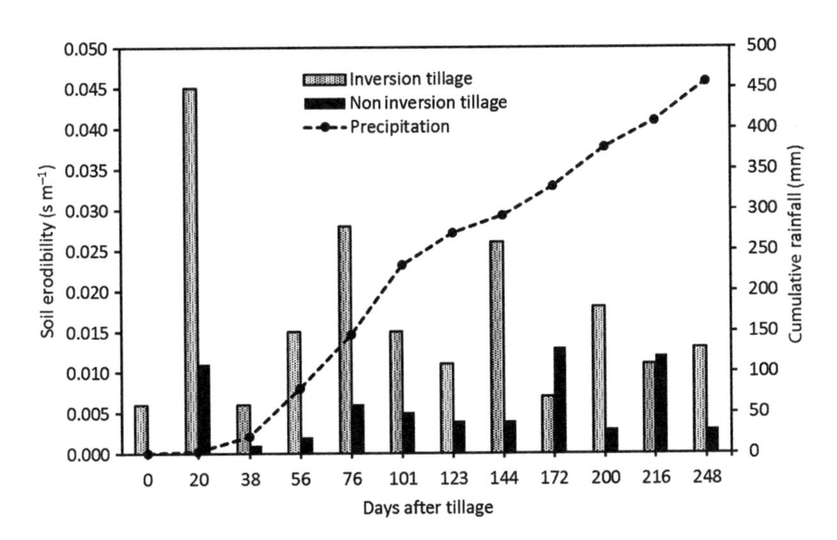

Figure 5.4
The concentrated flow soil erodibility factor increases after inversion tillage compared to noninversion tillage because of greater soil aggregate disturbance of a Belgian silt loam. *Adapted from Knapen, A., Poesen, J., 2006. Enhancing the soil's erosion resistance during concentrated runoff by reducing tillage disturbance. In: Horn, R., Fleige, H., Peth, S., Peng, X. (Eds.), Soil Management for Sustainability. Catena Verlag GmbH, Reiskirchen, Germany (Knapen and Poesen, 2006).*

The greatest negative effect of CAPS on soil chemistry is not only the intense use and introduction of herbicides and insecticides in the soil ecosystem, but the lack of crop rotations and cover crops which are known for adding organic carbon (C) and biological diversity to the soil system. The tillage systems used in most CAPS result in substantial soil disturbance and generally a negative impact on soil organic matter (West and Post, 2002). Aggressively tilling the soil leads to a considerable incorporation of air, which causes a spike in microbial activity and subsequently accelerates soil organic matter breakdown (Alvarez et al., 1995). Since soil organic matter is key to good soil health, any reductions in soil organic matter will obviously reduce soil health.

It has been shown that excessive introduction of certain N fertilizers into the soil ecosystem results in soil acidification. This added acidity further degrades soil by affecting soil primary particles, namely clay minerals. This was demonstrated by Barak et al. (1997) for glacially deposited soils in south central Wisconsin, United States, where they found a significant increase in soil acidity caused decreases in cation exchange capacity (CEC), base saturation, and exchangeable calcium (Ca) and magnesium (Mg). Some of the degradations in soil properties were considerable, e.g., clays in this soil showed a 16% loss of CEC. These degradations came from application of anhydrous ammonia as a N fertilizer in long-term fertility experiments. These results highlight the impact that CAPS management practices can have on soil health and productivity.

In addition to fertilizer applications, soil chemistry is altered by pesticide applications. When pesticides, especially insecticides are applied to the soil they sometimes change the soil chemical environment to the detriment of soil biological populations (Johnsen et al., 2001). Good soil quality is characterized as having a vast range of soil microorganisms (Dick, 1994). While soil-applied insecticides have been proven to generally kill target insects, they will also destroy "beneficial" organisms (Moorman, 1989).

When excessive pesticides are used, which is generally the case in CAPS, not only do we see degradation of the soil ecosystem, but the underlying groundwater system often becomes polluted. It was previously noted that soil chemistry is altered by leaching of chemical substances by water. Chemical leaching is not limited to natural soil chemical elements but synthetic chemicals applied to soil, such as pesticides and fertilizers, leach as well. This is especially true for sandy soils and soils with little organic matter content. Most pesticides are strongly retained by soil organic matter and clay minerals. Sandy soils are often low in organic matter and CEC, thus providing ideal conditions for pesticides and nitrate from organic and synthetic N fertilizers to leach. In addition, sandy soils tend to be well drained, thus water can move rapidly through them. Groundwater contamination is often reported in areas where the water table is close to the soil surface (Kraft and Stites, 2003; Bero et al., 2015).

5.2.2 Conventional Agricultural Production Systems' Impact on Soil Chemistry and Biology, and Organic Matter Interactions

As previously noted, natural soil chemistry is generally dictated by the soil parent material. For example, when soil is derived from rock the chemistry of this soil is a direct reflection of the rock, including the minerals in the rock. In most cases rocks do not impart organic matter to soil, yet organic matter is one of the most important measures of soil health. Organic matter is infused into soil over time, often considerable time, by decaying plants and biological organisms. In addition to the direct contributions from plants and microorganisms to soil organic matter, they promote soil health because they contribute to soil formation. Microorganisms produce organic chemicals that help dissolve rock, and plants aid in rock fracturing and weathering when their roots grow into rock fractures and pores. Therefore, these same biological organisms that contribute to the soil with organic matter formation can also help degrade (weather) rock into soil-size particles. Thus, this biological activity leads to the formation of a rich composition of minerals and organic matter making for a dynamic natural soil ecosystem.

While this has been indicated before, because of its importance it bears repeating, soil organic matter is the greatest biochemical property influencing soil health. Thus, healthy soil is considered to have large quantities of organic matter and a highly diversified biological population. Yet both the organic matter content and the biological population can be altered by the type of cropping system used. A single gram of healthy soil contains billions of diverse bacteria/microorganisms (Kennedy and Smith, 1995).

Given the amount of synthetic chemicals that are used in most CAPS for control of unwanted pests (i.e., weeds, insects, and diseases), soil chemistry and organic matter are bound to be negatively impacted. Thus, soil health is often degraded by such systems.

Organic matter can be replenished in soil by addition of plant and/or animal organic sources (Arriaga and Lowery, 2003). The question is how much material needs to be added and over what time period will it need to be applied to bring a soil back to a condition close to its natural or original state? In addition to animal waste applications, soil organic matter can be restored under natural vegetation such as prairie plants. Reconstructed prairies have been shown to increase soil organic matter (Guzman and Al-Kaisi, 2010; Olson et al., 2014).

5.2.3 Conventional Agricultural Production Systems' Impact on Soil Physical Properties

While good soil health is often described as being directly related to soil biology and its chemistry, especially abundance of organic matter, it should also be understood that ideal soil physical properties are also key to good soil health. Some of the key physical properties

Figure 5.5
Example of a soil that has been abused by compaction from heavy equipment during harvest of previous crop while soil was very wet. Note how the soil structure has been destroyed and subsequent tillage has pulled the B-horizon to the soil surface. *Courtesy of Richard P. Wolkowski, Personal Communication, Emeritus Research Scientist, University of Wisconsin-Madison.*

necessary for good soil health include low bulk density which in turn means high porosity (given that bulk density and porosity are inversely related/proportional), good infiltration and drainage, and rapid water movement under saturated and unsaturated conditions. It is generally a given that all these physical conditions are found in soil with an abundance of organic matter, but even soil with adequate organic matter can have poor physical properties or can be degraded to the point that its physical properties are not optimal. An example of this is when soil with good levels of organic matter becomes compacted, which leads to soil structure degradation (Fig. 5.5), decreases in porosity, and water flow is restricted. Granted for this to happen to soils with adequate organic matter content, it is likely the result of bad soil management practices such as the increased use of heavy machinery and improper management (e.g., machine traffic and cultivating during wet soil conditions), but this is typically the case with CAPS. This is not to say that all other agricultural systems such as organic agriculture systems will be managed any better.

The ratio of soil solids to soil pores, or void spaces, is key for accessing soil health because good or ideal soil condition is considered to have about 50% pores and 50% solids (Fig. 5.6). For good soil health a large soil pore volume, or porosity, is desired. In order to assess soil porosity one must first understand soil bulk density. The mass of dry soil solids per unit volume of soil (volume of solids + voids) is a definition of soil bulk density. Since bulk density and porosity are inversely proportional, the smaller the value of a soil's bulk density the greater will be its volume of pores. It should be noted that soil water and gas (air) reside in the pores and both are needed for proper plant function to occur. Soil bulk density is a finite value that can range between slightly less than 1000 and 2650 $kg\,m^{-3}$ (the density of

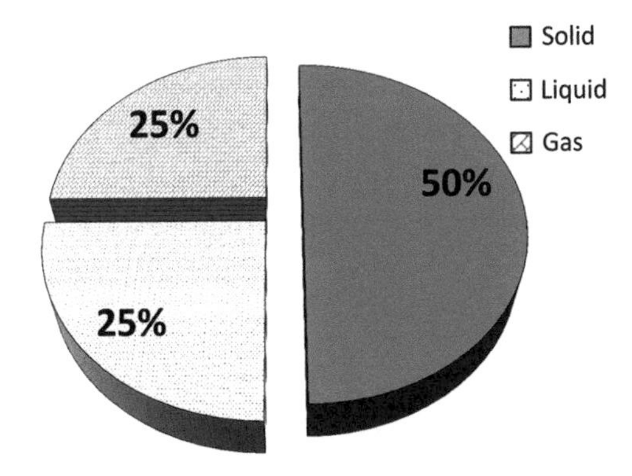

Figure 5.6
The ratio of soil solids to soil pores, or void spaces, in an ideal soil is considered to be about 50% each. The solid phase of an ideal soil is comprised of minerals (~45%) plus organic matter (~5%). The proportion of air and water that fills the soil pores (porosity) is dynamic, or always changing. An ideal soil with the characteristics described here would provide a good soil bulk density, which along with soil porosity are key for soil health and plant growth.

average soil particles) for an average mineral soil. Since organic matter has a much smaller density ($800\,kg\,m^{-3}$) when compared to mineral particles (i.e., quartz $2650\,kg\,m^{-3}$), increases in soil organic matter result in a decrease in soil bulk density.

The total porosity of a soil is important, but the distribution of pore sizes is also key to good soil quality. That is a soil with a very high porosity (i.e., low bulk density) is not necessarily a good-quality soil if the total porosity is mainly composed of small pores. A soil with mainly small pores will have a high capacity to retain or hold water but will have a poor drainage capacity. Such conditions will often result in anaerobic soil environments that limit microbial activities. Thus an ideal soil with good soil health should have a good pore size distribution. It has been shown that the addition of organic matter to such soil will improve the soil physical condition including bulk density, pore size distribution, infiltration, and hydraulic conductivity because soil aggregation is enhanced (Wei et al., 1985). These improvements will also likely promote good soil health.

Soil structure is defined as the combination and arrangement of primary (individual) soil particles into secondary structural units that form aggregates. The aggregation of primary soil particles (i.e., sand, silt, and clay) is accomplished by the necessary binding agents. Binding agents in the soil include certain chemicals such as oxides, clay minerals, and most of all soil organic matter (Six et al., 1998). These aggregates are separated from adjoining aggregates by planes of weakness, creating larger and typically more continuous pores that improve water infiltration and redistribution within a soil profile. Soils with good structure will likely have small bulk density values and thus high porosity. Good structure also means large distribution of pores and other desirable physical properties. Management practices that increase soil organic matter are typically associated with improved soil aggregation.

Soil structure is significantly impacted by some conventional agricultural systems in that structure in these systems tends to be poor, which means greater bulk density and compaction. This all combines to make for poor drainage and water-logged conditions. This is obviously soil-specific as sandy soil is an exception since sands are subject to compaction, but even under compacted conditions sand will likely have sufficient drainage. In most cases a sandy soil will not have good structure below surface soil layers. By nature, sands tend to have single grains in the subsoil.

Conventional agricultural production systems often require more intensive tillage and greater inputs of external nutrients (especially N) as compared to systems which use cropping rotations and cover crops to control pest and weeds. This often leads to degradation of soil structure because of declines in soil organic matter (Havlin et al., 1990). Consequently, it has been reported in several studies that there is a rapid decline in water-holding capacity and poor soil aeration associated with compaction, and reductions in soil organic matter, which in the long term results in yield reductions (Doran, 2002; Liebig et al., 2004). Such reductions in soil structural aggregate stability and water-holding capacity will have further ramifications on the resiliency of vegetable and grain production systems when considering the potential effects of climate change, where many studies and models have predicted an increase in frequency of extreme rainfall events and droughts globally (Pryor et al., 2014). Subsequently, many grain producers apply organic amendments to the soil, such as animal manures, sewage sludge, and household waste, to minimize the depletion of soil organic matter and improve soil structure and stability (Six et al., 1999; Bationo et al., 2007; Fließbach et al., 2007).

Good drainage and water movement through the soil (hydraulic conductivity) is key to soil health. In addition to good drainage, soil water retention is also important to soil health. There must however be a good balance between water retention and soil air for maximum plant growth and a healthy mix of soil microorganisms (Doran et al., 1996; Doran, 2002).

Soil compaction is a reduction in soil porosity often caused by human activities. Soil compaction is undesirable for agricultural plant production because it decreases infiltration and may physically restrict root growth (Hilfiker and Lowery, 1988; Lowery and Schuler, 1991). It also causes aeration problems. Examples of the impact of compaction on soil properties can be borrowed from engineering applications, such as (1) increasing soil strength for reducing settling of roadbeds and foundations, and (2) decreasing hydraulic conductivity in earth-fill dams and other water-retaining structures by increasing soil bulk density. Engineers achieve these goals by removing soil with high organic matter (i.e., topsoil) and destroying soil aggregates via excavation, tillage, and compaction. In agricultural settings, tillage using moldboard plow forms a compacted "plow sole" at the bottom of the furrow if plow depth is not varied from year-to-year. Tillage pans are not limited to moldboard plowing as most other tillage implements form a subsurface compacted zone; this includes disk-harrows, subsoilers, and disk-plows. Tillage also reduces a soil's resilience against soil compaction in the soil surface, since soil aggregates are degraded and loose friable soil is left (Lal, 1993). This reduction of soil aggregation in the long term will ultimately result in settling of smaller soil particles closer together, resulting in greater soil compaction. It has been shown that reduced and controlled traffic will reduce the amount of soil subjected to compaction in a given field area (Batey, 2009). Additionally, management practices such as diverse cropping rotations and cover crops can reduce the effect of field traffic, and can also help reduce the use of pesticides and/or fertilizers (Lal et al., 1994a; Unger and Jones, 1998).

5.3 Tillage Systems

Tillage is the mechanical manipulation of soil for weed control, seedbed preparation, and a good porous root environment. Intensive tillage practices common to CAPS, such as moldboard plowing and disk-harrow, tend to promote dry soil conditions. This drying is associated with the lack of surface plant residue (Johnson et al., 1984), disruption of soil aggregates, and reduction in organic matter with some of these systems. It has been reported that the lack of plant residue on the soil surface increases the warming of soil (Johnson and Lowery, 1985; Al-Darby and Lowery, 1987), causes more evaporation, less water storage (Johnson et al., 1984), plant stress and thus low production, which means less CO_2 fixed and low net primary production (Al-Darby and Lowery, 1986, 1987; Al-Darby et al., 1987).

Tillage, especially moldboard and chisel plowing, tends to reduce soil organic matter because it incorporates air into the soil, which promotes increased biological activity (Lou and Zhou, 2010). This increase in microbial activity, or respiration, is caused by the addition of oxygen which is often limiting for soil microbial activity. The C in organic matter serves as a "food" source for microorganisms, leading to the loss of soil organic matter as CO_2. Although organic matter is crucial for optimum soil function and improved soil quality, maintaining aggregate integrity over time is also of importance.

In some situations, tillage can have a greater impact on soil properties and health than organic matter additions to soil. An assessment of chemical, biological, and physical properties indicative of soil health compared two tillage practices (plow tillage and no-tillage) and corn stover residue harvest (residue returned to the soil and residue harvested) after 32 years, concluded that tillage practice had a greater impact on soil quality than the return of crop residues to the soil (Moebius-Clune et al., 2008). They observed negative impacts on eight out of 25 soil quality indicators studied after 32 years of corn stover residue harvest, but tillage had a greater impact on 15 of the 25 soil health indicators than stover harvest. Interestingly, soil health indicators that were affected by residue harvest were equally or greater affected by tillage (Fig. 5.7). Although organic residues are paramount for soil health, this type of results underscore the negative impact of tillage on soil aggregate disruption and that it can be as important as the organic matter loss from residue removal. Tillage disrupts

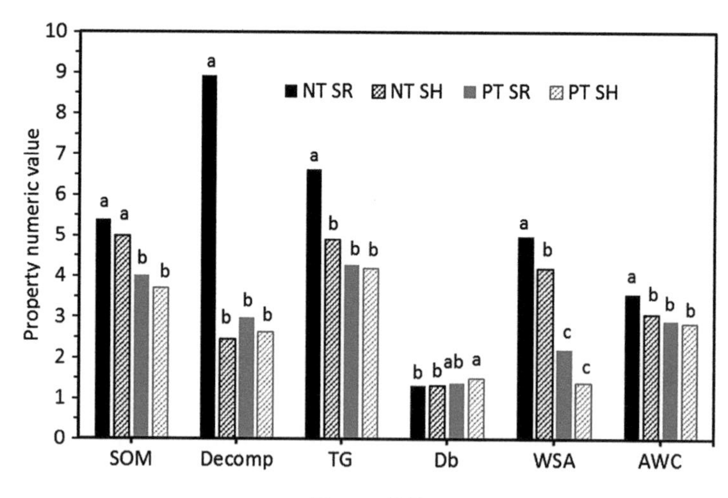

Figure 5.7

Soil health indicators affected by 32 years of corn stover harvest (SH) versus corn stover returned (SR) under no-tillage (NT) and inversion tillage (PT) management. Soil health parameters presented here include: soil organic matter (SOM; in %), cellulose decomposition rate (Decomp in; % $week^{-1}$), total glomalin concentration (TG; in mg g^{-1} dry soil), soil bulk density (Db; in Mg m^{-3}), water-stable aggregates (WSA; g g^{-1}×10), and available water capacity (AWC; m^3 m^{-3}×10). Different letters within a soil property indicate a significant difference at $\alpha = 0.10$ level. *Adapted from Moebius-Clune, B., van Es, H.M., Idowu, O.J., Schindelbeck, R.R., Moebius-Clune, D.J., Wolfe, D.W., et al., 2008. Long-term effects of harvesting maize stover and tillage on soil quality. Soil Sci. Soc. Am. J. 72, 960–969.*

aggregate formation and reduces the stabilization of new soil organic matter (Six et al., 1999). Therefore, it is not surprising that no-tillage will at least maintain soil organic matter over time (Reganold et al., 1987; Olson et al., 2014).

5.4 Crop Rotations

Crop rotation is defined as growing a different crop on a given land area every growing/planting cycle and season. Crops are rotated for different reasons, but one reason is to break disease and pest cycles. In some areas of the world conventional agriculture means crops are rotated for a given land area either seasonally or yearly. However, in some parts of the world monoculture of crops is often the norm for CAPS.

While crop rotation might not be a generally used practice, it is a practice that can promote soil health. Barley-soybean rotations with ryegrass or red clover increased the concentration of organic C in 1–2 mm and 4.7–9.5 mm aggregates after just two cycles of the rotation in Canada (Carter and Kunelius, 1993). Further, the use of ryegrass increased microaggregation. However, other times crop rotations might not positively affect soil organic matter accumulation. Varvel and Wilhelm (2008) reported an increase in soil organic matter with continuous corn when compared to a corn-soybean rotation and continuous soybean under irrigated conditions in a semiarid region in Nebraska. Similarly, corn and soybean yields were not different under continuous crop (monoculture) than those in crop rotation (i.e., corn-soybean rotation) in Piedmont and Coastal Plain soils of North Carolina (Wagger and Denton, 1992). Nevertheless, corn grain yield was 27% and 4% greater with no-tillage when compared to inversion tillage in the Piedmont and Coastal Plain soils, respectively. Similarly, soybean yields were 5% greater with no-tillage in the Piedmont, but there was no yield difference in the Coastal Plain. This increase in yield with no-tillage was attributed to greater soil water availability.

Although crop rotations might not enhance crop yields under all scenarios, crop rotations can provide other benefits. Including a leguminous crop in a rotation system can help reduce fertilizer applications and nitrate leaching to groundwater. In a 5-year study conducted in Pennsylvania, nitrate leaching was reduced by 75–80% when alfalfa was included in rotation with corn versus continuous corn (Toth and Fox, 1998). This reduction in nitrate leaching was attributed to greater organic N pools in the soil from the alfalfa grown in rotation.

5.5 Conventional Systems and Soil Erosion

Soil erosion is the process by which soil primary particles and aggregates are removed and lost from their point of origin by wind or water, or even mass wasting from gravitational forces (Troeh et al., 2004). However, wind and water erosion are generally most commonly associated with human activity with respect to agricultural practices. Soil erosion is common with human manipulation of soil, but in agriculture this is often great under many CAPS. In addition to erosion by water and wind, tillage erosion has been identified as significant with some agricultural managements systems (Lindstrom et al., 1992; Heckrath et al., 2005).

It has been shown that soil removal by erosional processes leads to significant reductions in soil quality, thereby resulting in poor soil health (Verity and Anderson, 1990; Lowery and Larson, 1995). Soil erosion by wind or water is a selective process where the organic-rich topsoil is lost (Lowery et al., 1995). Soil erosion is likely the greatest degrader of soil quality by natural forces because when soil is lost by erosion it is generally not replaced by natural weathering processes within a human life time (Alewell et al., 2014). Erosion, being a selective process of topsoil removal, leads to significant amounts of soil organic matter loss. Researchers have demonstrated that some of the organic matter lost by erosion can be restored. Arriaga and Lowery (2003) used animal manure to improve soil properties including organic matter. However, unlike erosional loss of soil organic matter, which has been shown to be reversible with proper management (Arriaga and Lowery, 2003), the eroded mineral soil itself cannot be replaced by typical agricultural management practices.

Agriculture-related erosion's main impact on soil quality is from the selective removal of the nutrient-enriched (C stocks) topsoil which generally reduces the soil production potential (productivity) (Arriaga and Lowery, 2005; Panagos et al., 2015). The eroded soil is often a source of pollution to surface water (Panagos et al., 2015). Further, Panagos et al. (2015) notes that soil erosion is one of the major threats to soils in the European Union.

Erosion is detrimental to soil physical properties and soil quality (Andraski and Lowery, 1992). When soil physical properties are degraded, soil is also degraded. Erosion causes the exposure of subsoil as the topsoil is lost, where subsoil is characterized by low organic matter and more clay content than surface soil. Soil plant-available water content, the most important soil physical property for crop production, is severely affected by erosion because of reductions in soil organic matter, decreased soil profile depth, and changes in soil particle size distribution (Adams, 1949; Langdale et al., 1979; Frye et al., 1982; Andraski and Lowery, 1992; Bakker et al., 2007). A good example of this effect can be found in the work of Lowery et al. (1995), where they reported a decrease in plant-available water capacity among 12 of the 13 Midwest soils they investigated.

When natural processes result in soil erosion this is referred to as geological erosion. When soil is eroded under these conditions it is likely not to be as the selective removal of topsoil with agriculture, but rather extensive loss including the subsoil. This kind of erosion is not unique to any farming practices. Thus, not only soil health is lost but the entire soil resources under geological erosion.

In general, conventional agriculture production systems result in greater soil erosion worldwide than would be expected under natural/geological conditions, thus agricultural activities generate more soil loss than geological erosion. This is not to say that conventional agriculture will result in more erosion than organic farming systems. In fact, some organic farming systems could generate more soil erosion than CAPS. The reason is most organic agriculture systems depend on tillage to control weeds and this excess tillage will result in soil erosion by water and possibly tillage erosion (Peigné et al., 2007). However, renewed interest in cover crops will help reduce erosion risks in both conventional and organic production systems.

5.6 Conclusions

Given the definition of conventional agriculture production systems and when it is compared to organic or sustainable cropping systems it is easy to see how conventional agriculture often results in a reduction in soil health. This need not be the case as some CAPS can maintain some, if not all, aspects of natural soil health. Extensive use of crop rotations and cover crops in combination with no-tillage will maintain and/or enhance soil health. No-tillage and other conservation tillage practices can reduce soil erosion. Other practices, such as control traffic, can be used in CAPS to reduce soil compaction. A management scheme that holds much promise is the use of high-residue cover crops, terminated with the aid of a roller crimper, and used in combination with reduce tillage or no-tillage practices, as this increases the number of plant species in a rotation, allows for a living root system to grow for longer periods of the year, and reduces the amount of soil disturbance.

References

Adams, W.E., 1949. Loss of topsoil reduced crop yield. J. Soil Water Conserv. 4, 130.

Al-Darby, A.M., Lowery, B., 1986. Evaluation of corn (*Zea mays* L.) growth with three conservation tillage systems. Agron. J. 78, 901–907.

Al-Darby, A.M., Lowery, B., 1987. Seed zone soil temperature and early corn growth with three conservation tillage systems. Soil. Sci. Soc. Am. J. 51, 768–774.

Al-Darby, A.M., Lowery, B., Daniel, T.C., 1987. Corn leaf water potential and water use efficiency under three conservation tillage systems. Soil Tillage Res. 9, 241–254.

Alewell, C., Egli, M., Meusburger, K., 2014. An attempt to estimate tolerable soil erosion rates by matching soil formation with denudation in Alpine grasslands. J. Soils Sediment. 15, 1383–1399. (http://dx.doi.org/10.1007/s11368-014-0920-6).

Alvarez, R., Diaz, R.A., Barbero, N., Santanatoglia, O.J., Blotta, L., 1995. Soil organic carbon, microbial biomass and CO_2-C production from three tillage systems. Soil Tillage Res. 33, 17–28.

Andraski, B.J., Lowery, B., 1992. Erosion effects on soil water storage, plant water uptake and corn growth. Soil. Sci. Soc. Am. J. 56, 1911–1919.

Arriaga, F.J., Lowery, B., 2003. Soil physical properties and corn yield of an eroded soil as affected by long-term cattle manure applications. Soil Sci. 168, 888–889.

Arriaga, F.J., Lowery, B., 2005. Spatial distribution of carbon over an eroded landscape in southwest Wisconsin. Soil Tillage Res. 81, 155–162.

Bakker, M.M., Govers, G., Jones, R.A., Rounsevell, M.D.A., 2007. The effect of soil erosion on Europe's crop yields. Ecosystems 10, 1209–1219.

Barak, P., Jobe, B.O., Krueger, A.R., Peterson, L.A., Laird, D.A., 1997. Effects of long-term soil acidification due to nitrogen fertilizer inputs in Wisconsin. Plant Soil 197, 61–69.

Batey, T., 2009. Soil compaction and soil management–a review. Soil Use Mgt. 25, 335–345.

Bationo, A., Kihara, J., Vanlauwe, B., Waswa, B., Kimetu, J., 2007. Soil organic carbon dynamics, functions and management in West African agro-ecosystems. Agric. Syst. 94, 13–25.

Bero, N.J., Ruark, M.D., Lowery, B., 2015. Bromide and chloride tracer application to determine sufficiency of plot size and well depth placement to capture preferential flow and solute leaching. Geoderma 262, 94–100.

Carter, M.R., Kunelius, H.T., 1993. Effect of undersowing barley with annual ryegrasses or red clover on soil structure in a barley-soybean rotation. Agric. Ecosyst. Environ. 43, 245–254.

Dick, R.P., 1994. Soil enzyme activities as indicators of soil quality. In: Doran, J.W., Coleman, D.C., Bezdicek, D.F., Stewart, B.A. (Eds.), Defining Soil Quality for a Sustainable Environment. Soil Science Society of America, Madison, Wisconsin.

Doran, J.W., 2002. Soil health and global sustainability: translating science into practice. Agric. Ecosyst. Environ. 88, 119–127.

Doran, J.W., Sarrantonio, M., Liebig, M., 1996. Soil health and sustainability. Adv. Agron. 56, 1–54.

Fließbach, A., Oberholzer, H.-R., Gunst, L., Mäder, P., 2007. Soil organic matter and biological soil quality indicators after 21 years of organic and conventional farming. Agric. Ecosyst. Environ. 118, 273–284.

Frye, W.W., Ebelhar, S.A., Murdock, L.W., Bleven, R.L., 1982. Soil erosion effects on properties and productivity of two Kentucky soils. Soil Sci. Soc. Am. J. 46, 1051–1055.

Guzman, J., Al-Kaisi, M., 2010. Soil carbon dynamics and soil carbon budget of newly reconstructed tall-grass prairies in south central Iowa. J. Environ. Qual. 39, 136–146.

Havlin, J., Kissel, D., Maddux, L., Claassen, M., Long, J., 1990. Crop rotation and tillage effects on soil organic carbon and nitrogen. Soil Sci. Soc. Am. J. 54, 448–452.
Heckrath, G., Djurhuus, J., Quine, T.A., Van Oost, K., Govers, G., Zhang, Y., 2005. Tillage erosion and its effect on soil properties and crop yield in Denmark. J. Environ. Qual. 34, 12–24.
Hilfiker, R.E., Lowery, B., 1988. Effect of conservation tillage systems on corn root growth. Soil Tillage Res. 12, 269–283.
Jitareanu, G., Ailincai, C., Bucur, D., 2006. Influence of tillage systems on soil physical and chemical characteristics and yield in a soybean and maize grown in the Moldavian Plain (North-Eastern Romania). In: Horn, R., Fleige, H., Peth, S., Peng, X. (Eds.), Soil Management for Sustainability. Catena Verlag GmbH, Reiskirchen, Germany.
Johnsen, K., Jacobsen, C.S., Torsvik, V., Sørensen, J., 2001. Pesticide effects on bacterial diversity in agricultural soils–a review. Biol. Fertil. Soils 33, 443–453.
Johnson, M.D., Lowery, B., 1985. Effect of conservation tillage practices on soil temperature and thermal properties. Soil. Sci. Soc. Am. J. 49, 1547–1552.
Johnson, M.D., Lowery, B., Daniel, T.C., 1984. Soil moisture regimes of three conservation tillage systems. Trans. ASAE 27, 1385–1390. 1395.
Kennedy, A., Smith, K., 1995. Soil microbial diversity and the sustainability of agricultural soils. Plant Soil 170, 75–86.
Knapen, A., Poesen, J., 2006. Enhancing the soil's erosion resistance during concentrated runoff by reducing tillage disturbance. In: Horn, R., Fleige, H., Peth, S., Peng, X. (Eds.), Soil Management for Sustainability. Catena Verlag GmbH, Reiskirchen, Germany.
Kraft, G.J., Stites, W., 2003. Nitrate impacts on groundwater from irrigated-vegetable systems in humid north-central US sand plain. Agric. Ecosyst. Environ. 100, 63–74.
Lal, R., 1993. Tillage effects on soil degradation, soil resilience, soil quality, and sustainability. Soil Tillage Res. 27, 1–8.
Lal, R., Mahboubi, A., Fausey, N., 1994a. Long-term tillage and rotation effects on properties of a central Ohio soil. Soil. Sci. Soc. Am. J. 58, 517–522.
Lal, R., Logan, T., Shipitalo, M., Eckert, D., Dick, W., Carter, M., 1994b. Conservation Tillage in the Corn Belt of the United States. Lewis Publishers Inc.
Landon, A.J., 2008. The "How" of the three sisters: the origins of agriculture in Mesoamerica and the human niche. Nebr. Anthropol 40, 110–124.
Langdale, G.W., Box Jr., J.E., Leonard, R.A., Barnett, A.P., Fleming, W.G., 1979. Crop yield reduction on eroded Southern Piedmont soils. J. Soil Water Conserv. 34, 226–228.
Liebig, M., Tanaka, D., Wienhold, B.J., 2004. Tillage and cropping effects on soil quality indicators in the northern Great Plains. Soil Tillage Res. 78, 131–141.
Lindstrom, M.J., Nelson, W.W., Schumacher, T.E., 1992. Quantifying tillage erosion rates due to moldboard plowing. Soil Tillage Res. 24, 243–255.
Lou, Y., Zhou, X., 2010. Soil Respiration and the Environment. Academic Press, Burlington, MA.
Lowery, B., Larson, W.E., 1995. Symposium: erosion impact on soil productivity; preamble. Soil Sci. Soc. Am. J. 59, 647–648.
Lowery, B., Schuler, R.T., 1991. Temporal effects of subsoil compaction on soil strength and plant growth. Soil Sci. Soc. Am. J. 55, 216–223.
Lowery, B., Swan, J., Schumacher, T., Jones, A., 1995. Physical properties of selected soils by erosion class. J. Soil Water Conserv. 50, 306–311.
Moebius-Clune, B., van Es, H.M., Idowu, O.J., Schindelbeck, R.R., Moebius-Clune, D.J., Wolfe, D.W., et al., 2008. Long-term effects of harvesting maize stover and tillage on soil quality. Soil Sci. Soc. Am. J. 72, 960–969.
Moorman, T., 1989. A review of pesticide effects on microorganisms and microbial processes related to soil fertility. J. Prod. Agric. 2, 14–23.

Olson, K.R., Al-Kaisi, M., Lal, R., Lowery, B., 2014. Experimental consideration, treatments, and methods in determining soil organic carbon sequestration rates. Soil Sci. Soc. Am. J. 78, 348–360.

Panagos, P., Borrelli, P., Poesen, J., Ballabio, C., Lugato, E., Meusburger, K., et al., 2015. The new assessment of soil loss by water in Europe. Environ. Sci. Policy 54, 438–447. http://dx.doi.org/10.1016/j.envsci.2015.08.012.

Peigné, J., Ball, B., Roger-Estrade, J., David, C., 2007. Is conservation tillage suitable for organic farming? A review. Soil Use Mgt. 23, 129–144.

Pryor, S.C., Scavia, D., Downer, C., et al., 2014. Midwest. Climate change impacts in the United States: the third national climate assessment. In: Melillo, J.M., Richmond, T.C., Yohe, G.W. (Eds.), National Climate Assessment Report. U.S. Global Change Research Program, Washington, DC, pp. 418–440.

Reganold, J.P., Elliott, L.F., Unger, Y.L., 1987. Long-term effects of organic and conventional farming on soil erosion. Nature 330, 370–372.

Six, J., Elliott, E., Paustian, K., Doran, J., 1998. Aggregation and soil organic matter accumulation in cultivated and native grassland soils. Soil Sci. Soc. Am. J. 62, 1367–1377.

Six, J., Elliott, E., Paustian, K., 1999. Aggregate and soil organic matter dynamics under conventional and no-tillage systems. Soil Sci. Soc. Am. J. 63, 1350–1358.

Toth, J.D., Fox, R.H., 1998. Nitrate losses from a corn-alfalfa rotation: lysimeters measurement of nitrate leaching. J. Environ. Qual. 27, 1027–1033.

Triplett, G., Dick, W.A., 2008. No-tillage crop production: a revolution in agriculture!. Agron. J. 100, 153–165.

Troeh, F.R., Hobbs, J.A., Donohue, R.L., 2004. Soil and Water Conservation: Productivity and Environmental Protection, fourth ed. Prentice-Hall, Inc, Upper Saddle River, NY.

Unger, P.W., Jones, O.R., 1998. Long-term tillage and cropping systems affect bulk density and penetration resistance of soil cropped to dryland wheat and grain sorghum. Soil Tillage Res. 45, 39–57.

Varvel, G.E., Wilhelm, W.W., 2008. Soil carbon levels in irrigated western corn belt rotations. Agron. J. 100, 1180–1184.

Verity, G., Anderson, D., 1990. Soil erosion effects on soil quality and yield. Can. J. Soil Sci. 70, 471–484.

Wagger, M.G., Denton, H.P., 1992. Crop and tillage rotations: grain yield, residue cover, and soil water. Soil Sci. Soc. Am. J. 56, 1233–1237.

Wei, Q.F., Lowery, B., Peterson, A.E., 1985. Effect of sludge application on physical properties of a silty clay loam soil. J. Environ. Qual. 14, 178–180.

West, T.O., Post, W.M., 2002. Soil organic carbon sequestration rates by tillage and crop rotation. Soil Sci. Soc. Am. J. 66, 1930–1946.

CHAPTER 6

Integration of Annual and Perennial Cover Crops for Improving Soil Health

Abbey Wick[1], Marisol Berti[1], Yvonne Lawley[2] and Mark Liebig[3]

[1]North Dakota State University, Fargo, ND, United States [2]University of Manitoba, Winnipeg, MB, Canada [3]United States Department of Agriculture, Agriculture Research Service, Northern Great Plains Research Laboratory, Mandan, ND, United States

6.1 Introduction

Inclusion of cover crops, whether annual or perennial, improves soil health as compared to leaving soils fallow (Wienhold et al., 2006). Annual cover crops (e.g., grasses, brassicas, and legumes) can fill gaps in rotations when soils may otherwise be fallow to provide important agronomic benefits including, but not limited to, erosion control and water management (Kaspar et al., 2001; De Baets et al., 2011; Chen et al., 2014; Lounsbury and Weil, 2015), nutrient capture and soil fertility (Thorup-Kristensen, 2001; Weil and Kremen, 2007; Dean and Weil, 2009; Thorup-Kristensen et al., 2009; Parr et al., 2014; Jani et al., 2015), weed suppression (Lawley et al., 2012), and disease control (Chew, 1988; Brown and Morra, 1995, 1997). Perennial grasses and legumes in rotation can provide significant ecosystem services by improving nutrient and water cycling efficiency (Walsh et al., 2014), enhance climate regulation (Lal et al., 2011; Franzluebbers et al., 2014), and also greatly improve soil health (Blanco-Canqui, 2010). Benefits to soil health parameters may include increasing nutrient retention, reducing erosion and fixing nitrogen (N) (Russelle and Birr, 2004; Congreves et al., 2015). Relative benefits of annual grasses, brassicas, and legumes and perennial grasses and legumes to selected soil health parameters are summarized in Table 6.1. The goal of this chapter is to discuss the use of annual and perennial cover crops in rotation in the context of physical, chemical, and biological soil health parameters.

6.2 Annual Cover Crops

6.2.1 Grasses

Warm- and cool-season grasses are used as cover crops, either as a volunteer following a harvested crop, interseeded into a standing crop, seeded postharvest as a monoculture, or as part of a mixture. In prevented plant situations, where climatic conditions inhibit the

Soil Health and Intensification of Agroecosystems.
DOI: http://dx.doi.org/10.1016/B978-0-12-805317-1.00006-3

Table 6.1: Relative effects of cover crops on soil health properties

	Annual Cover Crops			Perennial Cover Crops	
Category/Soil Property	**Grass**	**Brassica**	**Legume**	**Grass**	**Legume**
		Physical			
Soil bulk density	⇩[a]		⇩	⇩	
Macroporosity		⇧		⇧	⇧
Aggregate stability	⇧		⇧	⇧	⇧
Available water-holding capacity	⇧				
Infiltration		⇧		⇧	⇧
Hydraulic conductivity				⇧	⇧
		Chemical			
Soil nitrate (uptake)	⇧	⇧	⇧	⇧	⇧
Soil nitrate (release)			⇧		
Soil P (uptake)	⇧		⇧		
Soil P (release)		⇧	⇧		⇧
Electrical conductivity				⇩	⇩
Soil organic matter	⇧			⇧	⇧
		Biological			
Microbial biomass				⇧	⇧
Arbuscular mycorrhizal fungi	⇧				⇧
Enzyme activity			⇧	⇧	⇧
Earthworm abundance				⇧	⇧

[a]Up and down arrows imply an increase or decrease in associated soil property, respectively.

planting of a cash crop, grasses may also be planted to remediate a field and prepare it for the following year's crop. The rapid growth, tolerance to a spectrum of soil conditions, and fibrous root systems are just some of the benefits of using a grass as a cover crop. Annual grass cover crops of interest include cereal rye (*Secale cereal*), annual ryegrass (*Lolium multiflorum* L.), barley (*Hordeum vulgare* L.), winter wheat (*Triticum aestivum* L.), triticale (× *Triticosecale* Whittman), forage oat (*Avena sativa* L.), and sorghum × sudangrass [*Sorghum bicolor* × *Sorghum sudanense* (L.) Moench.].

Environmental services provided by grasses include erosion control (Kaspar et al., 2001; De Baets et al., 2011) and nutrient scavenging (Thorup-Kristensen, 2001; Thorup-Kristensen et al., 2009), both of which lead to improvements in water quality (Brandi-Dohrn et al., 1997; Ritter et al., 1998; Coale et al., 2001; Dabney et al., 2001; Strock et al., 2004). Additionally, use of cover crops in general may mitigate climate change through the transformation of carbon dioxide into soil organic matter (Reicosky and Forcella, 1998). Cereal rye is one of the most widely used cover crops, especially in the Corn Belt of the United States, because of winter hardy characteristics and versatility (Singer et al., 2007).

6.2.1.1 Benefits to soil physical properties

Use of cool-season grasses in rotation leads to soil physical property improvements primarily in the surface soils. Water-stable aggregation is greater in surface soils when a winter cover crop, such as cereal rye, is included (Hermawan and Bomke, 1997; Villamil et al., 2006; Steele et al., 2012). Prolific root establishment, organic matter contributions, and microbial stimulation under annual grasses also build soil aggregation (Haynes and Beare, 1997). Barley has less of an effect on aggregate formation in the macroaggregate size class (>2 mm) compared with cereal and annual ryegrass (Haynes and Beare, 1997). This is largely a result of shortened growth with winter-kill, whereas the cereal and annual ryegrass overwintered and had more influence on soil properties (Liu et al., 2005). Polysaccharides, an indicator of an active soil-binding agent for aggregate formation, were higher under annual ryegrass compared with cereal rye or barley (Liu et al., 2005).

Cover crops are also known for reducing compaction through the development of biopores in a process termed "biodrilling" (Cresswell and Kirkegaard, 1995). Though grasses are effective for remediating compacted soil layers, brassicas penetrate compacted layers better than grasses (Materechera et al., 1992; Chen and Weil, 2010). Increases in aggregation and rooting can also lead to decreases in bulk density and increases in total porosity in surface soils (Villamil et al., 2006). For example, crop rotations, which include cereal rye (in some mix combination), showed a 7% decrease in bulk density relative to rotations that did not include cereal rye (Villamil et al., 2006). Variability in pore size distribution from stimulated earthworm activity (Lamande et al., 2003; Schmidt et al., 2003) and root penetration results in better aeration and increased water retention (Villamil et al., 2006).

6.2.1.2 Benefits to soil chemical properties

Soil chemical properties can be influenced by cover crop inclusion in rotation, specifically from the perspective of nutrient capture and fertility. Cool-season grasses are often used as "catch crops" to retain excess nutrients remaining in the soil from the preceding crop and to supplement the fertility demands of the following cash crop (Vyn et al., 2000; Aronsson et al., 2016). The ability to capture nutrients, particularly nitrate, is related to biomass of the cover crop and the development of the root system (Thorup-Kristensen, 2001), making timing of planting and termination important (Komatsuzaki and Wagger, 2015). Phosphorus (P) uptake by cover crops and incorporation into aboveground biomass is also of interest, but less is known (Liu et al., 2014, 2015). Capturing excess nutrients reduces potential water quality issues from nitrate leaching (Strock et al., 2004) and can also reduce fertilizer inputs for the following crop (Shipley et al., 1992; Staver and Brinsfield, 1998).

A variety of cover crops can be used to capture nutrients to build soil fertility and protect water resources. Brassicas are commonly used because of their deep tap root (up to 2 m deep after 4 months growth) (Thorup-Kristensen et al., 2009) and rapid growth (root growth of

2 mm day^{-1} °C^{-1}) (Thorup-Kristensen, 2001), allowing for capture of nitrate from deep in the soil profile and movement back towards the soil surface and in biomass. Winter wheat, barley, and cereal rye, though they exhibit shallower rooting depths and slower growth than brassicas (Thorup-Kristensen, 2001), have been found to capture significant amounts of N between 0.90 and 1.5 m depths (Kuhlmann et al., 1989; Strebel and Duynisveld, 1989). However, nitrate-N (NO_3-N) capture was higher under brassica cover crops than grasses. Concentrations in groundwater at the 1.0–1.5 m depth was 119 μg nitrate L^{-1} under "no cover crop" treatments, 23 μg nitrate L^{-1} under rye, and 1.5 μg nitrate L^{-1} under radish (*Raphanus sativus* L.) (Thorup-Kristensen, 2001). The key to maximizing nitrate capture is establishing a cover crop immediately after harvest for sufficient fall growth and adequate growth in the spring prior to termination if winter annuals are used (Komatsuzaki and Wagger, 2015).

Though grasses may not be as deep-rooted and nitrate uptake can be less than brassicas, cereal rye, winter wheat, and triticale are important cover crops for nitrate capture because of tolerance to cold climatic conditions enabling them to overwinter and grow in the spring for additional nitrate capture (Martin et al., 1976; Komatsuzaki and Wagger, 2015). Winter wheat can reach rooting depths twice that of spring wheat because of the ability to grow in the fall, overwinter, and grow again in early spring (Thorup-Kristensen et al., 2009). In an 8-year study, use of cereal rye between cash crops reduced annual nitrate leaching by 80% in the mid-Atlantic coastal plain (Staver and Brinsfield, 1998), primarily as a result of its growth characteristics. Winter annual cover crops following a cash crop can be especially effective in years where climatic conditions, such as drought, limit cash crop uptake of fertilizers applied to the soil (Staver and Brinsfield, 1990).

6.2.1.3 Benefits to soil biological properties

One of the primary benefits of annual grass cover crops, but likely the least understood, pertains to their influence on soil biological processes. Arbuscular mycorrhizal fungi (AMF) are an important component of the soil biological system, assisting with crop P uptake, reduced pathogen pressures on host crops, and tolerance to environmental stresses (Rillig, 2004; Lehman et al., 2012). In the absence of a host crop, AMF populations will decline (Brundette, 2002). Use of cover crops which form associations with AMF, such as forage oat, cereal rye, and winter wheat, have been found to increase propagule and biomass numbers to a level which benefitted the following corn (*Zea mays* L.) crop (Boswell et al., 1998; Kabir and Koide, 2002; White and Weil, 2010; Karasawa and Takebe, 2012; Lehman et al., 2012). Forage oats promote higher AMF propagules than cereal rye or winter wheat or soils without a cover crop (Karasawa and Takebe, 2012; Lehman et al., 2012).

Fertility benefits of a cover crop to the following cash crop are dependent upon nutrient capture efficiency and mineralization potential of the cover crop, demand of the cash crop, environmental conditions, and management practices including tillage and timing of termination of the cover crop (Kamh et al., 1999; Vyn et al., 2000). Plant material quality becomes an important consideration when selecting cover crops to improve soil fertility. Grasses generally have an abundance of carbon (C) relative to N, while brassicas are known to have an abundance of N relative to C (SARE, 2012). Priming of microbial communities, which assists with crop uptake of nutrients, is also an important benefit of cover crops for fertility. Sorghum × sudangrass assists with P uptake of following crops (specifically wheat) by stimulating fungal populations that solubilize P (Karasawa and Takehashi, 2015).

Winter grasses used in rotation can increase soil organic matter. Including cereal rye in corn-soybean [*Glycine max* (L.) Merr.] rotations under no-tillage can significantly increase soil organic matter in surface soils, for both whole soil and readily available C pools (Fae et al., 2009; Sequiera and Alley, 2011; Moore et al., 2014). Annual ryegrass is known to have greater root biomass than cereal rye, resulting in greater potential for building organic matter and stimulating microbial communities (Liu et al., 2005). Annual ryegrass also has a greater ability to build total soil C versus cereal rye or barley (Liu et al., 2005).

6.2.2 Brassicas

Multiple characteristics have made plants in the *Brassicaceae* family useful cover crops, including small seeds size, ease of establishment, rapid and high biomass production (Thorup-Kristensen, 2001), ability to suppress weeds (Lawley et al., 2012), tap root structure (Chen and Weil, 2010), and production of secondary metabolites for biofumigation (Chew, 1988; Brown and Morra, 1995, 1997). Some of the most commonly used species for cover crops are rapeseed and canola (*Brassica napus* L. or *Brassica rapa* L.), brown and oriental mustard (*Brassica juncea* L.), yellow mustard (*Sinapis alba* L.), forage or oilseed or Daikon radish, turnip [*Brassica rapa* L. *var. rapa* (L.) Thell.], and leaf turnip Pasja (*Brassica campestris* x *Brassica napus*). Brassica species that are of growing interest for use as cover crops include kale [*Brassica oleracea* L. convar. *acephala* (DC)], camelina [*Camelina sativa* (L.) Crantz.], and pennycress (*Thlaspi arvense* L.).

6.2.2.1 Benefits to soil physical properties

Large tap roots are a feature of most species in the *Brassicaceae* family, though there is a broad range of shapes and rooting depths (Weil et al., 2009). Brassica tap roots can quickly grow deep in the soil (Gruver et al., 2012), at a rate of 2 mm day^{-1} $°C^{-1}$ (Thorup-Kristensen, 2001) to a depth of 2 m after 4 months (Thorup-Kristensen et al., 2009). Species such as radish and turnip have large fleshy tap roots that can grow up and out of the soil. The fine roots at the tip of these tap roots grow through root-restricting layers when cover crops are grown at a time of year when soils are wet (Chen and Weil, 2010). When the small roots senesce and decompose, the remaining root channels can act as tunnels for subsequent crop roots, allowing them to grow through these compacted root zones (Williams and Weil, 2004). This allows subsequent cash crops to access deep soil moisture and nutrients that might otherwise be unavailable (Chen and Weil, 2011).

Higher root counts in forage radish and rapeseed have been found in and below the compaction soil zone (5–25 cm depth) compared to the fibrous roots of cereal rye (Chen and Weil, 2010). Most importantly, root channels created by the forage radish and rapeseed persisted under the subsequent corn crop, allowing for greater root penetration of corn beyond the compacted soil layer. Under these conditions, corn yields were greater following all cover crop treatments, than following the no-cover crop control (Chen and Weil, 2010). Partnering the frequent use of cover crops with conservation tillage practices to conserve root channels may be an alternative strategy to deep tillage strategies currently being used to restore soil function in compacted soils.

Winterkilled brassica cover crops with large swollen tap roots, such as forage or oilseed radish, can leave large macropores at the soil surface the following spring (White and Weil, 2011). Intuitively, these macropores increase surface water infiltration. Surface water spring runoff decreased (likely as a result of macropores; though not measured) in production fields with forage radish cover crops compared with neighboring fields without cover crops (Weil et al., 2009). Differing soil moisture trends in surface soils (5 cm and 20 cm depths) indicate more rapid infiltration of surface water under radish than those without cover crops (Lounsbury and Weil, 2015). Similarly, air permeability is also increased following radish and rapeseed cover crops relative to cereal rye and no cover crops (Chen et al., 2014). This points to an increase in soil macroporosity that benefits plants by increasing water and air movement in the soil profile following brassica cover crops.

6.2.2.2 Benefits to soil chemical properties

The rapid emergence and growth of brassica cover crops makes them an ideal cover crop to scavenge N from a soil profile following a cash crop (Thorup-Kristensen, 2001). Brassica cover crops have been studied extensively in the United States Mid-Atlantic where they are being grown to reduce nutrient losses from agriculture in the Chesapeake Bay watershed (Weil and Kremen, 2007). N uptake in the fall by brassica cover crops (forage radish, oilseed radish, and rapeseed; 36–171 kg N ha^{-1}) was equal to or greater than cereal rye (42–112 kg N ha^{-1}; Dean and Weil, 2009). Rapeseed was the only brassica that overwintered and its spring N uptake was equal to that of cereal rye ranging from 41 to 118 kg N ha^{-1}.

Patterns in cover crop uptake of N were matched by soil N measurements in the soil profile to a depth of 105 cm (Dean and Weil, 2009). Brassica and cereal rye cover crops depleted soil N down the soil profile relative to a no-cover crop control. N release patterns differed between the two radish species and rapeseed cover crops because of their different termination patterns. Rapeseed overwintered due to mild winters in the study area (Maryland) and had N release patterns similar to cereal rye (a winter annual). Forage radish and oilseed radish winterkilled between December and January in the study area and the N release from the cover crop was detected in March and April, especially in sandy soils. Early-seeded crops are essential to take advantage of that early release of N. For example, early-season planting of spinach (*Spinacia oleracea* L.) into forage radish cover crop residues led to higher spring soil nitrate levels (+33 kg N ha^{-1}) 20 days after planting following forage radish (Lounsbury and Weil, 2015), resulting in significantly higher spinach yields relative to oat cover crop and no cover crop treatments.

6.2.2.3 Benefits to soil biological properties

Plants in the *Brassicaceae* family are known for their biofumigation potential. When brassica cover crops are terminated mechanically through mowing or tillage, or when killed by cold temperatures, secondary metabolites including glucosinolates are released. Glucosinolates are broken down by plant enzymes into other more toxic compounds, including isothiocyanates, that have bioactive properties on seeds and soilborne organisms including bacteria and fungi that can benefit or harm subsequent crop plants (Chew, 1988; Brown and Morra, 1995, 1997; Haramoto and Gallandt, 2004). The management of brassica cover crops can be optimized to exploit or minimize these biofumigant properties. Species and variety selection based on gulcosinolate content of vegetative plant tissues, higher soil fertility status, along with cover crop termination at or before flowering can be used to maximize glucosinolates in brassica cover crops (Haramoto and Gallandt, 2004).

Plants in the *Brassicaceae* family are nonmycorrhizal and have evolved mechanisms to increase their uptake of soil P by lowering rhizosphere pH and extruding organic acids through their roots (Hedley et al., 1982; Hoffland et al., 1989). Thus, brassica cover crops can increase P availability to subsequent crops by releasing the labile and recalcitrant P after uptake. However, their nonmycorrhizal status can potentially limit P uptake by subsequent crops that are very reliant on mycorrhizae for soil P uptake, such as corn. Mycorrhizal colonization of young (V4) corn roots was reduced following forage radish relative to corn following cereal rye (White and Weil, 2010). However, mycorrhizal colonization following forage radish was equivalent to colonization following a no-cover crop control treatment in four of six site years (White and Weil, 2010). Treatment differences in mycorrhizal colonization did not persist into later vegetative growth stages (V8) and differences in young (V4) corn did not always result in differences in corn P status at early stages.

6.2.3 Legumes

Annual legume cover crops are an important tool to increase soil N, reduce nitrate leaching, and reduce N fertilization, while improving soil health. Legume cover crops improve grain yield and can replace entirely or partially the N fertilizer required by cereals. There are many annual legumes used as cover crops, but among the most common are crimson clover (*Trifolium incarnatum* L.), pea (*Pisum sativum* L.), and hairy vetch (*Vicia villosa* Roth).

6.2.3.1 Benefits to soil physical properties

A positive response in soil aggregation and stability of soil aggregates has been observed with many annual legume cover crops (McVay et al., 1989; Villamil et al., 2006; N'Dayegamiye et al., 2015). Improved aggregate stability can be explained by the higher root activity of legumes, associated with higher microbial biomass than cereal crops. Aggregate formation is stimulated by microbial activity and contributions of organic material in the rooting zone, where organic matter acts as a nucleus for aggregate formation, polysaccharides secreted by bacteria act as glues to hold the soil particles together, and fungal hyphae and fine roots physically hold the soil particles together (Tisdall and Oades, 1982; Jastrow and Miller, 1998). Compared with a control, the highest increases of the 5-mm aggregate proportion (+28% to +39%) were recorded for monocropped legumes (hairy vetch and crimson clover) and for hairy vetch/wheat (N'Dayegamiye et al., 2015).

Legume cover crops, such as hairy vetch, also can reduce soil bulk density in surface soils (0–10 cm) and significantly increase total soil porosity (Villamil et al., 2006). These changes are correlated with an increase in soil organic matter content, root number, and abundance of earthworms.

6.2.3.2 Benefits to soil chemical properties

Improvements to N cycling efficiency associated with legumes are recognized as a primary benefit to agronomic systems. N credit to cash crops from the use of an annual legume cover crop are on average 72 kg N ha^{-1} (Hargrove, 1986), but varies greatly with the species, location, and seeding date. Legumes first use available N in the soil before forming microbial associations which fix N_2. Only a portion of N in the cover crop biomass is mineralized and available for the next crop, thus the N credit from the cover crop is 50% or less than the total N content in the plant. Also, mineralization of legume residue takes longer in dry years, if left on the soil surface, than in years with adequate moisture to drive decomposition of plant materials (Dabney et al., 2001). Crimson clover and hairy vetch released 50% of the N in the root biomass in 1 week, mainly from fine roots. But, the N in coarse roots was released in a period of 8 weeks, increasing inorganic N in the soil (Jani et al., 2015). N in crimson clover was immobilized and not available to the corn crop until 42 days after roller-crimper termination in the spring (Parr et al., 2014). Reports from other researchers indicate no crop yield response following crimson clover (Touchton et al., 1982; Bair et al., 2008). The method of termination influences the mineralization rate, where chemical desiccation has shown much faster mineralization of crimson clover biomass N and yield response (Wagger, 1989; Ranells and Wagger, 1996; Wagger et al., 1998).

N content of 88 kg N ha^{-1} was reported for hairy vetch, but only 20 kg N ha^{-1} of the N in the plant was available to the subsequent crop (Bollero and Bullock, 1994). N mineralization of hairy vetch aboveground biomass is fast and N release starts as early as 14 days after termination of the crop in the spring (Parr et al., 2014). N in the biomass of forage pea, Austrian winter pea, and hairy vetch was 116, 85, and 87 kg N ha^{-1}, respectively. Forage pea had approximately 45 kg N ha^{-1} from biological N-fixation during a 60-day growing period (Samarappuli et al., 2014). About 80–90% of N_2 fixed by crimson clover and hairy vetch is retained in the aboveground biomass (Shipley et al., 1992). Crimson clover and balansa clover (*Trifolium balansae* L.) scavenge N from soil and air (N_2 fixing) in amounts enough to provide the N requirements for several vegetable crops (Dabney et al., 2001). Crimson clover in pure stands or in a mixture with annual ryegrass provided enough N to sustain a corn crop (Kramberger et al., 2014).

Legumes, faba bean (*Vicia faba* L.), hairy vetch, and pea have higher P uptake and P mobilization potential (lower soil pH, higher organic acids, and phosphatase activity) than cereals (Kamh et al., 2002; Nuruzzaman et al., 2006; Li et al., 2007; Maltais-Landry, 2015). Inclusion of a legume in a crop sequence can reduce soil P concentrations as a result of immobilization of P in the cover crop biomass (Villamil et al., 2006). Thus, using legume cover crops alone or in mixture with cereal rye might reduce P runoff losses (Villamil et al., 2006).

Soil organic matter increases when legume cover crops are added to a corn-soybean-based crop rotation (Villamil et al., 2006). Also, mixing a legume, such as hairy vetch, with cereal rye improves the ability of the cereal rye to decompose and become part of the soil organic matter pool. This process is facilitated by the mixing of the low C:N ratio of legumes (generally <20) with the higher C:N ratio of cereals. Legume cover crops increase soil organic matter slower than cereals, but when combined, soil organic matter increases rapidly (Touchton et al., 1982; Ebelhar et al., 1984).

6.2.3.3 Benefits to soil biological properties

Legume green manures improve bacterial and fungal populations and enzyme activities (Biederbeck et al., 2005). Dehydrogenase and alkaline phosphatase activities are indicators of soil microbial activity. The greatest increase in these enzymes' activities was recorded for hairy vetch, and hairy vetch/wheat followed by crimson clover compared with continuous maize, indicating a more metabolically active microbial population (N'Dayegamiye et al., 2015). Austrian winter pea and crimson clover affect ß-glucosidase and ß-glucosaminidase activities more positively than hairy vetch. Austrian winter peas enhance nitrification potential more than crimson clover and hairy vetch, despite its lower biomass yield (Liang et al., 2014). Differences in chemical composition, C:N ratio, lignin and cellulose contents among legumes account for substantial changes in soil microbial properties. Usually a lower C:N ratio and lignin content promotes microbial activity and nutrient availability (Liang et al., 2014).

Increased soil microbial activity improves many soil physical properties and indirectly offers many ecosystem services. Microbes and mycorrhizae produce substances with gluing effect increasing soil aggregate stability. A soil with better soil aggregate stability is less prone to erosion, nutrient leaching, and runoff. Additionally, this improves soil water retention.

6.3 Perennial Cover Crops

Integration of perennial plants into annual cropping systems can provide significant ecosystem service benefits to agricultural landscapes through improvements to nutrient and water cycling efficiency, enhanced climate regulation, wildlife habitat, and increased aesthetic, educational, and recreational opportunities (Franzluebbers et al., 2014). Many ecosystem service benefits associated with perennial plants are derived from changes to soil properties (Blanco-Canqui, 2010). Changes to soil properties under perennials are driven mainly by limited soil disturbance, coupled with increased organic matter inputs from roots and rhizodeposits in comparison to annual crops (Franzluebbers, 2015). Understanding how perennial plants alter soil properties, and, in turn, how such alterations affect ecosystem services is essential for the development and adoption of improved management practices to facilitate a transition toward multifunctional agricultural landscapes (Sanderson and Adler, 2008).

6.3.1 Grasses

Several excellent literature reviews have contributed to general inferences of soil health improvement from perennial grasses on agricultural land (Anderson-Teixeira et al., 2009; Blanco-Canqui, 2010; Franzluebbers et al., 2014, 2015). While the reviews have been helpful in conveying general outcomes, one must take care to not extend this inference to all grasses and soils independent of stand duration. With over 10,000 species worldwide, perennial grasses differ considerably in their capacity to provide surface cover and produce above- and belowground biomass (Sanderson et al., 2009). Root morphology and maximum rooting depths also vary among perennial grasses (Weaver and Darland, 1949), thereby contributing to differential effects across the soil profile. Moreover, soil responses to perennial grasses can occur in the short term (<2 years) or require over a decade to be resolved, depending on the property (Blanco-Canqui, 2010). Such considerations need to be taken into account when reviewing perennial grass effects on soil properties and ecosystem services.

6.3.1.1 Benefits to soil physical properties

Among categories of soil attributes, soil physical properties are frequently cited to improve following inclusion of perennial grasses in annual crop production systems. Perennial grasses, with greater root biomass and rhizodeposits compared to annual crops, coupled with limited soil disturbance during the perennial grass phase, foster a biophysical environment favoring larger and more stable soil aggregates, increased soil pore space and continuity, and improved water entry and flow through the soil profile. In an extensive review of energy crop effects on soil and the environment (Blanco-Canqui, 2010), perennial grasses were found to enhance soil structural properties compared to cropland across a range of soil types and growing conditions. Specifically, perennial grasses decreased soil bulk density and increased water-stable aggregates, macroporosity, and saturated hydraulic conductivity (Blanco-Canqui, 2010). Similar improvements in soil physical properties under perennial grasses have been observed in grass-cropland contrasts not involving bioenergy production (Karlen et al., 1999; Franzluebbers et al., 2014).

Changes to soil physical properties under perennial grasses can confer significant benefits to agroecosystem function, such as decreased erosion risk, increased resistance to soil compaction, and improved air and water transfer in soil (Katsvairo et al., 2007; Blanco-Canqui, 2010). Such changes can alter hydrologic outcomes at larger spatial scales, as shown by studies evaluating perennial grass buffer strip effects on runoff and erosion (Gilley et al., 2000; Dabney et al., 2006). Grass buffer strips slow the movement of runoff and increase the potential for temporary ponding, thereby promoting sediment deposition and water infiltration into soil with improved structure (Dabney, 1998; Blanco-Canqui, 2010; Blanco-Canqui et al., 2014). As a result, perennial grass strips, if of sufficient width, can serve to significantly reduce water runoff and soil loss from cropland. Moreover, grass strips have been found effective at removing available N and P in runoff (Eghball et al., 2000; Mann and Tolbert, 2000; Blanco-Canqui et al., 2004), thereby enhancing water purification and nutrient cycling services on agricultural land.

6.3.1.2 Benefits to soil chemical properties

Accrual of soil organic carbon (SOC) under perennial grasses has been documented in numerous studies throughout the world (Anderson-Teixeira et al., 2009; Blanco-Canqui, 2010; Franzluebbers et al., 2014), reflecting contributions to climate regulation and nutrient cycling. Perennial grasses have been found to contribute to SOC accumulation in both near-surface (0–10 cm) and subsurface (30–90 cm) soil depths (Anderson-Teixeira et al., 2009; Culman et al., 2010; Schmer et al., 2011). Soil organic C accrual at subsoil depths by perennial grasses is particularly important in the context of permanence of climate regulation, as C stored below 30 cm is less susceptible to mineralization and loss by soil disturbance (Schmer et al., 2015).

In contrast to no-tillage crop production systems, SOC accrual can be 10-fold greater under warm-season perennial grasses [e.g., switchgrass (*Panicum virgatum* L.)] grown for bioenergy production (Liebig et al., 2008, 2009). Even under short-term perennial grass phases (<4 years), SOC can increase (Bremer et al., 2002; Ghosh et al., 2009) or be maintained (Janzen, 1987) within annual crop production systems. In an evaluation of 2-year perennial grass and mixed grass/legume leys, Persson et al. (2008) found replacing annual crops with leys in the crop rotation reduced losses of SOC without decreasing annual crop yields. Intermediately labile soil C pools are also enhanced by perennial grasses (Franzluebbers, 2012). In the southern Great Plains, Potter and Derner (2006) observed greater particulate organic matter in restored grasslands than cropped soils in the 0–5 cm depth. Bremer et al. (2002) found much of the SOC gain under perennial grasses was accounted for in easily mineralizable and light fraction C pools, implying that SOC gain would be readily lost upon conversion to annual cropping. To this point, Dupont et al. (2010) observed conversion of perennial vegetation to annual crops reduced root biomass by 43%, and was associated with significantly lower easily oxidizable soil C under annually cropped no-tillage management

3 years after conversion compared to perennial vegetation. Accordingly, conversion of perennial phases to annual cropping can have dramatic effects on climate regulation services derived from soil, and underscore the importance of identifying appropriate management practices to mitigate SOC loss during the transition to annual cropping.

Both perennial grasses and legumes have the ability to modulate hydrological attributes within agricultural landscapes, thereby providing a means to reduce the potential for salinity. Perennial pastures have been observed to reduce soil water content and decrease the amount of deep drainage compared to annual pastures owing to greater maximum rooting depth in the former (White et al., 2000; Ward et al., 2002). This ability for greater water uptake by perennial pastures limits salt accumulation in near-surface soil depths by capillary rise. Effective control of salinity with perennials in crop production systems, however, requires farmer adoption at large spatial scales to be effective (Roberts et al., 2009).

Similar mechanisms that contribute to dryland salinity control with perennial grasses apply to potential reductions in nutrient loss. Perennial grasses are effective at assimilating soil nitrate before it leaches below the root zone, resulting in much smaller levels of nitrate in the soil profile and concomitant decreases in nitrate lost from agricultural land compared to annual crops (Randall and Mulla, 2001; Karlen et al., 1999). Reduced nitrate loss with perennials may also mitigate soil acidification. In southern Australia, reduced soil nitrate loss from perennial pastures compared to annual pastures suggested soil acidification under perennials was reduced by 1 kmol H^+ ha^{-1} yr^{-1} (White et al., 2000).

6.3.1.3 Benefits to soil biological properties

Among soil attributes evaluated under perennial grasses included in cropping systems, soil biological properties are among the least studied. In general, increased root biomass and rhizodeposits from perennial grasses, coupled with attenuated fluctuations in soil temperature and water content, serve to create an environment suitable for the growth and activity of soil biota (Blanco-Canqui, 2010). Microbial biomass has been observed to be 10–64% greater under perennial grasses compared to cropped or fallow sites (Karlen et al., 1999; Acosta-Martinez et al., 2004; Al-Kaisi and Grote, 2007; Ghosh et al., 2009), implying a greater inherent capacity to cycle nutrients in the former. Similarly, fatty acid methyl ester profiles were found to be two times greater under bermudagrass (*Cynodon dactylon* L.) compared to foxtail millet [*Setaria italica* (L.) P. Beauv.]/cotton (*Gossypium hirsutum* L.) and continuous cotton cropping systems under semiarid conditions (Davinic et al., 2013). Moreover, among all vegetation types evaluated by Davinic et al. (2013), bermudagrass tended to have the greatest levels of AMF and microbial biomass. Enzyme activities have also been observed to be greater under perennial grasses than annual crops, with glucosidase (C cycling) and alkaline phosphatase (P cycling) levels greater under a perennial warm-season grass pasture compared to continuous cotton (Acosta-Martinez et al., 2010).

Among soil macrofauna, earthworm population densities were found to be greater in a sod rotation including bahiagrass (*Paspalum notatum* Fluggé) compared with a traditional peanut (*Arachis hypogaea* L.)/cotton cropping system (Katsvairo et al., 2007). In the same study, water infiltration was positively correlated to earthworm population densities, thereby linking the role of perennial grasses and earthworms to affect water regulation (Blouin et al., 2013). Earthworm biomass has been found to be greater under grass-white clover mixtures compared to grass only swards, suggesting the quality of litter (C:N ratio of root biomass) as an important variable regulating earthworm abundance (Van Eekeren et al., 2009).

Collectively, perennial grasses alter soil properties to facilitate decreases in erosion, nutrient loss, salinization, and acidification, while concurrently enhancing soil biological habitat and greenhouse gas mitigation potential. Such outcomes serve to improve production efficiencies, buffer climate-induced stressors, and increase agroecosystem resilience (Lal et al., 2011).

6.3.2 Legumes

Perennial legumes have the capacity to fix N_2 (Russelle and Birr, 2004), increase nutrient retention, reduce soil erosion, improve soil health (Congreves et al., 2015), and provide many other ecosystem services.

6.3.2.1 Alfalfa

Alfalfa (*Medicago sativa* L.) has a high capacity for symbiotic N_2 fixation (Russelle, 2008) and water use, is deeply rooted (Jiang et al., 2015), enhances soil physical properties and organic matter levels (Gregorich et al., 2001), and reduces loss of N. Alfalfa N_2 fixation ranges from 20 to 200 kg N ha^{-1} annually in the Mississippi River Basin in the United States (Russelle and Birr, 2004). Alfalfa increases the amount of soil N each year of production and in only two growing seasons may provide sufficient N for the following crop (Kelner et al., 1997), contributing all or most of the N required by a subsequent corn crop (Lory et al., 1995; Lawrence et al., 2008) and partially provide N requirements for a second-year corn or other crops (Yost et al., 2015). Alfalfa provides most of the N required by wheat (Entz et al., 2002), and canola (Mahli et al., 2007) and can substantially lower N fertilization and associated economic cost, and also provide energy savings from reduced on-farm fuel usage and fertilizer manufacture. Monocultures of alfalfa have been found to return 134 kg N ha^{-1} in the first year of establishment with the N uptake in corn following alfalfa being 115 kg N ha^{-1} (N'Dayegamiye et al., 2015). In addition, yields of succeeding crops can increase even when N supply is sufficient (Entz et al., 2002). This non-N effect is likely due to the increased soil porosity, infiltration, recycling of other nutrients, increased organic matter, and enhanced soil biodiversity.

6.3.2.1.1 Benefits to soil physical properties

Alfalfa increases soil infiltration rate, saturated hydraulic conductivity, macroporosity (Rasse and Smucker, 1998; Rasse et al., 2000), aggregate structure, and stability (N'Dayegamiye et al., 2015), and reduces soil compaction (Jiang et al., 2015). Macropore flow has been found to be more dominant and soil water distribution more uniform in soil with a 3-year alfalfa stand than in soil with corn on conventional tillage (Jiang et al., 2015). Bulk density increased with depth in surface soils (0–50 cm) under conventional corn, while bulk density remained similar between the surface and 50-cm depth under alfalfa (Jiang et al., 2015), an outcome possibly associated with the macropores left by decaying alfalfa roots. Alfalfa in rotation influences root distribution of corn roots the following season, where the proportion of corn roots recolonizing root-induced alfalfa macropores was 41%, and recolonization of corn-induced macropores by subsequent corn was 18% (Rasse and Smucker, 1998).

Alfalfa also improves soil water supply by trapping snow (Leep et al., 2001), and reduces development of soil salinity (Miller et al., 1981), a major problem in semiarid areas of the world. Although alfalfa can fix large amounts of N_2 symbiotically, it also absorbs N from the soil and reduces nitrate leaching (Entz et al., 2001; Syswerda et al., 2012). Thus, alfalfa can capture subsoil nitrate (Russelle et al., 2001; Ferchaud and Mary, 2016) and can utilize N from in-season manure applications (Ceotto and Spallacci, 2006). Nitrate leaching from conventional corn-soybean rotations is much higher than in alfalfa (Syswerda and Robertson, 2014). Nitrate concentrations surrounding the root zone of alfalfa typically are very low (Rasse et al., 1999). Thus, growing alfalfa is an effective way to reduce nitrate leaching (Entz et al., 2001) and recycle deep soil profile N.

6.3.2.1.2 Benefits to soil chemical properties

Soil C levels are greater for rotations containing 3 or more years of alfalfa than with conventional and no-tillage corn-soybean-wheat rotation (Syswerda and Robertson, 2014) or 2 year successions (corn-wheat and sugarbeet (*Beta vulgaris* L.)-wheat), continuous corn, and continuous wheat (Triberti et al., 2016). Soil C:N ratios were lowest for alfalfa when compared to annual crops (Syswerda and Robertson, 2014; N'Dayegamiye et al., 2015), with corresponding increases in soil alkaline phosphatase and dehydrogenase activities (N'Dayegamiye et al., 2015). Including perennial legumes and reduced tillage in a crop rotation increases soil organic matter. A rotation including 3 years of alfalfa followed by corn had higher soil particulate organic C (POM-C) than that of corn in monoculture (Cates et al., 2016).

P eroded from agricultural fields is contributing to eutrophication of lakes and rivers. Alfalfa in rotation with corn and oat fertilized every year with 67 P_2O_5 kg ha^{-1} had less P losses through tile drainage than continuous Kentucky bluegrass (*Poa annua* L.) or corn (Zhang et al., 2015). However, cover crops including perennial legumes such as alfalfa and red clover (*Trifolium pratense* L.) could contribute to significant amounts of inorganic P surface runoff losses. Plant biomass frozen and then thawed can release significant amount of inorganic soluble P, and this seems to be higher in more easily degradable annual cover crops with lower C:N ratios (Riddle and Bergstrom, 2013; Liu et al., 2014).

6.3.2.1.3 Additional ecosystem services

Other ecosystem services offered by alfalfa in the rotation include reduced greenhouse gas emissions, increased plant species richness (Syswerda and Robertson, 2014), reduced pest populations (Chen et al., 2006), and increased soil biodiversity and abundance (Crotty et al., 2015). Total soil microbial biomass was 12% greater in soils under alfalfa compared with soils under orchardgrass (*Dactylis glomerata* L.) production (Min et al., 2003). Alfalfa reduced soybean cyst nematode (SCN) (*Heterodera glycines* Ichinohe) populations by 55% in studies conducted in Minnesota where SCN populations were high (Chen et al., 2006).

6.3.3 Perennial Clovers

Red clover can fix N and match N uptake by the crop to reduce nitrate leaching through the soil profile (Vyn et al., 2000). Comparing a soybean-winter wheat-corn rotation with and without cover crop, with red clover frost-seeded into winter wheat in March and cereal rye seeded after corn, resulted in the increase of 8 out of 11 ecosystem services compared with the same rotation without cover crop. The supporting and regulating services included biomass production, N supply, soil C storage, nitrate retention, erosion control, weed suppression, AMF colonization, and beneficial insect conservation. The two services that were not different than the no-cover crop system were insect pest suppression and N_2O reduction (Fig. 6.1) (Schipanski et al., 2014).

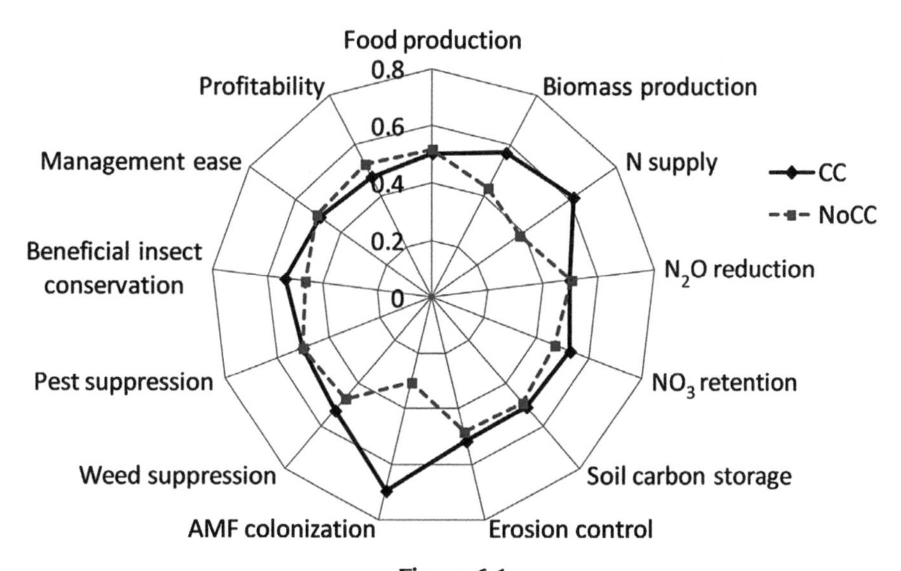

Figure 6.1
Normalized values for 11 ecosystem services and two economic metrics averaged across the 3-year rotation of cropping systems with (CC) and without (NoCC) cover crops (Schipanski et al., 2014).

Addition of red clover and cereal rye cover crops increased many supporting and regulating services including biomass production, N supply, soil C, nitrate retention, erosion control, weed suppression, mycorrhizal colonization, insect biodiversity (Schipanski et al., 2014), and SCN suppression (Chen et al., 2006). An overall greater abundance of earthworms was observed in soils under white clover (*Trifolium repens* L.) compared with red clover, perennial ryegrass (*Lolium perenne* L.), and chicory (*Chicorium intybus* L.) (Crotty et al., 2015).

Corn-red clover systems produce the highest and most stable yields of silage corn (Grabber and Jokel, 2013), with N credits to corn estimated at $168\,kg\,N\,ha^{-1}$ (Schipanski et al., 2014). This is probably due to the rapid N mineralization of fall-terminated red clover (Grabber et al., 2014) and the increase in the N use efficiency in the soil (Gaudin et al., 2015). Additionally, red clover interseeded into corn can have an N credit of $30–48\,kg\,N\,ha^{-1}$ (Gentry et al., 2013). Red clover reduces nitrate leaching even when manure is applied (Grabber et al., 2014; Thilkarathna et al., 2015). Reduced quantities of residual nitrate at a 120-cm soil profile depth on manured red clover stands were reported (Grabber et al., 2014). Lower nitrate leaching losses were also observed with white clover pastures compared with pastures without legumes (Malcolm et al., 2014).

Kura clover (*Trifolium ambiguum* M. Bieb.) is a rhizomatous winter hardy perennial legume, with high persistence. Kura clover mobilizes C to roots and rhizomes during establishment, resulting in limited nodulation and N_2 fixation in the seeding year (Laberge et al., 2005). Once established, kura clover can last for many years and has been proposed as a permanent living mulch. Kura clover has the potential to reduce corn N fertilization by increasing nitrate retention in the soil profile, while maintaining corn productivity (Sawyer et al., 2010), especially for silage corn (Ochsner et al., 2010). If kura clover is not suppressed, it will compete with corn, resulting in reduced corn yield (Sawyer et al., 2010). Ochsner et al. (2010) and Qi et al. (2011a) observed reduced nitrate concentrations in the soil solution, but not in drainage effluent.

6.4 Conclusions

Annual and perennial cover crops in rotation provide environmental services and benefit to soil health. Environmentally, cover crops can fill gaps in the rotation where the field may otherwise be fallow, reducing soil erosion. Additionally, using cover crops as a nutrient capture tool can reduce nutrient losses and loading in water resources. From a soil health perspective, extending the growing season with live plant cover has the potential to improve soil physical, chemical, and biological properties.

Grasses and brassicas improve water infiltration into soils by facilitating aggregation and creating biopores and preferential flow for water movement from the surface into the soil profile. Both grasses and brassicas can penetrate compacted layers to improve rooting depth of subsequent cash crops. Annual and perennial legumes make nutrient cycling more efficient through additions of N-rich organic material to the soil, potentially reducing inorganic fertilizer inputs required to support cash crops. Biologically, cover crop associations with mycorrhizal fungi provide benefits for nutrient cycling and uptake of both the cover crop and also the following cash crop through soil priming. Root and organic matter additions from cover crops supporting microbial communities benefit soil health; however, this is an area that requires continuous research.

The scientific and agricultural communities continue to recognize the importance of including cover crops within cash crop rotations. Based on this review of the literature, motivations to using cover crops for improving soil health appear to be justified.

References

Acosta-Martínez, V., Upchurch, D.R., Schubert, A.M., Porter, D., Wheeler, T., 2004. Early impacts of cotton and peanut cropping systems on selected soil chemical, physical, microbiological and biochemical properties. Biol. Fertil. Soils 40, 44–54.

Acosta Martinez, V., Bell, C., Morris, B., Zak, J., Allen, V.G., 2010. Long-term soil microbial community and enzyme activity responses to an integrated cropping-livestock system in a semi-arid region. Agric. Ecosys. Environ. 137 (3–4), 231–240.

Al-Kaisi, M.M., Grote, J.B., 2007. Cropping systems effects on improving soil carbon stocks of exposed subsoil. Soil Sci. Soc. Am. J. 71, 1381–1388.

Anderson-Teixeira, K.J., Davis, S.C., Masters, M.D., Delucia, E.H., 2009. Changes in soil organic carbon under biofuel crops. Glob. Change Biol. Bioenerg. 1, 75–96.

Aronsson, H., Hansen, E.M., Thomsen, I.K., Liu, J., Ogaard, A.F., Kankanen, H., et al., 2016. The ability of cover crops to reduce nitrogen and phosphorous losses from arable land in southern Scandinavia and Finland. J. Soil Water Conserv. 71 (1), 41–55.

Bair, K.E., Stevens, R.G., Davenport, J.R., 2008. Release of available nitrogen after incorporation of a legume cover crop in Concord grape. Hort. Sci. 43, 875–880.

Biederbeck, V.O., Zentner, R.P., Campbell, C.A., 2005. Soil microbial populations and activities as influenced by legume green fallow in a semiarid climate. Soil Biol. Biochem 37, 1775–1784.

Blanco-Canqui, H., 2010. Energy crops and their implications on soil and environment. Agron. J. 102, 403–419.

Blanco-Canqui, H., Gantzer, C.J., Anderson, S.H., Alberts, E.E., Thompson, A.L., 2004. Grass barriers and vegetative filter strip effectiveness in reducing runoff, sediment, and nutrient loss. Soil Sci. Soc. Am. J. 68, 1670–1678.

Blanco-Canqui, H., Gilley, J.E., Eisenhauer, D.E., Jasa, P.J., Boldt, A., 2014. Soil carbon accumulation under switchgrass barriers. Agron. J. 106, 2185–2192.

Blouin, M., Hodson, M.E., Delgado, E.A., Baker, G., Brussard, L., Butt, K.R., et al., 2013. A review of earthworm impact on soil function and ecosystem services. Eur. J. Soil Sci. 64, 161–182.

Bollero, G.A., Bullock, D.G., 1994. Cover cropping systems for the central corn belt. J. Prod. Agric. 7, 55–58.

Boswell, E.P., Koide, R.T., Shumway, D.L., Addy, H.D., 1998. Winter wheat cover cropping, VA mycorrhizal fungi and maize growth and yield. Agric. Ecosys. Environ. 67, 55–65.

Bradi-Dohrn, F.M., Dick, R.P., Hess, M., Kauffman, S.M.M., Hemphill, D.D., Selker, J.S., 1997. Nitrate leaching under cereal rye cover crop. J. Environ. Qual. 26, 181–188.

Bremer, E., Janzen, H.H., McKenzie, R.H., 2002. Short-term impact of fallow frequency and perennial grass on soil organic carbon in a Brown Chernozem in southern Alberta. Can. J. Soil Sci. 82, 481–488.

Brown, P.D., Morra, M.J., 1995. Glucosinolate-containing plant tissues as bioherbidies. J. Agric. Food Chem. 43, 3070–3074.

Brown, P.D., Morra, M.J., 1997. Control of soil-borne plant pests using glucosinolate-containing plants. Adv. Agron. 61, 167–231.

Brundette, M.C., 2002. Tansley review no. 134. Coevolution of roots and mycorrhizas of land plants. New Phytol. 154, 275–304.

Cates, A.M., Ruark, M.D., Hedtckec, J.L., Posnera, J.L., 2016. Long-term tillage, rotation and perennialization effects on particulate and aggregate soil organic matter. Soil Till. Res. 155, 371–380.

Ceotto, E., Spallacci, P., 2006. Pig slurry applications to alfalfa: productivity, solar radiation utilization, N and P removal. Field Crops Res. 95, 135–155.

Chen, G., Weil, R.R., 2010. Penetration of cover crop roots through compacted soils. Plant Soil 331, 31–43.

Chen, G., Weil, R.R., 2011. Root growth and yield of maize as affected by soil compaction and cover crops. Soil Till. Res. 117, 17–27.

Chen, G., Weil, R.R., Hill, R., 2014. Effects of compaction and cover crops on soil least limiting water range and air permeability. Soil Till. Res. 136, 61–69.

Chen, S.Y., Wyse, D.L., Johnson, G.A., Porter, P.M., Stetina, S.R., Miller, D.R., et al., 2006. Effect of cover crops alfalfa, red clover, and perennial ryegrass on soybean cyst nematode population and soybean and corn yields in Minnesota. Crop Sci. 46, 1890–1897.

Chew, F.S., 1988. Biological effects of glucosinolates. ACS Symp. Ser. 380, 155–181.

Coale, F.J., Costa, J.M., Bollero, G.A., Schlosnagle, S.P., 2001. Small grain winter cover crops for conservation of residual soil nitrogen in the mid-Atlantic Coastal Plain. Am. J. Altern. Agric. 16, 66–72.

Congreves, K.A., Hayes, A., Verhallen, E.A., Van Eerd, L.L., 2015. Long-term impact of tillage and crop rotation on soil health at four temperate agroecosystems. Soil Till. Res. 152, 17–28.

Cresswell, H.P., Kirkegaard, J.A., 1995. Subsoil amelioration by plant roots – the process and the evidence. Aust. J. Soil Res. 33, 221–239.

Crotty, F.V., Fychan, R., Scullion, J., Sanderson, R., Marley, C.L., 2015. Assessing the impact of agricultural forage crops on soil biodiversity and abundance. Soil Biol. Biochem. 91, 119–196.

Culman, S.W., DuPont, S.T., Glover, J.D., Buckley, D.H., Fick, G.W., Ferris, H., et al., 2010. Long-term impacts of high-input annual cropping and unfertilized perennial grass production on soil properties and belowground food webs in Kansas, USA. Agric. Ecosys. Environ. 137, 13–24.

Dabney, S.M., 1998. Cover crop impacts on watershed hydrology. J. Soil Water Conserv. 53, 207–213.

Dabney, S.M., Delgado, J.A., Reeves, D.W., 2001. Using winter cover crops to improve soil and water quality. Comm. Soil Sci. Plant Anal. 32, 1221–1250.

Dabney, S.M., Moore, M.T., Locke, M.A., 2006. Integrated management of in-field, edge-of-field, and after-field buffers. J. Am. Water Res. Assoc. 42 (1), 15–24.

Davinic, M., Moore-Kucera, J., Acosta-Martínez, V., Zak, J., Allen, V., 2013. Soil fungal distribution and functionality as affected by grazing and vegetation components of integrated crop-livestock agroecosystems. Appl. Soil Ecol. 66, 61–70.

Dean, J.E., Weil, R.R., 2009. *Brassica* cover crops for N retention in the Mid-Atlantic coastal plain. J. Environ. Qual. 38, 520–528.

De Baets, S., Poesen, J., Meersmans, J., Serlet, L., 2011. Cover crops and their erosion-reducing effects during concentrated flow erosion. Catena 85, 237–244.

DuPont, S.T., Culman, S.W., Ferris, H., Buckley, D.H., Glover, J.D., 2010. No-tillage conversion of harvested perennial grassland to annual cropland reduces root biomass, decreases active carbon stocks, and impacts soil biota. Agric. Ecosys. Environ. 137, 25–32.

Ebelhar, S.A., Frye, W.W., Blevins, R.L., 1984. Nitrogen from legume cover crops for no-tillage corn. Agron. J. 76, 51–55.

Eghball, B., Gilley, J.E., Kramer, L.A., Moorman, T.B., 2000. Narrow grass hedge effects on phosphorus and nitrogen in runoff following manure and fertilizer application. J. Soil Water Cons. 55 (2), 172–176.

Entz, M.H., Bullied, W.J., Forster, D.A., Gulden, R., Vessey, J.K., 2001. Extraction of subsoil nitrogen by alfalfa, alfalfa-wheat, and perennial grass systems. Agron. J. 93, 495–503.

Entz, M.H., Baron, V.S., Carr, P.M., Meyer, D.W., Smith Jr., S.R., McCaughey, W.P., 2002. Potential of forages to diversify cropping systems in the Northern Great Plains. Agron. J. 94, 240–250.

Fae, G.S., Sule, R.M., Barker, D.J., Dick, R.P., Eastridge, M.L., Lorenz, N., 2009. Integrating winter annual forages into a no-till corn silage systems. Agron. J. 101, 1286–1296.

Ferchaud, F., Mary, B., 2016. Drainage and nitrate leaching assessed during 7 years under perennial and annual bioenergy crops. BioEnergy Res., 1–15.

Franzluebbers, A.J., 2012. Grass roots of soil carbon sequestration. Carbon Manage 3, 9–11.

Franzluebbers, A.J., 2015. Farming strategies to fuel bioenergy demands and facilitate essential soil services. Geoderma 259–260, 251–258.

Franzluebbers, A.J., Sawchik, J., Taboada, M.A., 2014. Agronomic and environmental impacts of pasture-crop rotations in temperate North and South America. Agric. Ecosys. Environ. 190, 18–26.

Gaudin, A.C.M., Janovicek, J., Deen, B., Hooker, D.C., 2015. Wheat improves nitrogen use efficiency of maize and soybean-based cropping systems. Agric. Ecosys. Environ. 210, 1–10.

Gentry, L.E., Snapp, S.S., Price, R.F., Gentry, L.F., 2013. Apparent red clover nitrogen credit to corn: evaluating cover crop introduction. Agron. J. 105, 1658–1664.

Ghosh, P.K., Saha, R., Gupta, J.J., Ramesh, T., Das, A., Lama, T.D., et al., 2009. Long-term effect of pastures on soil quality in acid soil of North-East India. Aust. J. Soil Res. 47, 372–379.

Gilley, J.E., Eghball, B., Kramer, L.A., Moorman, T.B., 2000. Narrow grass hedge effects on runoff and soil loss. J Soil Water Conserv. 55, 190–196.

Grabber, J.H., Jokela, W.E., 2013. Off-season groundcover and runoff characteristics of perennial clover and annual grass companion crops for no-till corn fertilized with manure. J. Soil Water Conserv. 68, 411–418.

Grabber, J.H., Jokela, W.E., Lauer, J.G., 2014. Soil nitrogen and forage yields of corn grown with clover or grass companion crops and manure. Agron. J. 106, 952–961.

Gregorich, E.G., Drury, C.F., Baldock, J.A., 2001. Changes in soil carbon under long-term maize in monoculture and legume-based rotation. Can. J. Soil Sci. 81, 21–31.

Gruver, J., Weil, R.R., White, C., Lawley, Y., 2012. Radishes – a new cover crop for organic farming systems. e-Organic e-Extension. Available at <http://www.extension.org/pages/64400/radishes-a-new-cover-crop-for-organic-farming-systems> (verified 6.03.16.).

Haramoto, E.R., Gallandt, E.R., 2004. *Brassica* cover cropping for weed management: a review. Renewable agric. Food Syst. 19, 187–198.

Hargrove, W.L., 1986. Winter legumes as nitrogen sources for no-tillage grain sorghum. Agron. J. 78, 70–74.

Haynes, R.J., Beare, M.H., 1997. Influence of six crop species on aggregate stability and some labile organic matter fractions. Soil Biol. Biochem 29, 1647–1653.

Hedley, M.J., Nye, P.H., White, R.E., 1982. Plant-induced changes in the rhizosphere of rape (*Brassica napus* var. Emerald) seedlings: II. Origin of the pH change. New Phytol. 91, 31–44.

Hermawan, B., Bomke, A.A., 1997. Effects of winter cover crops and successive spring tillage on soil aggregation. Soil Till. Res. 44, 109–120.

Hoffland, E., Findenegg, G.R., Nelemans, J.A., 1989. Solubilization of rock phosphate by rape: II. Local root exudation of organic acids as a response to P-starvation. Plant Soil 113, 161–165.

Jani, A.D., Grossman, J.M., Smyth, T.J., Hu, S., 2015. Influence of soil inorganic nitrogen and root diameter size on legume cover crop root decomposition and nitrogen release. Plant Soil 393, 57–68.

Janzen, H.H., 1987. Soil organic matter characteristics after long-term cropping to various spring wheat rotations. Can. J. Soil Sci. 67, 845–856.

Jastrow, J.D., Miller, R.M., 1998. Soil aggregate stabilization and carbon sequestration: feedbacks through organiomineral associations. In: Lal, R., Kimble, J.M., Follett, R.F., Stewart, B.A. (Eds.), Soil Processes and the Carbon Cycle. CRC Press, Boca Raton, pp. 207–223.

Jiang, X., Liu, X., Wang, E., Li, X.G., Sun, R., Shi, W., 2015. Effects of tillage pan on soil water distribution in alfalfa-corn crop rotation systems using a dye tracer and geostatistical methods. Soil Till. Res. 150, 68–77.

Kabir, Z., Koide, R.T., 2002. Effect of autumn and winter mycorrhizal cover crops on soil properties, nutrient uptake and yield of sweet corn in Pennsylvania, USA. Plant Soil 238, 205–215.

Kamh, M., Horst, W.J., Amer, F., Mostafa, H., Maier, P., 1999. Mobilization of soil and fertilizer phosphate by cover crops. Plant Soil 211, 19–27.

Kamh, M., Abdou, M., Chude, V., Wiesler, F., Horst, W.J., 2002. Mobilization of phosphorus contributes to positive rotational effects of leguminous cover crops on maize grown on soils from northern Nigeria. J. Plant Nutr. Soil Sci 165, 566–572.

Karasawa, T., Takebe, M., 2012. Temporal or spatial arrangements of cover crops to promote arbuscular mycorrhizal colonization and P uptake of upland crops grown after non-mycorrhizal crops. Plant Soil 353, 355–366.

Karasawa, T., Takahashi, S., 2015. Introduction of various cover crop species to improve soil biological P parameters and P uptake of the following crops. Nutr. Cycl. Agroecosyst. 103, 15–28.

Karlen, D.L., Rosek, M.J., Gardner, J.C., Allan, D.L., Alms, M.J., Bezdicek, D.F., et al., 1999. Conservation reserve program effects on soil quality indicators. J. Soil Water Conserv. 54, 439–444.

Kaspar, T.C., Radke, J.K., Laflen, J.M., 2001. Small grain cover crops and wheel traffic effects on infiltration, runoff and erosion. J. Soil Water Conserv. 56, 160–164.

Katsvairo, T.W., Wright, D.L., Marois, J.J., Hartzog, D.L., Balkcom, K.B., Wiatrak, P.J., et al., 2007. Cotton roots, earthworms and infiltration characteristics in peanut/cotton cropping systems. Agron. J. 99, 390–398.

Kelner, D.J., Vessey, J.K., Entz, M.H., 1997. The nitrogen dynamics of 1-, 2-, and 3-year stands of alfalfa in a cropping system. Agric. Ecosys. Environ. 64, 1–10.

Komatsuzaki, M., Wagger, M.G., 2015. Nitrogen recovery by cover crops in relation to time of planting and growth termination. J. Soil Water Conserv. 70 (6), 385–398.

Kramberger, B., Gselman, A., Kristl, J., Lesnik, M., Sustar, V., Mursec, M., et al., 2014. Winter cover crop: the effects of grass–clover mixture proportion and biomass management on maize and the apparent residual N in the soil. Eur. J. Agron. 55, 63–71.

Kuhlmann, H., Barraclough, P.B., Weir, A.H., 1989. Utilization of mineral nitrogen in the subsoil by winter wheat. Z Pflanzenernaehr Bodenkd 152, 291–295.

Laberge, G., Seguin, P., Peterson, P.R., Sheaffer, C.C., Ehlke, N.J., 2005. Forage yield and species composition in years following kura clover sod-seeding into grass swards. Agron. J. 97, 1352–1360.

Lal, R., Delgado, J.A., Groffman, P.M., Millar, N., Dell, C., Rotz, A., 2011. Management to mitigate and adapt to climate change. J. Soil Water Conserv. 66 (4), 276–285.

Lamande, M., Hallaire, V., Curmi, P., Peres, G., Cluzeau, D., 2003. Changes in pore morphology, infiltration and earthworm community in a loam soil under different agricultural managements. Catena 54, 637–649.

Lawley, Y., Weil, R., Teasdale, J., 2012. The mechanism for weed suppression by a forage radish cover crop. Agron. J. 104 (2), 205–214.

Lawrence, J.R., Ketterings, Q.M., Cherney, J.H., 2008. Effect of nitrogen application on yield and quality of first year corn. Agron. J. 100, 73–79.

Leep, R.H., Andresen, J.A., Jeranyama, P., 2001. Fall dormancy and snow depth effects on winterkill of alfalfa. Agron. J. 93, 1142–1148.

Lehman, R.M., Taheri, W.I., Osborne, S.L., Buyer, J.S., Douds Jr., D.D., 2012. Fall cover cropping can increase arbuscular mycorrhizae in soils supporting intensive agricultural production. Appl. Soil Ecol. 61, 300–304.

Li, L., Li, S.M., Sun, J.H., Zhou, L.L., Bao, X.G., Zhang, H.G., et al., 2007. Diversity enhances agricultural productivity via rhizosphere phosphorus facilitation on phosphorus-deficient soils. PNAS 104, 11192–11196.

Liang, S., Grossman, J., Shi, W., 2014. Soil microbial responses to winter legume cover crop management during organic transition. Eur. J. Soil Biol. 65, 15–22.

Liebig, M.A., Schmer, M.R., Vogel, K.P., Mitchell, R., 2008. Soil carbon storage by switchgrass grown for bioenergy. Bioenergy Res. 1, 215–222.

Liebig, M.A., Mikha, M.M., Potter, K.N., 2009. Management of dryland cropping systems in the U.S. Great Plains: effects on soil organic carbon. SSSA Special Publ. In: Lal, R., Follett, R.F. (Eds.), Soil carbon sequestration and the greenhouse effect, second ed. SSSA Spec. Publ. 57. ASA-CSSA-SSSA, Madison, WI, pp. 97–113.

Liu, A., Ma, L., Bomke, A.A., 2005. Effects of cover crops on soil aggregate stability, total organic carbon, and polysaccharides. Soil Sci. Soc. Am. J. 69, 2041–2048.

Liu, J., Ulen, B., Bergkvist, G., 2014. Freezing-thawing effects on phosphorus leaching from catch crops. Nutr. Cycl. Agroecosyst. 99, 17–30.

Liu, J., Bergkvist, G., Ulen, B., 2015. Biomass production and phosphorous retention by catch crops on clayey soils in southern and central Sweden. Field Crops Res. 171, 130–137.

Lory, J.A., Randall, G.W., Russelle, M.P., 1995. Crop sequence effects on response of corn and soil inorganic nitrogen to fertilizer and manure nitrogen. Agron. J. 87, 876–883.

Lounsbury, N.P., Weil, R.R., 2015. No-till seeded spinach after winterkilled cover crops in an organic production system. Renew. Agric. Food Syst. 30, 473–485.

Mahli, S.S., Johnston, A.M., Lopeky, H., Vera, C.L., Beckie, H.J., Bandara, P.M.S., 2007. Immediate Effects of Time and Method of Alfalfa Termination on Soil Mineral Nitrogen, Moisture, Weed Control, and Seed Yield, Quality, and Nitrogen Uptake. J. Plant Nutr. 30 (7), 1059–1081.

Malcolm, B.J., Cameron, K.C., Di, H.J., Edwards, G.R., Moir, J.L., 2014. The effect of four different pasture species compositions on nitrate leaching losses under high N loading. Soil Use Manag. 30, 58–68.

Maltais-Landry, G., 2015. Legumes have a greater effect on rhizosphere properties (pH, organic acids and enzyme activity) but a smaller impact on soil. Plant Soil 394, 139–154.

Mann, L., Tolbert, V., 2000. Soil sustainability in renewable biomass plantings. Ambio 29, 492–498.

Martin, J.H., Leonard, W.H., Stamp, D.L., 1976. Rye and Triticale. Principles of Field Crop Production, third ed. Macmillan Publishing Co., Inc., New York, NY.

Materechera, S.A., Alston, A.M., Kirby, J.M., Dexter, A.R., 1992. Influence of root diameter on the penetration of seminal roots into a compacted subsoil. Plant Soil 144, 297–303.

McVay, K., Radcliffe, D., Hargrove, W., 1989. Winter legume effects on soil properties and nitrogen fertilizer requirements. Soil Sci. Soc. Am. J. 53, 1856–1862.

Miller, M.R., Brown, P.L., Donovan, J.J., Bergatino, R.N., Sonderegger, J.L., Schmidt, F.A., 1981. Saline seep development and control in the North American Great Plains - Hydrogeological aspects. Agric. Water Manag. 4, 115–141.

Min, D.H., Islam, K.R., Vough, L.R., Weil, R.R., 2003. Dairy manure effects on soil quality properties and carbon sequestration in alfalfa-orchardgrass systems. Comm. Soil Sci. Plant Anal. 34 (5–6), 781–799.

Moore, E.B., Wiedenhoeft, M.H., Kaspar, T.C., Cambardella, C.A., 2014. Rye cover crop effects on soil quality in no-till corn silage-soybean cropping systems. Soil Sci. Soc. Am. J. 78, 968–976.

N'Dayegamiye, N., Whalen, J.K., Tremblay, G., Nyiraneza, J., Grenier, M., Drapeau, A., et al., 2015. The benefits of legume crops on corn and wheat yield, nitrogen nutrition, and soil properties improvement. Agron. J. 107 (5), 1653–1665.

Nuruzzaman, M., Lambers, H., Bolland, M.D.A., Veneklaas, E.J., 2006. Distribution of carboxylates and acid phosphatase and depletion of different phosphorus fractions in the rhizosphere of a cereal and three grain legumes. Plant Soil 281, 109–120.

Ochsner, T.E., Albrecht, K.A., Schumacher, T.W., Baker, J.M., Berkevich, R.J., 2010. Water balance and nitrate leaching under corn in kura clover living mulch. Agron. J. 102, 1169–1178.

Parr, M., Grossman, J.M., Chris Reberg-Horton, S., Brinton, C., Crozier, C., 2014. Roller-crimper termination for legume cover crops in North Carolina: impacts on nutrient availability to a succeeding corn crop. Comm. Soil Sci. Plant Anal. 45, 1106–1119.

Persson, T., Bergkvist, G., Katterer, T., 2008. Long-term effects of crop rotations with and without perennial leys on soil carbon stocks and grain yields of winter wheat. Nutr. Cycl. Agroecosys. 81, 193–202.

Potter, K.N., Derner, J.D., 2006. Soil carbon pools in Central Texas: prairies, restored grasslands, and croplands. J. Soil Water Conserv. 61, 124–128.

Qi, Z., Helmers, M.J., Christianson, R.D., Pederson, C., 2011a. Nitrate-nitrogen losses through subsurface drainage under various agricultural land covers. J. Environ. Qual. 40, 1578–1585.

Qi, Z., Helmers, M.J., Kaleita, A.L., 2011b. Soil water dynamics under various agricultural land covers on a subsurface drained field in north-central Iowa, USA. Agric. Water Manag. 98, 665–674.

Randall, G.W., Mulla, D.J., 2001. Nitrate nitrogen in surface waters as influenced by climatic conditions and agricultural practices. J. Environ. Qual. 30, 337–344.

Ranells, N.N., Wagger, M.G., 1996. Nitrogen release from grass and legume cover crop monocultures and bicultures. Agron. J. 88, 777–782.

Rasse, D.P., Smucker, A.J.M., 1998. Root recolonization of previous root channels in corn and alfalfa rotations. Plant Soil 204, 203–212.

Rasse, D.P., Smucker, A.J.M., Santos, D., 2000. Alfalfa root and shoot mulching effects on soil hydraulic properties and aggregation. Soil Sci. Soc. Am. J. 64, 725–731.

Rasse, D.P., Smucker, A.J.M., Schabenberger, O., 1999. Modifications of soil nitrogen pools in response to alfalfa root systems and shoot mulch. Agron. J. 91, 471–477.

Reicosky, D.C., Forcella, F., 1998. Cover crop and soil quality interactions in agroecosystems. J. Soil Water Conserv. 53, 224–229.

Riddle, M.U., Bergström, L., 2013. Phosphorus leaching from two soils with catch crops exposed to freeze-thaw cycle. Agron. J. 105, 803–811.

Rillig, M.C., 2004. Arbuscular mycorrhizae and terrestrial ecosystems processes. Ecol. Lett. 7, 740–754.

Ritter, W.F., Scarborough, R.W., Chirnside, A.E.M., 1998. Winter cover crops as best management practice for reducing nitrate leaching. J. Contam. Hydrol. 34, 1–15.

Roberts, A.M., Helmers, M.J., Fillery, I.R.P., 2009. The adoptability of perennial-based farming systems for hydrologic and salinity control in dryland farming systems in Australia and the United States of America. Crop Pasture Sci. 60, 83–99.

Russelle, M.P., 2008. Biological dinitrogen fixation in agriculture In: Schepers, J.S. Raun, W.R. (Eds.), Nitrogen in Agricultural Systems, vol. 69. Agron Monogr, Madison, WI, pp. 281–359. Am Soc Agron, Crop Sci Soc Am, and Soil Sci Soc Am.

Russelle, M.P., Birr, A.S., 2004. Large-scale assessment of symbiotic dinitrogen fixation by crops: soybean and alfalfa in the Mississippi River Basin. Agron. J. 96, 1754–1760.

Russelle, M.P., Lamb, J.F.S., Montgomery, B.R., Eisenheimer, D.W., Miller, B.S., Vance, C.P., 2001. Alfalfa rapidly remediates excess inorganic nitrogen at a fertilizer spill site. J. Environ. Qual. 30, 30–36.

Samarappuli, D., Johnson, B.L., Kandel, H.H., Berti, M.T., 2014. Biomass yield and nitrogen content of annual energy/forage crops preceded by cover crops. Field Crops Res. 167, 31–39.

Sanderson, M.A., Adler, P.R., 2008. Perennial forages as second generation bioenergy crops. Int. J. Mol. Sci. 9 (5), 768–788.

Sanderson, M.A., Wedin, D., Tracy, B., 2009. Grassland: definition, origins, extent, and future. In: Wedin, W.F., Fales, S.L. (Eds.), Grassland: Quietness and Strength for a New American Agriculture. ASA-CSSA-SSSA, Madison, WI, pp. 57–74.

Sawyer, J.E., Pedersen, P., Barker, D.W., Ruiz-Diaz, D.A., Albrecht, K., 2010. Intercropping corn and kura clover: response to nitrogen fertilization. Agron. J. 102, 568–574.

Schipanski, M.E., Barbercheck, M., Douglas, M.R., Finney, D.M., Haider, K., Haye, J.P., et al., 2014. A framework for evaluating ecosystem services provided by cover crops in agro-ecosystems. Agric. Syst. 125, 12–22.

Schmer, M.R., Liebig, M.A., Vogel, K.P., Mitchell, R.B., 2011. Field-scale soil property changes under switchgrass managed for bioenergy. Glob. Change Biol. Bioenergy 3, 439–448.

Schmer, M.R., Jin, V.L., Wienhold, B.J., 2015. Sub-surface soil carbon changes affects biofuel greenhouse gas emissions. Biomass Bioenergy 81, 31–34.

Schmidt, O., Clements, R.O., Donaldson, G., 2003. Why do cereal-legume intercrops support large earthworm populations? Appl. Soil Ecol. 22, 181–190.

Sequiera, C.H., Alley, M.M., 2011. Soil organic matter fractions as indices of soil quality changes. Soil Sci. Soc. Am. J. 75, 1766–1773.

Shipley, P.R., Meisinger, J.J., Dekker, A.M., 1992. Conserving residual corn fertilizer nitrogen with winter cover crop. Agron. J. 84, 869–876.

Singer, J.W., Nusser, S.M., Alf, C.J., 2007. Are cover crops being used in the U.S. Corn Belt? J. Soil Water Conserv. 62, 353–358.

Staver, K.W., Brinsfield, R.B., 1990. Patterns of soil nitrate availability in corn production systems: Implications for reducing groundwater contamination. J. Soil Water Conserv. 45, 318–322.

Staver, K.W., Brinsfield, R.B., 1998. Using cereal grain winter cover crops to reduce groundwater contamination in the mid-Atlantic coastal plains. Soil Water Cons. 53 (3), 230–240.

Steele, M.K., Coale, F.J., Hill, R.L., 2012. Winter annual cover crop impacts on no-till soil physical properties and organic matter. Soil Sci. Soc. Am. J. 76, 2164–2173.

Strebel, O., Duynisveld, W.H.M., 1989. Nitrogen supply to cereals and sugar beet by mass flow and diffusion on a silty loam soil. Z Pflanzenernaehr Bodenkd 152, 135–141.

Strock, J., Porter, P., Russelle, M., 2004. Cover cropping to reduce nitrate loss through subsurface drainage in the northern UC Corn Belt. J. Environ. Qual. 33, 1010–1016.

Sustainable Agriculture Research and Education (SARE), 2012. Managing Cover Crops Profitably, third ed. SARE, Maryland.

Syswerda, S.P., Robertson, G.P., 2014. Ecosystem services along a management gradient in Michigan (USA) cropping systems. Agric. Ecosys. Environ. 189, 28–35.

Syswerda, S.P., Basso, B., Hamilton, S.K., Tausig, J.B., Robertson, G.P., 2012. Long-term nitrate loss along an agricultural intensity gradient in the Upper Midwest USA. Agric. Ecosys. Environ. 149, 10–19.

Thilakarathna, M.S., Serran, S., Lauzon, J., Janovicek, K., Deen, B., 2015. Management of manure nitrogen using cover crops. Agron. J. 107, 1295–1607.

Thorup-Kristensen, K., 2001. Are differences in root growth of nitrogen catch crops important for their ability to reduce soil nitrate-N content, and how can this be measured? Plant Soil 230, 185–195.

Thorup-Kristensen, K., Cortasa, M., Loges, R., 2009. Winter wheat roots grow twice as deep as spring wheat roots, is this important for N uptake and N leaching losses? Plant Soil 322, 101–114.

Tisdall, J.M., Oades, J.M., 1982. Organic matter and water stable aggregates in soils. J. Soil Sci. 33, 141–163.

Touchton, J.T., Gardner, W.A., Hargrove, W.L., Duncan, R.R., 1982. Reseeding crimson clover as a N source for no-tillage grain sorghum production. Agron. J. 74, 283–287.

Triberti, L., Nastri, A., Baldoni, G., 2016. Long-term effects of crop rotation, manure and mineral fertilization on carbon sequestration and soil fertility. Eur. J. Agron. 74, 47–55.

Van Eekeren, N., van Liere, D., de Vries, F., Rutgers, M., de Goede, R., Brussaard, L., 2009. A mixture of grass and clover combines the positive effects of both plant species on selected soil biota. Appl. Soil Ecol. 42, 254–263.

Villamil, M.B., Bollero, G.A., Darmody, R.G., Simmons, F.W., Bullock, D.G., 2006. No-till corn/soybean systems including winter cover crops: effects on soil properties. Soil Sci. Soc. Am. J. 70, 1936–1944.

Vyn, T.J., Faber, J.G., Janovicek, K.J., Beauchamp, E.G., 2000. Cover crop effects on nitrogen availability to corn following wheat. Agron. J. 92, 915–924.

Wagger, M.G., 1989. Time of desiccation effects on plant composition and subsequent nitrogen release from several winter annual cover crops. Agron. J. 81, 236–241.

Wagger, M.G., Cabrera, M.L., Ranells, N.N., 1998. Nitrogen and carbon cycling in relation to cover crop residue quality. J. Soil Water Conserv. 53, 214–218.

Walsh, J., Wuebbles, D., Hayhoe, K., Kossin, J., Kunkel, K., Stephens, G., et al., 2014. Our changing climate. In: Melillo, J.M., Richmond, T.C., Yohe, G.W. (Eds.), Climate Change Impacts in the United States: The Third National Climate Assessment. U.S. Global Change Research Program, pp. 19–67. http://dx.doi.org/10.7930/JOD798BC.

Ward, P.R., Dunin, F.X., Micin, S.F., 2002. Water use and root growth by annual and perennial pastures and subsequent crops in a phase rotation. Agric. Water Manage. 53, 83–97.

Weaver, J.E., Darland, R.W., 1949. Soil-root relationships of certain native grasses in various soil types. Ecol. Monogr. 19, 303–338.

Weil, R., Kremen, A., 2007. Thinking across and beyond disciplines to make cover crops pay. J. Sci. Food Agric. 87, 551–557.

Weil, R.R., White, C., Lawley, Y., 2009. Forage Radish: New Multi-Purpose Cover Crop for the Mid-Atlantic. Maryland Cooperative Extension, College Park, MD.

White, C.M., Weil, R.R., 2010. Forage radish and cereal rye cover crop effects on mycorrhizal fungus colonization of maize roots. Plant Soil 328, 507–521.

White, C.M., Weil, R.R., 2011. Forage radish cover crops increase soil test phosphorus surrounding holes created by radish taproots. Soil Sci. Soc. Am. J. 75, 121–130.

White, R.E., Helyar, K.R., Ridley, A.M., Chen, D., Heng, L.K., Evans, J., et al., 2000. Soil factors affecting the sustainability and productivity of perennial and annual pastures in the high rainfall zone of south-eastern Australia. Aust. J. Exp. Agric. 40, 267–283.

Wienhold, B.J., Pikul Jr., J.L., Liebig, M.A., Mikha, M.M., Varvel, G.E., Doran, J.W., et al., 2006. Cropping systems effects on soil quality in the Great Plains: Synthesis from a regional project. Renew. Agric. Food Syst. 21, 49–59.

Williams, S.M., Weil, R.R., 2004. Crop cover root channels my alleviate soil compaction effects on soybean crop. Soil Sci. Soc. Am. J. 68, 1403–1409.

Yost, M.A., Russelle, M.P., Coulter, J.A., Schmitt, M.A., Sheaffer, C.C., Randall, G.W., 2015. Stand age affects fertilizer nitrogen response in first-year corn following alfalfa. Agron. J. 107, 486–494.

Zhang, T.Q., Tan, C.S., Zheng, Z.M., Drury, C.F., 2015. Tile drainage phosphorus loss with long-term consistent cropping systems and fertilization. J. Environ. Qual. 44, 503–511.

CHAPTER 7

Perennial-Based Agricultural Systems and Livestock Impact on Soil and Ecological Services

John Hendrickson and Matt Sanderson

USDA-Agricultural Research Service, Mandan, ND, United States

7.1 Introduction

The integration of perennial grass and forages, as a distinct component of cropping systems, is fairly new in the 8000 plus years of agriculture. A significant shift to using perennials in agricultural systems occurred in the Middle Ages. Bubonic plague (1350–1480) caused massive human mortality and significantly shrank the number of farm workers (Thirsk, 1997). Because of the lack of labor, some land remained uncultivated and therefore was often seeded to grasses or clovers for livestock production (Thirsk, 1997). Eventually farmers realized the benefits of perennial sods as a means of rejuvenating soils. This realization led to the development of new crop rotations (i.e., a 4-year rotation with grasses and clovers) that increased grain yields and also freed up labor for the industrial revolution (Sinclair and Sinclair, 2010).

As recently as 1930, approximately 26 million ha of farmland was used to feed horses and mules but, as farms mechanized, this land could be used for crop instead of forage production (Olmstead and Rhode, 2001). Besides reducing the need for perennials on farms, mechanization and other technological advances changed the structure of agriculture. Farms became more specialized (i.e., crops only or concentrated feeding operations) (Brummer, 1998; Russelle et al., 2007; O'Donoghue et al., 2011) and cropping systems became less diversified (Aguilar et al., 2015), which further reduced the amount of perennials in agricultural systems.

These concerns over the decoupling of crop and livestock production (Brummer, 1998) have led to renewed interest in incorporating perennials into agricultural systems (Asbjornsen et al., 2014). Perennials are still important in many agricultural systems around the world. In the Buenos Aires province of Argentina, e.g., 4 years of cash crops are often rotated with 3–5

Soil Health and Intensification of Agroecosystems.
DOI: http://dx.doi.org/10.1016/B978-0-12-805317-1.00007-5

years of pasture (Studdert et al., 1997), and in Uruguay, cropland–pasture rotation is practiced on 20% of the land (García-Préchac et al., 2004). Incorporating alfalfa (*Medicago sativa* L.) into crop rotations may address rising water tables and salinity issues in Australia (Ward et al., 2006; Verburg et al., 2008). The inclusion of perennial forages also reduced some annual weeds in European organic agriculture (van Elsen, 2000).

In this chapter, perennial-based agriculture is defined as the intentional integration of perennial grass and other forages into agricultural systems (FAO, 2011). This chapter will focus on the impacts, both positive and negative, of incorporating perennials and associated livestock grazing on soil properties and ecosystem services. Perennial-based systems and livestock grazing have different impacts on soil properties, as well as soil carbon (C), nitrogen (C), phosphorus (P), and microbial biomass concentrations than annual cropping systems. These systems can also enhance or provide different ecosystem services compared to annual systems. However, understanding how perennials and livestock impact these variables is crucial in developing management scenarios.

7.2 Background

7.2.1 Livestock Integration in Agriculture

Crop diversity in the United States has decreased (Aguilar et al., 2015) during the past 30 years and livestock production has become larger and more concentrated. A potential solution may be full integration of livestock into agricultural systems either within the same land base or at a regional level to potentially reduce environmental and economic problems associated with specialized agriculture (Naylor et al., 2005; Sulc and Tracy, 2007; Hendrickson et al., 2008). Integrated crop/livestock producers traditionally raise a greater diversity of crops, encouraging crop rotation (Honeyman, 1996). Table 7.1 details the benefits of using forages, diverse crops, and livestock in a rotation which includes reducing inputs (Brummer, 1998; Entz et al., 2002; Schiere et al., 2002), increasing crop yield (Entz et al., 1995, 2002; Tracy and Zhang, 2008), enhancing nutrient cycling (Brummer, 1998; Schiere et al., 2002), reducing plant disease (Krupinsky et al., 2002), and improving soil quality (Krall and Schuman, 1996). High-density livestock operations have been a source of N and P runoff (Carpenter et al., 1998), but incorporating livestock into cropping systems can reduce livestock density and crop residues or failed crops can provide forage for livestock (Oltjen and Beckett, 1996).

7.2.2 Changes in the Role of Livestock and Perennial Systems

Interest in ecosystem services and agriculture's role in potentially providing ecosystem services has caused a rethinking of the primary role of agriculture from production to ecological functioning (Boody et al., 2005; Power, 2010). Agriculture provides and consumes ecosystem services (Power, 2010) but the potential for agriculture to contribute ecosystem

Table 7.1: Benefits of incorporating forages, diverse crops, and livestock into agricultural systems

Metric	Impact	References
Crop yield increase	71% of respondents reported increased yield when annual crops followed forages	Entz et al. (1995)
	Corn in integrated crop–livestock systems yielded 1 $Mg\,ha^{-1}$ more than continuous corn	Tracy and Zhang (2008)
Enhance nutrient cycling	Moving from high external input system to integrated crop–livestock system tightens within system nutrient cycling	Schiere et al. (2002)
Reduce plant disease	Rotating between cereal and noncereal crops reduces soil- and residue-borne diseases Including a pulse crop in the rotation enhances beneficial soil organisms and decreases root diseases	Krupinsky et al. (2002)
Improve soil quality	Australian wheat-sheep and Alberta perennial legume-wheat systems are examples of integrated systems that improve soil quality	Krall and Schuman (1996)

services beyond the traditional thinking of "provisioning" services provides an opportunity and a challenge (Swinton et al., 2006). Increasing diversity in both pasture and cropping systems can maintain productivity while also providing additional ecosystem services (Sanderson et al., 2013), which suggests increasing diversity in agricultural systems could be done without significant public costs (Boody et al., 2005).

This change from focusing on production to considering other ecological functions can also alter how perennials are viewed within the context of farming systems. For example, perennial vegetation may be incorporated into agricultural systems by targeting strategic positions within the landscape (Asbjornsen et al., 2014). Perennial vegetation can also be used to modify farming systems to more closely align with natural systems (Scherr and McNeely, 2008). But, for these systems to be effective, critical parts of the landscape need to be identified (Asbjornsen et al., 2014), appropriate management strategies need to be used (Powers, 2010), and quantification of the economic and environmental indicators is needed (Sanderson et al., 2013).

7.3 The Impact of Livestock on Soil Properties Within a Perennial-Based System

Perennial systems often impact soil properties differently than annual-based systems. There are multiple reasons for this but this chapter will be focused mainly on the livestock impact in perennial-based systems. The impact that perennials have on soil properties is discussed in Chapter 6, Integration of Annual and Perennial Cover Crops for Improving Soil Health.

7.3.1 Physical Properties

Livestock impact soil physical properties just by their presence on a landscape. Greenwood and McKenzie (2001) indicated cattle treading on a pasture can exert pressures of 98–192 kPa at the soil surface and this force can increase if the livestock are climbing a slope (Trimble and Mendel, 1995). In comparison, wheeled tractors or tracked vehicles exert a force of 74–81 kPa or less at the point of soil contact (Blunden et al., 1994). With time, livestock treading can affect a large area, whereas tractors with properly inflated tires subject less than 25% of the soil surface to pressures greater than 75 kPa (Raper et al., 1995). Table 7.2 shows soil bulk density data from a series of studies from Australasia, Africa, and North America that evaluated the impact of increased stocking rate on soil bulk density. In general, soil bulk density increased with an increase in grazing pressure or the presence of grazing livestock. Grazing livestock can increase soil compaction resulting in decreased water infiltration rates and increased surface runoff. Sheep-grazed pastures in Western Australia had increased bulk density and water infiltration decreased with increased stocking rate (Willatt and Pullar, 1984). Trimble and Mendel (1995) reviewed the impact of cattle grazing and suggested heavy cattle grazing in upland areas compacts the soil and reduces water infiltration. Increases in bulk density with increased cattle grazing pressure have also been seen on relatively flat native ranges in North Dakota (Wienhold et al., 2001).

Management of stocking rate (i.e., the number of animals per unit area of land per unit time) is critical in moderating livestock impact on soil physical properties. Data from the Edwards Plateau region in Texas, United States, suggest that heavy stocking rate and drought were the major contributing factors to reducing infiltration rate and increasing interrill erosion in rangeland (Thurow et al., 1988). Liebig et al. (2006) evaluated long-term (>70 years) moderately and heavily grazed native pasture and a fertilized crested wheatgrasss [*Agropyron cristatum* (L.) Gaertn.] in the northern Great Plains and found bulk densities were not great enough to restrict root growth. Cattle grazing crop residues in North Dakota did not significantly increase soil bulk density when compared to cropland that was hayed or ungrazed (Liebig et al., 2012). The strong freeze–thaw cycle in the northern Great Plains may alleviate negative grazing effects on soil bulk density. Even in more temperate areas, such as Illinois, United States, cattle presence on cropland may have increased soil compaction in some years, but grain yields were not impacted (Tracy and Zhang, 2008). Livestock and farm machinery may differ in how they affect soil compaction. In general, livestock compaction is limited to the 5–20 cm of the soil surface while farm machinery compaction is usually deeper and affects the zone from 10 to 60 cm (Hamza and Anderson, 2005). Farm implements usually restrict their compaction zone to a relatively small area (Raper et al., 1995), although with repeated passes, up to 100% of the soil surface may be impacted (Hamza and Anderson, 2005).

Table 7.2: Bulk Density for sheep and cattle at different animal densities. depth refers to depth of soil sample and animal density is recorded in number of head per hectare

Location	Soil Depth (cm)	Soil Bulk Density (Mg m^{-3})				References
		Sheep				
New South Wales, Australia	0–5	0 sheep ha^{-1} 1.17	10 sheep ha^{-1} 1.26	15 sheep ha^{-1} 1.25	20 sheep ha^{-1} 1.23	Greenwood et al. (1997)
Southland, New Zealand	0–5	0 sheep ha^{-1} 0.8			1800 sheep ha^{-1} 0.9	Drewry et al. (1999)
Sadoré Niger	0–7	0 sheep ha^{-1} 1.59		125 kg wt ha^{-1} 1.58	Uncontrolled mixed grazing 1.65	Hiernaux et al., 1999
		Cattle				
Mandan, North Dakota, USA	0–5	0 steers ha^{-1} 0.39		0.4 steers ha^{-1} 0.48	1.1 steers ha^{-1} 0.61	Wienhold et al. (2001)
Nunn, Colorado, USA	0–5		0.6 steers ha^{-1} 1.37	0.8 steers ha^{-1} 1.42	1.4 steers ha^{-1} 1.45	Van Haveren (1983)

7.3.2 Soil Carbon

Globally, livestock grazing does not have a clear impact on soil C (Milchunas and Lauenroth, 1993). However, in ecosystems with a strong evolutionary history of livestock grazing, grazing appears to increase soil C (Conant et al., 2001). Livestock grazing may change vegetative species composition in grasslands. For example, interactions between fire, atmospheric CO_2 increases, and grazing can increase woody plant establishment in post oak savannahs (Archer et al., 1995), although woody plant establishment may occur equally well in grazed or ungrazed areas (Jurena and Archer, 2003). Shrubs (Springsteen et al., 2010) and bunchgrasses (Derner et al., 1997) can accumulate soil organic C directly underneath individual plants creating nutrient islands. Frank et al. (1995) evaluated long-term (>70 years) grazed and ungrazed pastures in North Dakota and found species changes caused by heavy grazing, primarily a shift to a short, shallow-rooted, grazing tolerant, grasses, compensated for potential C losses caused by grazing.

C fluxes from rangelands and pastures have also been measured. An evaluation of eight North American rangelands indicated that a majority were typically sinks for CO_2; although, two rangelands in the desert Southwest were sources for CO_2 (Svejcar et al., 2008). Livestock grazing does not appear to impact carbon dioxide effluxes from grazing lands (Risch and Frank, 2006; Liebig et al., 2013). Understanding dormant season CO_2 flux is critical in determining whether grazing lands will be a source or sink for CO_2 (Frank, 2002; Frank et al., 2002). Frank et al. (2002) indicated soil temperature accounted for 65% of the variation in CO_2 flux. Their data indicate that dormant season CO_2 fluxes still occurred even though soil temperatures were as low as −12°C (Fig. 7.1).

7.3.3 Soil Nitrogen

Globally soil N is similar to soil C in that it exhibits both positive and negative effects due to grazing (Milchunas and Lauenroth, 1993). Inorganic N was similar across long-term (>70 years) grazing treatments at shallow depths (0–5 cm) but N-NO_3 was greater under heavy grazing compared to moderately or ungrazed pastures (Wienhold et al., 2001). Surface soils of ungrazed (0–7.5 cm) rangelands had lower soil N than surface soils of either continuous season-long or rotational grazing systems in Wyoming, United States (Manley et al., 1995). A comparison of ungrazed, light, and heavy grazed rangelands in Wyoming found no differences in total N mass in the 0–60 cm soil profile, but N distribution within the profile was changed (Schuman et al., 1999). In North Dakota, however, ungrazed rangeland had more soil N than did moderate or heavy grazed rangelands (Frank et al., 1995) and similarly, total soil N decreased with increased grazing intensity on a meadow steppe in Inner Mongolia (Han et al., 2008).

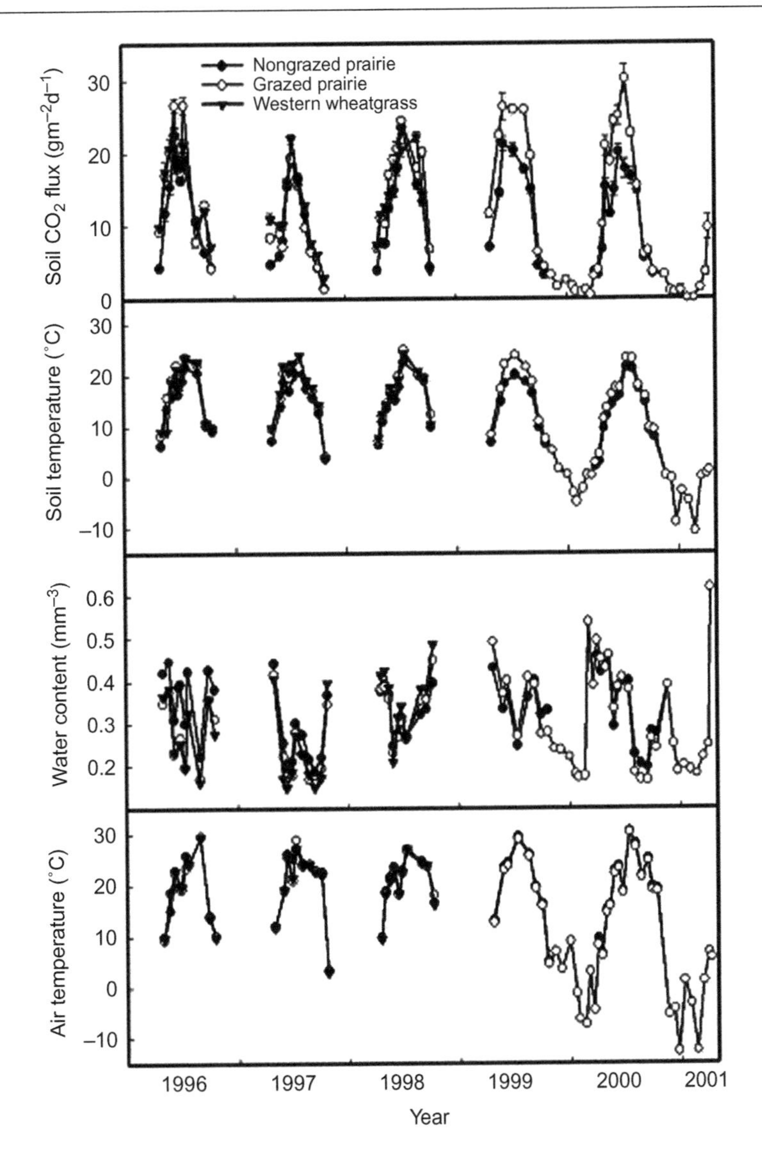

Figure 7.1
Soil temperature, soil water content, air temperature, and soil CO_2 fluxes for western wheatgrass, nongrazed native range, and grazed native range from 1996 to 2001. Fluxes were measured only in the western wheatgrass from 1996 to 98 and during this time dormant season fluxes were not measured. *After Frank, A.B., Liebig, M.A., Hanson, J.D., 2002. Soil carbon dioxide fluxes in northern semiarid grasslands. Soil Biol. Biochem. 34, 1235–1241.*

Soil organic N increased in a grazed coastal bermudagrass [*Cynodon dactylon* (L.) Pers.] and annual ryegrass (*Lolium multiflorum* Lam.) pasture as stand age increased (Wright et al., 2004). Soil N-NO_3 was greater under a fertilized crested wheatgrass pasture than ungrazed native rangeland (Wienhold et al., 2001). Grazing management did not impact extractable N in the 0–15 cm soil depth in a Florida, United States, bermudagrass pasture, but N concentrations were greater when shade, water, or supplement were closer (Mathews et al., 1994). N concentrations are lower in hayed compared to grazed pastures (Mathews et al., 1994; Franzluebbers et al., 2000) primarily because of the removal of nutrients under hay production (Mathews et al., 1994).

7.3.4 Phosphorus Concentration

The impact of livestock grazing on soil P concentration centers on losses of P off-site rather than alterations of P on-site. Irish pastures grazed by cattle had increased organic P concentration in the overland flow than ungrazed pastures (Kurz et al., 2006). Soil P losses in New Zealand pastures were linked to soil physical changes and potential soil erosion and surface runoff caused by livestock traffic (Sharpley and Syers, 1979; Smith and Monaghan, 2003).

Neff et al. (2005) compared ungrazed and historically grazed native rangelands in Utah, United States and found losses of P under grazing. However, the authors implied the losses in P were primarily due to soil losses to erosion. The flooded Pampas in Argentina also had lower soil P when grazed than when left ungrazed (Lavado et al., 1996). A related study suggested livestock consumption of forage may remove the P in the biomass from the system, which resulted in lower P levels (Chaneton et al., 1996). In Florida, United States, grasslands grazing did not impact soil P in the short term (Mathews et al., 1994).

7.3.5 Soil Microbial Biomass

Livestock grazing is closely linked to soil microbial biomass. An evaluation of *Poa pratensis* L. invaded grassland, in Yellowstone National Park, Wyoming, United States, indicated grazing caused roots to release C, which was used by the soil microbial community and subsequently increased in available N promoting regrowth (Hamilton and Frank, 2001). Another report from the same region suggested that topography influenced the size of the soil microbial biomass component while grazing increased microbial activity as measured by N fluxes (Tracy and Frank, 1998). However, other reports indicate that soil microbial populations in native grasslands in Australia and the United States exhibited decline under even moderate grazing because of losses in soil organic C (Northup et al., 1999; Ingram et al., 2008). In both cases these declines may be linked to a drought–grazing interaction.

Livestock grazing also influences soil microbial communities. Bardgett et al. (2001) found that bacterial-based organisms were dominant in intensively grazed grasslands in the UK, while less intensively grazed or unmanaged grasslands have decomposer changes dominated by fungi. Both nematode (Wang et al., 2006) and microbial communities (Clegg, 2006) were altered by cattle grazing. These changes must be caused by changes in nutrient cycling or plant community composition since cattle waste does not seem to increase soil microbial biomass unless it is incorporated into the soil (Lovell and Jarvis, 1996).

7.3.6 Summary

Livestock grazing exhibits a variety of effects on physical, nutrient, and biological properties of soil. Impacts depend on the evolutionary history of region of the system, level of grazing, and climatic conditions prior to and during the grazing event. However, including livestock grazing as part of an integrated crop/livestock system appears to alleviate some of the negative impacts of annual cropping on the environment (Hendrickson et al., 2008) by enhancing soil physical and biological properties (Acosta-Martínez, 2004), while having little if any impact on crop yields (Tracy and Zhang, 2008).

7.4 Perennial-Based Agricultural Systems and Livestock Impact on Soil and Ecological Services

7.4.1 Review of Ecosystem Goods and Services

Ecosystem goods and services include the "benefits human populations derive, directly or indirectly, from ecosystem functions" (Costanza et al., 2014). Several functional characteristics of perennial-based agricultural systems contribute to ecosystem services of benefit to humans (Table 7.3). In addition to the demand for provisioning ecosystem services (e.g., food, feed, fiber, and fuel), society is expecting agriculture to provide regulating (e.g., pollination), cultural (e.g., recreation), and supporting (e.g., water cycling) services. Meeting the demands for multiple services requires a unique management approach. For example, diversifying agricultural systems via increasing plant species in pastureland and integrating crops and livestock is a multifunctional approach that can benefit production (provisioning service), efficient use of natural resources (regulating, supporting, and cultural services), and potentially improve economic returns. Livestock are a key component of perennial-based systems because they can efficiently use the forage resource from both high-quality land and marginal lands not suited to intensive management.

7.4.2 Positive Impacts on Ecosystem Services

Provisioning services from perennial systems have a significant economic benefit. The forage and livestock production from perennial hayland and pastureland alone accounts for about

Table 7.3: Selected ecosystem services of perennial systems, their potential benefit to humans, and potential tradeoffs

Ecosystem Service	Potential Benefit	Potential Tradeoffs
Herbage production	Production is positively related to food supply and economic return	High production increases nutrient loss, pesticide use, and greater greenhouse gas losses
Sustain pollinators	Native pollinators are needed for reproductive success of some native plants; pollinator habitat has high positive value for honey bee production	Difficulty and cost of establishing and maintaining diverse flora; potential loss of forage production and quality
Biodiversity	Biodiversity has positive benefits, such as providing vegetation cover and habitat for wildlife. Greater plant diversity may stabilize forage production over time	Benefits often are inversely related to other factors, e.g., low production and high biodiversity at low soil fertility
Social values (aesthetics)	Grasslands have positive value because a well-managed pasture "looks good"	Aesthetic values are in the eye of the beholder
Water quality	High water quality from grasslands has high environmental value. Vegetation cover minimizes soil erosion	Land taken out of production for filter strips or fenced riparian areas
Sustain wildlife	Keeping natural trophic levels intact; provide hunting opportunities	Competition and interference between wildlife and livestock, e.g., nest trampling, disease transmission, predation of livestock
Carbon sequestration	Carbon sequestration has a positive benefit from increased soil organic matter	Carbon sequestration has a moderate negative value because stored carbon= "lost" production

Source: Adapted from Sanderson, M.A., Watzold, F., 2010. Balancing tradeoffs in ecosystem functions and services in grassland management. Grassland Sci. Eur. 15, 639–648, Sanderson and Watzold (2010).

$45 billion of agricultural receipts in the United States, with about $19 billion from the 23 million ha of hayland (Sanderson et al., 2012). Forage and grazing lands supply feedstuffs for approximately 90 million head of cattle on US farms (USDA-NASS, 2014).

Perennial grasslands (both natural and managed) support pollinator services via diverse and abundant populations of native pollinators such as bumblebees, hoverflies, and butterflies along with introduced honey bees (Spivak et al., 2011; Kimoto et al., 2012a). For example, switchgrass (*Panicum virgatum* L.) monocultures and plantings of mixed prairie species provided more ecosystem services in the form of pollinator habitat, grassland bird habitat, and pest suppression than annual maize-ethanol production systems (Werling et al., 2014). Diverse floral resources (e.g., forbs including legumes; shrubs and trees) present on grasslands provide pollen and nectar but also provide an undisturbed habitat for ground-nesting bees. The native pollinators are not only needed for pollinating native plants in grasslands, but can also enhance pollination services for nearby agricultural crops (Black et al., 2011). Commercial beekeeping operations rely on perennial forage and grazing lands for bee pasture. For example, a survey of 320 apiaries across North Dakota revealed that about 80% of surveyed apiaries were located in or adjacent to hayland, pastureland, or rangeland (Sanderson, 2015).

Perennial vegetation provides critical conservation ecosystem services to mitigate livestock impacts. Managed grasslands are used as one of soil and water conservation practices, such as windbreaks for reducing wind erosion, vegetative barriers for trapping sediment, and reducing water erosion, filter strips for slowing surface water runoff, and buffer strips for protecting riparian zones (Sollenberger et al., 2012).

Perennial systems can be a source of cultural services in the form of recreation, vistas (e.g., the wide open spaces of the US West), and aesthetic values (Sayre et al., 2012). Perennial lands can provide significant habitat for wildlife such as grassland birds (USDA-NRCS, 2010) and wild ungulates (e.g., bison, deer, and elk). In some instances, farmers or ranchers may actively manage perennial lands to produce wildlife either for meat or other products or to provide hunting opportunities and income (Sayre et al., 2012). In addition to those services, grassland ecosystems provide significant enhancement in soil ecosystem services and soil health indicators that can include soil organic matter, soil microbial community, and water and nutrient cycling.

7.4.3 Negative Impacts

Livestock-based perennial systems may also compromise ecosystem services, especially on marginal lands if management is lax and intensity of use crosses a threshold. Intensification of perennial system production (e.g., greater livestock densities on pasture; greater external inputs to stimulate biomass production) can lead to negative effects on ecosystem services such as diminished pollinator habitat and degraded water quality.

Grazing livestock may affect populations of native pollinators differently. Increased grazing intensity (0–60% utilization of forage) tended to reduce the abundance and diversity of native bumble bees (*Bombus* sp.) but did not affect sweat bee (*Lasioglossum* sp.) populations on native bunchgrass prairie in northeastern Oregon, United States (Kimoto et al., 2012b). Improperly timed or extremely intensive grazing by livestock can eliminate some flowering plants, reducing the food supply for pollinators and trampling and destroying nest sites (Sugden, 1985; Black et al., 2011).

Improper livestock management can lead to degradation of the hydrological functions of perennial systems. Improperly managed grazing may reduce vegetative cover, increase bare ground, and compact the soil. With soil stability compromised, the land is then at greater risk of soil erosion, surface water runoff, and contamination of surface and ground waters

(Sollenberger et al., 2012). In arid regions, the same unstable soil conditions can exacerbate wind erosion and thereby degrade air quality (Pellant et al., 2005).

7.4.4 Tradeoffs

Ideally, farmers and other land managers would prefer to benefit economically from provisioning services (i.e., production of agricultural products) as much as possible while at the same time satisfying society's preference for environmental ecosystem services. Unfortunately, it is not always possible simultaneously to realize or maximize all ecosystem services. Ultimately, compromises in management must be made, resulting in economic or environmental tradeoffs.

Perennial grassland systems generally accumulate soil organic matter and may reduce greenhouse gas (GHG) emissions, a valuable ecosystem service. The ability of perennial systems to be a sink for atmospheric C, however, depends on management and climate. Liebig et al. (2010) estimated net global warming potential of native rangeland in North Dakota and concluded that these grasslands could significantly mitigate GHG emissions but required a trade off in lower stocking rates and reduced beef production per hectare. Precision measurements of C flux of pastures in Pennsylvania, United States demonstrated that mature (>40 years old) pastures were significant C sinks for a brief period in April and May but after that they became a C source at typical forage utilization levels (Skinner, 2008). Delaying forage harvest by cutting or grazing as late as possible during spring could maximize the C sink; however, that would involve an economic tradeoff in forage quality and animal performance.

Establishing and maintaining pollinator habitat in perennial systems may require a change in management to favor certain plant species, such as flowering forbs, at a significant economic cost and a potential tradeoff in forage productivity and forage quality for livestock. Both the social and economic context of the ecosystem service must be considered when attempting to estimate its value in tradeoff analysis. Commercial beekeepers or wildlands managers may value the perennial landscape as a pollen and nectar source for pollinators, whereas a rancher may value the same landscape for its ability to supply forage for grazing livestock. In this instance, it may be possible to estimate the potential loss in forage production and propose a payment to offset the rancher's costs. It could be argued, however, that healthy pollinator populations resulting from maintaining a diversity of forbs, grasses, and flowering shrubs could enhance the reproductive success and persistence of some rangeland plants and could also provide spillover pollination services to nearby crops.

7.5 Management

A challenge to agriculture involves achieving multifunctionality or the joint production of agricultural commodities and ecosystem services. Complex ecosystems provide multiple benefits but require multiple species (Duffy, 2008; Isbell et al., 2011). This then leads to the need for multifunctional management or manipulating agroecosystems to realize more than one output or ecosystem service.

7.5.1 Multienterprise (Multifunctional) Management

Maintaining diversity in natural systems or agricultural systems adds resilience and enables systems to withstand shocks such as climate extremes (Darnhoffer et al., 2010). In managed grasslands, plant species diversity can be enhanced by planting diverse mixtures of grasses, legumes, and forbs. Several studies on managed grasslands have demonstrated increased forage production and greater resistance to weed invasion with greater plant diversity in the system (e.g., Deak et al., 2007; Bonin and Tracy, 2012). Equally important, however, is the species composition of the vegetation diversity. Including species with specific functional traits, such as drought tolerance, N-fixation, or photosynthetic pathway can be more effective than simply planting an ad hoc mixture of species (Goslee et al., 2013). Including multiple species of grazing livestock in perennial systems through complementary stocking methods or mixed species grazing offers potential for sustainable intensification of botanically diverse grasslands (Muir et al., 2015). Managing a diverse array of grass, legume, and forb species in agricultural grasslands can also improve the availability of floral resources for enhanced pollinator populations (Woodcock et al., 2014).

Managing grazing animal distributions and timing of grazing to result in greater landscape heterogeneity of floral and nesting resources was suggested as a tool to maintain pollinator diversity and pollination services on working rangelands (Kimoto et al., 2012b). Mowing and prescribed burning can also be used to create habitat heterogeneity, but no more than one-third of the habitat should be mowed or burned at one time (Black et al., 2011).

Integrated crop–livestock systems offer a strategy for sustainable intensification of perennial systems. These systems combine management of livestock, crops, and soils to produce cash crops, livestock feed, and enhance nutrient cycling. On the northern plains of the United States, integrating livestock into an annual cropping system by grazing crop residues dramatically lowered winter feeding costs (Karn et al., 2005).

On the southern plains of the United States, grazing cattle on cover crops during the winter on cropland used to grow cotton (*Gossypium hirsutum* L.) or peanuts (*Arachis hypogea* L.) during the summer increased farm income (Allen et al., 2012). Nearly 25% less irrigation water was used in an integrated cotton-forage-beef production system on the Texas High Plains compared with a cotton monoculture. The integrated system also had less soil erosion, required fewer chemical and N fertilizer inputs, and had greater rainfall infiltration than the nonintegrated system.

Adding cover crops to integrated crop–livestock systems provides a valuable source of forage and conservation benefits such as reduced soil erosion, better soil structure, greater water infiltration, and enhanced soil biology (Snapp et al., 2005; Martens and Entz, 2012). Potential drawbacks to the use of cover crops include seed and machinery costs along with slower soil warming and drying in the spring and potential interference with cash crops that may follow in the rotation.

7.5.2 Management Intensity Versus Input Intensity

Agriculture is one of the major land uses globally (Tilman et al., 2001) and in the continental United States, croplands occupy 22% of the land base (Aguilar et al., 2015). Agriculture has increased productivity but has had a corresponding increase in input use (Matson et al., 1997; Tilman et al., 2001). While reintegrating crop–livestock systems may provide a way to reduce negative impacts of agriculture (Brummer, 1998; Naylor et al., 2005), these systems often require more management input. Hendrickson et al. (2008) devised a hierarchy of agricultural systems based on management intensity (Fig. 7.2). They found more complex agricultural systems can decrease input use but management intensity increases.

The degree of management in complex agricultural systems requires adaptability and seamless communication between research and agriculturalists (Tanaka et al., 2002) to maximize knowledge and enhance the decision-making process. These more complex agricultural systems can provide a mechanism for greater sustainable intensification of agriculture, but they also need to incorporate flexibility so emerging issues, that were unrecognized when the systems were designed, can be addressed. The need for future

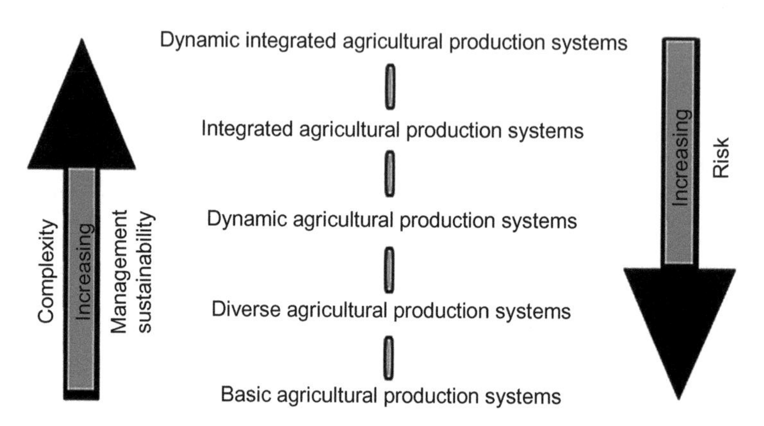

Figure 7.2

Hierarchical ranking of agricultural systems. As system complexity increases, e.g., moving from a basic agricultural production system to an integrated system, management intensity increases but so does sustainability. *After Hendrickson, J.R., Hanson, J.D., Tanaka, D.L., Sassenrath, G., 2008. Principles of integrated agricultural systems: introduction to processes and definition. Renew. Agric. Food Syst. 23, 265–271.*

flexibility further challenges producers and researchers to be adaptable and engage in productive two-way communication so new challenges and opportunities can be addressed as they arise.

7.6 Research Needs

7.6.1 Evaluating Multiple Ecosystem Functions Simultaneously

Multiple drivers are forcing agriculture to transform (Jordan and Warner, 2010) and part of this transformation is realizing agricultural landscapes can fulfill a variety of ecosystem functions (Boody et al., 2005). Incorporating perennials, livestock, and crops into the same landscape can help fulfill multiple ecosystem functions.

However, the number of integrated producers is decreasing (Hendrickson et al., 2008) and few research projects are focused on developing integrated crop–livestock systems (Tanaka et al., 2008). While there are many potential reasons for the lack of integrated crop–livestock system research, it is important to remember that this type of research is complex, challenging, and requires a multiple discipline scientific team, relatively low replication, and large land areas (Tanaka et al., 2008). In addition, in designing projects, it is important that the farmer's, as well as society's, goals are considered (Russelle et al., 2007) and these goals are considered important for properly functioning agroecosystems.

7.6.2 Research Networks to Scale Up (e.g., Long-Term Agroecosystems Research)

As agriculture makes the shift from a production focus to including concerns about its environmental footprint (Walbridge and Shafer, 2011) and potential multifunctionality (Boody et al., 2005), there needs to be a simultaneous shift in how research is conducted. Traditionally, agricultural research has been conducted at a small plot level (Robertson et al., 2008); however, with historical challenges emerging, there is a need to scale up both temporally and spatially. Adding up system responses on smaller scales is inadequate to understanding ecosystem functions at a larger scale (Carpenter et al., 2006) and changing environmental conditions and other long-term concerns indicate that research timeframes longer than the current 2–3 years are needed (Robertson et al., 2008). Establishing these long-term, large-scale experiments can provide benefits not originally envisioned in their original establishment. For example, a 100-year-old grazing experiment on the Northern Great Plains Research Laboratory near Mandan, ND, was originally set up to evaluate stocking rate but more recent contributions have focused on GHG emissions (Sanderson et al., 2015).

Because of the benefits of large-scale, long-term research in agricultural systems, there have been calls to initiate a Long-Term Agroecosystems Research (LTAR) network (Robertson et al., 2008). The USDA has proposed and developed an LTAR network (Walbridge and

Shafer, 2011) and there are currently 18 sites in the network (Karlen et al., 2014). These sites provide information on watersheds (Steiner et al., 2014; Sadler et al., 2015); soil management (Karlen et al., 2014); grazing systems (Sanderson et al., 2015), and cropping systems. The development of the LTAR network should provide a means to scale temporally and spatially the impacts of inclusion of perennials and livestock grazing in agricultural systems.

7.7 Conclusions

This chapter discussed how livestock grazing and inclusion of perennials can impact agricultural systems. This discussion highlighted the fact that incorporating perennials and livestock into agricultural systems can have both positive and negative impacts. For example, enhancing plant diversity and integrating crops and livestock into an agricultural operation can provide multiple ecosystem services but may increase overland P flow. Successfully incorporating perennials and livestock into existing agricultural systems requires increased management intensity.

Understanding how livestock grazing and inclusion of perennials impact agricultural systems becomes more relevant as agriculture shifts from focusing on productivity to consideration of multiple ecosystem functions. In particular, the development of integrated crop–livestock systems presents opportunities to minimize the environmental footprint of agriculture while providing multiple ecosystem services. However, development and adoption of these systems is not easy. Within these systems, input reduction is often achieved by increasing management intensity. Crop–livestock system research requires multidisciplinary teams and large land bases and scientific compromise. Still, the development of large spatial and temporal research networks suggests (1) an interest in developing multifunctional systems and (2) a means for the research to be accomplished.

References

Acosta-Martinez, V., Zobeck, T.M., Allen, V., 2004. Soil microbial, chemical and physical properties in continuous cotton and integrated crop–livestock systems. Soil Sci. Soc. Amer. J. 68, 1875–1884.

Aguilar, J., Gramig, G.G., Hendrickson, J.R., Archer, D.W., Forcella, F., Liebig, M.A., 2015. Crop species diversity changes in the United States: 1978–2012. PLoS One 10, e0136580.

Allen, V.G., Brown, C.P., Kellison, R., Green, P., Zilverberg, C.J., Johnson, P., et al., 2012. Integrating cotton and beef production in the Texas southern high plains: I. Water use and measures of productivity. Agron. J. 104, 1625–1642.

Archer, S., Schimel, D.S., Holland, E.A., 1995. Mechanisms of shrubland expansion: land use, climate or CO_2? Clim. Change 29, 91–99.

Asbjornsen, H., Hernandez-Santana, V., Liebman, M., Bayala, J., Chen, J., Helmers, M., et al., 2014. Targeting perennial vegetation in agricultural landscapes for enhancing ecosystem services. Renew. Agric. Food Syst. 29, 101–125.

Bardgett, R.D., Jones, A.C., Jones, D.L., Kemmitt, S.J., Cook, R., Hobbs, P.J., 2001. Soil microbial community patterns related to the history and intensity of grazing in sub-montane ecosystems. Soil Biol. Biochem. 33, 1653–1664.

Black, S.H., Shepherd, M., Vaughn, M., 2011. Rangeland management for pollinators. Rangelands, 9–13.
Blunden, B.G., McBride, R.A., Daniel, H., Blackwell, P.S., 1994. Compaction of an earthy sand by rubber tracked and tired vehicles. Soil Res. 32, 1095–1108.
Bonin, C.L., Tracy, B.F., 2012. Diversity influences forage yield and stability in perennial prairie plant mixtures. Agri. Ecosyst. Environ. 162, 1–7.
Boody, G., Vondracek, B., Andow, D.A., Krinke, M., Westra, J., Zimmerman, J., et al., 2005. Multifunctional agriculture in the United States. BioScience 55, 27–38.
Brummer, E.C., 1998. Diversity, stability, and sustainable American agriculture. Agron. J. 90, 1–2.
Carpenter, S.R., Caraco, N.F., Correll, D.L., Howarth, R.W., Sharpley, A.N., Smith, V.H., 1998. Nonpoint pollution of surface waters with phosphorus and nitrogen. Ecol. Appl. 8, 559–568.
Carpenter, S.R., De Fries, R., Dietz, T., Mooney, H.A., Polasky, S., Reid, W.V., et al., 2006. Millennium ecosystem assessment: research needs. Science 314, 257–258.
Chaneton, E.J., Lemcoff, J.H., Lavado, R.S., 1996. Nitrogen and phosphorus cycling in grazed and ungrazed plots in a temperate subhumid grassland in Argentina. J. Appl. Ecol. 33, 291–302.
Clegg, C.D., 2006. Impact of cattle grazing and inorganic fertiliser additions to managed grasslands on the microbial community composition of soils. Appl. Soil Ecol. 31, 73–82.
Connant, R.T., Paustian, K., Elliot, E.T., 2001. Grassland management and conversion into grassland: effects on soil C. Ecol. Appl. 11, 343–355.
Costanza, R., de Groot, R., Sutton, P., ver der Ploeg, S., Anderson, S.J., Kubiszewski, I., et al., 2014. Changes in the global value of ecosystem services. Global Environ. Change 26, 152–158.
Darnhoffer, I., Fairweather, J., Moller, H., 2010. Assessing a farm's sustainability: insights from resilience thinking. Int. J. Sustain Agric. 8, 186–198.
Deak, A., Hall, M.H., Sanderson, M.A., 2007. Production and nutritive value of grazed simple and complex forage mixtures. Agron. J. 99, 814–821.
Derner, J.D., Briske, D.D., Boutton, T.W., 1997. Does grazing mediate soil carbon and nitrogen accumulation beneath C4, perennial grasses along an environmental gradient? Plant Soil 191, 147–156.
Drewry, J.J., Lowe, J.A.H., Paton, R.J., 1999. Effect of sheep stocking intensity on soil physical properties and dry matter production on a Pallic Soil in Southland. New Zeal. J. Agric. Res. 42, 493–499.
Duffy, J.E., 2008. Why biodiversity is important to the functioning of real-world ecosystems. Front Ecol. Environ. 7, 437–444.
Entz, M.H., Bullied, W.J., Katepa-Mupondwa, F., 1995. Rotational benefits of forage crops in Canadian prairie cropping systems. J. Prod. Agric. 8, 521–529.
Entz, M.H., Baron, V.S., Carr, P.M., Meyer, D.W., Smith, S., McCaughey, W.P., 2002. Potential of forages to diversify cropping systems in the Northern Great Plains. Agron. J. 94, 240–250.
FAO, 2011. Perennial Agriculture: Landscape resilience for the future. <http://www.fao.org/fileadmin/templates/agphome/documents/scpi/PerennialPolicyBrief.pdf > (accessed 2.05.16).
Frank, A.B., 2002. Carbon dioxide fluxes over a grazed prairie and seeded pasture in the Northern Great Plains. Environ. Pollut. 116, 397–403.
Frank, A.B., Tanaka, D.L., Hofmann, L., Follett, R.F., 1995. Soil carbon and nitrogen of Northern Great Plains grasslands as influenced by long-term grazing. J. Range Manage. 48, 470–474.
Frank, A.B., Liebig, M.A., Hanson, J.D., 2002. Soil carbon dioxide fluxes in northern semiarid grasslands. Soil Biol. Biochem. 34, 1235–1241.
Franzluebbers, A.J., Stuedemann, J.A., Schomberg, H.H., Wilkinson, S.R., 2000. Soil organic C and N pools under long-term pasture management in the Southern Piedmont USA. Soil Biol. Biochem. 32, 469–478.
García-Préchac, F., Ernst, O., Siri-Prieto, G., Terra, J.A., 2004. Integrating no-till into crop–pasture rotations in Uruguay. Soil Tillage Res. 77, 1–13.
Goslee, S.C., Veith, T.L., Skinner, R.H., Comas, L.H., 2013. Optimizing ecosystem function by manipulating pasture community composition. Basic Appl. Ecol. 14, 630–641.
Greenwood, K.L., McKenzie, B.M., 2001. Grazing effects on soil physical properties and the consequences for pastures: a review. Anim. Prod. Sci. 41, 1231–1250.

Greenwood, K.L., MacLeod, D.A., Hutchinson, K.J., 1997. Long-term stocking rate effects on soil physical properties. Anim. Prod. Sci. 37, 413–419.

Hamilton III, E.W., Frank, D.A., 2001. Can plants stimulate soil microbes and their own nutrient supply? Evidence from a grazing tolerant grass. Ecology 82, 2397–2402.

Hamza, M.A., Anderson, W.K., 2005. Soil compaction in cropping systems: a review of the nature, causes and possible solutions. Soil Tillage Res. 82, 121–145.

Han, G., Hao, X., Zhao, M., Wang, M., Ellert, B.H., Willms, W., et al., 2008. Effect of grazing intensity on carbon and nitrogen in soil and vegetation in a meadow steppe in Inner Mongolia. Agric. Ecosyst. Environ. 125, 21–32.

Hendrickson, J.R., Hanson, J.D., Tanaka, D.L., Sassenrath, G., 2008. Principles of integrated agricultural systems: introduction to processes and definition. Renew. Agric. Food Syst. 23, 265–271.

Hiernaux, P., Bielders, C.L., Valentin, C., Bationo, A., Fernandez-Rivera, S., 1999. Effects of livestock grazing on physical and chemical properties of sandy soils in Sahelian rangelands. J. Arid. Environ. 41, 231–245.

Honeyman, M.S., 1996. Sustainability issues of US swine production. J. Anim. Sci. 74, 1410–1417.

Ingram, L.J., Stahl, P.D., Schuman, G.E., Buyer, J.S., Vance, G.F., Ganjegunte, G.K., et al., 2008. Grazing impacts on soil carbon and microbial communities in a mixed-grass ecosystem. Soil Sci. Soc. Amer. J. 72, 939–948.

Isbell, F., Calcagno, V., Hector, A., Connolly, J., Harpole, W.S., Reich, P.B., et al., 2011. High plant diversity is needed to maintain ecosystem services. Nature 477, 199–203.

Jordan, N., Warner, K.D., 2010. Enhancing the multifunctionality of US agriculture. BioScience 60, 60–66.

Jurena, P.N., Archer, S., 2003. Woody plant establishment and spatial heterogeneity in grasslands. Ecology 84, 907–919.

Karlen, D.L., Peterson, G.A., Westfall, D.G., 2014. Soil and water conservation: our history and future challenges. Soil Sci. Soc. Amer. J. 78, 1493–1499.

Karn, J.F., Tanaka, D.L., Liebig, M.A., Ries, R.E., Kronberg, S.L., Hanson, J.D., 2005. An integrated approach to crop/livestock systems: wintering beef cows on swathed crops. Renew. Agric. Food Syst. 20, 232–242.

Kimoto, C., DeBano, S.J., Thorp, R.W., Taylor, R.V., Schmalz, H., DelCurto, T., et al., 2012a. Short-term responses of native bees to livestock and implications for managing ecosystem services in grasslands. Ecosphere 3, 1–19.

Kimoto, C., DeBano, S.J., Thorp, R.W., Rao, S., Stephen, W.P., 2012b. Investigating temporal patterns of a native bee community in a remnant North American bunchgrass prairie using blue vane traps. J. Insect. Sci. 12, 108. Available online: <http://ww.insectscience.org/12.108>.

Krall, J.M., Schuman, G.E., 1996. Integrated dryland crop and livestock production systems on the Great Plains: extent and outlook. J. Prod. Agric. 9, 187–191.

Krupinsky, J.M., Bailey, K.L., McMullen, M.P., Gossen, B.D., Turkington, T.K., 2002. Managing plant disease risk in diversified cropping systems. Agron. J. 94, 198–209.

Kurz, I., O'Reilly, C.D., Tunney, H., 2006. Impact of cattle on soil physical properties and nutrient concentrations in overland flow from pasture in Ireland. Agric. Ecosyst. Environ. 113, 378–390.

Lavado, R.S., Sierra, J.O., Hashimoto, P.N., 1996. Impact of grazing on soil nutrients in a Pampean grassland. J. Range Manage. 49, 452–457.

Liebig, M.A., Gross, J.R., Kronberg, S.L., Hanson, J.D., Frank, A.B., Phillips, R.L., 2006. Soil response to long-term grazing in the northern Great Plains of North America. Agric. Ecosyst. Environ. 115, 270–276.

Liebig, M.A., Gross, J.R., Kronberg, S.L., Phillips, R.L., 2010. Grazing management contributions to net global warming potential: a long-term evaluation in the northern Great Plains. J. Environ. Qual. 39, 799–809.

Liebig, M.A., Tanaka, D.L., Kronberg, S.L., Scholljegerdes, E.J., Karn, J.F., 2012. Integrated crops and livestock in central North Dakota, USA: agroecosystem management to buffer soil change. Renew. Agric. Food Syst. 27, 115–124.

Liebig, M.A., Kronberg, S.L., Hendrickson, J.R., Dong, X., Gross, J.R., 2013. Carbon dioxide efflux from long-term grazing management systems in a semiarid region. Agric. Ecosyst. Environ. 164, 137–144.

Lovell, R.D., Jarvis, S.C., 1996. Effect of cattle dung on soil microbial biomass C and N in a permanent pasture soil. Soil Biol. Biochem. 28, 291–299.

Manley, J.T., Schuman, G.E., Reeder, J.D., Hart, R.H., 1995. Rangeland soil carbon and nitrogen responses to grazing. J. Soil Water Conserv. 50, 294–298.

Martens, J.R.T., Entz, M.H., 2012. Integrating green manure and grazing systems: a review. Can. J. Plant Sci. 81, 811–824.

Matson, P.A., Parton, W.J., Power, A.G., Swift, M.J., 1997. Agricultural intensification and ecosystem properties. Science 277, 504–509.

Mathews, B.W., Sollenberger, L.E., Nair, V.D., Staples, C.R., 1994. Impact of grazing management on soil nitrogen, phosphorus, potassium, and sulfur distribution. J. Environ. Qual. 23, 1006–1013.

Milchunas, D.G., Lauenroth, W.K., 1993. Quantitative effects of grazing on vegetation and soils over a global range of environments. Ecol. Monogr. 63, 327–366.

Muir, J.P., Pitman, W.D., Foster, J.L., Dubeux Jr., J.C., 2015. Sustainable intensification of cultivated pastures using multiple herbivore species. Afr. J. Range Forage Sci. 32, 97–112.

Naylor, R., Steinfeld, H., Falcon, W., Galloway, J., Smil, V., Bradford, E., et al., 2005. Losing the links between livestock and land. Science 310, 1621–1622.

Neff, J.C., Reynolds, R.L., Belnap, J., Lamothe, P., 2005. Multi-decadal impacts of grazing on soil physical and biogeochemical properties in southeast Utah. Ecol. Appl. 15, 87–95.

Northup, B.K., Brown, J.R., Holt, J.A., 1999. Grazing impacts on the spatial distribution of soil microbial biomass around tussock grasses in a tropical grassland. Appl. Soil Ecol. 13, 259–270.

O'Donoghue, E., MacDonald, J., Vasavada, U., Sullivan, P., 2011. USDA Economic Research Service-Changing Farming Practices. USDA-Economic Research Service. <http://www.ers.usda.gov/amber-waves/2011-december/changing-farming-practices.aspx?keepThis=true&TB_iframe=true&height=650&width=850&caption=Publications+-+USDA+-+Economic+Research+Service#.Vre9aFK_Zj8> (accessed 02.7.16.).

Olmstead, A.L., Rhode, P.W., 2001. Reshaping the landscape: the impact and diffusion of the tractor in American Agriculture, 1910-1960. J. Econ. Hist. 61, 663–698.

Oltjen, J.W., Beckett, J.L., 1996. Role of ruminant livestock in sustainable agricultural systems. J. Anim. Sci. 74, 1406–1409.

Pellant, M., Shaver, P., Pyke, D.A., Herrick, J.E., 2005. Interpreting Indicators of Rangeland Health, Version 4. Technical Reference 1734-6. U.S. Department of the Interior, Bureau of Land Management, National Science and Technology Center, Denver, CO122., BLM/WO/ST-00/001+ 1734/REV05.

Power, A.G., 2010. Ecosystem services and agriculture: tradeoffs and synergies. Philos. Trans. R. Soc. B 365, 2959–2971.

Raper, R.L., Bailey, A.C., Burt, E.C., Way, T.R., Liberati, P., 1995. Inflation pressure and dynamic load effects on soil deformation and soil-tire interface stresses. Trans ASAE 38, 685–689.

Risch, A.C., Frank, D.A., 2006. Carbon dioxide fluxes in a spatially and temporally heterogeneous temperate grassland. Oecologia 147, 291–302.

Robertson, G.P., Allen, V.G., Boody, G., Boose, E.R., Creamer, N.G., Drinkwater, L.E., et al., 2008. Long-term agricultural research: a research, education, and extension imperative. BioScience 58, 640–645.

Russelle, M.P., Entz, M.H., Franzluebbers, A.J., 2007. Reconsidering integrated crop–livestock systems in North America. Agron. J. 99 (2), 325–334.

Sadler, E.J., Lerch, R.N., Kitchen, N.R., Anderson, S.H., Baffaut, C., Sudduth, K.A., et al., 2015. Long-term agroecosystem research in the Central Mississippi River Basin: introduction, establishment, and overview. J. Environ. Qual. 44, 3–12.

Sanderson, M.A., 2015. Agroecosystem diversity and pollinator ecosystem services on the northern Great Plains. Agronomy Abstracts paper No. 91924.

Sanderson, M.A., Watzold, F., 2010. Balancing tradeoffs in ecosystem functions and services in grassland management. Grassland Sci. Eur. 15, 639–648.

Sanderson, M.A., Jolley, L.M., Dobrowolski, J.P., 2012. Pastureland and hayland in the USA. In: Nelson, C.J. (Ed.), Environmental Outcomes of Conservation Practices Applied to Pasture and Hayland in the U.S: The Pastureland Conservation Effects Assessment Project (CEAP). Allen Press, Lawrence, KS, USA, pp. 25–40.

Sanderson, M.A., Archer, D., Hendrickson, J., Kronberg, S., Liebig, M., Nichols, K., et al., 2013. Diversification and ecosystem services for conservation agriculture: outcomes from pastures and integrated crop–livestock systems. Renew. Agric. Food Syst. 28, 129–144.

Sanderson, M.A., Liebig, M.A., Hendrickson, J.R., Kronberg, S.L., Toledo, D., Derner, J.D., et al., 2015. Long-term agroecosystem research on northern Great Plains mixed-grass prairie near Mandan, North Dakota. Can. J. Plant Sci. 95, 1101–1116.

Sayre, N.F., Carlisle, L., Huntsinger, L., Fisher, G., Shattuck, A., 2012. The role of rangelands in diversified farming systems: innovations, obstacles, and opportunities in the USA. Ecol. Soc. 17 (4), 43. http://dx.doi.org/10.5751/ES-04790-170443.

Scherr, S.J., McNeely, J.A., 2008. Biodiversity conservation and agricultural sustainability: towards a new paradigm of 'ecoagriculture' landscapes. Philos. Trans. R. Soc. B 363 (1491), 477–494.

Schiere, J.B., Ibrahim, M.N.M., Van Keulen, H., 2002. The role of livestock for sustainability in mixed farming: criteria and scenario studies under varying resource allocation. Agric. Ecosyst. Environ. 90, 139–153.

Schuman, G.E., Reeder, J.D., Manley, J.T., Hart, R.H., Manley, W.A., 1999. Impact of grazing management on the carbon and nitrogen balance of a mixed-grass rangeland. Ecol. Appl. 9, 65–71.

Sharpley, A.N., Syers, J.K., 1979. Loss of nitrogen and phosphorus in tile drainage as influenced by urea application and grazing animals. New Zeal. J. Agric. Res 22, 127–131.

Sinclair, T.R., Sinclair, C.J., 2010. Bread, Beer and the Seeds of Change: Agriculture's Imprint on World History. CABI, Cambridge, MA.

Skinner, R.H., 2008. High biomass removal limits carbon sequestration of mature temperate pastures. J. Environ. Qual. 37, 1319–1326.

Smith, L.C., Monaghan, R.M., 2003. Nitrogen and phosphorus losses in overland flow from a cattle-grazed pasture in Southland. New Zeal. J. Agric. Res. 46, 225–237.

Snapp, S.S., Swinton, S.M., Labarta, R., Mutch, D., Black, J.R., Leep, R., et al., 2005. Evaluating cover crops for benefits and costs and performance within cropping system niches. Agron. J. 97, 322–332.

Sollenberger, L.E., Agouridis, C.T., Vanzant, E.S., Franzluebbers, A.J., Owens, L.B., 2012. Prescribed grazing in pasturelands. In: Nelson, C.J. (ed.), Environmental Outcomes of Conservation Practices Applied to Pasture and Hayland in the U.S: The Pastureland Conservation Effects Assessment Project (CEAP).

Spivak, M., Mader, E., Vaughan, M., Euliss jr., N.H., 2011. The plight of bees. Environ. Sci. Technol. 45, 34–38.

Springsteen, A., Loya, W., Liebig, M., Hendrickson, J., 2010. Soil carbon and nitrogen across a chronosequence of woody plant expansion in North Dakota. Plant Soil 328, 369–379.

Steiner, J.L., Starks, P.J., Garbrecht, J.D., Moriasi, D.N., Zhang, X., Schneider, J.M., et al., 2014. Long-term environmental research: the upper Washita river experimental watersheds, Oklahoma, USA. J. Environ. Qual. 43, 1227–1238.

Studdert, G.A., Echeverria, H.E., Casanovas, E.M., 1997. Crop-pasture rotation for sustaining the quality and productivity of a Typic Argiudoll. Soil Sci. Soc. Am. J 61, 1466–1472.

Sugden, E.A., 1985. Pollinators of *Astragalus monoensis* Barneby (Fabaceae): New host records; Potential impact of sheep grazing. Great Basin Nat. 45, 299–312.

Sulc, R.M., Tracy, B.F., 2007. Integrated crop–livestock systems in the US Corn Belt. Agron. J. 99, 335–345.

Svejcar, T., Angell, R., Bradford, J.A., Dugas, W., Emmerich, W., Frank, A.B., et al., 2008. Carbon fluxes on North American rangelands. Range Ecol. Manage. 61, 465–474.

Swinton, S.M., Lupi, F., Robertson, G.P., Landis, D.A., 2006. Ecosystem services from agriculture: looking beyond the usual suspects. Am. J. Agric. Econ. 88, 1160–1166.

Tanaka, D.L., Krupinsky, J.M., Liebig, M.A., Merrill, S.D., Ries, R.E., Hendrickson, J.R., et al., 2002. Dynamic cropping systems. Agronomy J. 94, 957–961.

Tanaka, D.L., Karn, J.F., Scholljegerdes, E.J., 2008. Integrated crop/livestock systems research: practical research considerations. Renew. Agric. Food Syst. 23, 80–86.

Thirsk, J., 1997. Alternative Agriculture: A History: From the Black Death to the Present Day: From the Black Death to the Present Day. Oxford University Press Inc., New York, NY, USA.

Thurow, T.L., Blackburn, W.H., Taylor Jr., C.A., 1988. Infiltration and interrill erosion responses to selected livestock grazing strategies, Edwards Plateau, Texas. J. Range Manage. 41, 296–302.

Tilman, D., Fargione, J., Wolff, B., D'Antonio, C., Dobson, A., Howarth, R., et al., 2001. Forecasting agriculturally driven global environmental change. Science 292, 281–284.

Tracy, B.F., Frank, D.A., 1998. Herbivore influence on soil microbial biomass and nitrogen mineralization in a northern grassland ecosystem: Yellowstone National Park. Oecologia 114, 556–562.

Tracy, B.F., Zhang, Y., 2008. Soil compaction, corn yield response, and soil nutrient pool dynamics within an integrated crop-livestock system in Illinois. Crop Sci. 48, 1211–1218.

Trimble, S.W., Mendel, A.C., 1995. The cow as a geomorphic agent—a critical review. Geomorphology 13, 233–253.
USDA National Agricultural Statistics Service, 2014. Cattle – final estimates 2009-2013. Statistical Bulletin No. 1034.
USDA Natural Resources Conservation Service, 2010. Management considerations for grassland birds in northeastern haylands and pasturelands. Wildlife Insight, Washington, DC.
van Elsen, T., 2000. Species diversity as a task for organic agriculture in Europe. Agric. Ecosyst. Environ. 77, 101–109.
Van Haveren, B.P., 1983. Soil bulk density as influenced by grazing intensity and soil type on a shortgrass prairie site. J. Range Manage. 36, 586–588.
Verburg, K., Bond, W.J., Hirth, J.R., Ridley, A.M., 2008. Lucerne in crop rotations on the Riverine Plains. 3. Model evaluation and simulation analyses. Crop Pasture Sci. 58, 1129–1141.
Walbridge, M.R., Shafer, S.R., 2011. A long-term agro-ecosystem research (LTAR) network for agriculture. In: C.N. Medley, G. Patterson, M.J. Parker (Eds.), Observing, Studying, and Managing Change: Proceedings of the Fourth Interagency Conference in the Watersheds. US Geological Survey Scientific Investigations Report # 2011-5169, pp. 26–30.
Wang, K.H., McSorley, R., Bohlen, P., Gathumbi, S.M., 2006. Cattle grazing increases microbial biomass and alters soil nematode communities in subtropical pastures. Soil Biol. Biochem. 38, 1956–1965.
Ward, P.R., Micin, S.F., Dunin, F.X., 2006. Using soil, climate, and agronomy to predict soil water use by lucerne compared with soil water use by annual crops or pastures. Crop Pasture Sci. 57, 347–354.
Werling, B.P., Dickson, T.L., Isaacs, R., Gaines, H., Gratton, C., Gross, K.L., et al., 2014. Perennial grasslands enhance biodiversity and multiple ecosystem services in bioenergy landscapes. Proc. Natl. Acad. Sci. 111, 1652–1657.
Wienhold, B.J., Hendrickson, J.R., Karn, J.F., 2001. Pasture management influences on soil properties in the Northern Great Plains. J. Soil Water Conserv. 56, 27–31.
Willatt, S.T., Pullar, D.M., 1984. Changes in soil physical properties under grazed pastures. Soil Res. 22, 343–348.
Woodcock, B.A., Savage, J., Bullock, J.M., Nowakowski, M., Orr, R., Tallowin, J.R.B., et al., 2014. Enhancing floral resources for pollinators in productive agricultural grasslands. Biol. Cons. 171, 44–51.
Wright, A.L., Hons, F.M., Rouquette, F.M., 2004. Long-term management impacts on soil carbon and nitrogen dynamics of grazed bermudagrass pastures. Soil Biol. Biochem. 36, 1809–1816.

CHAPTER 8

Intensified Agroecosystems and Their Effects on Soil Biodiversity and Soil Functions

Mathew E. Dornbush[1] and Adam C. von Haden[2]

[1]*University of Wisconsin-Green Bay, Green Bay, WI, United States* [2]*University of Wisconsin-Madison, Madison, WI, United States*

8.1 Introduction

Human modification and the intensified use of lands and seas is a defining signature of the last 300 years (Meyer and Turner, 1992; Ellis and Ramankutty, 2008). For much of the terrestrial biosphere, these changes have come in the form of an increased extent of, and intensive use for, agricultural production (Matson et al., 1997; Tscharntke et al., 2012). To the benefit of humanity, these efforts have significantly increased crop yields and reduced human hunger globally (Borlaug, 2007). Nevertheless, the collective "tyranny of small decisions" (Odum, 1982) associated with intensified agriculture has homogenized landscapes (Brown and Schulte, 2011), both physically and biologically, placed significant stress upon soils used for agricultural production (Anonymous, 2004), and unintentionally altered the ecosystem services provided to humanity (Foley et al., 2005). This chapter defines agricultural intensification and ecological services, highlights various ways by which agricultural intensification of row crop agriculture has occurred historically, and discusses how these approaches have affected selected soil parameters to the detriment of soil health, soil biodiversity, and the ecosystem services supplied by soil ecosystems. To provide new insight into these well-documented patterns, the chapter describes the implications of these changes to both the mean level and heterogeneity of soil parameters, rather than focusing solely on average changes to soil parameters. The chapter ends with a discussion of how a seeming societal transition toward a greater valuation of noncommodity services is promoting the application of ecological intensification to restore diversity, system heterogeneity, and soil ecosystem health.

Agricultural intensification is generally defined as an increase in agricultural commodity production per unit area and time (Donald et al., 2001). Agricultural intensification in row

Soil Health and Intensification of Agroecosystems.
DOI: http://dx.doi.org/10.1016/B978-0-12-805317-1.00008-7

crop systems has been most pronounced in developed countries, with significant increases in crop yields occurring over the last half century (Rudel et al., 2009; Neumann et al., 2010). While the expansion of agricultural intensification into new regions has contributed to increased global yield during this period, yield increases far surpassed land conversion rates, stressing that yield gains came primarily from increased production on existing agricultural lands (Rudel et al., 2009). However, the mechanisms responsible for increased yields, and thus the potential impact that agricultural intensification has on ecosystems and associated ecosystem services, are quite distinct depending upon both the agricultural sector and process of interest.

Matson et al. (1997) emphasized the increased use of high-yielding crop varieties, fertilization, irrigation, pesticides, and mechanization in modern industrial agriculture (Fig. 8.1). The widespread adoption of genetically modified organisms (GMOs) now warrants mention when defining agricultural intensification, as the planted acreage of GMO crops increased 100-fold between 1996 and 2012 alone (Clive, 2012). The cumulative trend associated with agricultural intensification is the replacement of human and animal labor with fossil-fuel-powered mechanization, a reduced dependence on natural and mechanical controls for weeds and pests in favor of agrochemicals, an exchange of organic for inorganic fertilizers, the replacement of traditional crop varieties for carefully bred and vetted hybrids, and a reduced dependence upon traditional breeding and species-specific gene pools (Phipps and Park, 2002).

An alternative and perhaps more beneficial perspective, is to view the intensification of row crop agriculture as the intentional application of external inputs (e.g., fossil fuels, engineering, horticulture, and biotechnology) into agricultural systems in an attempt to close the *yield gap* between the *potential yield*, set by climatic limitations from solar radiation and temperature, and the *realized yield* obtained by producers (Bommarco et al., 2013). As is observed with many intentional manipulations of ecosystem services (e.g., water quality treatment), initial external investments provide large initial improvements in services (e.g., yield), but returns on investment diminish through time as one seeks to continually enhance a given service to a higher and higher level (Cunningham and Cunningham, 2015). Viewing intensification within this framework identifies a reality of decreasing yield gains relative to economic inputs as maximum potential yields are approached (Cassman, 1999). Matson et al. (1997) highlight another important pattern associated with this transition; because agricultural intensification has focused almost exclusively on maximizing a sole ecosystem service, yield, there has been an inherent undervaluing of and associated decline in many of the other services historically provided by agricultural soil ecosystems (Fig. 8.1). The wide-ranging impacts that these changes have on both the probability of sustaining current yields into the future and the unintended costs to society spurred by these changes is reflected in the large and emerging body of work focused on reducing the environmental impacts of modern agriculture (Robertson et al., 2014; Rey et al., 2015).

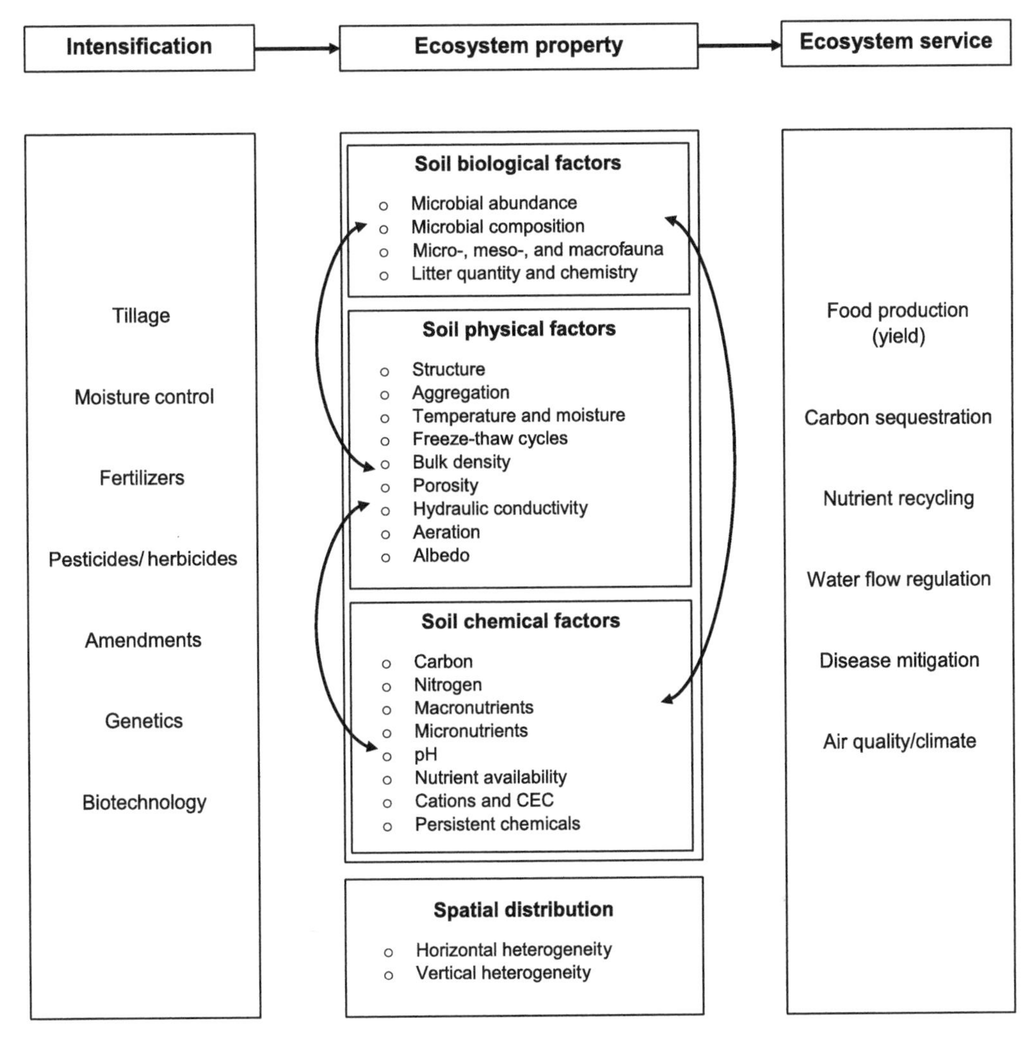

Figure 8.1
Relationships among select components of agricultural intensification, the soil properties they affect, and the resulting ecosystem service outputs.

8.2 Ecosystem Services Defined

Ecosystem services are defined simply as the benefits provided to humans by ecosystems (MEA, 2005). It is intentional that this definition includes both managed and unmanaged ecosystems, as human-managed ecosystems cover more than 75% of the earth's ice-free land area (Ellis and Ramankutty, 2008). This statistic emphasizes that services historically provided by "natural" areas are currently, or increasingly, unlikely to meet human needs. The remaining option is a growing dependence on the delivery of multiple services from the agroecosystems within which most humans live.

Ecosystem services are generally divided into four subcategories: provisioning services, regulating services, cultural services, and supporting services (MEA, 2005). *Provisioning services* include the products people acquire from ecosystems, such as food, biofuels, fiber, freshwater, or genetic resources (MEA, 2005). These services generally have a clear and recognized market value, and thus are often already well represented within existing economic systems. *Regulating services* provide the renewal processes needed to maintain basic human needs, such as clean air and water, erosion control, disease regulation, or climate mitigation (MEA, 2005). *Cultural services* include nonmaterial benefits to humanity (MEA, 2005). In his very forward-looking paper, Janzen (1988) justifies the necessity of restoring cleared tropical forests as akin to the need for maintaining the cultural assets of the world's great museums, libraries, or galleries; to serve as societal centers for spurring intellectual wonder, growth, and innovation. This example captures the heart of cultural services, those services providing humans with spiritual enrichment, recreational opportunities, or aesthetic enjoyment (MEA, 2005). Finally, *supporting services* are needed to produce all other services (MEA, 2005). Nutrient cycling and soil formation are two commonly used examples of supporting services. In this chapter, it is emphasized that, to date, intensified agriculture has most closely reflected historic markets by emphasizing provisioning services (i.e., yield) over all others. In fact, the low emphasis placed on nonprovisioning services largely accounts for the documented, yet unintended, degradation of other essential services available from agroecosystems.

8.3 Intensification of Row Crop Systems

8.3.1 Tillage Effects on Vertical Distributions of Soil Organic Carbon and Microorganisms

Intensification of row crop systems is often associated with changes in tillage practices. Perhaps the most widely told story relating to the unintended effects of tillage on soil ecosystems describes the large and widespread loss of organic matter following conversion of native grasslands of the Midwest United States into row crop agriculture. Specifically, soil organic carbon (SOC) losses of 40–60% relative to historic levels have been estimated for this region (Parton et al., 2005). However, the degree to which tillage practices impact soil properties depends upon the intensity of tillage system, not simply its presence or absence. Using a broad brush, soil tillage represents soil disturbance, which is not in itself an unnatural process. When viewed along a continuum, soil disturbance in natural systems is dominated by bioturbation from organisms ranging in size from wallows created by large grazers (e.g., bison [*Bos bison* L.]) to burrowing by pocket gophers (*Geomys bursarius* Shaw), earthworms (*Lumbricus terrestric* L.), or other invertebrates (Wilkinson et al., 2009). Within annual cropping systems, no-tillage agriculture most closely mimics this "natural" condition, followed by reduced or conservation tillage practices, and finally conventional tillage practices associated with the largest physical disruption to the soil system. While exact definitions are a bit elusive, conservation tillage generally includes a suite of practices that retain 30% or greater residue cover on the soil surface (Baker et al., 2007). Implicit in this definition is an acknowledgment that the end location and associated environment where aboveground litter decomposes also differs along the tillage intensity gradient. These changes have profound impacts on both the physical and biological attributes of the soil system.

Aboveground litter in natural systems, and to a lesser extent in no-tillage agricultural systems, is deposited and ultimately decomposes on the soil surface, producing a characteristic litter layer above the mineral soil. While the depth of the litter layer varies among ecosystem types (e.g., forests verses grasslands), and for various reasons (e.g., fire, grazing), soil organic matter formation within the soil profile of natural systems is consistently driven by root, not aboveground litter, inputs (Jobbágy and Jackson, 2000; Rasse et al., 2005). While some select soils (e.g., vertisols) and animal activities provide for a limited, and spatially heterogeneous, natural movement of surface litter downward into the soil profile, conventional tillage appears to represent a unique disturbance in its temporal consistency, its capacity for moving large quantities of surface litter downward, and in its spatially extensive signature across the landscape. Agroecosystem-specific differences in the type (roots vs. roots plus aboveground litter) and depth (only roots at depth vs. roots and aboveground litter at depth) of organic-C inputs produce notable and associated differences in vertical SOC distributions. For example, SOC concentrations are proportionally greater at shallower depths within no-tillage systems, and proportionally greater at depth within conventional tillage systems because of the mixing effect (Baker et al., 2007). When viewed within the context of soil heterogeneity, conventional tillage appears to reduce vertical SOC heterogeneity, while systems with reduced disturbances, including perennial natural systems, tend to concentrate SOC near the surface, and thus increase vertical SOC heterogeneity. These patterns appear to hold even in vertisols (Dalal et al., 1991), again emphasizing the uniqueness of tillage relative to natural vertical mixing.

From an ecosystem services perspective, differences in the relative location and source of SOC between natural, no-tillage, and conventional tillage systems are important, as organic C affects a wide suite of soil processes, from water infiltration rates to soil erosion, soil nutrient cycling, habitat structure for microorganisms, and biogeochemical processes with global implications. For example, using phospholipid fatty acid analysis, Fierer et al. (2003) documented increases in Gram-positive bacteria and *Actinomycetes* and decreases in Gram-negative bacteria, fungi, and protozoa with depth, which they concluded were driven by differences in SOC availability and quality across the soil profile (Table 8.1). Vertical patterns of total and specific microbial enzyme activity follow similar declining patterns with depth, suggesting a link between SOC availability, microbial community structure, and potential SOC turnover rates (Schnecker et al., 2015). For example, Kandeler et al. (1999) reported greater nitrogen (N) mineralization and nitrification potentials within the 0–10 cm depth

Table 8.1: The vertical distribution of soil microbial groups (mean and standard error) within a terrace soil profile

Sampling Depth (cm)	Gram-Positive Bacteria (mol%)	Gram-Negative Bacteria (mol%)	Actinomycetes (mol%)	Protozoa (mol%)	Fungi (mol%)
0–5	22 (0.95)	24 (0.98)	4.7 (1.4)	0.70 (0.026)	2.9 (0.16)
5–15	25 (0.59)	20 (0.29)	11 (0.082)	0.62 (0.086)	1.8 (0.36)
15–25	27 (0.40)	18 (0.33)	12 (0.46)	0.50 (0.15)	1.1 (0.074)
50	32 (0.88)	13 (0.11)	17 (1.2)	0[a]	1.1 (0.21)
100	24 (4.1)	13 (3.6)	13 (0.86)	0[a]	1.4 (0.33)
200	28 (3.2)	5.6 (0.44)	10 (2.1)	0[a]	0[a]

Adapted from Fierer, N., Schimel, J.P., Holden, P.A., 2003. Variations in microbial community composition through two soil depth profiles. Soil Biol. Biochem. 35, 167–176.

[a] Mol% below the detection limit of the instrument.

in no-tillage than conventional tillage fields, but this trend was reversed at the 20–30 cm depth. Kandeler et al. (1999) attributed this pattern, in part, to differences in the vertical distribution of microbial biomass between these two systems. Thus, tillage-induced changes in the vertical distribution of SOC alter the vertical distribution of microbial abundance and structure, thereby altering the depth distribution of many other soil ecosystem processes.

While fields converted from conventional tillage into both no-tillage row crops and perennial grasslands are characterized by similar patterns of proportionally larger near-surface C pools and near-surface SOC accrual rates (Kucharik, 2007), perennial systems benefit from significantly deeper root systems with larger, and temporally more consistent organic-C inputs to the soil than no-tillage systems (Tufekcioglu et al., 1999). Thus, while there are many profile similarities between no-tillage systems and restored perennial grasslands, larger and deeper belowground inputs in perennial systems appear to increase the potential for supporting larger, more active microbial communities, for maintaining SOC increases for longer periods of time, and for accumulating SOC to larger maximum levels (Rasse et al., 2005). An underappreciation of this nuanced difference in root inputs and vertical SOC distributions appears to have contributed in part to overly optimistic expectations regarding the merits of expanding no-tillage agriculture alone as a promising strategy for sequestering C within agricultural soils (Baker et al., 2007). Differences in SOC accrual patterns between no-tillage and perennial grasslands may be partially mitigated through increased cover crop use to supplement the size, depth, and temporal consistency of belowground inputs in row crop systems (Sainju et al., 2002). Irrespective, from an ecosystem service perspective, it may still be preferable to concentrate SOC in near-surface soil horizons, even if the total quantity of SOC is identical, or slightly less, between no-tillage and conventional tillage systems. Essentially, because the majority of water infiltration and nutrient cycling processes occur near the soil surface (Schnecker et al., 2015), and because these processes are strongly influenced by soil C levels (Zak et al., 1994), soils with greater concentrations of C near the surface may be preferable to soils with similar total C quantities spread across the soil profile. This suggestion deserves greater attention, particularly when also considering whether shallow or deeper SOC differs in its resistance to changes in management and climate (i.e., C security).

8.3.2 Tillage Effects on Soil Organisms, Soil Structure, and Aggregation

Tillage generally reduces total soil microbial biomass, fungal biomass, and the ratio of fungi to bacteria (Fig. 8.2) (Beare et al., 1992; Frey et al., 1999). These changes may initially occur via the direct destruction of fungal hyphal networks by physical disturbance (Frey et al., 1999). Relative decreases in soil fungi under tillage may also result indirectly from the spatial homogenization of organic substrates or changes in moisture availability (Beare et al., 1992; Frey et al., 1999). Fungi tend to have higher carbon (C) use efficiencies than bacteria

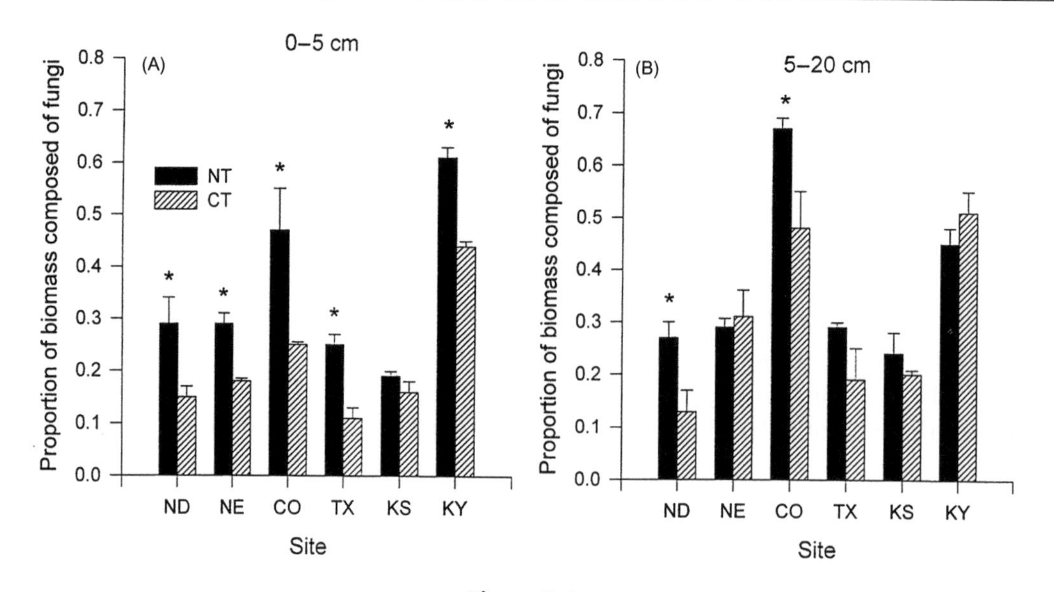

Figure 8.2

Fungal dominance of no-tilled (NT) verses conventionally tilled (CT) agricultural soils among six sites at the (A) 0–5 cm and (B) 5–20 cm depths (Frey et al., 1999).

(Keiblinger et al., 2010), and therefore a reduction in relative fungal biomass increases the proportion of C respired relative to C allocated to microbial growth. Since most SOC is microbially derived (Gleixner, 2013), a reduction in microbial substrate use efficiency in tilled systems may diminish the overall potential for SOC formation (Cotrufo et al., 2013).

Whether a direct or indirect effect, tillage also affects soil faunal groups, with particularly negative effects upon those with larger body sizes (Postma-Blaauw et al., 2010). Although soil fauna represent a small portion of soil organism biomass (Fierer et al., 2009), they play significant roles as soil ecosystem engineers (Lavelle et al., 1997). Less frequent tillage can promote diversity of larger soil organisms such as oribatid mites, collembolans, and earthworms (Tsiafouli et al., 2015), thereby leading to improvements in soil structure and nutrient cycling potential (Chan, 2001). Overall, increased tillage intensity reduces soil heterogeneity at small spatial scales, with apparent reductions to the overall complexity of the soil community.

Tillage and other soil disturbances also directly alter soil structure, with significant effects on soil physical, chemical, and biological properties and subsequent ecosystem processes. Soil structural stability decreases along a disturbance gradient from native ecosystems to no-tillage agriculture to conventionally tilled systems (Six et al., 1998). Decreased soil stability following tillage does not necessarily result from lower total soil aggregation, per se (Verchot et al., 2011). Rather, soil tillage preferentially destroys macroaggregates (>250 μm), as

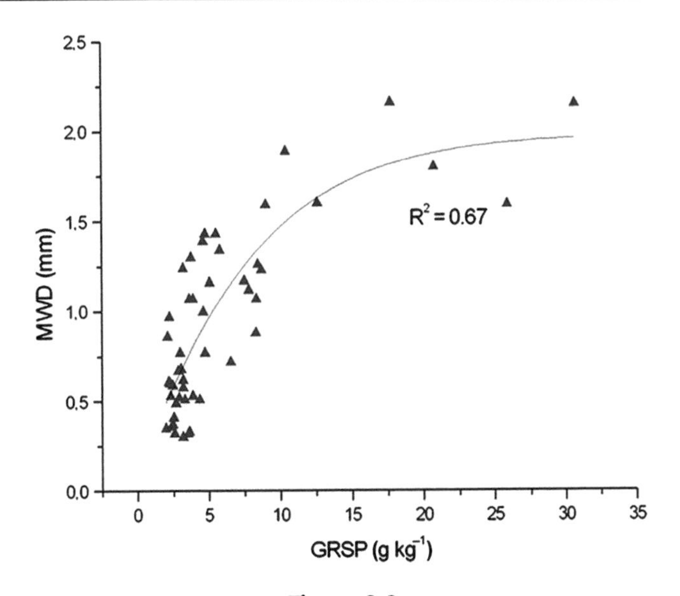

Figure 8.3
Relationship between glomalin-related soil protein (GRSP) and aggregate mean weighted diameter (MWD) (Spohn and Giani, 2010).

evidenced by shifts toward smaller aggregate size distribution and smaller aggregate mean weighted diameter in tilled soils (Grandy and Robertson, 2006). Reduction in soil fungi under tillage decreases the production of glomalin binding agents, hindering the reformation of soil macroaggregates, and thus reduces the potential for SOC stabilization within aggregates (Fig. 8.3) (Jastrow et al., 1998; Six et al., 2006; Spohn and Giani, 2010).

The ecological ramifications of tillage-induced decreases in soil aggregation are far-reaching. For example, reduced aggregation leaves soils more vulnerable to soil loss via water erosion (LeBissonnais, 1996). Decreased soil aggregation can reduce the capillary pore space and subsequently limit soil water storage potential and plant water availability (Ball-Coelho et al., 1998). A shift towards smaller aggregate size distribution may decrease the probability of intra-aggregate anaerobic zones, potentially leading to lower denitrification rates (Sexstone et al., 1985). For example, Doran (1980) found that tillage typically decreased denitrifiers near the soil surface by 85% and increased nitrifiers deeper in the soil by about 55%, thus leading to overall higher nitrification and lower denitrification potential within tilled systems. Soil aggregation also plays a central role in SOC dynamics, with major implications for nutrient cycling and C sequestration. By physically separating organic substrates from the microbial community and providing less favorable environmental conditions, soil aggregates protect SOC from decomposition (Six et al., 2002) and thus conventional tillage generally leads to SOC loss (West and Post, 2002).

8.3.3 Soil Moisture Control and Soil Microorganisms

Fundamental to reducing the agricultural yield gap is the delivery of reliable, adequate, and relatively stable soil moisture content (Jackson et al., 2001). Globally, humans now appropriate approximately 50% of surface freshwater (Postel et al., 1996; Jackson et al., 2001), with agriculture accounting for roughly 33% of consumptive uses (Jackson et al., 2001). While discussions often focus on these supply-side demands for crop production, in rainfed agricultural regions one of the most notable water-related impacts on ecosystem services is the observed increase in surface runoff and transport of sediment and nutrients into adjacent waterbodies (Bennett et al., 2001). Viewed within a soil disturbance continuum of perennial to conventional tillage systems, changes (generally decreases) in maximum and seasonally averaged leaf area index (LAI) and surface litter mass are associated with changes (again generally decreases) in water interception storage and evapotranspiration (Chapin et al., 2011). Combined reductions to LAI and surface litter reduce the vertical space available to store precipitation, thereby increasing water inputs to the mineral soil surface, increasing water available for either infiltration or runoff (Chapin et al., 2011). Reductions in the canopy and litter layer also increase the energy associated with raindrops as they hit the soil surface, or as water moves across soil, increasing the potential for soil crusting and erosion (Chapin et al., 2011). The discussion that follows focuses on less-explored impacts of agricultural intensification on the soil system, those occurring below the canopy and litter layer, emphasizing feedbacks to the soil system.

As discussed previously, tillage is a homogenizing practice, both within the soil profile and across the soil surface. Specifically, tillage notably reduces microtopographic differences in surface elevation, thereby reducing surface water storage across the landscape. For example, natural forest ecosystems often possess extensive *pit-and-mound topography* caused by the uprooting of wind-thrown trees (Beatty, 2003). *Mima mounds*, circular soil hillocks approximately 1 m in height, are found within natural landscapes on all continents except Antarctica (Gabet et al., 2003), and *gilgai*, a complex of circular soil mounds and linear ridges and depressions, are common to undisturbed shrink-swell soils, such as Vertisols (Kishné et al., 2014). These and like formations tend to be abundant, often dominant features of pretillage, or unmanaged, landscapes that collectively create undulating soil surfaces supporting a heterogeneous, juxtaposition of drained soils and surface ponding. For example, Ulanova (2000) reported that pit-and-mound topography covered 7–12% of the soil surface in a natural, uneven age spruce forest, and Kishné et al. (2014) estimated that gilgai microtopography in a Texas, United States, pasture held approximately 0.024 m^3 of water m^{-2}, summing to a capture of 43.74 m^3 of water within their 1800 m^2 study area. While relatively few attempts have been made to quantify the impact of these features on surface water storage and runoff, the evidence to date suggests that the homogenizing effects of tillage on the soil surface result in an associated reduction in surface water storage, and a permanent increase in surface runoff.

Historically, widespread surface ponding also occurred within recently glaciated regions across slightly larger spatial scales. For example, the Midwestern United States agricultural states of Illinois, Iowa, Minnesota, and Wisconsin averaged from 11 to 28% wetland coverage in their preagricultural states (Dahl, 1990). These wetlands are generally too large to expect any significant alteration to hydrology with tillage alone, requiring a more intentional and energy-intensive approach to controlling soil moisture levels. Moisture control within these landscapes is accomplished through the installation of tile drains and manmade ditches. Mitsch et al. (2001) estimated that the Mississippi River Basin lost more than 30 million ha of wetlands during the 20th century, with 19 million ha lost from the extensively tile-drained states of Indiana, Illinois, Iowa, Missouri, Ohio, and Wisconsin alone. Tiles and ditches lower water tables (Schilling et al., 2006), increase the rate by which water is transported from uplands to streams, and serve as direct conduits for leached nutrients into adjacent water bodies (Bolton et al., 1970; Sims et al., 1998), effectively shifting many microbial-mediated nutrient cycling processes from terrestrial to aquatic systems.

Soil moisture and oxygen availability are important biophysical constraints on soil communities and microbial-driven soil biogeochemical processes. For example, over prolonged periods, wetter soils typically support larger populations of herbaceous nematodes, oribatid mites, and collembolans (Todd et al., 1999; Tsiafouli et al., 2005), and wetland drainage results in a relatively rapid and large loss of SOC (Euliss et al., 2006). Reciprocally, historic changes to drainage must also be acknowledged for their effect on potential SOC recovery, as evidenced by the strong relationship between SOC accrual rates and soil moisture levels (O'Brien et al., 2010). The relationship between soil moisture and SOC accrual is likely a function of reduced microbial activity in water-saturated soils. For example, von Haden and Dornbush (2014) found that soil moisture was negatively related to root decomposition and positively related to SOC, thus linking soil moisture, decomposition, and SOC accrual potential. Similar to the effects of tillage, soil-moisture control strategies by intensive agriculture in rainfed systems appear to have a homogenizing effect on soil C pools and processes, as evidenced by greater SOC variability (i.e., coefficient of variation) in remnant grasslands compared to younger, restored grasslands (Kucharik et al., 2006). Thus, the effects of intensified agriculture on soil moisture levels have likely contributed to reduced SOC heterogeneity and associated distinct C-hotspot processing points within the landscape, and places underappreciated constraints on SOC accrual potential.

Challenges to identifying the effects of altered hydrology on microbial processes result in part from the high plasticity with which microbes and microbial communities respond to changing soil moisture conditions. For example, over short time scales, facultative anaerobic microbes can cope with changing soil moisture conditions by switching from aerobic respiration when oxygen is available to anaerobic metabolisms such as denitrification when oxygen becomes limiting (Zumft, 1997). Thus, irrigation and rain events typically trigger immediate increases in nitrous oxide emissions (Scheer et al., 2012). The ratio of fungi to bacteria can also decrease within hours of soil wetting (Burger et al., 2005), as bacterial populations are typically more responsive than fungi to changes in soil moisture (Barnard et al., 2013). Changes in the relative abundances of bacterial groups are also commonly observed in response to soil moisture changes. For example, Barnard et al. (2013) reported consistent decreases in *Actinobacteria* and increases in *Acidobacteria* when soils were wetted. These changes in the microbial community can be persistent in regularly irrigated soils (Williams and Rice, 2007), but may be short-lived if soils are allowed to redry (Burger et al., 2005). As a result of soil rewetting and subsequent release of microbial cytoplasmic osmolytes (Fierer and Schimel, 2003), bacterial growth and C respiration rapidly increase by as much as two- and tenfold, respectively (Iovieno and Bååth, 2008). However, the ramifications of water-induced changes to the microbial community are largely dependent on the frequency of wet and dry cycles. For example, Tiemann and Billings (2011) showed that less frequently watered soils had greater enzyme activities and lower C use efficiency than soils that were watered more frequently with the same total amount of water. Consequently, total C mineralization was greater in the less frequently watered soils in which soil moisture was temporally more variable (Tiemann and Billings, 2011). These dynamics help to explain the SOC sequestration that is typically observed in irrigated compared to nonirrigated systems, particularly in arid environments (Trost et al., 2013).

8.3.4 Fertilization and Soil Microorganisms

The Green Revolution of the mid-1900s represents the most dramatic shift in agricultural practices in human history. As a result of crop genetic improvements, and increased fertilizer and irrigation use, cereal crop production has tripled since the 1960s (Pingali, 2012). During that same time period, the efficiency of fertilizer uptake by plants rapidly declined, indicating that increased fertilizer use outpaced yield improvements (Tilman et al., 2002). At modern fertilizer application rates, only 30–50% of N is recovered by the crop, indicating that 50–70% of applied N remains unused (Ladha et al., 2005). On a regional basis, intensification of livestock production tends to separate grain and livestock production, favoring a shift toward industrial-scale farms (e.g., concentrated animal feeding operations), and often disconnecting crop and animal nutrient cycles (Gerber et al., 2005). With the intensification and concentration of animal agriculture, the manure application intensity (amount per area) subsequently increases (Key et al., 2011). The excess of both organic and inorganic fertilizers has short- and long-term implications for ecosystem processes both within the agricultural system itself and within downstream aquatic systems (Carpenter et al., 1998).

Considering that soil microbes drive many soil biogeochemical processes and are often limited by N and phosphorus (P) availability (Hartman and Richardson, 2013), fertilization has major effects on ecosystem services via changes in microbial structure and function.

Although inorganic fertilizers, particularly urea, typically decrease total microbial biomass in agricultural systems (Lu et al., 2011b), organic fertilizers (e.g., manure) usually increase soil microbial biomass (Böhme et al., 2005). Soil microbial diversity may decrease slightly with fertilizer additions (Zeng et al., 2016), but this effect is apparently contingent upon ecosystem type (Leff et al., 2015). More importantly, soil microbial communities change in a predictable, functionally relevant manner under fertilization. When fertilizer is applied, soil bacterial communities shift from those with oligotrophic (slow metabolism) to copiotrophic (fast metabolism) life histories (Ramirez et al., 2012; Leff et al., 2015). Common soil bacterial taxa *Alphaproteobacteria* and *Actinobacteria* typically increase, whereas *Deltaproteobacteria* and *Acidobacteria* usually decrease under fertilization (Leff et al., 2015). Mycorrhizal fungi and methanogenic archaea both decrease with N and P fertilization (Leff et al., 2015). The relative abundances of microbial functional groups involved with N processing may increase with fertilization, thereby resulting in greater nitrification, denitrification, and nitrous oxide production (Lu et al., 2011a; Clark et al., 2012; Sun et al., 2015). For example, in a global metaanalysis, Lu et al. (2011a) reported that N fertilization induced a 25% increase in N mineralization, a 154% increase in nitrification, an 84% increase in denitrification, and stimulated gaseous losses of N_2O and N leachate losses by 134% and 461% relative to control treatments, respectively. Fertilizer-induced shifts in the microbial community result both directly from increased nutrient availability, and indirectly through changes in soil C, pH, or other soil properties (Fig. 8.4) (Fierer et al., 2009; Zeng et al., 2016).

Intuitively, relief from nutrient limitation may be expected to increase microbial-driven organic matter decomposition, but neutral or negative effects of fertilizer on microbial respiration and decomposition rates are often observed (Ramirez et al., 2012). This phenomenon could result from microbial scaling of C use efficiency in conjunction with the availability of nutrients (Manzoni et al., 2010), or via a shift from oligotrophic to copiotrophic organisms with increasing nutrient availability (Ramirez et al., 2012). In both cases, nutrient-limited microbes "mine" nutrients from low-quality organic matter by being C-inefficient and metabolizing high quantities of C per unit of growth, thereby maintaining appropriate C-to-nutrient stoichiometry (Manzoni et al., 2010). Thus, a population of C-efficient microbes with adequate nutrient supply can maintain similar or slightly lower decomposition rates as a population of nutrient-limited, C-inefficient microbes. In agreement, Ramirez et al. (2012) reported β-glucosidase, acid phosphatase, and peroxidase enzyme activities decreased by 24%, 36%, and 25%, respectively, when N was added at $125\,g\,N\,m^{-2}$. Reduced extracellular enzyme activities and greater C use efficiency in fertilized systems reduce net C mineralization (Ramirez et al., 2012; Cotrufo et al., 2013) and thus help to explain the 3.5% greater SOC levels typically found in fertilized agricultural soils (Lu et al., 2011b).

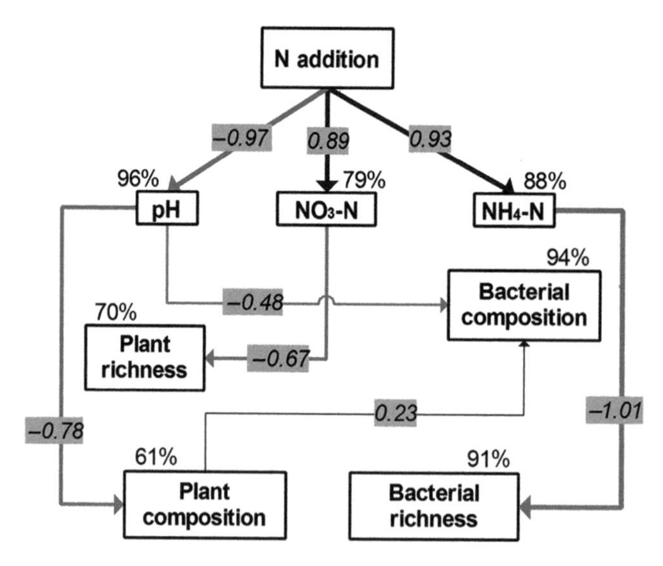

Figure 8.4
Structural equation model showing the direct and indirect relationships between N addition, soil properties, and the soil microbial community. The arrow line widths are in proportion to the strength of the relationships, and the numbers on the lines are correlation coefficients. The percentages are the variance accounted for by the structural equation model (Zeng et al., 2016).

8.3.5 Pesticides and Soil Organisms

In 2007, global pesticide use exceeded 2.3 Tg, with the United States accounting for about 20% of worldwide use (Grube et al., 2011). Hundreds of pesticide compounds are used in the United States (Goldman, 2007), and approximately 80% of total pesticide use is attributable to the agricultural sector (Grube et al., 2011). Because the biological toxicity mechanism varies among pesticide types (Jacobsen and Hjelmsø, 2014), it is difficult to assess the overall impact of pesticides on soil organisms and function. A metaanalysis by Jänsch et al. (2006) indicated that most soil invertebrate groups are minimally impacted by most pesticide classes. However, lumbricid earthworms and Enchytraeidae species are usually affected by fungicides, whereas arachnids, Collembola, and lumbricid earthworms are generally impacted by insecticides (Jänsch et al., 2006). The effect of pesticides on the soil microbial community is highly variable, with potential changes to community structure, diversity, and richness, or in other cases, no detectable effects (Imfeld and Vuilleumier, 2012). In the short term, pesticides can decrease microbial biomass, but these changes are typically short-lived (Schuster and Schröder, 1990). Inhibition of dehydrogenase and stimulation of cellulose enzymes within the soil are commonly reported for many pesticides (Riah et al., 2014), but these effects do not necessarily translate into measureable effects to microbial processes. For example, Hart and Brookes (1996) found no short- or long-term effects of a wide array of pesticides on C mineralization rates. Thus, we currently lack a comprehensive understanding of the implications of pesticide use on soil functions (Imfeld and Vuilleumier, 2012).

8.4 The Path Forward with Ecological Intensification

Historically, agricultural intensification focused almost entirely on yield delivery, largely neglecting the cost of detrimental effects to broader ecosystem service provisioning. More recently, focus has turned to finding solutions within the confines of a dual necessity for maintaining high yields and broadening services through the intensified use of ecological knowledge (Robertson et al., 2014; Rey et al., 2015). Many compelling arguments have emphasized that these objectives must be met without sole reliance on either the further expansion of agricultural acreage or natural-area preservation in place of existing working lands (Fischer et al., 2008; Phalan et al., 2011). *Ecological intensification*, defined as the management of organisms to enhance ecosystem services (Bommarco et al., 2013), provides a means to reduce the negative ramifications of agricultural intensification without compromising crop yields.

At scales relevant for fields, there is a well-developed literature documenting the benefits of increased plant diversity on ecosystem processes (e.g., Isbell et al., 2011), a relationship ripe to benefit agriculture through the use of cover crops. Specifically, cover crop mixtures diversify plant canopies, plant rooting zones, and the timing and quantity of plant growth and C inputs into the soil (Dabney et al., 2001; Snapp et al., 2005). As plant species differ in their ability to suppress weeds and pests (Shrestha et al., 2002), in the size and composition of their mycorrhizal communities (Eom et al., 2000), in their production of root-derived phosphatases (Shen et al., 2011), and in their ability to fix atmospheric N (Ebelhar et al., 1984), diverse cropping systems should also deliver a broader suite of ecological services (Altieri, 1999). While economic, climatic, and implementation challenges remain (e.g., Singer et al., 2007), cover crop use stimulates soil communities, increases soil quality (Altieri, 1999; Sainju et al., 2002), reduces chemical inputs, and maintains agricultural yields (Robertson et al., 2014). Ecological service delivery may also benefit from the use of soil additives such as biochars (Atkinson et al., 2010), which stimulate soil microbial biomass and activity, and improve overall soil quality (Kolb et al., 2009; Lehmann et al., 2011).

At the watershed scale, riparian buffers have long been effective at reducing off-site impacts of intensive agriculture (Mayer et al., 2007). Chapter 14, Agroforestry Practices and Soil Ecosystem Services, addresses the impact of agroforestry practices integration in agricultural production systems and their ecological services. New modifications to this approach place perennial grassland strips in targeted locations throughout the watershed, significantly improving field and watershed-scale ecological services with a minimal loss of existing farm acreage. For example, Helmers et al. (2012) found that the conversion of 10% of existing watershed cropland into perennial grassland reduced soil sediment loss by 96% over 4 years.

In a similar study, Pérez-Suárez et al. (2014) reported 100% and 37% increases in soil N and C levels under prairie vegetation buffers within 5 years. Such approaches may be essential to recovering the benefits of historic topographic variation and the associated landscape-level water storage and heterogeneous biogeochemical process points provided in untilled landscapes.

In the end, a combination of both watershed and within-field management changes will be required to meet ever-increasing demands for both agricultural production and environmental quality. Irrespective of the final combination that gains widespread adoption, recent gains in no-tillage farming, cover crop use, biochar amendments, and the adoption of perennial microhabitats across the landscape all suggest an increasing incorporation of, and dependence upon, ecological knowledge to meet societal expectations for ecosystem service delivery. This final transition represents the long-term goal of conservation in the age of agricultural intensification, to match global societal benefits of yield gain with the principles of sustainability needed to support local communities.

8.5 Conclusions

Agroecosystems are increasingly expected to provide a gamut of ecosystem services including food production, C sequestration, nutrient recycling, and climate regulation. Simultaneously, agricultural intensification through the use of tillage, fertilizers, and pesticides alters soil physical, chemical, and biological properties, and thus increases yields at the cost of other ecosystem services. Tillage typically reduces mean SOC content, but also tends to homogenize the horizontal and vertical distribution of SOC. Since most soil biological communities are dependent upon SOC substrates, the spatial distribution of biologically mediated soil ecosystem services is impacted by agricultural practices that alter SOC. Fertilizer application has dramatic and predictable effects on the soil microbial community, generally leading to increased nutrient losses. Ecological intensification seeks to improve ecosystem service provisioning within agroecosystems by adding or modifying biological components of the system. Examples of ecological intensification include the use of cover crops, perennial plant species, and biochar soil additions. In the end, agricultural intensification must be combined with ecological knowledge and practices in order to balance crop yields with other critical ecosystem services.

References

Altieri, M.A., 1999. The ecological role of biodiversity in agroecosystems. Agr. Ecosyst. Environ. 74, 19–31.
Anonymous, 2004. Soil and trouble. Science 304, 1614–1615.
Atkinson, C.J., Fitzgerald, J.D., Hipps, N.A., 2010. Potential mechanisms for achieving agricultural benefits from biochar application to temperate soils: a review. Plant Soil. 337, 1–18.
Baker, J.M., Ochsner, T.E., Venterea, R.T., Griffis, T.J., 2007. Tillage and soil carbon sequestration – what do we really know? Agr. Ecosyst. Environ. 118, 1–5.

Ball-Coelho, B., Roy, R., Swanton, C., 1998. Tillage alters corn root distribution in coarse-textured soil. Soil Tillage Res 45, 237–249.
Barnard, R.L., Osborne, C.A., Firestone, M.K., 2013. Responses of soil bacterial and fungal communities to extreme desiccation and rewetting. ISME J. 7, 2229–2241.
Beare, M.H., Parmelee, R.W., Hendrix, P.F., Cheng, W., Coleman, D.C., Crossley, D.A., 1992. Microbial and faunal interactions and effects on litter nitrogen and decomposition in agroecosystems. Ecol. Monogr. 62, 569–591.
Beatty, S.W., 2003. Habitat heterogeneity and maintenance of species in understory communities. In: Gilliam, F.S., Roberts, M.R. (Eds.), The Herbaceous Layer in Forests of Eastern North America. Oxford University Press, New York, pp. 177–197.
Bennett, E.M., Carpenter, S.R., Caraco, N.F., 2001. Human impact on erodible phosphorous and eutrophication: a global perspective. Bioscience 51, 227–234.
Böhme, L., Langer, U., Böhme, F., 2005. Microbial biomass, enzyme activities and microbial community structure in two European long-term field experiments. Agr. Ecosyst. Environ. 109, 141–152.
Bolton, E.F., Aylesworth, J.W., Hore, F.R., 1970. Nutrient losses through tile drains under three cropping systems and two fertility levels on a Brookston clay soil. Can. J. Soil Sci. 50, 275–279.
Bommarco, R., Kleijn, D., Potts, S.G., 2013. Ecological intensification: harnessing ecosystem services for food security. Trends Ecol. Evol. 28, 230–238.
Borlaug, N., 2007. Feeding a hungry world. Science 318, 359. 359.
Brown, P.W., Schulte, L.A., 2011. Agricultural landscape change (1937–2002) in three townships in Iowa, USA. Landscape Urban Plan. 100, 202–212.
Burger, M., Jackson, L.E., Lundquist, E.J., Louie, D.T., Miller, R.L., Rolston, D.E., et al., 2005. Microbial responses and nitrous oxide emissions during wetting and drying of organically and conventionally managed soil under tomatoes. Biol. Fert. Soils 42, 109–118.
Carpenter, S.R., Caraco, N.F., Correll, D.L., Howarth, R.W., Sharpley, A.N., Smith, V.H., 1998. Nonpoint pollution of surface waters with phosphorus and nitrogen. Ecol. Appl. 8, 559–568.
Cassman, K.G., 1999. Ecological intensification of cereal production systems: yield potential, soil quality, and precision agriculture. Proc. Natl. Acad. Sci. U. S. A. 96, 5952–5959.
Chan, K., 2001. An overview of some tillage impacts on earthworm population abundance and diversity – implications for functioning in soils. Soil Tillage Res 57, 179–191.
Chapin III, F.S., Matson, P.A., Vitousek, P.M., 2011. Principles of Terrestrial Ecosystem Ecology, second ed. Springer, New York.
Clark, I.M., Buchkina, N., Jhurreea, D., Goulding, K.W.T., Hirsch, P.R., 2012. Impacts of nitrogen application rates on the activity and diversity of denitrifying bacteria in the Broadbalk Wheat Experiment. Philos. T. Roy. Soc. B. 367, 1235–1244.
Clive, J., 2012. Global status of commercialized biotech/GM crops: 2012. ISAAA Brief, 44.
Cotrufo, M.F., Wallenstein, M.D., Boot, C.M., Denef, K., Paul, E., 2013. The Microbial Efficiency-Matrix Stabilization (MEMS) framework integrates plant litter decomposition with soil organic matter stabilization: do labile plant inputs form stable soil organic matter? Glob. Change Biol. 19, 988–995.
Cunningham, W.P., Cunningham, M.A., 2015.. In: Thirteenth, Environmental Science: A Global Concern. McGraw-Hill Education, New York.
Dabney, S.M., Delgado, J.A., Reeves, D.W., 2001. Using winter cover crops to improve soil and water quality. Commun. Soil Sci. Plan. 32, 1221–1250.
Dahl, T.E., 1990. Wetlands losses in the United States, 1780's to 1980's. Report to the Congress (No. PB-91-169284/XAB). National Wetlands Inventory, St. Petersburg, FL, USA.
Dalal, R.C., Henderson, P.A., Glasby, J.M., 1991. Organic Matter and Microbial Biomass in a Vertisol After 20 yr of Zero-Tillage. Soil Biol. Biochem. 23, 435–441.
Donald, P.F., Green, R.E., Heath, M.F., 2001. Agricultural intensification and the collapse of Europe's farmland bird populations. Proc Roy. Soc, B—Biol. Sci. 268, 25–29.
Doran, J.W., 1980. Soil microbial and biochemical changes associated with reduced tillage. Soil Sci. Soc. Am. J. 44, 765–771.

Ebelhar, S.A., Frye, W.W., Belvins, R.L., 1984. Nitrogen from legume cover crops for no-tillage corn. Agron. J. 76, 51–55.

Ellis, E.C., Ramankutty, N., 2008. Putting people in the map: anthropogenic biomes of the world. Front. Ecol. Environ. 6, 439–447.

Eom, A.H., Hartnett, D.C., Wilson, G.W.T., 2000. Host plant species effects on arbuscular mycorrhizal fungal communities in tallgrass prairie. Oecologia 122, 435–444.

Euliss, N.H., Gleason, R.A., Olness, A., McDougal, R.L., Murkin, H.R., Robarts, R.D., et al., 2006. North American prairie wetlands are important nonforested land-based carbon storage sites. Sci. Total Environ. 361, 179–188.

Fierer, N., Schimel, J.P., 2003. A proposed mechanism for the pulse in carbon dioxide production commonly observed following the rapid rewetting of a dry soil. Soil Sci. Soc. Am. J. 67, 798–805.

Fierer, N., Schimel, J.P., Holden, P.A., 2003. Variations in microbial community composition through two soil depth profiles. Soil Biol. Biochem. 35, 167–176.

Fierer, N., Strickland, M.S., Liptzin, D., Bradford, M.A., Cleveland, C.C., 2009. Global patterns in belowground communities. Ecol. Lett. 12, 1238–1249.

Fischer, J., Brosi, B., Daily, G.C., Ehrlich, P.R., Goldman, R., Goldstein, J., et al., 2008. Should agricultural policies encourage land sparing or wildlife-friendly farming? Front. Ecol. Environ. 6, 380–385.

Foley, J.A., DeFries, R., Asner, G.P., Barford, C., Bonan, G., Carpenter, S.R., et al., 2005. Global consequences of land use. Science 309, 570–574.

Frey, S.D., Elliott, E.T., Paustian, K., 1999. Bacterial and fungal abundance and biomass in conventional and no-tillage agroecosystems along two climatic gradients. Soil Biol. Biochem. 31, 573–585.

Gabet, E.J., Reichman, O.J., Seabloom, E.W., 2003. The effects of bioturbation on soil processes and sediment transport. Annu. Rev. Earth Pl. Sci. 31, 249–273.

Gerber, P., Chilonda, P., Franceschini, G., Menzi, H., 2005. Geographical determinants and environmental implications of livestock production intensification in Asia. Bioresour. Technol. 96, 263–276.

Gleixner, G., 2013. Soil organic matter dynamics: a biological perspective derived from the use of compound-specific isotopes studies. Ecol. Res. 28, 683–695.

Goldman, L.R., 2007. Managing pesticide chronic health risks. J. Agromedicine 12, 67–75.

Grandy, A.S., Robertson, G.P., 2006. Aggregation and organic matter protection following tillage of a previously uncultivated soil. Soil Sci. Soc. Am. J. 70, 1398–1406.

Grube, A., Donaldson, D., Kiely, T., Wu, L., 2011. Pesticide's Industry Sales and Usage: 2006–2007 Market Estimates. US Environmental Protection Agency, Washington (DC), Report No. EPA-733-R-11-001.

Hart, M.R., Brookes, P.C., 1996. Soil microbial biomass and mineralisation of soil organic matter after 19 years of cumulative field applications of pesticides. Soil Biol. Biochem. 28, 1641–1649.

Hartman, W.H., Richardson, C.J., 2013. Differential nutrient limitation of soil microbial biomass and metabolic quotients (qCO2): is there a biological stoichiometry of soil microbes? PLoS One 8, e57127.

Helmers, M.J., Zhou, X.B., Asbjornsen, H., Kolka, R., Tomer, M.D., Cruse, R.M., 2012. Sediment removal by prairie filter strips in row-cropped ephemeral watersheds. J. Environ. Qual. 41, 1531–1539.

Imfeld, G., Vuilleumier, S., 2012. Measuring the effects of pesticides on bacterial communities in soil: a critical review. Eur. J. Soil. Biol. 49, 22–30.

Iovieno, P., Bååth, E., 2008. Effect of drying and rewetting on bacterial growth rates in soil. FEMS Microbiol. Ecol. 65, 400–407.

Isbell, F., Calcagno, V., Hector, A., Connolly, J., Harpole, W.S., Reich, P.B., et al., 2011. High plant diversity is needed to maintain ecosystem services. Nature 477, 199–202.

Jackson, R.B., Carpenter, S.R., Dahm, C.N., McKnight, D.M., Naiman, R.J., Postel, S.L., et al., 2001. Water in a changing world. Ecol. Appl. 11, 1027–1045.

Jacobsen, C.S., Hjelmsø, M.H., 2014. Agricultural soils, pesticides and microbial diversity. Curr. Opin. Biotech. 27, 15–20.

Jänsch, S., Frampton, G.K., Römbke, J., Van den Brink, P.J., Scott-Fordsmand, J.J., 2006. Effects of pesticides on soil invertebrates in model ecosystem and field studies: a review and comparison with laboratory toxicity data. Environ. Toxicol. Chem. 25, 2490–2501.

Janzen, D.H., 1988. Tropical ecological and biocultural restoration. Science 239, 243–244.
Jastrow, J., Miller, R., Lussenhop, J., 1998. Contributions of interacting biological mechanisms to soil aggregate stabilization in restored prairie. Soil Biol. Biochem. 30, 905–916.
Jobbágy, E.G., Jackson, R.B., 2000. The vertical distribution of soil organic carbon and its relation to climate and vegetation. Ecol. Appl. 10, 423–436.
Kandeler, E., Tscherko, D., Spiegel, H., 1999. Long-term monitoring of microbial biomass, N mineralisation and enzyme activities of a Chernozem under different tillage management. Biol. Fert. Soils 28, 343–351.
Keiblinger, K.M., Hall, E.K., Wanek, W., Szukics, U., Hammerle, I., Ellersdorfer, G., et al., 2010. The effect of resource quantity and resource stoichiometry on microbial carbon-use-efficiency. FEMS Microbiol. Ecol. 73, 430–440.
Key, N., McBride, W.D., Ribaudo, M., Sneeringer, S., 2011. Trends and Developments in Hog Manure Management: 1998–2009. US Department of Agriculture (Economic Research Service), Washington, DC.
Kishné, A.S., Morgan, C.L.S., Neely, H.L., 2014. How much surface water can gilgai microtopography capture? J. Hydrol. 513, 256–261.
Kolb, S.E., Fermanich, K.J., Dornbush, M.E., 2009. Effect of charcoal quantity on microbial biomass and activity in temperate soils. Soil Sci. Soc. Am. J. 73, 1173–1181.
Kucharik, C.J., 2007. Impact of prairie age and soil order on carbon and nitrogen sequestration. Soil Sci. Soc. Am. J. 71, 430–441.
Kucharik, C.J., Fayram, N.J., Cahill, K.N., 2006. A paired study of prairie carbon stocks, fluxes, and phenology: comparing the world's oldest prairie restoration with an adjacent remnant. Glob. Change Biol. 12, 122–139.
Ladha, J.K., Pathak, H., Krupnik, T.J., Six, J., van Kessel, C., 2005. Efficiency of fertilizer nitrogen in cereal production: retrospects and prospects. Adv. Agron. 87, 85–156.
Lavelle, P., Bignell, D., Lepage, M., Wolters, V., Roger, P., Ineson, P., et al., 1997. Soil function in a changing world: the role of invertebrate ecosystem engineers. Eur. J. Soil Biol. 33, 159–193.
LeBissonnais, Y., 1996. Aggregate stability and assessment of soil crustability and erodibility. 1. Theory and methodology. Eur. J. Soil Sci. 47, 425–437.
Leff, J.W., Jones, S.E., Prober, S.M., Barberán, A., Borer, E.T., Firn, J.L., et al., 2015. Consistent responses of soil microbial communities to elevated nutrient inputs in grasslands across the globe. Proc. Natl. Acad. Sci. U.S.A. 112, 10967–10972.
Lehmann, J., Rillig, M.C., Thies, J., Masiello, C.A., Hockaday, W.C., Crowley, D., 2011. Biochar effects on soil biota – a review. Soil Biol. Biochem. 43, 1812–1836.
Lu, M., Yang, Y.H., Luo, Y.Q., Fang, C.M., Zhou, X.H., Chen, J.K., et al., 2011a. Responses of ecosystem nitrogen cycle to nitrogen addition: a meta-analysis. New Phytol. 189, 1040–1050.
Lu, M., Zhou, X.H., Luo, Y.Q., Yang, Y.H., Fang, C.M., Chen, J.K., et al., 2011b. Minor stimulation of soil carbon storage by nitrogen addition: a meta-analysis. Agr. Ecosyst. Environ. 140, 234–244.
Manzoni, S., Trofymow, J.A., Jackson, R.B., Porporato, A., 2010. Stoichiometric controls on carbon, nitrogen, and phosphorus dynamics in decomposing litter. Ecol. Monogr. 80, 89–106.
Matson, P.A., Parton, W.J., Power, A.G., Swift, M.J., 1997. Agricultural intensification and ecosystem properties. Science 277, 504–509.
Mayer, P.M., Reynolds Jr., S.K., McCutchen, M.D., Canfield, T.J., 2007. Meta-analysis of nitrogen removal in riparian buffers. J. Environ. Qual. 36, 1172–1180.
Meyer, W.B., Turner, B.L., 1992. Human population growth and global land-use/cover change. Annu. Rev. Ecol. Syst. 23, 39–61.
Millenium Ecosystem Assessment (MEA), 2005. Millenium Ecosystem Assessment, Ecosystems and Human Well-Being. World Resources Institute, Washington, DC.
Mitsch, W.J., Day, J.W., Gilliam, J.W., Groffman, P.M., Hey, D.L., Randall, G.W., et al., 2001. Reducing nitrogen loading to the Gulf of Mexico from the Mississippi River Basin: strategies to counter a persistent ecological problem. Bioscience 51, 373–388.
Neumann, K., Verburg, P.H., Stehfest, E., Mueller, C., 2010. The yield gap of global grain production: a spatial analysis. Agr. Syst. 103, 316–326.

O'Brien, S.L., Jastrow, J.D., Grimley, D.A., Gonzalez-Meler, M.A., 2010. Moisture and vegetation controls on decadal-scale accrual of soil organic carbon and total nitrogen in restored grasslands. Glob. Change Biol. 16, 2573–2588.

Odum, W.E., 1982. Environmental degradation and the tyranny of small decisions. Bioscience 32, 728–729.

Parton, W.J., Gutmann, M.P., Williams, S.A., Easter, M., Ojima, D., 2005. Ecological impact of historical land-use patterns in the great plains: a methodological assessment. Ecol. Appl. 15, 1915–1928.

Pérez-Suáreza, M., Castellano, M.J., Kolka, R., Asbjornsen, H., Helmers, M., 2014. Nitrogen and carbon dynamics in prairie vegetation strips across topographical gradients in mixed Central Iowa agroecosystems. Agr. Ecosyst. Environ. 188, 1–11.

Phalan, B., Onial, M., Balmford, A., Green, R.E., 2011. Reconciling food production and biodiversity conservation: land sharing and land sparing compared. Science 333, 1289–1291.

Phipps, R.H., Park, J.R., 2002. Environmental benefits of genetically modified crops: global and European perspectives on their ability to reduce pesticide use. J. Anim. Feed Sci. 11, 1–18.

Pingali, P.L., 2012. Green revolution: impacts, limits, and the path ahead. Proc. Natl. Acad. Sci. U. S. A. 109, 12302–12308.

Postel, S.L., Daily, G.C., Ehrlich, P.R., 1996. Human appropriation of renewable fresh water. Science 271, 785–788.

Postma-Blaauw, M.B., de Goede, R.G.M., Bloem, J., Faber, J.H., Brussaard, L., 2010. Soil biota community structure and abundance under agricultural intensification and extensification. Ecology 91, 460–473.

Ramirez, K.S., Craine, J.M., Fierer, N., 2012. Consistent effects of nitrogen amendments on soil microbial communities and processes across biomes. Glob. Change Biol. 18, 1918–1927.

Rasse, D., Rumpel, C., Dignac, M.-F., 2005. Is soil carbon mostly root carbon? Mechanisms for a specific stabilisation. Plant Soil 269, 341–356.

Rey, F., Cécillon, L., Cordonnier, T., Jaunatre, R., Loucougaray, G., 2015. Integrating ecological engineering and ecological intensification from management practices to ecosystem services into a generic framework: a review. Agron. Sustain. Dev. 35, 1335–1345.

Riah, W., Laval, K., Laroche-Ajzenberg, E., Mougin, C., Latour, X., Trinsoutrot-Gattin, I., 2014. Effects of pesticides on soil enzymes: a review. Environ. Chem. Lett. 12, 257–273.

Robertson, G.P., Gross, K.L., Hamilton, S.K., Landis, D.A., Schmidt, T.M., Snapp, S.S., et al., 2014. Farming for ecosystem services: an ecological approach to production agriculture. Bioscience 64, 404–415.

Rudel, T.K., Schneider, L., Uriarte, M., Turner II, B.L., DeFries, R., Lawrence, D., et al., 2009. Agricultural intensification and changes in cultivated areas, 1970–2005. Proc. Natl. Acad. Sci. 106, 20675–20680.

Sainju, U.M., Singh, B.P., Whitehead, W.F., 2002. Long-term effects of tillage, cover crops, and nitrogen fertilization on organic carbon and nitrogen concentrations in sandy loam soils in Georgia, USA. Soil Till. Res. 63, 167–179.

Scheer, C., Grace, P.R., Rowlings, D.W., Payero, J., 2012. Nitrous oxide emissions from irrigated wheat in Australia: impact of irrigation management. Plant. Soil 359, 351–362.

Schilling, K.E., Li, Z., Zhang, Y.-K., 2006. Groundwater – surface water interaction in the riparian zone of an incised channel, Walnut Creek, Iowa. J. Hydrol. 327, 140–150.

Schnecker, J., Wild, B., Takriti, M., Eloy Alves, R.J., Gentsch, N., Gittel, A., et al., 2015. Microbial community composition shapes enzyme patterns in topsoil and subsoil horizons along a latitudinal transect in Western Siberia. Soil Biol. Biochem. 83, 106–115.

Schuster, E., Schröder, D., 1990. Side-effects of sequentially-applied pesticides on non-target soil microorganisms: field experiments. Soil Biol. Biochem. 22, 367–373.

Sexstone, A.J., Revsbech, N.P., Parkin, T.B., Tiedje, J.M., 1985. Direct measurement of oxygen profiles and denitrification rates in soil aggregates. Soil Sci. Soc. Am. J. 49, 645–651.

Shen, J.B., Yuan, L.X., Zhang, J.L., Li, H.G., Bai, Z.H., Chen, X.P., et al., 2011. Phosphorus dynamics: from soil to plant. Plant. Physiol. 156, 997–1005.

Shrestha, A., Knezevic, S.Z., Roy, R.C., Ball-Coelho, B.R., Swanton, C.J., 2002. Effect of tillage, cover crop and crop rotation on the composition of weed flora in a sandy soil. Weed Res. 42, 76–87.

Sims, J.T., Simard, R.R., Joern, B.C., 1998. Phosphorus loss in agricultural drainage: historical perspective and current research. J. Environ. Qual. 27, 277–293.

Singer, J.W., Nusser, S.M., 2007. Are cover crops being used in the US corn belt? J. Soil Water Conserv. 62, 353–358.

Six, J., Elliott, E.T., Paustian, K., Doran, J.W., 1998. Aggregation and soil organic matter accumulation in cultivated and native grassland soils. Soil Sci. Soc. Am. J. 62, 1367–1377.

Six, J., Conant, R.T., Paul, E.A., Paustian, K., 2002. Stabilization mechanisms of soil organic matter: implications for C-saturation of soils. Plant Soil 241, 155–176.

Six, J., Frey, S.D., Thiet, R.K., Batten, K.M., 2006. Bacterial and fungal contributions to carbon sequestration in agroecosystems. Soil Sci. Soc. Am. J. 70, 555–569.

Snapp, S.S., Swinton, S.M., Labarta, R., Mutch, D., Black, J.R., Leep, R., et al., 2005. Evaluating cover crops for benefits, costs and performance within cropping system niches. Agron. J. 97, 322–332.

Spohn, M., Giani, L., 2010. Water-stable aggregates, glomalin-related soil protein, and carbohydrates in a chronosequence of sandy hydromorphic soils. Soil Biol. Biochem. 42, 1505–1511.

Sun, R., Guo, X., Wang, D., Chu, H., 2015. Effects of long-term application of chemical and organic fertilizers on the abundance of microbial communities involved in the nitrogen cycle. Appl. Soil Ecol. 95, 171–178.

Tiemann, L.K., Billings, S.A., 2011. Changes in variability of soil moisture alter microbial community C and N resource use. Soil Biol. Biochem. 43, 1837–1847.

Tilman, D., Cassman, K.G., Matson, P.A., Naylor, R., Polasky, S., 2002. Agricultural sustainability and intensive production practices. Nature 418, 671–677.

Todd, T.C., Blair, J.M., Milliken, G.A., 1999. Effects of altered soil–water availability on a tallgrass prairie nematode community. Appl. Soil Ecol. 13, 45–55.

Trost, B., Prochnow, A., Drastig, K., Meyer-Aurich, A., Ellmer, F., Baumecker, M., 2013. Irrigation, soil organic carbon and N2O emissions. A review. Agron Sustain. Dev. 33, 733–749.

Tscharntke, T., Clough, Y., Wanger, T.C., Jackson, L., Motzke, I., Perfecto, I., et al., 2012. Global food security, biodiversity conservation and the future of agricultural intensification. Biol. Conserv. 151, 53–59.

Tsiafouli, M.A., Kallimanis, A.S., Katana, E., Stamou, G.P., Sgardelis, S.P., 2005. Responses of soil microarthropods to experimental short-term manipulations of soil moisture. Appl. Soil Ecol. 29, 17–26.

Tsiafouli, M.A., Thebault, E., Sgardelis, S.P., de Ruiter, P.C., van der Putten, W.H., Birkhofer, K., et al., 2015. Intensive agriculture reduces soil biodiversity across Europe. Glob. Change Biol. 21, 973–985.

Tufekcioglu, A., Raich, J.W., Isenhart, T.M., Schultz, R.C., 1999. Fine root dynamics, coarse root biomass, root distribution, and soil respiration in a multispecies riparian buffer in Central Iowa, USA. Agrofor. Syst. 44, 163–174.

Ulanova, N.G., 2000. The effects of windthrow on forests at different spatial scales: a review. For. Ecol. Manag. 135, 155–167.

Verchot, L.V., Dutaur, L., Shepherd, K.D., Albrecht, A., 2011. Organic matter stabilization in soil aggregates: understanding the biogeochemical mechanisms that determine the fate of carbon inputs in soils. Geoderma 161, 182–193.

von Haden, A.C., Dornbush, M.E., 2014. Patterns of root decomposition in response to soil moisture best explain high soil organic carbon heterogeneity within a mesic, restored prairie. Agr. Ecosyst. Environ. 185, 188–196.

West, T.O., Post, W.M., 2002. Soil organic carbon sequestration rates by tillage and crop rotation: a global data analysis. Soil Sci. Soc. Am. J. 66, 1930–1946.

Wilkinson, M.T., Richards, P.J., Humphreys, G.S., 2009. Breaking ground: pedological, geological, and ecological implications of soil bioturbation. Earth-Sci. Rev. 97, 254–269.

Williams, M.A., Rice, C.W., 2007. Seven years of enhanced water availability influences the physiological, structural, and functional attributes of a soil microbial community. Appl. Soil Ecol. 35, 535–545.

Zak, D.R., Tilman, D., Parmenter, R.R., Rice, C.W., Fisher, F.M., Vose, J., et al., 1994. Plant production and soil microorganisms in late-successional ecosystems: a continental-scale study. Ecology 75, 2333–2347.

Zeng, J., Liu, X.J., Song, L., Lin, X.G., Zhang, H.Y., Shen, C.C., et al., 2016. Nitrogen fertilization directly affects soil bacterial diversity and indirectly affects bacterial community composition. Soil Biol. Biochem. 92, 41–49.

Zumft, W.G., 1997. Cell biology and molecular basis of denitrification. Microbiol. Mol. Biol. Rev 61, 533–616.

CHAPTER 9

Intensified Agroecosystems and Changes in Soil Carbon Dynamics

Abdullah Alhameid[1], Colin Tobin[1], Amadou Maiga[1,2], Sandeep Kumar[1], Shannon Osborne[3] and Thomas Schumacher[1]

[1]South Dakota State University, Brookings, SD, United States [2]University of Sciences, Technics and Technologies of Bamako, Mali [3]USDA-ARS, Brookings, SD, United States

9.1 Introduction

Agroecosystems comprise 30% of the Earth's surface (Altieri, 1991), and according to Swift et al. (1996) can be defined as, "the ecosystems in which humans have exerted a deliberate selectivity on the composition of the biota i.e., the crops and the livestock maintained by the farmer, replacing to a greater or lesser degree the natural flora and fauna of the site." Agroecosystems provide various ecosystem services, and management practices used in the agroecosystems determine the state of the global environment (Tilman et al., 2002). However, most agroecosystems are disturbed more frequently and with greater intensity than natural ecosystems resulting in reduced biological diversity (Gliessman, 2015). Poor land management practices within many agroecosystems result in reduced soil organic carbon (SOC) and crop production, and increased erosion. About 50% of the global arable land is already under mechanical and chemical intensive agriculture (Cohen, 1995), which requires high inputs of nutrient, energy, and water (Krishna, 2010). In this chapter, alternatives to mechanical and chemical intensification of agriculture using an approach called sustainable intensification of agroecosystems based on the concept of ecological intensification described by Gaba et al. (2014) will be addressed. Sustainable intensification often reduces mechanical and chemical inputs and increases biological diversification of agroecosystems. Examples include the use of diverse crop rotations, cover crops, no-tillage (NT), and the integration of livestock onto cropped lands. The adoption of diversification in agroecosystems has several advantages including building SOC, reducing insects/pests and diseases, and over time may result in improved crop productivity and other ecological services. However, as stated by Gaba et al. (2014) "manipulating biotic interactions is not necessarily gentler than conventional agriculture and may also have undesirable effects." Sustainable intensification based on diversification of agroecosystems also includes a number of concerns such as higher

Soil Health and Intensification of Agroecosystems.
DOI: http://dx.doi.org/10.1016/B978-0-12-805317-1.00009-9

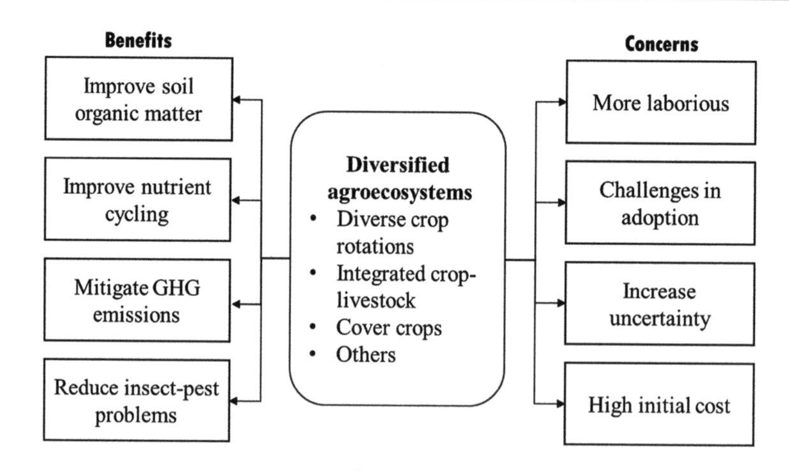

Figure 9.1
Benefits and concerns of diversified agroecosystems. *Source: FAO AQUASTAT (2009).*

initial cost, increased management skills, and an increase in uncertainty during the transition from mechanical and chemical intensification to sustainable intensification (Fig. 9.1). Gliessman (2015) defined diversification in agroecosystems as "Diversity is at once a product, a measure, and foundation of system's complexity—and therefore, of its ability to support sustainable functioning. From this perspective, ecosystem diversity comes about as a result of ways that different living and nonliving components of the system are organized and interact. From another perspective, diversity as manifested by the complex of biogeochemical cycle and the variety of living organisms—is what makes the organization and interaction of the system possible." Diversification can generate greater employment opportunities and higher incomes for producers (Ghosh et al., 2014). A few examples of diversified farming systems include complex crop rotations, cover crops, and integrated crop-livestock (ICL) systems. This chapter focuses on the impacts of diversified agroecosystems on soil carbon (C) dynamics.

9.2 Examples of Sustainable Intensification

The economy for almost 80% of the world's population is based on agriculture (FAO, 2003). In West Africa, about 80% of the cropped land is estimated to be under mixed cultivation (Steiner, 1982). Agricultural development strategies around the world are primarily based on a few major crops and include crops such as rice (*Oryza sativa*), wheat (*Triticum aestivum* L), corn (*Zea mays* L.), cotton (*Gossypium hirsutum L*), coffee (*Coffea arabica L*), cocoa (*Theobroma cacao*), and groundnuts (*Arachis hypogaea*). These crops have facilitated development within the agricultural sector for several decades in countries such as North America, Burkina Faso, Mali, Togo, Benin, Senegal, Ghana, and Ivory Coast. There has

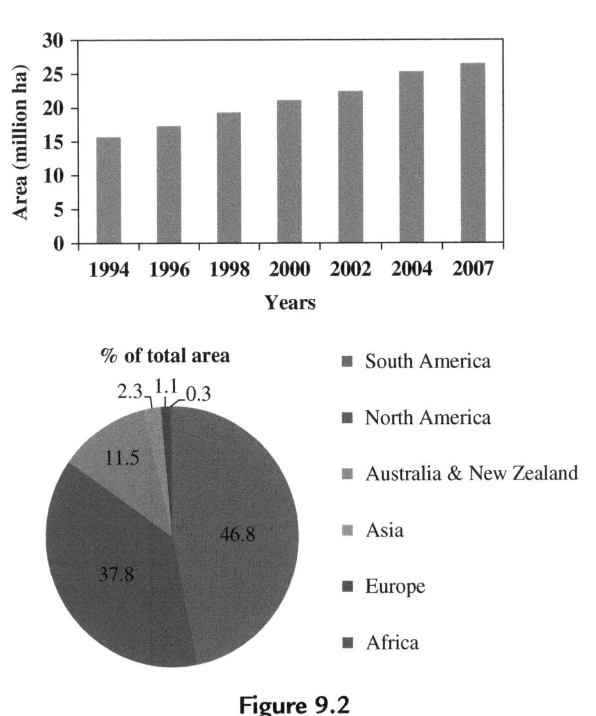

Figure 9.2

The area under no-tillage farming in the United States (top) and globally (bottom) (countries with >100,000 ha) in 2008–09. *Source: FAO AQUASTAT (2009).*

been increased emphasis in many parts of the world on reducing mechanical and chemical inputs into major cropping systems. NT is becoming widely adopted in many parts of the world. In order to increase the effectiveness of NT systems in agroecosystems, it needs to be coupled with the use of diversified crop rotations, increased use of cover crops, and increased integration of livestock-cropping systems.

Tillage practices have been considered important for crop production in most of the world for millennia (Carr et al., 2012). However, due to the negative consequences of tillage on erosion and soil health, producers have developed an increased interest in reducing tillage practices in many regions such as the North Great Plains of North America (Carr et al., 2012). NT management has become one of the most adopted farming techniques around the world. The data in Fig. 9.2 show the acreage (above 100,000 ha) under NT farming across the world and the United States. The data in this figure show that the area under NT in North and South America has increased. In the United States, the area under NT farming was increased by about 10 million ha from 1994 to 2007. This type of farming has emerged as one of the main practices to overcome negative impacts of tillage by reducing soil erosion and increasing SOC. The United States Department of Agriculture (USDA) data show that approximately 35.5% of United States cropland area was under NT farming systems in 2009 (Horowitz et al., 2010).

Table 9.1: Literature collected from previous published papers across the world showing the impacts of tillage management on soil organic carbon

Author	Location	Precipitation (Annual)	Soils	Tillage	Duration (Years)	SOC Range
Kumar et al. (2012)	Ohio, USA	1000 mm	Silt loam	[a]NT, CT, PT	49	Higher in 0–20 cm under NT and RT were by 26% than the MP and CT
Mikha et al. (2013)	Colorado, USA	418 mm	Silt loam	CT, MP, NT, RT	39	Higher in 0–30 cm under NT (32 $Mg\,ha^{-1}$) and RT (31.6 $Mg\,ha^{-1}$) than MP (26.1 $Mg\,ha^{-1}$) and CT (26.5 $Mg\,ha^{-1}$)
Dolan et al. (2006)	Minnesota, USA	–	Silt loam	MB, CH, NT	23	Higher under NT by 30% in 0–20 cm than MB and CH
Wright and Hons (2005)	Texas, USA	980 mm	Silty clay loam	CT, NT	23	Higher in 0–5 cm by 72% than CT, no-tillage effect at the 5–15 cm depth
Álvaro-Fuentes et al. (2008)	Lleida, Spain	430 mm	Silt loam	CT, ST, RT, NT	21	Higher in 0–5 cm, under NT (12.8 $Mg\,ha^{-1}$) than RT (9.1 $Mg\,ha^{-1}$), ST (7.7 $Mg\,ha^{-1}$), CT (5.6 $Mg\,ha^{-1}$)
Chivenge et al. (2007)	Harare, Zimbabwe	800–1000 mm	Clay	CT, CR, TR, MC	19	Higher under TR (20.4 $mg\,g^{-1}$) compared to MC (17.2 $mg\,g^{-1}$) CR (16.8 $mg\,g^{-1}$), CT (14.9 $mg\,g^{-1}$)
Zhao et al. (2015)	Hebei, China	480.7 mm	Silt loam	NT, PT, RT	14	Higher under NT by 12% in 0–5 cm, 17% in 5–10 cm, and 5.6% in 10–20 cm than RT and PT

[a]*NT*, no-tillage; *PT*, plow tillage; *CT*, conventional tillage; *ST*, subsoil tillage; *CR*, clean ripping; *TR*, tied ridging; *MB*, moldboard plow; *CH*, chisel plow; *RT*, rotary tillage; *MC*, mulch ripping.

Table 9.1 shows the benefits of NT systems in improving SOC across the world. However, the benefits of NT farming in increasing SOC are generally observed near the surface (Table 9.1). NT farming systems when combined with diverse crop rotations can increase SOC compared to conventional tillage (CT) systems. However, in short-term studies significant impacts on SOC and crop performance are sometimes insignificant (Ibrahim et al., 2015).

ICL systems provide an additional example of diversification that improves SOC, nutrient cycling, minimizing nitrogen (N) losses, and benefiting the environment (Lemaire et al., 2014). ICL systems cover about 25 million km^2 of the cultivated area in the world (De Haan et al., 1997), and land area under this type of system has been increasing in recent years because of its rare ability to supply society demands by producing food and positively affecting the environment at the same time (Thornton, 2010). Furthermore, livestock systems are valued at about $1.4 trillion (Thornton, 2010), and these integrated systems have been promoted across the world to improve farm production and food security. The use of cover crops in ICL systems improves soil properties and reduces spring nitrate-N concentrations in the soil (Meisinger et al., 1991; Reese et al., 2014).

9.3 Crop Rotations in Row Crops

Crop rotation is defined as a "system of growing different kinds of crops in recurrent succession on the same land" (Martin et al., 1976). Rotating different crops year after year adds various economic and environmental benefits. In addition, crop rotation is helpful in long-term soil and farm management. Rotating different crops can break pest cycles and add extra nutrients to the soil. Crop rotations build soil fertility, preserve the environment, control weeds, diseases, and insects, and add to crop and market diversity (Baldwin, 2006). Some examples of crop rotations include: corn-soybean (*Glycine max* L.)-wheat/red clover (*Trifolium pretense* L.), corn–soybean, corn–soybean–winter wheat, rice–wheat, and other potential rotations. However, these crop rotations are not universally common, rather they depend on different environmental and soils conditions. In the Midwest United States, corn–soybean is the dominating biannual crop rotation (Plourde et al., 2013), whereas, in Asia, rice–wheat is the dominating rotation (Mishra and Singh, 2012). The rice–wheat crop rotation occupies about 18 million hectare area in Asia, of which 75% are in the Indo-Gangetic Plains of India (10 million ha), Pakistan (2.2 million ha), Bangladesh (0.8 million ha), and Nepal (0.5 million ha) (Mishra and Singh, 2012). Both rice and wheat crops feed about 20% of the world population (Mishra and Singh, 2012). Soybean and corn summer crops are widely grown in rotation with Italian ryegrass (*Lolium multiflorum* Lam) in the subtropical South American regions of Brazil, Argentina, Uruguay, and Paraguay (Neto et al., 2014). In Europe, planting of wheat after rapeseed (*Brassica oleracea* L.), and wheat after small-grain cereals is very common (Peyraud et al., 2014). Small-grain cereals, grain and forage maize, rapeseed, and sunflower (*Helianthus annus* L.) accounted for 75% of crops preceding wheat in 2001 (Peyraud et al., 2014).

The use of long-term diverse rotations helps in improving grain yield as well. Baldwin (2006) documented that the grain producers of the United States believe that the use of 2- and 3-year crop rotations produce higher crop yields compared to those grown as monoculture. In Africa, the sole cropping system, crop rotation, and intercropping of legumes and cereals are the dominant cultural practices, and crop rotation has been indicated as the most superior, followed by monoculture, and lastly the intercropping system. Consequently, rotational cropping involving legumes and cereals were reported to be a more sustainable system for increasing food production in Africa than intercropping (Dakora and Keya, 1997).

Mixed cropping systems are favored because they reduce the risk of crop failure caused by unfavorable rainfall patterns as well as pests and disease pressures. There are various benefits to soil health in a diversified cropping system (Sauerborn et al., 2000). Most notably is the improvement of water use efficiency especially in arid and semiarid areas such as West Asia and North Africa. The latter region covers almost about 1.7 billion ha (Kassam, 1981) and is recognized as dryland with a rainfall range of 100–600 mm $year^{-1}$ (Dennett, 1987). The effectiveness of crop rotation has been studied in another arid area, Saudi Arabia, where rotations of cereals and alfalfa crops managed with center pivot systems showed an improvement in soil hydrological properties and subsequent yield improvement (Al-Rumikhani, 2002). Another long-term study in Syria conducted by Jones and Singh (2000) showed increase in crop yield of a barley-legume rotation compared to continuous barley. A recent study in Egypt showed the ability of crop rotation in decreasing nematode population in the root zone (Ahlam et al., 2015). Therefore, based on the studies conducted across different parts of the world, it has been demonstrated that rotating crops every other year has various economic and environmental benefits.

9.4 Crop Rotation Diversification and Cover Crop Impacts on Soil Carbon Dynamics

9.4.1 Soil Organic Carbon

Soil C dynamics is a key component of crop production system and provides various ecosystem services. A discussion of soil C dynamics has been included in previous sections of this chapter and other chapters of this book. However, the impacts of crop rotations and cover crops on SOC dynamics will be presented in this section. Ecosystem C dynamics is the soil component that involves SOC storage through the C sequestration process and potential release of C as carbon dioxide (CO_2) into the atmosphere (Heimann and Reichstein, 2008). Therefore, soil C dynamics relate to gain and loss of C by the soil. SOC can be recycled through the aboveground assimilatory processes (plant photosynthesis) and belowground heterotrophic respiratory processes through decomposition by soil microbial activity and respiration by animal and other organisms in the soil. It is evident that the above- and belowground processes are intimately linked, constituting a complex and dynamic system with non-negligible interactions (Heimann and Reichstein, 2008). The living part of soil organic matter (SOM), known as soil microbial biomass, acts as an important ecological indicator and is responsible for the decomposition and mineralization of plant and animal residues present in the soil (Marinari et al., 2006). The large portion of SOM that persists for decades to centuries is critical for long-term ecosystem services, stability, and the global C cycle. These dynamics and processes are essential for building soil health and functioning system.

Crop rotations and management practices impact soil C dynamics by impacting C sources and sinks. However, the rate of C input and loss depend on soil type, cropping system, environmental conditions, and duration of management. For example, research conducted by Zhu et al. (2014), in the Yangtze River Delta of China, showed that under short-term duration, in rice–wheat rotation systems, SOC was increased because of the additions of crop residues produced from rice and wheat straw in the soil. The use of NT management when used under diverse crop rotations is beneficial in increasing SOC compared to that when used in monoculture systems. NT with high residue left on the soil surface can help in building up the SOM. A study conducted by Wegner et al. (2015) reported that the use of an NT system with low crop residue removal rate can help in improving SOC under a corn–soybean rotation compared to that with high residue removal rate. Further, they observed that cover crop also improved SOC content, but only in one of the 4 years of sampling, and this may be due to the fact that cover crop had not been used long enough to show some positive impacts on SOC. Two studies conducted at Ridgetown (Canada) by Van Eerd et al. (2014) evaluated the long-term (11 and 15 years) effects of tillage and crop rotation on SOC and total nitrogen (TN). The results from these studies showed that SOC and TN were greater with NT than plow tillage (PT) systems. Further, crop rotations with winter wheat–soybean, and soybean–winter wheat–corn were compared with rotations without winter wheat. In both long-term trials, NT had 21 $Mg\,ha^{-1}$ more or 14% greater SOC than PT system in the 0–100-cm soil profile. Research conducted by Kibet et al. (2016) to estimate the long-term tillage impacts on SOM in a Sharpsburg silty clay loam, Nebraska (United States), under NT, double disc (DD), chisel plow (CP) and PT with corn–soybean rotation. They reported that SOC content at the 0–10 cm depth, in NT and DD were greater by 1.2 times than that with PT.

Diverse crop rotations can impact the amount of C in the soil. A 42-year long-term study in Iowa, United States that involved comparing two crop rotations (2-year, corn–soybean vs. 4-year, corn–oat [*Avena sativa*]-meadow-meadow) showed that there was an increase in SOC under four crops rotation compared to two crops rotation (Robinson et al., 1996). Similar findings were reported from a 26-year long-term experiment in Brazil under tropical weather conditions (Campos et al., 2011). The results from this study illustrated that a five-crop rotation consisting of black oat (*Avena strigose*)-soybean-black, oat–corn–oilseed radish (*Raphanus sativus*) increased SOC compared to four- and two-crop rotation systems on a clay soil. A study conducted in eastern Australia by Liu et al. (2016) showed that SOC changes under crop-pasture rotation varies from northern to southern sites. These researchers reported that temperature and rainfall play an important role and noted that this drove the dynamics of SOC and its interactions among farming management practices. According to a study conducted in Illinois, United States, on a silty clay loam soil by Zuber et al. (2015) after 15 years the corn–soybean–wheat (3-year) rotation increased SOC content compared to that under continuous corn (2-year) or continuous soybean (2-year). However, in some short-term studies, benefits of crop rotation in improving SOC have been also observed. A study conducted by Zhu et al. (2014) under a rice–wheat rotation system in China, where residues of rice and wheat were added into the soil improved SOM. Data from this study showed that rice–wheat crop rotation built-up SOM because of the left over residue. In the findings from a 49-year long-term study in Ohio, Kumar et al. (2012) reported that SOC stocks under continuous corn and corn–soybean rotation were almost the same. Another study of 14 years in China under winter wheat–summer corn crop rotation showed increase in SOC by 12% in 0–5 cm, 17% in 5–10 cm, and 6% in 10–20 cm, and decrease in SOC by 7% in 20–30 cm soil depth affected by crop rotations and tillage systems (Zhao et al., 2015). Shrestha et al. (2013) conducted a study from 1995 to 2005 in semiarid southwestern Saskatchewan, Canada to compare SOC stocks and rate of SOC change under one continuous crop and four 3-year fallow-containing crop rotations managed with an NT system, and two fallow-containing crop rotations under minimum-tillage (MT). After 11 years, they observed that the SOC (0- to 15-cm depth) was $0.2\,Mg\,C\,ha^{-1}$ greater under continuous crop compared with the fallow-containing systems. These researchers concluded that there were no significant differences in SOC and rate of SOC change among fallow-containing rotations or between MT and NT. A study by Aziz et al. (2011) on Van Meter farm in Ohio, United States, to evaluate the impact of crop rotations on soil quality showed that corn–soybean–wheat (2002–07) showed a significant impact of these crops on SOC and TN contents. Table 9.2 shows the data collected from various studies conducted across the world that involves diverse crop rotations and their impacts on SOC, N, and crop production. Data reported in Table 9.2 show that diverse crop rotations have significant impacts on improving SOC across various locations around the world.

9.4.2 Soil Surface Greenhouse Gas Fluxes

Agriculture contributes a significant amount of anthropogenic emissions of greenhouse gas (GHG) to an extent of 10–12% of worldwide emissions (Lehuger et al., 2011). These GHG emissions include mainly carbon dioxide (CO_2), nitrous oxide (N_2O), and methane (CH_4). This GHG can lead to global warming and climate variability. Greenhouse gas emissions from soils are produced from various sources such as CO_2, which is mostly produced from decomposition of SOM, while N_2O is produced mainly during denitrification processes, and CH_4 emissions are produced mostly by methanogenic bacteria or from waterlogged conditions such as rice fields.

Table 9.2: Literature collected from previously published papers across the world showing the impacts of diverse crop rotation systems on soil organic carbon and crop yield

Authors	Location	Precipitation (Annual)	Soil Type	Crop Rotation	Duration (Years)	[a]SOC Range
Kumar et al. (2012)	Wooster, USA	1000 mm	Well-drained (silt loam)	Corn–corn (R1), corn–soybean (R2)	49	Higher in 0–10 cm under R1 (16.2 g kg^{-1}) than R2 (15.7 g kg^{-1}), nonsignificant
Robinson et al. (1996)	Iowa, USA	711 mm	Clay loam	Corn–soybean (R1), Corn–oat–meadow–meadow (R2)	42	Higher under R2 (32.3 g kg^{-1}) than R1 (28.2 g kg^{-1})
Campos et al. (2011)	Cruz Alta, Brazil	1774 mm	Clay	Wheat–soybean (R1), wheat–soybean–black oat-soybean (R2), black oat–soybean–black oat–maize–oilseed radish (R3)	26	Higher in 0–5 cm in order R3 (6.2 Mg ha^{-1}), R2 (5.12 Mg ha^{-1}) than R1 (4.03 Mg ha^{-1})
Varvel (2006)	North Dakota, USA	-	Silty clay loam	Oat-grain sorghum–soybean–corn (R1), corn–soybean (R2)	20	Higher under R1 (15.92 Mg ha^{-1}) compared to R1 (13.88 Mg ha^{-1})
Barbera et al. (2012)	Sicily, Italy	481 mm	Clay	Wheat–wheat (R1), wheat–faba bean (R2)	19	Higher under R2 (21.2 g kg^{-1}) than R1 (18.2 g kg^{-1})
Zuber et al. (2015)	Illinois, USA	978 mm	Silty clay loam	Corn–corn (R1), soybean–soybean (R2), corn–soybean–wheat (R3)	19	Higher in 0–20 cm depth under R3 (47.1 Mg ha^{-1}) than R1 (44.6 Mg ha^{-1}) and R2 (41.2 Mg ha^{-1})
Shrestha et al. (2013)	Saskatchewan, CA	Semiarid	Silt loam	Fallow–canola–wheat (R1), wheat-oilseed-pulse (R2)	17	Higher under R2 (30.8 Mg ha^{-1}), than R1 (29.3 Mg ha^{-1}) under NT

[a]*SOC*, soil organic carbon; *R*, rotation system; *NT*, no-tillage; *N*, nitrogen.

Soil and crop management impacts soil surface GHG emissions. Research has shown that GHG emissions can be affected by changes in land use practices. Many studies have been conducted and results show that the use of United States croplands for corn production increases soil surface GHG emissions (e.g., Searchinger et al., 2008). However, mixed crop rotations can help in reducing the GHG emissions. A study conducted by Campbell et al. (2014) using two long-term experimental sites in Ohio, United States, to study the effects of crop rotation on GHG emissions using continuous corn (C–C) and corn–soybean (C–S) rotation reported that emissions of CO_2 and N_2O were significantly greater in C–C rotation than that of C–S rotation. It was concluded that the greater emissions of CO_2 and N_2O in C–C rotation were due to a greater amount of crop residue produced in C–C rotation compared to C–S rotation, where the C was emitted as CO_2. However, the greater amount of N fertilizer used in the C–C system as compared to corn after soybean in C–S rotation, caused the greater emission rate of N_2O. Another study was carried out by Sainju et al. (2012) in North Dakota, United States, to assess the impacts of crop rotation, irrigation, and tillage and N fertilization on GHG emissions. In this study, treatments consist of varying cropping systems such as conventionally tilled malt barley with N fertilizer (CT-N), conventionally tilled malt barley with no N fertilizer (CT-C), NT malt barley-pea with N fertilizer (NT-PN), NT malt barley with N fertilizer (NT-N), and NT malt barley with no N fertilizer (NT-C). The findings showed that CO_2 and N_2O emissions were greater in CT-N, and CH_4 was greater in NT-PN. The GHG emissions are influenced by climate condition as demonstrated by the findings of a study conducted to estimate the influence of lupin–wheat and wheat–wheat rotations on N_2O emissions in a semiarid soil of Australia (Barton et al., 2013). It was found that including a grain legume in a cropping rotation in such conditions decreased GHG emissions. Ma et al. (2013) studied the net global warming potential and GHG intensity on crop seasonal scale for two cycles of rice–wheat rotations in China and reported that soil surface CH_4 flux was significantly greater during rice-growing season and negligible during the wheat-growing season. It showed that net CH_4 fluxes were increased after transplantation and decreased during midseason drainage. These researchers further reported that N_2O was emitted during the wheat-growing season and small emissions peaked during the rice-growing season. Seasonal and cumulative CO_2 emission rates in this study were found to be different among cultivation practices. A study carried out by Carvalho et al. (2014) to quantify soil GHG emissions due to conversion of four different land uses: native vegetation (NV), pasture (PA), crop-succession (CS), and crop-pasture rotation in Brazilian Cerrado showed that the implementation of crop-pasture rotation is a good strategy to reduce soil GHG emissions. Research conducted by Zhang et al. (2015) to estimate the effects of tillage practices on GHG emissions in rice–wheat cropping systems in Central China reported that NT significantly reduced annual CH_4 emissions by 7.5%, but N_2O emissions were not affected much during rice and wheat seasons. Management practices that include crop rotation, tillage system, N fertilizer use, and seasonal variability can influence GHG emissions as demonstrated by many studies (Zhang et al., 2015; Liu et al., 2016).

9.4.3 Management Practices Effects on Nutrients Cycling

Nutrients are added to soil from different sources such as chemical fertilizers, manure, leguminous crops, or crop residue (Eltz and Norton, 1997). Some other factors such as weather, soil and crop types, and tillage management associated with cropping systems can greatly impact nutrient cycling in the soil profile. The choice of crop rotation along with management practices can influence economic returns (Martin-Rueda et al., 2007). Also, crop rotation and soil management practices choice influence SOM and nutrient pools in the soil (Zhou et al., 2014). A study conducted by Edwards et al. (1992) in Crossville, Alabama, United States, showed that implementing different crop rotations increased the availability of K compared with planting the same crop continuously under same conditions. The study illustrated how continuous use of a specific crop, such as corn, can greatly decrease the availability of certain nutrients such as phosphorus (P), calcium (Ca), and magnesium (Mg). However, in a long-term (24-year) rotation study conducted by Bünemann et al. (2006) in the southern region of Australia, it was found that an increase in total P as extractable inorganic P in a wheat–lupin rotation that included stubble burning was comparable to continuous wheat.

Soil fertility of a land in a particular management system can be maintained when nutrients are efficiently recycled through the soil food web and soil–plant–animal system. Soil microorganisms such as bacteria, fungi, and other decomposers play a significant role in transforming the organic matter pool of C compounds to simpler compounds, and release of nutrients in available forms for root uptake. Crop roots and residues play an essential role in improving soil fertility by stimulating soil microbial communities and improving soil aggregation, where SOM acts as binding agent for aggregation. However, SOM is often affected by a number of crop and soil management practices. Crop rotations and residue management are some of these practices that can influence SOM and contribute to soil fertility pool and nutrient availability through increasing nutrient input from roots and crop residues decomposition (Hatfield et al., 2001). Nutrient availability, including N, is affected by many factors such as soil type, tillage management, crop rotation, crop residue, and precipitation (Hatfield et al., 2001). Wood et al. (1991) showed that topography is the main factor of N movement in the soil system without significant impact on plant N uptake. When N content exceeds plant needs, N moves freely in the form of nitrate (NO_3^-) in the soil system to ground water and rivers and impacts the environment by reducing water quality (Ibrahim et al., 2015). The use of legumes in cropping systems can have nutritional and economic benefits from N fixing and reduction of N application. Legume crops can be used strategically in regular crop rotations for improving the soil fertility and meeting the needs of N-demanding crops. Types of crops and amount of residue also highly impact N content of soil system. A study conducted by Halvorson et al. (1999) showed that N content in dry lands under cropping systems was impacted by crop type and the amount of residue returned to the soil system. An 8-year experiment in Brooking, South Dakota, United States, by

Carpenter-Boggs et al. (2000) found that diverse crop rotation (corn–soybean–wheat–alfalfa) increased N in the soil profile compared to continuous corn and corn–soybean rotation systems on a clay loam soil (Peterson and Varvel, 1989). The inclusion of legumes in crop rotation adds significant benefits by capturing atmospheric N and fixes it into forms available to plants. Specialized bacteria (*Rhizobium* spp.) associated with the roots of legumes convert atmospheric N (N_2 gas) into plant-available N. The amount of N fixed by this association between bacteria and legumes varies with plant species and variety, soil type, climate, crop management, and the length of time the crop is grown. A research study conducted in Burkina Faso, West Africa, by Bado (2002) showed that crop rotations with peanut or cowpea increased the amount of available N for the next crop, resulting in a considerable increase in the sorghum yield. In a study conducted by Dakora and Keya (1997) on N dynamics in crop rotation systems including cowpea (*Vigna unguiculata* L. Walp) corn and groundnuts it was revealed that the recovery rate of N was 27% and 60% from cowpea and groundnut vestiges, respectively. Similar studies conducted in African agroecological systems also showed that the N fixation has played a significant role in crop productivity by contributing to soil fertility (Okorogbona and Adebisi, 2012). Diverse crop rotations with drainage management are helpful in improving soils and crop production. Overwinter cover crops are beneficial in reducing surface water flow, enhancing soil water evapotranspiration, and extracting residual soil N left over in the crop root zone after harvest (Drury et al., 2014). Constantin et al. (2010) reported from a long-term study conducted in France that cover crops (white mustard, *Sinapis alba* L.; Italian ryegrass, *Lolium perenne* L. ssp. *multiflorum* (Lam.) Husnot; radish, *Raphanus* L.; and winter wheat) increased soil storage of organic N and decreased residual soil NO_3^- concentrations in the root zone by 34–52%, and reduced NO_3^- leaching below the root zone by 36–62%.

9.5 Integrated Crop-Livestock Systems

ICL systems can increase ecological interactions among land use systems which can improve the efficiency of agricultural ecosystems in cycling nutrients, enhancing soil quality and preserving natural resources and the environment. Key ICL systems are common throughout the world, and have been increasing in recent years because of its various economic and environmental benefits (Thornton, 2010). ICL systems are an example of diversification that increases SOC (Snapp et al., 2005; Lemaire et al., 2014).

Greater demand for grains with increased population has shifted land use away from animal production systems to crop production systems, and the shift in land management has begun to deteriorate productive land to less fertile land. Specifically, in the United States ICL systems have been on the rise in recent years because of lower commodity prices, high land rent prices, and the limited amount of grazing land for animals. ICL systems throughout the world are somewhat similar to those used in the United States. Some ICL systems include: the use of large and small ruminants for weed control and manure application under palms in Malaysia; grazing crop residues by ruminants in Asia (Devendra and Thomas, 2002); and grazing after cropping and during fallow in Africa (Smith et al., 1997). The examples of crop-livestock systems within the United States include: planting and grazing of cover crops; grazing of crop residue after harvest; and grazing of annual crops swathed for winter feed only (Liebig et al., 2011). However, there are other types of crop-livestock integrated systems that are being adopted in the United States, and the most commonly used are grass-based crop rotation, livestock grazing of cover crops within cash-crop rotation, grazing of crop residues, grass intercropping, dual-purpose cereal crops, and agroforestry (Sulc and Franzluebbers, 2014).

9.5.1 Integrated Systems Effects on SOC and Nutrient Cycling

Grazing systems develop complex pathways for the C and N in soil causing highly localized concentrations of available C and N. In a study near Lubbock, Texas, United States an ICL system was studied by Acosta-Martinez et al. (2004) to determine soil C dynamics changes to a Pullman clay soil. These researchers found that microbial biomass C and N contents were greater in the ICL system compared to that of continuous cotton for the top 15-cm soil depth. Different stocking rates of grazing under livestock management systems have a strong influence on soil C dynamics. The SOC under rotational grazed systems was greater than in nongrazed, light stocking rate continuous grazing, and heavy continuous grazing systems (Teague et al., 2011). Excessive grazing under continuous grazing systems removes crop biomass and litter that cause soil exposure and soil degradation (Teague et al., 2013). In Brazil, under a clayey Oxisol soil with corn–soybean rotation in NT followed by summer grazing of black oat and Italian ryegrass, moderate grazing intensities (20–40 cm shoot height) led to SOC levels similar to those of nongrazed areas compared to high grazing intensity (10 cm shoot height) (Assmann et al., 2014). In the same study, moderate grazing intensities, with sward pasture heights between 20 and 40 cm, and a long period of a crop–livestock integration under NT, increased total particulate and mineral-associated organic C and N stocks similar to nongrazed areas with NT system (Assmann et al., 2014). A study located near New Deal, Texas, United States, under a Pullman clay loam soil with 0–1% slope, reported that SOC increased by of 22% during a 13-year ICL rotation of Old World Bluestem, and in an NT cotton-wheat-fallow-rye compared to continuous cotton (Fultz et al., 2013).

Integrating livestock into arable cropping systems helps in improving nutrient cycling and reducing N losses. These integrated systems enhance soil fertility and C sequestration, as the nutrients in the forage consumed by the livestock are applied back to the soils through manure deposition (Russelle et al., 2007). In north central United States, winter grazing is a commonly used practice that farmers have been using for a long time. In a study near Mandan, North Dakota, Liebig et al. (2011) reported that winter grazing of annual crops showed minor effects on near-surface soil properties. Further, it was noted that soil bulk density had an increase of 0.1 $Mg\,m^{-3}$ between the fall of 2007 and the spring of 2008 because of animal hoof-induced traffic during the grazing period of 2007. These researchers reported that soil nutrients such as available P, SOC, and total N increased between 2005 and 2008 in the high-traffic (HT) zone, and this could be partially attributed to the increased accumulation of manure from cattle in the HT zone because of the relatively close proximity of the zone to the winter shelter and water source (Liebig et al., 2011). Grazing in ICL systems may alter soil P dynamics. The understanding of P dynamics is important because it can impact on soil health. Research conducted on an oxisol in Brazil under soybeans rotated to a winter cover crop mixture of black oat and Italian ryegrass managed with an NT system showed that after 6 years of ICL systems, the total P was greater in the 0–5 cm depth in grazed areas due to intensified P-cycling compared to nongrazed, whereas, the nongrazed treatments had higher P above the surface because of biomass accumulation (Costa et al., 2014).

9.5.2 Integrated Systems Effects on Root Growth

9.5.2.1 Grazing systems

Grazing creates compaction and impacts root growth and soil properties. Root growth is essential in influencing soil aggregation through the addition of SOM and fungi growth (Oades, 1984; Six et al., 2000). The incorporation of perennial pastures into cropping systems can improve soil aggregation. The extensive deep root systems of perennial grasses contribute to SOM and improve soil aggregation. The fibrous root systems of these perennial grasses act as a binding agent in soil aggregation. Well-aggregated soils are capable of maintaining their structure and have more permanent macro- and mesopores, which provide more continuous paths for root growth than soils with less aggregation (Maughan et al., 2009). A well-managed grazed pasture (rotationally grazed) improves root growth and soil aggregation compared to that intensively grazed. A study conducted by Kumar et al. (2010) reported that intensive grazing reduced total root length and total surface areas compared to those of a rotational grazing system. The lower root length under intensive grazing systems was partially attributed to the lower soil C in this system (Kumar et al., 2010). Poorly managed grazed areas can lead to increased soil bulk density, which subsequently creates conditions of poor aeration in the soil and hence restricted root growth, which affects the uptake of nutrients from the soil (Lipiec and Hatano, 2003).

9.5.2.2 Cropping systems

Growing a variety of crops in sequence has many positive effects on the soil environment. Differences in plant rooting patterns including root density and root branching at different soil depths also result in more efficient extraction of nutrients from all soil layers when a series of different crops are grown (Ma et al., 2013). The long-term research conducted by Congreves et al. (2015) in Canada to evaluate the impact of tillage and crop rotation on soil health of four sites in Ontario (Ridgetown, Delhi, Elora, and Ottawa) showed that crop rotation significantly affected the soil attributes including root health. This study showed that soil aggregate stability is related to root health, sand content, and extractable P, which highlights the interdependence of aggregate stability to root growth and penetration, erosion control, soil compaction, and aeration. Perennial crops increase plant residue and hence the C input into the soil. Perennial energy crops could increase SOC stocks by 15–20 $Mg\,ha^{-1}$ compared to annual energy crops in conventional arable systems (crop rotation with plow system) according to the research conducted over 11 years in Germany (Gauder et al., 2016).

The design of low-input cropping systems including legumes in the crop rotation could be a key parameter to reduce C and N losses. An experiment conducted by Plaza-Bonilla et al. (2016) that included different levels of legumes in 3-year rotations showed that rotations significantly affected the amount of C and N inputs, and SOC and SON, and helped in mitigating the losses of C and N. Raphael et al. (2016) conducted a study on changes in SOM concentrations and quality as a result of crop rotation including grasses and a legume grown in the fall/winter and spring under NT and showed that SOM was affected by spring crops. The effects of diversified cropping system on SOC and soil health parameters can also be shown in the development and growth of root systems.

9.6 Conclusions

Diversification in agriculture systems is helpful in improving soil resilience and increasing long-term crop productivity. Diverse crop rotations, application of ICL systems, and use of cover crops are a few examples of diversification that can improve soil C dynamics and nutrient cycling. The material in this chapter is focused mainly on the impacts of diverse crop rotations, cover crops, and ICL systems on soil C dynamics and nutrient cycling.

The use of cover crops can reduce nitrate-N concentrations, and NT management reduces SOC turnover, resulting in reduced GHG emissions. Cover crops can reduce nitrate-N concentrations in soil and may be responsible for decreased emissions of nitrous oxide (N_2O) from soils. These cover crops used in crop rotations help in improving SOC, however, the significant improvement in C storage due to cover crops can generally be observed after adopting diverse crop rotations for longer durations. These crops enhance soil fertility by stimulating soil microbial communities and improve soil aggregation because organic matter acts as a binding agent for soil aggregation. Cover crops supply nutrients to crops, supporting rapid nutrient cycling through microbial biomass. The use of cover crops in ICL systems is also helpful in improving soil C dynamics, and nutrient availability. Livestock, when integrated into a cropping system, can improve nutrient cycling, minimize N losses, and benefit the environment.

Soil management practices need to be designed and adapted for local conditions to reduce mechanical and chemical inputs, and increase organic matter inputs into the soil. The successful implementation of sustainable intensification strategies in agroecosystems to improve soil C dynamics will require a rational integration of a diverse set of management practices into current world cropping systems.

Acknowledgments

The authors acknowledge funding support from South Dakota Agriculture Experiment Station, Fulbright scholarship, Saudi Arabian Cultural Mission (SACM), Natural Resources Conservation Service (NRCS), and United States Department of Agriculture grant. In addition, the authors would like to thank Ambika Karkee for her comments in improving the chapter.

References

Acosta-Martinez, V., Zobeck, T., Allen, V., 2004. Soil microbial, chemical and physical properties in continuous cotton and integrated crop-livestock systems. Soil Sci. Soc. Am. J. 68, 1875–1884.

Ahlam, M., El-Nagdi, W.M., Youssef, M., 2015. Seasonal variations of population density of root knot nematode, Meloidogyne incognita as affected by different cropping sequences. Scientia. 10, 35–37.

Al-Rumikhani, Y.A., 2002. Effect of crop sequence, soil sample location and depth on soil water holding capacity under center pivot irrigation. Agric. Water Manage. 55, 93–104.

Altieri, M.A., 1991. How best can we use biodiversity in agroecosystems? Outlook. Agric. 20 (1), 15–21.

Álvaro-Fuentes, J., López, M., Cantero-Martinez, C., Arrúe, J., 2008. Tillage effects on soil organic carbon fractions in Mediterranean dryland agroecosystems. Soil Sci. Soc. Am. J. 72, 541–547.

Assmann, J.M., Anghinoni, I., Martins, A.P., de Andrade, S.E.V.G., Cecagno, D., Carlos, F.S., et al., 2014. Soil carbon and nitrogen stocks and fractions in a long-term integrated crop–livestock system under no-tillage in southern Brazil. Agric. Ecosyst. Environ. 190, 52–59.

Aziz, I., Ashraf, M., Mahmood, T., Islam, K., 2011. Crop rotation impact on soil quality. Pak. J. Bot. 43, 949–960.

Bado, B.V., 2002. Rôle des légumineuses sur la fertilité des sols ferrugineux tropicaux des zones Guinéenne et Soudanienne du Burkina Faso-Thèse de Doctorat. Université de Laval, Québec, 184 p.

Baldwin, K.R., 2006. Crop Rotations on Organic Farms. North Carolina Cooperative Extension Service College of Agriculture and Life Sciences, NC State University, Rayleigh.1–16.

Barbera, V., Poma, I., Gristina, L., Novara, A., Egli, M., 2012. Long-term cropping systems and tillage management effects on soil organic carbon stock and steady state level of C sequestration rates in a semiarid environment. Land Degradat. Dev. 23, 82–91.

Barton, L., Murphy, D.V., Butterbach-Bahl, K., 2013. Influence of crop rotation and liming on greenhouse gas emissions from a semi-arid soil. Agric. Ecosyst. Environ. 167, 23–32.

Bünemann, E., Heenan, D., Marschner, P., McNeill, A., 2006. Long-term effects of crop rotation, stubble management and tillage on soil phosphorus dynamics. Soil Res. 44, 611–618.

Campbell, B., Chen, L., Dygert, C., Dick, W., 2014. Tillage and crop rotation impacts on greenhouse gas fluxes from soil at two longterm agronomic experimental sites in Ohio. Soil Water Conserv. 69, 543–552.

Campos, B.-H. C. d, Amado, T.J.C., Bayer, C., Nicoloso, R. d S., Fiorin, J.E., 2011. Carbon stock and its compartments in a subtropical oxisol under long-term tillage and crop rotation systems. Revista Brasileira de Ciência do Solo 35, 805–817.

Carpenter-Boggs, L., Pikul, J.L., Vigil, M.F., Riedell, W.E., 2000. Soil nitrogen mineralization influenced by crop rotation and nitrogen fertilization. Soil Sci. Soc. Am. J. 64, 2038–2045.

Carr, P.M., Mäder, P., Creamer, N.G., Beeby, J.S., 2012. Editorial: overview and comparison of conservation tillage practices and organic farming in Europe and North America. Renewable Agric. Food Syst. 27, 2–6.

Carvalho, J.L.N., Raucci, G.S., Frazão, L.A., Cerri, C.E.P., Bernoux, M., Cerri, C.C., 2014. Crop-pasture rotation: a strategy to reduce soil greenhouse gas emissions in the Brazilian Cerrado. Agric. Ecosyst. Environ. 183, 167–175.
Chivenge, P., Murwira, H., Giller, K., Mapfumo, P., Six, J., 2007. Long-term impact of reduced tillage and residue management on soil carbon stabilization: Implications for conservation agriculture on contrasting soils. Soil Tillage Res. 94, 328–337.
Cohen, J.E., 1995. How many people can the earth support? Sciences (New York). 35 (6), 18–23.
Congreves, K.A., Hayes, A., Verhallen, E.A., Van Eerd, L.L., 2015. Long-term impact of tillage and crop rotation on soil health at four temperate agroecosystems. Soil Tillage Res. 152, 17–28.
Constantin, J., Mary, B., Laurent, F., Aubrion, G., Fontaine, A., Kerveillant, P., et al., 2010. Effects of catch crops, no till and reduced nitrogen fertilization on nitrogen leaching and balance in three long-term experiments. Agric. Ecosyst. Environ. 135, 268–278.
Costa, S.E.V.G.A., Souza, E.D., Anghinoni, I., Carvalho, P.C.F., Martins, A.P., Kunrath, T.R., et al., 2014. Impact of an integrated no-till crop–livestock system on phosphorus distribution, availability and stock. Agric. Ecosyst. Environ. 190, 43–51.
Dakora, F.D., Keya, S.O., 1997. Contribution of legume nitrogen fixation to sustainable agriculture in sub-Saharan Africa. Soil. Biol. Biochem. 29 (5/6), 809–917.
De Haan, C., Steinfeld, H., Blackburn, H., 1997. Livestock & the environment: Finding a balance. European Commission Directorate-General for Development, Development Policy Sustainable Development and Natural Resources, Rome, Italy.
Dennett, M., 1987. Variation of rainfall-the background to soil and water management in dryland regions. Soil Use Manage. 3, 47–51.
Devendra, C., Thomas, D., 2002. Crop–animal interactions in mixed farming systems in Asia. Agric. Syst. 71, 27–40.
Dolan, M., Clapp, C., Allmaras, R., Baker, J., Molina, J., 2006. Soil organic carbon and nitrogen in a Minnesota soil as related to tillage, residue and nitrogen management. Soil Tillage Res. 89, 221–231.
Drury, C., Tan, C., Welacky, T., Reynolds, W., Zhang, T., Oloya, T., et al., 2014. Reducing nitrate loss in tile drainage water with cover crops and water-table management systems. J. Environ. Qual. 43 (2), 587–598.
Edwards, J., Wood, C., Thurlow, D., Ruf, M., 1992. Tillage and crop rotation effects on fertility status of a Hapludult soil. Soil Sci. Soc. Am. J. 56, 1577–1582.
Eltz, F.L.F., Norton, L.D., 1997. Surface roughness changes as affected by rainfall erosivity, tillage, and canopy cover. Soil Sci. Soc. Am. J. 61, 1746–1755.
FAO, 2003. Synthesis report on the role of agriculture on environmental externalities. Food and Agriculture Organization., http://ftp.fao.org/es/esa/roa/pdf/2.../Environment_MaliNA.pdf.
Fultz, L.M., Moore-Kucera, J., Zobeck, T.M., Acosta-Martínez, V., Allen, V.G., 2013. Aggregate carbon pools after 13 years of integrated crop-livestock management in semiarid soils. Soil Sci. Soc. Am. J. 77, 1659–1666.
Gaba, S., Bretagnolle, F., Rigaud, T., Philippot, L., 2014. Managing biotic interactions for ecological intensification of agroecosystems. Front. Ecol. Evol. 2, 29. http://dx.doi.org/10.3389/fevo.2014.00029.
Gauder, M., Billen, N., Zikeli, S., Laub, M., Graeff-Hönninger, S., Claupein, W., 2016. Soil carbon stocks in different bioenergy cropping systems including subsoil. Soil Tillage Res. 155, 308–317.
Ghosh, M., Sarkar, D., Roy, B.C., 2014. Diversification of Agriculture in Eastern India (India Studies in Business and Economics). Springer.
Gliessman, S.R., 2014. Agroecology: The Ecology of Sustainable Food Systems, Third ed. CRC Press.
Halvorson, A.D., Reule, C.A., Follett, R.F., 1999. Nitrogen fertilization effects on soil carbon and nitrogen in a dryland cropping system. Soil Sci. Soc. Am. J. 63, 912–917.
Hatfield, J.L., Sauer, T.J., Prueger, J.H., 2001. Managing soils to achieve greater water use efficiency. Agron. J. 93, 271–280.
Heimann, M., Reichstein, M., 2008. Terrestrial ecosystem carbon dynamics and climate feedbacks. Nature. 451, 289–292.
Horowitz, J., Ebel, R., Ueda, K., 2010. No-till" farming is a growing practice. Economic Information Bulletin No. 70. USDA Economic Research Service, Washington, DC.
Ibrahim, M., Alhameid, A., Kumar, S., Chintala, R., Sexton, P., 2015. Long-term tillage and crop rotation impacts on a Northern Great Plainsmollisol. Adv. Crop. Sci. Tech. 3, 2.

Jones, M., Singh, M., 2000. Long-term yield patterns in barley-based cropping systems in northern Syria. 2. The role of feed legumes. J. Agric. Sci. 135, 237–249.

Kassam, A., 1981. Climate, soil and land resources in North Africa and West Asia. Soil Water and Nitrogen in Mediterranean-Type Environments, 1–29.

Kibet, L.C., Blanco-Canqui, H., Jasa, P., 2016. Long-term tillage impacts on soil organic matter components and related properties on a Typic Argiudoll. Soil Tillage Res. 155, 78–84.

Krishna, K.R., 2010. Agroecosystems of South India: Nutrient Dynamics, Ecology and Productivity. Universal Publishers.

Kumar, S., Udawatta, R., Anderson, S., 2010. Root length density and carbon content of agroforestry and grass buffers under grazed pasture systems in a Hapludalf. Agrofor. Syst. 80, 85–96.

Kumar, S., Kadono, A., Lal, R., Dick, W., 2012. Long-term no-till impacts on organic carbon and properties of two contrasting soils and corn yields in Ohio. Soil Sci. Soc. Am. J. 76, 1798–1809.

Lehuger, S., Gabrielle, B., Laville, P., Lamboni, M., Loubet, B., Cellier, P., 2011. Predicting and mitigating the net greenhouse gas emissions of crop rotations in Western Europe. Agric. For. Meteorol. 151, 1654–1671.

Lemaire, G., Franzluebbers, A., de Faccio Carvalho, P.C., Dedieu, B., 2014. Integrated crop–livestock systems: strategies to achieve synergy between agricultural production and environmental quality. Agric. Ecosyst. Environ. 190, 4–8.

Liebig, M.A., Tanaka, D.L., Kronberg, S.L., Scholljegerdes, E.J., Karn, J.F., 2011. Soil hydrological attributes of an integrated crop-livestock agroecosystem: Increased adaptation through resistance to soil change. Appl. Environ. Soil Sci. 2011, 1–7.

Lipiec, J., Hatano, R., 2003. Quantification of compaction effects on soil physical properties and crop growth. Geoderma. 116, 107–136.

Liu, S.M., Qi, X.H., Li, X., Ye, H.R., Wu, Y., Ren, J.L., et al., 2016. Nutrient dynamics from the Changjiang (Yangtze River) estuary to the East China Sea. J. Mar. Syst. 154, 15–27.

Ma, Y.C., Kong, X.W., Yang, B., Zhang, X.L., Yan, X.Y., Yang, J.C., et al., 2013. Net global warming potential and greenhouse gas intensity of annual rice–wheat rotations with integrated soil–crop system management. Agric. Ecosyst. Environ. 164, 209–219.

Marinari, S., Mancinelli, R., Campiglia, E., Grego, S., 2006. Chemical and biological indicators of soil quality in organic and conventional farming systems in Central Italy. Ecol. Indicators 6, 701–711.

Martin-Rueda, I., Munoz-Guerra, L., Yunta, F., Esteban, E., Tenorio, J., Lucena, J., 2007. Tillage and crop rotation effects on barley yield and soil nutrients on a Calciortidic Haploxeralf. Soil Tillage Res. 92, 1–9.

Martin, J., Leonard, W., Stamp, D., 1976. Principles of Feld Crop Production. Macmillan publishing Co, Inc, New York. 898–932.

Maughan, M.W., Flores, J.P.C., Anghinoni, I., Bollero, G., Fernández, F.G., Tracy, B.F., 2009. Soil quality and corn yield under crop–livestock integration in Illinois. Agron. J. 101, 1503–1510.

Meisinger, J., Hargrove, W., Mikkelsen, R., Williams, J., Benson, V., 1991. Effects of cover crops on groundwater quality Cover Crops for Clean Water Soil and Water Conservation Society, Ankeny, Iowa. 266, 793–799.

Mikha, M.M., Vigil, M.F., Benjamin, J.G., 2013. Long-term tillage impacts on soil aggregation and carbon dynamics under wheat-fallow in the central Great Plains. Soil Sci. Soc. Am. J. 77, 594–605.

Mishra, J., Singh, V., 2012. Tillage and weed control effects on productivity of a dry seeded rice–wheat system on a Vertisol in Central India. Soil Tillage Res. 123, 11–20.

Neto, A.B., Savian, J.V., Schons, R.M.T., Bonnet, O.J.F., do Canto, M.W., de Moraes, A., et al., 2014. Italian ryegrass establishment by self-seeding in integrated crop–livestock systems: effects of grazing management and crop rotation strategies. Eur. J. Agron. 57, 77–83.

Okorogbona, A.O.M., Adebisi, L.O., 2012. Animal Manure for Smallholder Agriculture in South Africa. In: Lichtfouse, E. (Ed.), Farming for Food and Water Security. Springer, Netherlands, Dordrecht, pp. 201–242. http://dx.doi.org/10.1007/978-94-007-4500-1-9.

Oades, J.M., 1984. Soil organic matter and structural stability: mechanisms and implications for management. Plant. Soil. 76, 319–337.

Peterson, T.A., Varvel, G., 1989. Crop yield as affected by rotation and nitrogen rate. I. Soybean. Agron. J. 81, 727–731.

Peyraud, J.-L., Taboada, M., Delaby, L., 2014. Integrated crop and livestock systems in Western Europe and South America: a review. Eur. J. Agron. 57, 31–42.

Plaza-Bonilla, D., Nolot, J.-M., Passot, S., Raffaillac, D., Justes, E., 2016. Grain legume-based rotations managed under conventional tillage need cover crops to mitigate soil organic matter losses. Soil Tillage Res. 156, 33–43.

Plourde, J.D., Pijanowski, B.C., Pekin, B.K., 2013. Evidence for increased monoculture cropping in the Central United States. Agric. Ecosyst. Environ. 165, 50–59.

Raphael, J.P.A., Calonego, J.C., Milori, D.M.B.P., Rosolem, C.A., 2016. Soil organic matter in crop rotations under no-till. Soil Tillage Res. 155, 45–53.

Reese, C.L., Clay, D.E., Clay, S.A., Bich, A.D., Kennedy, A.C., Hansen, S.A., et al., 2014. Winter cover crops impact on corn production in semiarid regions. Agron. J. 106, 1479–1488.

Robinson, C., Cruse, R., Ghaffarzadeh, M., 1996. Cropping system and nitrogen effects on Mollisol organic carbon. Soil Sci. Soc. Am. J. 60, 264–269.

Russelle, M.P., Entz, M.H., Franzluebbers, A.J., 2007. Reconsidering integrated crop–livestock systems in North America. Agron. J. 99, 325–334.

Sainju, U.M., Stevens, W.B., Caesar-TonThat, T., Liebig, M.A., 2012. Soil greenhouse gas emissions affected by irrigation, tillage, crop rotation, and nitrogen fertilization. J. Environ. Qual. 41 (6), 1774–1786.

Sauerborn, J., Sprich, H., Mercer-Quarshie, H., 2000. Crop rotation to improve agricultural production in sub-Saharan Africa. J. Agron. Crop Sci. 184, 64–72.

Searchinger, T., Heimlich, R., Houghton, R.A., Dong, F., Elobeid, A., Fabiosa, J., et al., 2008. Use of U.S. croplands for biofuels increases greenhouse gases through emissions from land-use change. Science. 319, 1238–1240.

Shrestha, B., McConkey, B., Smith, W., Desjardins, R., Campbell, C., Grant, B., et al., 2013. Effects of crop rotation, crop type and tillage on soil organic carbon in a semiarid climate. Can. J. Soil Sci. 93, 137–146.

Six, J., Elliott, E., Paustian, K., 2000. Soil macroaggregate turnover and microaggregate formation: a mechanism for C sequestration under no-tillage agriculture. Soil Biol. Biochem. 32, 2099–2103.

Smith, J., Naazie, A., Larbi, A., Agyemang, K., Tarawali, S., 1997. Integrated crop-livestock systems in sub-Saharan Africa: an option or an imperative? Outlook on Agriculture-ici limited then pergamon press 26, 237–246.

Snapp, S., Swinton, S., Labarta, R., Mutch, D., Black, J., Leep, R., et al., 2005. Evaluating cover crops for benefits, costs and performance within cropping system niches. Agron. J. 97, 322–332.

Steiner K.G., Intercropping in tropical smallholder agriculture with special reference to West Africa, 1982, Schriftenreihe der Deutsches Gesellschaft für Technische Zusammenarbeit (137).

Sulc, R.M., Franzluebbers, A.J., 2014. Exploring integrated crop–livestock systems in different ecoregions of the United States. Eur. J. Agron. 57, 21–30.

Swift, M., Vandermeer, J., Ramakrishnan, P., Anderson, J., Ong, C., Hawkins, B., 1996. Biodiversity and agroecosystem function. In: Mooney, H., Cushman, J., Medina, E., Sala, O. (Eds.), Functional Roles of Biodiversity: A Global Perspective. John Wiley and Sons, New York, NY, pp. 261–298.

Teague, R., Provenza, F., Kreuter, U., Steffens, T., Barnes, M., 2013. Multi-paddock grazing on rangelands: why the perceptual dichotomy between research results and rancher experience? J. Environ. Manage. 128, 699–717.

Teague, W.R., Dowhower, S.L., Baker, S.A., Haile, N., DeLaune, P.B., Conover, D.M., 2011. Grazing management impacts on vegetation, soil biota and soil chemical, physical and hydrological properties in tall grass prairie. Agric. Ecosyst. Environ. 141, 310–322.

Thornton, P.K., 2010. Livestock production: recent trends, future prospects. Philos. Trans. R. Soc. B: Biol. Sci. 365, 2853–2867.

Tilman, D., Cassman, K.G., Matson, P.A., Naylor, R., Polasky, S., 2002. Agricultural sustainability and intensive production practices. Nature. 418, 671–677.

Van Eerd, L.L., Congreves, K.A., Hayes, A., Verhallen, A., Hooker, D.C., 2014. Long-term tillage and crop rotation effects on soil quality, organic carbon, and total nitrogen. Can. J. Soil Sci. 94, 303–315.

Varvel, G.E., 2006. Soil organic carbon changes in diversified rotations of the western Corn Belt. Soil Sci. Soc. Am. J. 70, 426–433.

Wegner, B.R., Kumar, S., Osborne, S.L., Schumacher, T.E., Vahyala, I.E., Eynarde, A., 2015. Soil response to corn residue removal and cover crops in Eastern South Dakota. Soil Sci. Soc. Am. J. 79, 1179–1187.

Wood, C., Peterson, G., Westfall, D., Cole, C., Willis, W., 1991. Nitrogen balance and biomass production of newly established no-till dryland agroecosystems. Agron. J. 83, 519–526.

Wright, A.L., Hons, F.M., 2005. Soil carbon and nitrogen storage in aggregates from different tillage and crop regimes. Soil Sci. Soc. Am. J. 69, 141–147.

Zhang, Z.S., Guo, L.J., Liu, T.Q., Li, C.F., Cao, C.G., 2015. Effects of tillage practices and straw returning methods on greenhouse gas emissions and net ecosystem economic budget in rice–wheat cropping systems in central China. Atmos. Environ. 122, 636–644.

Zhao, X., Xue, J.-F., Zhang, X.-Q., Kong, F.-L., Chen, F., Lal, R., et al., 2015. Stratification and storage of soil organic carbon and nitrogen as affected by Tillage practices in the North China Plain. PLoS One 10 (6), e0128873.

Zhou, G., Yin, X., Verbree, D.A., 2014. Residual effects of potassium to cotton on corn productivity under no-tillage. Agron. J. 106, 893–903.

Zhu, L., Hu, N., Yang, M., Zhan, X., Zhang, Z., 2014. Effects of different tillage and straw return on soil organic carbon in a rice–wheat rotation system. PLoS ONE. 9, e88900.

Zuber, S.M., Behnke, G.D., Nafziger, E.D., Villamil, M.B., 2015. Crop rotation and tillage effects on soil physical and chemical properties in illinois. Agron. J. 107, 971–978.

CHAPTER 10

Agroecosystem Net Primary Productivity and Carbon Footprint

Jose Guzman[1] and Mohammad H. Golabi[2]

[1]The Ohio State University, Columbus, OH, United States [2]University of Guam, Mangilao, Guam

10.1 Introduction

The intensification of agroecosystems for food and fiber production and improved soil health are the primary focus of this book. In this regard, this chapter will be devoted to land management practices which increase net primary productivity (NPP) in agroecosystems, while simultaneously lowering their carbon (C) footprint (i.e., lower carbon dioxide (CO_2) emissions) and improving soil health. NPP is the amount of net "biomass" (vegetation) produced by plants in a particular geographic location per year. Total global NPP values in croplands and pasturelands are approximately 6.8 and 5.3 Pg C $year^{-1}$, respectively (Janzen, 2004). These NPP values depend on soil, water, and biological processes and vary by regional climate. At the same time, global NPP and agriculture systems play a major role in the global C cycle, which has as a significant effect on climate change.

Many scientists have concluded that if increases in anthropogenic atmospheric CO_2 continue to increase, that this would have many detrimental effects to our world's climate and food security (Lal, 2004). Recently, the Earth's atmosphere has surpassed the 848 Pg C or 400 ppm CO_2, and is increasing at a rate of 4 Pg C $year^{-1}$ (IPCC, 2013). Changes in atmospheric C are directly linked to the Earth's terrestrial C cycles (forest, cropland, pastureland, urban, and unsuitable land), gas exchanges between the ocean, and anthropogenic emissions from fossil fuel burning and manufacture production such as cement. Focusing on agroecosystems' impact on atmospheric C, global croplands and pasturelands comprise of approximately 1.43 billion ha (11%) and 3.51 billion ha (27%) of the total global land use, respectively (FAOSTAT, 2015). These two major land uses contain about 765 Pg of the Earth's terrestrial C stocks (total of 2500 Pg to a 1 m depth) (IPCC, 2013). Most of that C (~90–98%) is stored in the soil rather than the plant biomass in cropland and pasturelands in temperate regions (Janzen, 2004). The potential of soil C sequestration under these soils if recommended conservation agricultural practices were adapted could be on average of 0.8 Pg C $year^{-1}$ (Lal, 2004). Soil C sequestration in this chapter is defined as "the capture and secure storage

Soil Health and Intensification of Agroecosystems.
DOI: http://dx.doi.org/10.1016/B978-0-12-805317-1.00010-5

(in soil) of atmospheric CO_2 that would otherwise be emitted to or remain in the atmosphere" (Olson et al., 2014).

The NPP will be discussed within the context of soil C budget as it relates to the potential for agroecosystems to act as a sink or source of atmospheric C. Land use and altered C dynamics are the two primary components of global C change, and the effect of land use management on C cycling is a critical biogeochemical factor at a regional scale. Previous studies (Vitousek, 1994; Houghton et al., 1999) have shown that climate conditions control NPP at regional scales, while agroecosystems can affect NPP at local scales. However, few researchers have attempted to quantify the effect of agroecosystem effects on NPP at a global scale (Haberl et al., 2007; Monfreda et al., 2008).

The objective of this chapter is therefore to improve the knowledge base that is required to understand and implement land management technologies that purposefully sequester soil C and reduce GHG generated from agricultural production practices. The views in this chapter are based on three exercises; first, a review of NPP in agroecosystems in selected regions, second on agroecosystem contributions to GHG emission, and lastly, linking NPP with changes in soil C sequestration and improved soil health.

10.2 Estimating Net Primary Productivity in Agroecosystems

Estimates of NPP are often done to evaluate ecosystem functions and structures in order to determine biomass production and C sequestration as they relate to the climate change (Deal, 2011). Specifically, at the agroecosystem level, cultivation practices, microclimate variables, and soil properties (soil structure, soil texture, nutrient availability, etc.) are the major factors affecting NPP. Woodwell and Whittaker (1968) have defined NPP as the following: "the NPP is the gross primary production (GPP), or the total amount of energy fixed by live plants, minus autotrophic respiration (R_A), and can also be estimated by measuring the dry mass of combined aboveground and belowground components of an agroecosystem during one year of growth," as shown in Eq. (10.1):

$$\mathrm{NPP} = \mathrm{GPP} - \mathrm{R_A} \tag{10.1}$$

Total new biomass is often measured at the end of the growing season in order to estimate the accumulated new biomass after the products of the photosynthesis are exhausted via R_A. The accumulated biomass is expressed in terms of biomass per unit of land surface per unit time (Mg biomass ha^{-1} $year^{-1}$). The NPP is measured within an ecosystem scale including the two main components, namely the aboveground net primary productivity (ANPP) (shoots, leaves, and litter), and the belowground net primary productivity (BNPP) (roots). Additionally, it is also important to note that the annual gross primary productivity (GPP) is defined as the amount of solar energy fixed by plants during photosynthesis (Woodwell and Whittaker, 1968) and it therefore represents the integral rate of photosynthesis on an annual basis (Gilmanov et al., 2003).

Agroecosystems' NPP is very different from that of natural ecosystems due to land disturbances by tools and machinery, as well as the amount of external input, such as inorganic fertilizers and other nutrient sources. Hence, NPP can be controlled by management strategies including crop selections and rotations, chemical inputs, tillage practices (conventional versus no-tillage), as well as other biophysical controls (e.g., irrigation and/or application of fertilizers) (Deal, 2011). Additionally, NPP can be negatively affected by the removal of aboveground biomass by grazing and postharvest removal for animal feed (e.g., alfalfa) or bioenergy (e.g., maize (*Zea mays*), miscanthus (*Miscanthus* × *giganteus*), and switchgrass (*Panicum virgatum*)). Several researchers have reported that the removal of crop residues after harvest (i.e., maize stover) can have negative impacts on the soil organic C, water conservation, and crop yields among others (Blanco-Canqui and Lal, 2007). Decreases in crop yields may not be apparent during the first few years of crop residue removal, but over the long term (>5 years), decreases in soil nutrients and physical properties are expected to reduce overall soil health and productivity (Guzman, 2013).

Intensively managed agroecosystems worldwide are often considered as a major sink for CO_2. However, the data available are scarce, site-specific, and sometimes not robust enough for an accurate assessment of the global sink potential for agroecosystems worldwide (Patel et al., 2011). The NPP assessment by remote sensing, which is based on the light use efficiency (LUE), is therefore used for this purpose, which also includes C assimilation of plants in an ecosystem (Patel et al., 2011) as described below:

$$NPP = \sum PAR \times fPAR \times \varepsilon \tag{10.2}$$

where NPP is the net primary productivity for a season and/or a year ($g\,C\,M^{-2}$), PAR is the photosynthetically active radiation in Mega Joule (MJ), and fPAR is the fractional absorbed PAR (unitless) which is quantified from remotely sensed vegetation indices, and ε is the LUE ($g\,C\,MJ^{-1}$) factor.

10.3 Management Practices Effects on Net Primary Productivity

Since a major source of atmospheric CO_2 absorption is through photosynthesis during growth stage and construction of plant tissues, determining NPP becomes a priority for evaluating global C budgets. However, few studies have assessed NPP in agroecosystems at a global scale (Haberl et al., 2007; Monfreda et al., 2008). In general, the largest agroecosystem NPP value estimates occur in the humid/subhumid regions where C_4 grass species crops are primarily planted, such as maize and sugarcane (*Saccharum officinarum*). In arid regions and areas where legumes (Fabaceae) and tuber crops (i.e., beans (*Phaseolus vulgaris*) and potatoes (*Solanum tuberosum*)) are planted, NPP is lowest. Table 10.1 lists the major food crops by climate region and potential NPP. However, the planting of perennial forage and grain crops in place of annual food crops could lead to higher NPP, primarily due to longer growing periods and extensive rooting systems (Post and Kwon, 2000).

In developing countries, aboveground crop residues after grain harvest are typically burned, removed, or grazed by livestock, lowering these agroecosystems' potential for sequestrating CO_2 from the atmosphere into the soil (Kane, 2015). Conversely, by increasing NPP through the use of cover crop, green fallow, and other cropping systems coupled with no-tillage farming can result in net gain of C in the soil (McDaniel et al., 2014; Kane, 2015). Additionally, increased crop residue input as the result of intensified cropping may stimulate microbial mineralization of soil organic matter (SOM), as residue and SOM pool size increases, thus enhancing nutrient cycling, all together improving overall soil health (Paustian et al., 2000).

Fertilization effects on soil C and plant growth are uncertain and complicated. Some studies have shown that extensive application of N fertilizers to the soil may actually decrease soil microbial activities, resulting in reduced soil respiration and CO_2 emission to the atmosphere (Al-Kaisi et al., 2008). Results from other studies have shown that the addition of inorganic fertilizers can increase soil C as a result of increased NPP (Neff et al., 2002; Guzman et al., 2015). Fertilization typically enhances grain production and ANPP accumulation, especially where nutrient N is limited for crop production (Högberg, 2007). In contrast to the positive response of ANPP to N addition, the response of BNPP is not as apparent. Several studies have supported that N fertilization increases the production of leaves, stems, and coarse roots, but the production of fine roots may, in some cases, decrease (Högberg, 2007; Dietzel et al., 2015). Nonetheless, BNPP can be greatly higher by improving interactions between roots and

Table 10.1: Major food crops by climate region and relative global net primary productivity

Climate Region	Major Food Crop	Net Primary Productivity
Warm, humid	Cassava (*Manihot esculenta*), yam (*Dioscorea*), maize (*Zea mays*), banana (*Musa*), rice (*Oryza sativa*), pineapple (*Ananas comosus*), perennial tree crops, sugar cane (*Saccharum officinarum*)	Very high
Warm, subhumid	Maize, soybean (*Glycine max*), cotton (*Gossypium*), wheat (*Triticum*), forage crops	Moderately high
Warm, arid	Sorghum (*Sorghum bicolor*), millet (*Pennisetum glaucum*), cow pea (*Vigna unguiculata*), ground nut (*Arachis hypogaea*), sweet potato (*Ipomoea batatas*), forage crops	Low
Cool, subhumid	Wheat, maize, vegetables, pulse crops	Moderate

soil moisture, temperature, and enhancing nutrient cycling by promoting microbial activities and diversity from additions of crop residues and other C inputs, such as animal manures (Paustian et al., 2000).

Increasing cropping intensity is an important management practice for increasing NPP, soil C inputs/accumulation, microbial activities, and overall soil health (Peterson et al., 1998; Acosta-Martinez et al., 2007). Cropping intensity can be defined as the amount of production (biomass or grain) per area cultivated per year. Cropping intensity can be increased when a higher number of crops are grown in a year, or by reducing the length or eliminating fallow periods. In rain-fed crop production systems where the growing season is sufficiently long (no or very short winter seasons), two to three crops can be planted per year; however, sufficient and timely amount of perception must occur during crop growth. Increasing cropping intensity is typically done to raise productivity and profitability, but using appropriate cropping sequences can also lead to improved weed, insect, and disease pest management, increases in SOM and nutrient status, and reduced wind/water erosion (Bullock, 1992). For instance, a cover or forage crop planted immediately after harvest of the primary grain crop, not only provides residue cover during the nongrowing season to reduce soil erosion, it also aids in elevating SOM and soil nutrient status (especially if legume is used) for the next crop that is planted. Crop rotations that incorporate nitrogen (N)-fixing legumes such as soybeans and alfalfa provide subsequent crops with substantial amounts of N (Edwards et al., 1988).

In semiarid regions, water can be a limiting factor not only for production of the crop but also for the residue decomposition following the harvest. Under semiarid conditions, wheat and other small grains are the major crops, and summer/fall fallow is commonly practiced to store water for the subsequent crop production season (Paustian et al., 2000). During fallow, soil warming due to the lack of canopy cover can increase SOM decomposition and CO_2 emissions. Additionally, practices like intensive tillage may increase SOM decomposition rates (Blair and Crocker, 2000). Studies in semiarid regions in Africa, Australia, and the United States have indicated that fallow periods in wheat-fallow systems can cause significant losses of soil C (Rasmussen and Smiley, 1997; Peterson et al., 1998).

In semiarid regions, C sequestration can be enhanced by reducing the net CO_2 emission from the soil through agricultural practices, such as no-tillage, that increase C input while reducing the decomposition rate (Post and Kwon, 2000). No-tillage management technique intensifies cropping systems by maximizing the use of water and nutrients, hence increasing crop yields, which can in turn result in greater soil C sequestration. Under no-tillage farming practices, crop residues remain on the soil surface after harvest, which provides a barrier between the soil and the atmosphere, reducing both evaporation and rapid SOM decomposition (Golabi et al., 1989). Additionally, the standing stubble after the harvest can potentially decrease the wind speed at the soil surface, hence reducing wind erosion in no-tillage systems. The standing stubble also lowers the risk of abrupt fluctuations in moisture and surface temperature regimes (Smika, 1983).

10.4 Effect of Climate Factors on Cropping Systems' Net Primary Productivity

Grain and NPP in cropping systems depend on climate factors such as light, temperature, precipitation, and CO_2 concentration (White and Howden, 2007). Temperature and precipitation have the largest impact on cropping systems, both of which can vary in response to climate change. Managing crops for a sustainable production system thus depends on how best to manage climate factors (risks) such as droughts and extreme rainfall events. For instance, adaptation strategies may be considered in crop production planning in order to cope with rising global temperature, which may cause lower soil moisture content as a result of reduced rainfall events in certain regions. In contrast, adaptation strategies may require coping with different rainfall pattern in different regions of the world where higher intensity and more frequent storm events may cause flooding and heavy runoffs (White and Howden, 2007). Those extreme weather events may require modifications of cropping systems by introducing different crop species, cropping sequence, and/or cultivars, which may also require different land use management and cropping practices. Those changes in cropping systems and land management can have an impact on NPP. Changes in climate may also impact cycles and durations of pests, weeds, and diseases that can influence NPP.

Changes in soil moisture and temperature are the major drivers for microbial activities, soil respiration, and SOM decomposition (Fig. 10.1). In general, an increase in soil temperature causes an increase in soil respiration rates accordingly (Lloyd and Taylor, 1994). Soils under conventional tillage practices, which leave the soil surface bare, will have higher temperatures during the day time than the soils under no-tillage management systems where soil surface is covered with residue. Removal of crop residues also results in higher soil temperatures at the surface due to lower albedo at the soil surface. Changes in soil surface conditions can alter microbial activity, where soil respiration rates are affected by soil moisture conditions. Optimum soil respiration occurs at moisture level of 35–40% (on volume basis), where the microbial activities are the highest (Orchard and Cook, 1983). Anaerobic conditions primarily occur at 40% soil moisture and above, resulting in lower microbial activities. As previously stated, no-tillage practices tend to retain higher soil moisture content by leaving crop residues on the soil surface after the harvest. The higher soil moisture content is also expected to moderate the soil temperature (Golabi et al., 1989).

10.5 Cropping Systems' Contribution to Greenhouse Gas Emissions

It is reported by the World Bank (2012) that agriculture contributes to 30% of global GHG emissions, including the deforestation caused by the expansion of crop and livestock production for food, fiber, and fuel. Despite the fact that agriculture is a large source of GHG emissions, agroecosystems can sequester a significant amount CO_2 out of the atmosphere by

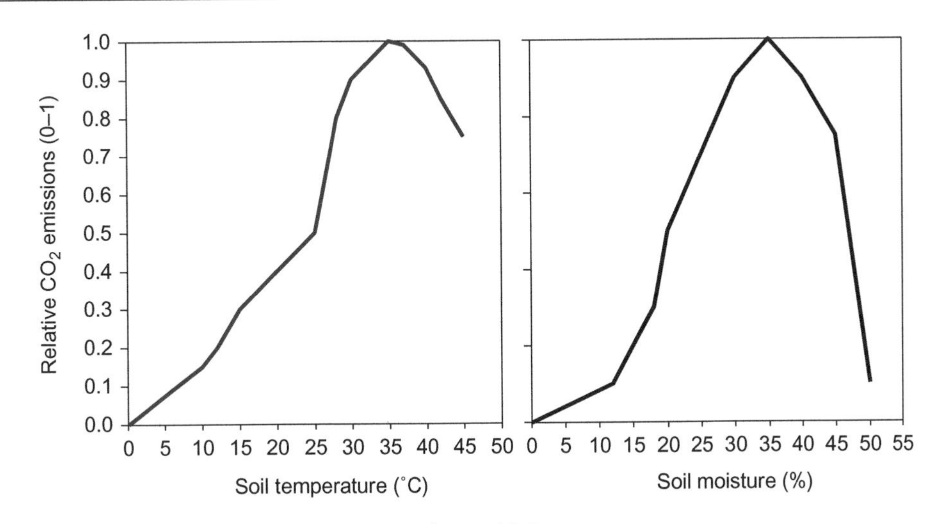

Figure 10.1
Conceptual relative soil respiration via CO_2 emissions from lowest (0) to highest (1) as affected by soil temperature and soil moisture (volume basis).

increasing NPP. In addition to CO_2 as one of the largest GHGs emitted from agroecosystems, methane (CH_4) and nitrous oxide (N_2O) are also emitted from these lands. According to World Bank (2015a,b) reports, 70% of the anthropogenic N_2O and 40% of CH_4 emissions are due to agricultural activities worldwide. However, neither of these two gases can be taken up by plants, nor can they be sequestered in the soil (i.e., changes in soil C) such as is the case for CO_2. The net annual mean contribution from CO_2 and N_2O equivalents under maize cropping systems varying in management practices in the mid-west United States is presented in Fig. 10.2.

The N_2O emissions are the greatest when the soil is saturated with water, creating anaerobic conditions and microbes are forced to utilize nitrate (NO_3-N) instead of oxygen (O_2) as an electron acceptor during the denitrification process (Smith et al., 2009). These processes are especially enhanced at high temperature levels in the soil (Linn and Doran, 1984). Cropping systems such as rice (*Oryza sativa*) production practices in Asia require not only flooding the field but also require a high rate of N application because of the loss of NO_3 as N_2O (off-gassed) due to the denitrification processes (Basche et al., 2014). Also, practices that could lead to increased N fixation in soil, such as legumes, can potentially become a source of N_2O under the saturated conditions caused by heavy rain and/or irrigation (Basche et al., 2014). Consideration of N sources and application rates is critical in managing or reducing excess N fertilizer application contributions to GHG emissions (Millar et al., 2010). For the third major source of GHG, CH_4, agricultural soils typically are a minor emitter or small sink for CH_4 under aerobic soil conditions (Bronson and Mosier, 1994). Other sources of CH_4 emissions from the agricultural sectors are enteric fermentation in livestock lots, especially with large

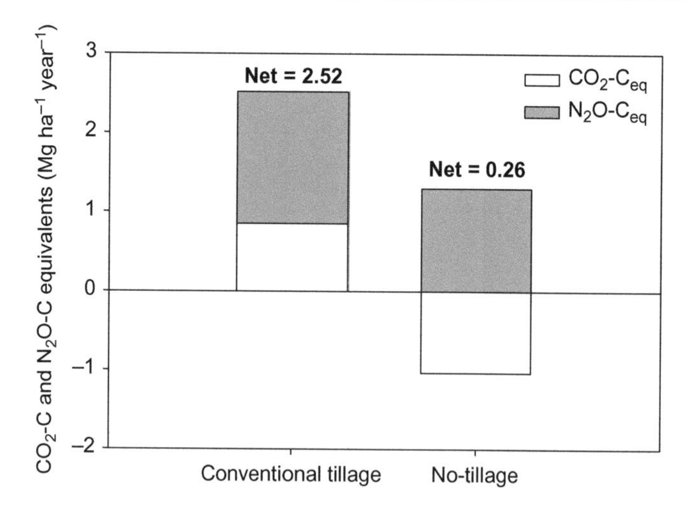

Figure 10.2

Net greenhouse gas emissions (CO_2-C and N_2O-C) in carbon equivalents in continuous maize varying by tillage system in mid-west United Sates. Positive carbon equivalents indicate sources of atmospheric carbon, while negative values indicate gains in soil carbon. *Data for these calculations were obtained from Ussiri, D.A., Lal, R., 2009a. Long-term tillage effects on soil carbon storage and carbon dioxide emissions in continuous corn cropping system from an alfisol in Ohio. Soil Tillage Res. 104 (1), 39–47; Ussiri, D.A., Lal, R., Jarecki, M.K., 2009b. Nitrous oxide and methane emissions from long-term tillage under a continuous corn cropping system in Ohio. Soil Tillage Res. 104 (2), 247–255 (Ussiri and Lal, 2009a,b) and Guzman, J.G., 2013. Evaluation of residue management practices effects on corn productivity, soil quality, and greenhouse gas emissions. Ph.D. Dissertation. Iowa State University, Ames, IA.*

animal husbandry operations, and where long periods of anaerobic soil conditions occur, such as in rice production (Basche et al., 2014).

10.6 Linking Net Primary Productivity with Soil Carbon

The intensification of agroecosystems can significantly affect soil C status. Under similar climate conditions, agroecosystems with high year-round NPP, such as grass pastures, tend to have greater soil C content at the surface than cultivated croplands, largely due to undisturbed soil resulting in slowly decomposing plant debris. On the contrary, tillage management increases the microbial oxidation of plant residue (above and below ground), which could have been stored as SOM (Al-Kaisi and Yin, 2005). Land-use changes from perennial grass systems into annual cropping systems have been shown to result in approximately 30–60% soil C losses (Ellert and Gregorich, 1996; Post and Kwon, 2000). In addition, tillage also reduces aggregate formation, shown to be critical for C accumulation (Six et al., 2002; Jones and Donnelly, 2004). Thus, cropland and disturbed land have been targeted as possibilities to sequester C and are considered a partial means for slowing further increases in GHG concentrations through soil C sequestration under no-tillage and organic management regimes (Lal et al., 1999; West and Post, 2002). For instance, many studies have shown increases in soil C content after establishing grasslands on previously cultivated cropland, especially over the first decade of grass establishment (Fig. 10.3).

10.6.1 Carbon Budgets

Soil C sequestration and the mitigation of GHG emissions into the atmosphere entail high NPP and increases in soil C pools protected from microbial activities as well as oxidation processes (i.e., tillage) in agroecosystems. The determination of whether cropping systems result in net sink or source of CO_2 and other GHGs requires accurate measurements in C budgets. The balance between C inputs (ANPP and BNPP) and losses via microbial respiration (R_h) has been used to reflect changes in soil C. Direct measurement of changes (i.e., $\pm$ 0.8 Mg C ha^{-1} year^{-1}) in soil C is difficult because these changes are relatively small compared to the natural variability of soil C stocks (i.e., $\pm$ 2.5 Mg C ha^{-1}) (Olson et al., 2014). Thus, it is suggested that assessing changes in soil C should be done in long-term studies (>10 years) (Conen et al., 2003). However, the calculation of a C budget from R_h

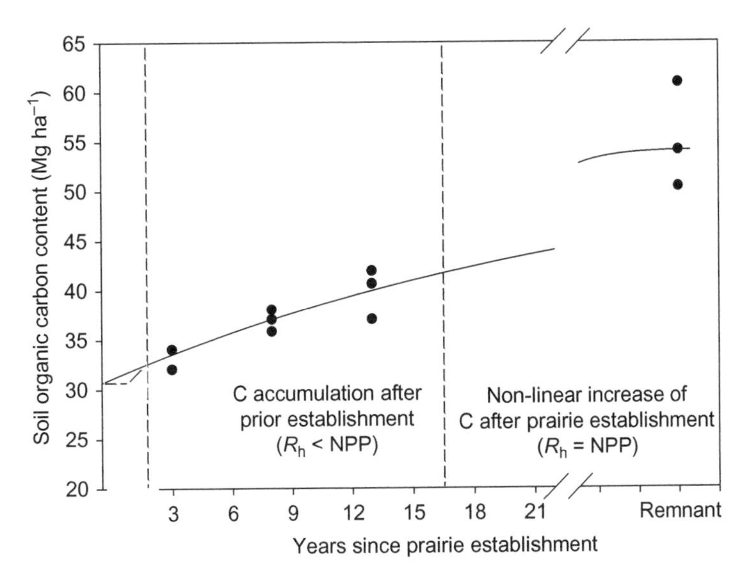

Figure 10.3
Soil organic carbon content over a 2-14-year period of reconstructed prairies and potential future increase. A remnant prairie site was included to identify the upper limits for soil organic carbon prairie ecosystem before conversion to cultivated cropland approximately 150 years ago. Each point represents one plot that was averaged from 2005 to 2007. R_h is heterotrophic respiration and NPP is net primary productivity. *Figure modified from Guzman J.G., Al-Kaisi M.M., Soil carbon dynamics and carbon budget of newly reconstructed tall-grass prairies in south central Iowa, J. Environ. Qual., 39(1), 2010, 136–46. With permission from Journal of Environmental Quality.*

measurements and C added from crop residues can give valuable insights into the processes contributing to changes in soil C induced by recent land-use changes because both C inputs and loses can be quantified at smaller time scales (>1 year). For example, a soil C budget can be estimated by measuring net ecosystem productivity (NEP) as done by Duiker and Lal (2000) and Kucharik et al. (2006). The potential soil C change is calculated as the difference between ANPP and BNPP C input, and the C losses through organic matter decomposition (i.e., R_h) for the entire year:

$$\text{Net soil carbon change} \left(\text{Mg C ha}^{-1}\text{yr}^{-1}\right) = \left(\text{ANPP}_c + \text{BNPP}_c\right) - R_h \tag{10.3}$$

where ANPP_c is potential C content input from aboveground plant biomass, BNPP_c is potential C content input from belowground root biomass, and R_h is C loss as CO_2 due to microbial respiration. An example of a NEP study is shown in Fig. 10.4.

The magnitude of soil C sequestration depends on inherent soil properties (i.e., mineralogy, texture, soil depth), factors limiting crop production (i.e., weather and variety selection), and management practices (i.e., tillage and fertilization). The amount of C in the system is determined by biomass C input, which is dependent on climate, and ANPP/BNPP ratio.

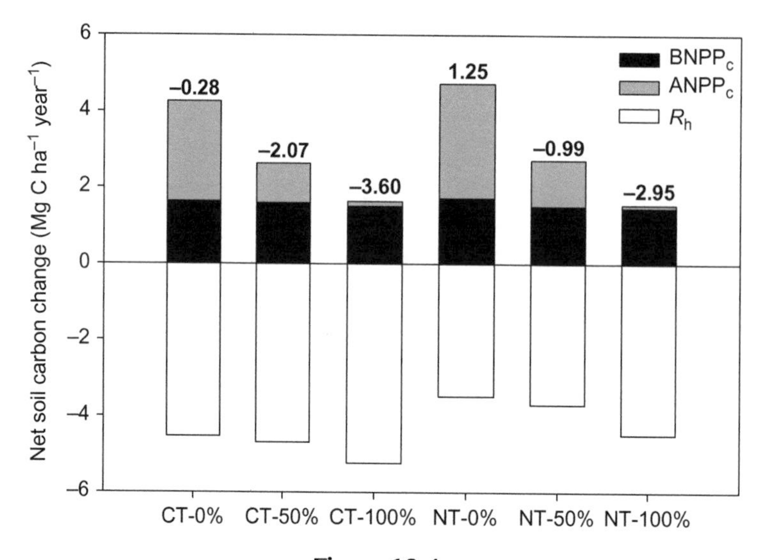

Figure 10.4

Net soil carbon change as estimated from carbon inputs in aboveground net primary productivity (ANPPc) and belowground (BNPPc) minus microbial respiration (R_h). Management practices included conventional tillage (CT), no-tillage (NT), and residue (maize stover) removal rates at 0, 50%, and 100%. Negative values indicate source of atmospheric carbon, and positive values indicate sinks of soil carbon. *Data for these calculations were obtained from Guzman, J.G., 2013. Evaluation of residue management practices effects on corn productivity, soil quality, and greenhouse gas emissions. Ph.D. Dissertation. Iowa State University, Ames, IA.*

Factors that result in loss of soil C include soil erosion, tillage, and residue removal, which determine the rate of SOC sequestration. Crop residues, being a C input, have a major influence on the level of soil C sequestration, although ANPP has a lesser impact on soil C than that from BNPP (Rasse et al., 2005). Nevertheless, microbial activities are the highest near soil surface, and any decrease in C input from the ANPP will adversely impact soil C and soil health (Blanco-Canqui and Lal, 2007). As mentioned earlier, management practices that lower soil temperature and increase soil moisture, such as no-tillage, usually result in an increase in soil C sequestration.

10.6.2 Soil Carbon Sequestration and Soil Health

There are numerous agricultural and natural resource management strategies that show promise for improving soil health and sequestering the major portion of C from being emitted into the atmosphere as CO_2 and other GHG. No-tillage cropping systems often increase soil C, however, because no-tillage technology requires specialized planting equipment and chemical/herbicide applications, this technology is primarily used in the United States, South America, and other developed countries with large-scale and mechanized agriculture. No-tillage farming can increase soil C particularly near the soil surface because of surface residue accumulation (West and Post, 2002). However, this system must remain continuously under no-tillage in order to restore soil C. Tilling soils can release C back to the atmosphere due to increased oxidation processes and microbial utilization with tillage (Grandy and Robertson, 2006). Some studies have cast doubt on the extent of C sequestration in no-tillage farming, especially when ignoring the lower depths where the differences between soil C stocks in no-tillage and conventionally tilled soils may be the same or lower (Baker et al., 2007; Syswerda et al., 2011; Powlson et al., 2014). With that mentioned, no-tillage works best in semiarid regions, since no-tillage systems were originally designed to conserve soil and water to increase grain and NPP. In soils where soil fertility and water are not limiting NPP, the addition of green fallows, cover crops, diversified crop rotation that uses semiperennial crops (i.e., alfalfa (*Medicago sativa*)) would be needed to increase the C sequestration in no-tillage soils.

There is a strong relationship between soil structure and soil C sequestration. Soil structure can be described as the arrangement of soil (clay, silt, and sand) particles and pore spaces. Soil aggregates are formed when individual soil particles are bonded together. The binding of soil particles results from chemical properties of organic carbon (OC) and inorganic binding agents (i.e., Ca^{2+}), biological factors (microorganisms, roots, and soil fauna), and/or physical (compaction) factors (Six et al., 2004). While organic and biological C compounds play a critical role in soil aggregation, aggregates can also provide protective mechanisms which result in soil C accumulation. The rate of soil C decomposition is slowed down due to the blocking off of the accessibility to OC compounds within the pores of aggregates, which microbes cannot reach (Jastrow et al., 2007). Management practices which enhance soil aggregation include those that reduce soil disturbance, such as in no-tillage systems and increase C inputs, such as animal manure and plant residue additions. Increasing soil aggregation not only results in higher soil C sequestration, but also improves soil health. Soil aggregation directly affects air to water ratios in the soil as well as water-holding capacity. This in turns affects soil biodiversity and their activities which significantly impact nutrient cycling and the rate of soil C accumulation.

10.7 Conclusions

The need for intensified cropping systems is driven by the increasing global demand for food, feed, fiber, and fuel, but also by the need to mitigate GHG emissions. Good understanding of determining NPP is of large importance when conducting C and GHG budgets on regional and/or global scales, since a major source of atmospheric CO_2 absorption is through photosynthesis, which is essential to soil C sequestration/retention and soil health outcome. The NPP is defined as the total amount of energy fixed by live plants (via photosynthesis), minus autotrophic respiration, and can be estimated by measuring the dry biomass of combined aboveground (shoots, leaves, and litter) and belowground (roots) biomass per unit of land surface per unit time (i.e., Mg biomass ha^{-1} $year^{-1}$). When evaluating NPP, agroecosystems differ from natural ecosystems due to disturbances (e.g., use of machinery) and external inputs (e.g., fertilizer) to improve cultivation of crops. Intensified management practices of an agroecosystem such as crop selections and rotation, application of fertilizers, tillage practices (conventional versus no-tillage), irrigation, and others can be implemented to increase NPP, yet may lead to negative C footprints by affecting certain soil health and ecosystem services. Some of these management practices' unintended consequences may include GHG emissions (i.e., CO_2, N_2O, and CH_4) and soil C loss which is essential for building soil health. Therefore, the assessment of whether current or intensified cropping systems result in a net sink or source of CO_2 is required, and can be accomplished by measuring NEP. Management practices that increase soil C sequestration through increase in NPP such as no-tillage (especially in semiarid regions) can have potential for sequestering a major portion of C from being emitted into the atmosphere as CO_2, while simultaneously increasing soil health (i.e., soil aggregation).

References

Acosta-Martinez, V., Mikha, M.M., Vigil, M.F., 2007. Microbial communities and enzyme activities in soils under alternative crop rotations compared to wheat–fallow for the Central Great Plains. Appl. Soil Ecol. 37 (1), 41–52.

Al-Kaisi, M.M., Yin, X., 2005. Tillage and crop residue effects on soil carbon and carbon dioxide emission in corn–soybean rotations. J. Environ. Qual. 34 (2), 437–445.

Al-Kaisi, M.M., Kruse, M.L., Sawyer, J.E., 2008. Effect of nitrogen fertilizer application on growing season soil carbon dioxide emission in a corn–soybean rotation. J. Environ. Qual. 37 (2), 325–332.

Baker, J.M., Ochsner, T.E., Venterea, R.T., Griffis, T.J., 2007. Tillage and soil carbon sequestration–what do we really know? Agric. Ecosyst. Environ. 118, 1–5.

Basche, A.D., Miguez, F.E., Kaspar, T.C., Castellano, M.J., 2014. Do cover crops increase or decrease nitrous oxide emissions? A meta-analysis. J. Soil Water Conserv. 69 (6), 471–482.

Blair, N., Crocker, G.J., 2000. Crop rotation effects on soil carbon and physical fertility of two Australian soils. Soil Res. 38 (1), 71–84.

Blanco-Canqui, H., Lal, R., 2007. Soil and crop response to harvesting corn residues for biofuel production. Geoderma 141, 355–362.

Bronson, K.F., Mosier, A.R., 1994. Suppression of methane oxidation in aerobic soil by nitrogen fertilizers, nitrification inhibitors, and urease inhibitors. Biol. Fertility of Soils 17, 263–268.

Bullock, D.G., 1992. Crop rotation. Crit. Rev. Plant Sci. 11 (4), 309–326.

Conen, F., Yakutin, M.V., Sambuu, A.D., 2003. Potential for detecting changes in soil organic carbon concentrations resulting from climate change. Global Change Biol. 9 (11), 1515–1520.

Deal, M.W., 2011. Net primary production in three bioenergy crop systems following land conversion. Thesis and Dissertation. The University of Toledo Digital Repository, The University of Toledo, Toledo, OH.

Dietzel, R., Jarchow, M.E., Liebman, M., 2015. Above-and belowground growth, biomass, and nitrogen use in maize and reconstructed prairie cropping systems. Crop Sci. 55 (2), 910–923.

Duiker, S.W., Lal, R., 2000. Carbon budget study using CO2 flux measurements from a no till system in central Ohio. Soil Tillage Res. 54 (1), 21–30.

Edwards, J.H., Thurlow, D.L., Eason, J.T., 1988. Influence of tillage and crop rotation on yields of corn, soybean, and wheat. Agron. J. 80 (1), 76–80.

Ellert, B.H., Gregorich, E.G., 1996. Storage of carbon, nitrogen, and phosphorus in cultivated and adjacent forested soils of Ontario. Soil Sci. 161 (9), 587–603.

FAOSTAT, 2015. Food and Agriculture Organization of the United Nations Statistics Division. FAO, Rome.

Gilmanov, T.G., Verma, S.B., Sims, P.L., Meyers, T.P., Bradford, J.A., Burba, G.G., et al., 2003. Gross primary production and light response parameters of four Southern Plains ecosystems estimated using long-term CO2-flux tower measurements. Global Biogeochem. Cycles 17 (2), 401–415.

Golabi, M.H., Radcliffe, D.E., Hargrove, W.L., Tollner, W., 1989. Influence of long-term no-tillage on the physical properties of an Ultisol. In: Proceedings of the Southern Conservation Tillage Conference, Athens, Georgia, August 1989.

Grandy, A.S., Robertson, G.P., 2006. Aggregation and organic matter protection following tillage of a previously uncultivated soil. Soil Sci. Soc. Am. J. 70 (4), 1398–1406.

Guzman, J.G., 2013. Evaluation of residue management practices effects on corn productivity, soil quality, and greenhouse gas emissions. Ph.D. Dissertation. Iowa State University, Ames, IA.

Guzman, J.G., Al-Kaisi, M.M., Parkin, T., 2015. Greenhouse gas emissions dynamics as influenced by corn residue removal in continuous corn system. Soil Sci. Soc. Am. J. 79 (2), 612–625.

Haberl, H., Erb, K.H., Krausmann, F., Gaube, V., Bondeau, A., Plutzar, C., et al., 2007. Quantifying and mapping the human appropriation of net primary production in earth's terrestrial ecosystems. Proc. Natl. Acad. Sci. 104 (31), 12942–12947.

Högberg, P., 2007. Environmental science: nitrogen impacts on forest carbon. Nature 447 (7146), 781–782.

Houghton, R.A., Hackler, J.L., Lawrence, K.T., 1999. The U.S. carbon budget: contributions from land-use change. Science 285, 574–578.

IPCC, 2013. Summary for policymakers. In: Stocker, T.F., Qin, D., Plattner, G.-K., Tignor, M., Allen, S.K., Boschung, J. (Eds.), Climate Change 2013: The Physical Science Basis. Working Group I Contribution to the Fifth Assessment Report of the Intergovernmental Panel on Climate Change. Cambridge University Press, Cambridge, UK and New York, pp. 1–27.

Janzen, H.H., 2004. Carbon-cycling in earth systems – a soil science perspective. Agric. Ecosyst. Environ. 104, 399–417.

Jastrow, J.D., Amonette, J.E., Bailey, V.L., 2007. Mechanisms controlling soil carbon turnover and their potential application for enhancing carbon sequestration. Clim. Change 80 (1-2), 5–23.

Jones, M.B., Donnelly, A., 2004. Carbon sequestration in temperate grassland ecosystems and the influence of management, climate and elevated CO_2. New Phytol. 164 (3), 423–439.

Kane, D., 2015. C-Sequestration potential on agricultural lands: a review of current science and available practices. In: Breakthrough, Strategies & Solutions. National Sustainable Agriculture Coalition. Online <http://sustainableagriculture.net/publications>.

Kucharik, C.J., Fayram, N.J., Cahill, K.N., 2006. A paired study of prairie carbon stocks, fluxes, and phenology: comparing the world's oldest prairie restoration with an adjacent remnant. Global Change Biol. 12 (1), 122–139.

Lal, R., 2004. Soil carbon sequestration impacts on global climate change and food security. Science 304 (5677), 1623–1627.

Lal, R., Follett, R.F., Kimble, J., Cole, C.V., 1999. Managing US cropland to sequester carbon in soil. J. Soil Water Conserv. 54 (1), 374–381.

Linn, D., Doran, J., 1984. Effect of water-filled pore-space on C-dioxide and nitrous oxide production in tilled and nontitle soils. Soil Sci. Soc. Am. J. 48 (6), 1267–1272.

Lloyd, J., Taylor, J.A., 1994. On the temperature dependence of soil respiration. Functional ecology 1, 315–323.

McDaniel, M.D., Tiemann, L.K., Grandy, A.S., 2014. Does agricultural crop diversity enhance soil microbial biomass and organic matter dynamic? A meta-analysis. Ecol. Appl. 24 (3), 560–570.

Millar, N., Robertson, G.P., Grace, P.R., Gehl, R.J., Hoben, J.P., 2010. Nitrogen fertilizer management for nitrous oxide (N_2O) mitigation in intensive corn (Maize) production: an emissions reduction protocol for US Midwest agriculture. Mitig. Adapt. Strateg. Glob. Change 15 (2), 185–204.

Monfreda, C., Ramankutty, N., Foley, J.A., 2008. Farming the planet: geographic distribution of crop areas, yields, physiological types, and net primary production in the year 2000. Global Biogeochem. Cycles 22 (1), GB1022.

Neff, J.C., Townsend, A.R., Gleixner, G., Lehman, S.J., Turnbull, J., Bowman, W.D., 2002. Variable effects of nitrogen additions on the stability and turnover of soil carbon. Nature 419 (6910), 915–917.

Olson, K.R., Al-Kaisi, M.M., Lal, R., Lowery, B., 2014. Experimental consideration, treatments, and methods in determining soil organic carbon sequestration rates. Soil Sci. Soc. Am. J. 78 (2), 348–360.

Orchard, V.A., Cook, F.J., 1983. Relationship between soil respiration and soil moisture. Soil Biol. Biochem. 15 (4), 447–453.

Patel, N.R., Dadhwal, V.K., Agrawal, S., Saha, S.K., 2011. Satellite driven estimation of primary productivity of agroecosystem in India. In: International Archives of the Photogrammetry, Remote Sensing, and Spatial Information Sciences, vol. XXXVIII-8/W20. ISPRS Bhopal 2011 workshop, November 8, 2011, Bopal, India.

Paustian, K., Six, J., Elliot, E.T., Hunt, H.W., 2000. Management options for reducing CO_2 emissions from agricultural soils. Biogeochemistry 48, 147–163.

Peterson, G.A., Halvorson, A.D., Havlin, J.L., Jones, O.R., Lyon, D.J., Tanaka, D.L., 1998. Reduced tillage and increasing cropping intensity in Great Plains conserves soil carbon. Soil Tillage Res. 47, 207–218.

Post, W.M., Kwon, K.C., 2000. Soil carbon sequestration and land-use change: processes and potential. Global Change Biol. 6 (3), 317–327.

Powlson, D.S., Stirling, C.M., Jat, M.L., Gerard, B.G., Palm, C.A., Sanchez, P.A., et al., 2014. Limited potential of "no-till" agriculture for climate change mitigation. Nat. Clim. Change 4 (8), 678–683.

Rasmussen, P.E., Smiley, R.W., 1997. Soil carbon and nitrogen change in long-term agricultural experiments at Pendleton, Oregon. In: Paul, E.A., Paustian, K., Elliott, E.T., Cole, C.V. (Eds.), Soil Organic Matter in Temperate Agro-ecosystems: Long-term Experiments in North America. CRC Press, Boca Raton, FL, pp. 353–360.

Rasse, D.P., Rumpel, C., Dignac, M.F., 2005. Is soil carbon mostly root carbon? Mechanisms for a specific stabilisation. Plant Soil 269, 341–356.

Six, J., Feller, C., Denef, K., Ogle, S., Sa, J.C.D.M., Albrecht, A., 2002. Soil organic matter, biota and aggregation in temperate and tropical soils-effects of no-tillage. Agronomie 22 (7-8), 755–775.

Six, J., Bossuyt, H., Degryze, S., Denef, K., 2004. A history of research on the link between (micro) aggregates, soil biota, and soil organic matter dynamics. Soil Tillage Res. 79 (1), 7–31.

Smika, D.E., 1983. Soil water change as related to position of wheat straw mulch on the soil surface. Soil Sci. Soc. Am. Proc. 47, 988–991.

Smith, P., Falloon, P., Kutsch, W.L., 2009. The role of soils in the Kyoto Protocol. In: Kutsch, K.L., Bahn, M., Heinemeyer, A. (Eds.), Soil Carbon Dynamics. Cambridge Publication, Cambridge, UK and New York, pp. 245–256.

Syswerda, S.P., Corbin, A.T., Mokma, D.L., Kravchenko, A.N., Robertson, G.P., 2011. Agricultural management and soil carbon storage in surface vs. deep layers. Soil Sci. Soc. Am. J. 75 (1), 92–101.

Ussiri, D.A., Lal, R., 2009a. Long-term tillage effects on soil carbon storage and carbon dioxide emissions in continuous corn cropping system from an alfisol in Ohio. Soil Tillage Res. 104 (1), 39–47.

Ussiri, D.A., Lal, R., Jarecki, M.K., 2009b. Nitrous oxide and methane emissions from long-term tillage under a continuous corn cropping system in Ohio. Soil Tillage Res. 104 (2), 247–255.

Vitousek, P.M., 1994. Beyond global warming: ecology and global change. Ecology 75, 1861–1876.

West, T.O., Post, W.M., 2002. Soil organic carbon sequestration rates by tillage and crop rotation: a global data analysis. Soil Sci. Soc. Am. J. 66 (6), 1930–1946.

White, D.H., Howden, S.M., 2007. Climate and its effect on crop productivity and management. In: Verheye, W.H. (Ed.), Soil, Plant Growth and Crop Production. Eolss Publishers, Oxford, UK, pp. 44–78.

Woodwell, G.M., Whittaker, R.H., 1968. Primary production in terrestrial ecosystems. Am. Zool. 8, 19–30.

World Bank, 2012. Carbon sequestration in agricultural soils. Economic and Sector Work. Report number: 67395-GLB. Online <ftp://ftp.fao.org/ag/agp/ca/CA_CoP_May12/ARD_ESW12_CarbonSeq_web[1].pdf>.

World Bank, 2015a. Agricultural nitrous oxide emissions. Online <http://data.worldbank.org/indicator/EN.ATM.NOXE.AG.ZS/countries/1W?display=graph> (accessed 01.06.16).

World Bank, 2015b. Agricultural methane emissions. Online <http://data.worldbank.org/indicator/EN.ATM.METH.AG.ZS/countries/1W?display=graph> (accessed 01.06.16).

CHAPTER 11

Nutrient Cycling and Soil Biology in Row Crop Systems under Intensive Tillage

Yucheng Feng[1] and Kipling S. Balkcom[2]

[1]Auburn University, Auburn, AL, United States [2]USDA-ARS National Soil Dynamics Laboratory, Auburn, AL, United States

11.1 Introduction

Intensive tillage, popularly known as conventional tillage, typically includes multiple surface tillage operations to bury surface residue. The 1996 Glossary of Soil Science terms defines conventional tillage as "primary and secondary tillage operations normally performed in preparing a seedbed and/or cultivating for a given crop grown in a given geographical area, usually resulting in <30% cover of the crop residues remaining on the surface after completion of the tillage sequence" (SSSA, 1997; Koller, 2003). Reasons for using this tillage system have varied from yield security, desire for a residue-free soil surface that promotes seedling establishment, incorporation of fertilizers and herbicides, and burying weed seed to promote weed control (Koller, 2003). However, energy requirements are high for this type of intensive tillage, and the subsequent bare soil surface that remains following these operations is quite susceptible to erosion. As a result, numerous research efforts can be found throughout the literature that document how tillage systems that minimize or eliminate inversion tillage and retain residues on the soil surface can maintain and/or stimulate agricultural productivity, while enhancing soil health.

11.2 Soil Characteristics of Intensive Tillage Systems

11.2.1 Physical Properties

Intensive tillage operations bury protective crop residues, which leaves the soil exposed to raindrop energy and highly susceptible to erosion. Mixing surface crop residues with soil creates a flush of microbial activity that rapidly decomposes soil organic matter (Novak et al., 2009). In contrast, residue remaining on the soil surface can protect it from excessive compaction and crusting, while limiting the breakdown and dispersion of soil aggregates (Blanco-Canqui and Lal, 2009). Increasing soil organic matter at the soil surface, typically

Soil Health and Intensification of Agroecosystems.
DOI: http://dx.doi.org/10.1016/B978-0-12-805317-1.00011-7

associated with noninversion tillage systems (e.g., conservation tillage), can significantly improve soil physical, as well as chemical and biological properties (Langdale et al., 1990).

Soil compaction, a condition that restricts air and water movement into and within the soil and plant root growth within the soil profile, is typically monitored by measuring soil bulk density and/or soil strength using cone index measurements. A compacted layer located within the plant rooting depth of the soil limits root exploration of the soil profile below this layer. This reduction in root growth will also limit nutrient and water uptake. Nutrient and water uptake limitations negatively impact yield and subsequent productivity. The reduced root volume available to plants may also enhance and/or promote short-term drought stress. Soil bulk density can be decreased and subsequent permeability of the soil can be increased with tillage in the tilled zone (Baumhardt et al., 1993), but eventually, intensive tillage breaks down the soil structure (So et al., 2009). Systems that minimize tillage and retain surface residues generally result in less consolidated soil over time. However, there are reports of increased bulk densities for no-tillage systems when soils are initially converted to these systems or limited surface residue is present, which can initially depress crop yields (Hill, 1990; Baumhardt et al., 1993; So et al., 2009).

Cover crops, an integral part of conservation tillage systems, particularly in the humid Southeast United States, are designed to complement existing surface residue to improve degraded soils prevalent across the region that are attributed to low organic matter contents (Balkcom et al., 2015). However, cone index measurements have identified the presence of a compacted zone in the soil profile, prior to spring tillage operations on Coastal Plain, United States, soils (Fig. 11.1A). In this region, some form of tillage, such as strip tillage is performed to maximize belowground disruption, but maintain surface residue by minimizing surface soil disturbance (Balkcom et al., 2013). This tillage operation can be very effective at eliminating compacted zones underneath the row within the soil profile (Fig. 11.1B). Fig. 11.1B highlights how shallow compaction can occur in the profile and the severity of the compaction based on a threshold of 2.0 MPa being identified as root-limiting (Taylor and Gardner, 1963). Although the tillage operation corrected the problem, gravimetric soil moisture differences at 30 cm, corresponding to each sample time, differed by 2 $kg\,kg^{-1}$, which accounts for some of the soil strength differences observed before and after tillage.

Surface crop residue can offset compaction by absorbing and dissipating wheel and animal traffic forces (Blanco-Canqui and Lal, 2009). These decomposed crop residues form soil organic matter that lowers the bulk density of the mineral fraction (Blanco-Canqui and Lal, 2009). In Australia, intensive tillage resulted in greater bulk densities compared to no-tillage after 14 years; however, no differences were detected between tillage systems during the early years of the study (So et al., 2009). In contrast, Hill (1990) reported higher bulk densities for no-tillage systems across all measured depths compared to conventional tillage for three sites with silt loam soils in Maryland, United States. As bulk density decreases, soil porosity

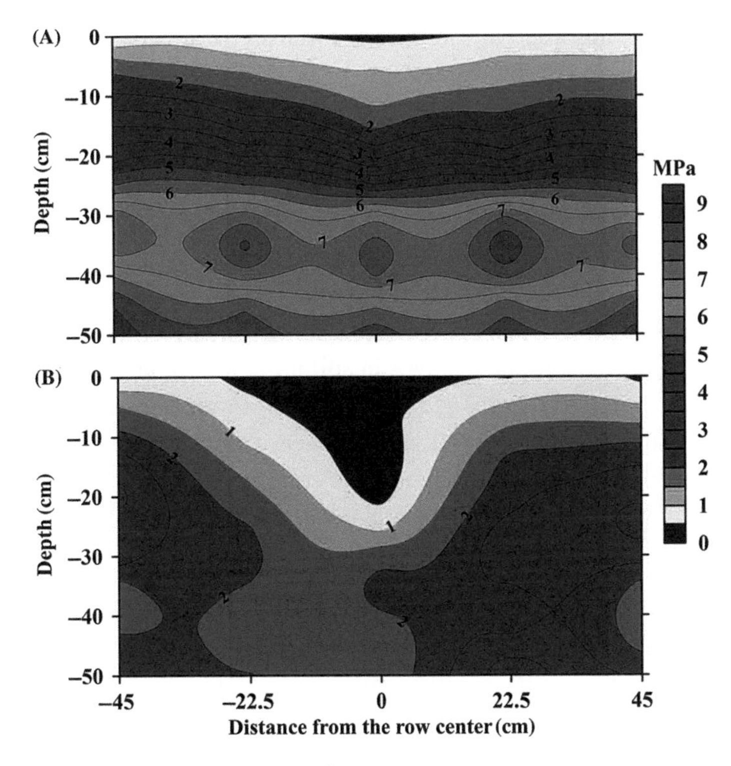

Figure 11.1
Soil strength measurements collected before tillage (A) and after tillage (B) for a Coastal Plain soil located in southeast Alabama, United States, that contained a rye (*Secale cereale* L.) cover crop. *Unpublished data from K.S. Balkcom.*

increases, likely resulting in greater water movement. Traditionally, this decrease in bulk density has been observed near the soil surface (Blanco-Canqui and Lal, 2009; So et al., 2009).

Porosity influences hydraulic properties of the soil and subsequent water availability to crops. Increases in porosity for no-tillage systems, particularly at the soil surface, promote greater infiltration of water compared to intensive tillage systems. Continuous intensive tillage breaks down the soil structure that tends to reduce soil porosity leading to compaction (So et al., 2009). The surface soil is especially sensitive to compaction and the thin, compacted layer formed on the soil surface is generally referred to as "soil crusting" or "surface sealing." Bare soil following intensive tillage absorbs rainfall energy that disperses soil aggregates and those soil particles move into and obstruct soil pores near the soil surface (Baumhardt et al., 2004; Blanco-Canqui and Lal, 2009). In addition, these crusts can also negatively affect crop establishment because seedlings must penetrate this compacted zone to emerge and maintain yield potential (Baumhardt et al., 2004). Once a crust forms, infiltration rates are significantly reduced and runoff increases, promoting soil erosion. Soil particles leaving the field through erosion may also contain adsorbed nutrients that diminish the fertility status of the soil, as well as promote pollution of nearby waterways.

11.2.2 Chemical Properties

Intensive tillage systems mechanically mix soil within the tilled zone. As a result, constituents that are present or added to the soil, such as residues, fertilizer, or lime are also subsequently mixed and presumably distributed throughout the tilled zone. This inherent activity associated with intensive tillage has always been perceived as an advantage of these systems, so much so that researchers began to examine changes in soil properties as research efforts began to focus on no-tillage systems (Blevins et al., 1977,1983; Franzluebbers and Hons, 1996). Tillage systems that eliminate inversion tillage practices (e.g., no-tillage) minimize surface soil disturbance that creates a different soil surface condition compared to intensive tillage (Blevins et al., 1983), which prompted concerns related to stratification.

Soil pH is one soil chemical property that was compared across different tillage systems. Soil pH is important because it affects root growth and microbial activity, phosphorus (P) and micronutrient availability, and herbicide activity (Franzluebbers and Hons, 1996). Blevins et al. (1977) observed that the surface soil becomes more acidic for no-tillage systems compared to intensive tillage systems, which was attributed to acidification associated with nitrification of surface-applied nitrogen (N) fertilizer. Dick (1983) observed lower soil pH values from no-tillage systems compared to conventional tillage systems to a depth of 22.5 cm for two Alfisols, the tillage systems had been in place for 18 years. Surface-applied lime can effectively neutralize surface soil acidity; however, soil acidity that occurs deeper in the soil profile cannot be neutralized effectively under no-tillage management because there is no mixing of the soil and lime (Dick, 1983).

Nutrient stratification prompted many research efforts related to concerns about differences in nutrient distribution between conventional and conservation tillage systems (Eckert, 1985; Franzluebbers and Hons, 1996; Crozier et al., 1999; Howard et al., 1999; Wright et al., 2007). These concerns led some regions to recommend shallow soil samples for lime and fertilizer requirements for tillage systems that limit soil mixing (Touchton and Sims, 1987; James and Wells, 1990). Balkcom et al. (2005) showed how stratification near the soil surface can be exacerbated with manure applications. No differences in nutrient concentrations were observed between tillage systems across depths using commercial fertilizer, but these results were observed after only 3 years (Fig. 11.2). Surface-applied dairy manure increased nutrient concentrations, particularly in the 0–5 cm depth, across tillage systems with the largest increase observed in the conservation tillage system (Fig. 11.2). Nutrient concentrations also increased in the 5–15 cm depth where dairy manure was applied to conventional tillage, which can be attributed to the soil mixing of this system (Fig. 11.2). The elevated nutrient concentrations for conventional tillage systems create potential environmental concerns.

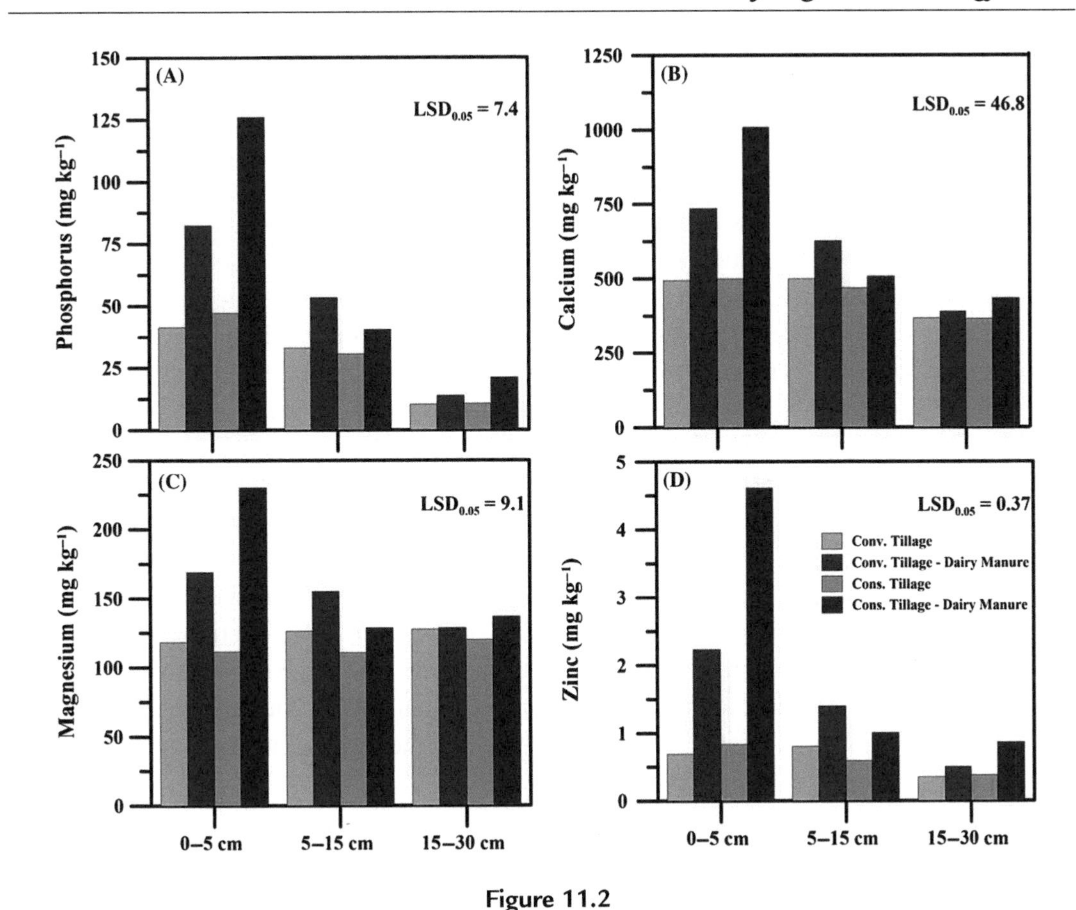

Figure 11.2

Phosphorus (A), calcium (B), magnesium (C), and zinc (D) concentrations measured across two tillage systems with and without manure for three depths in central Alabama, United States (Conv. = Conventional, Cons. = Conservation). *Adapted from Balkcom, K.S., Terra, J.A., Shaw, J.N., Reeves, D.W., Raper, R.L., 2005. Soil management system and landscape position interactions on nutrient distribution in a Coastal Plain field. J. Soil Water Conserv. 60, 431–437.*

Soil erosion associated with these systems could move soil containing adsorbed nutrients to nearby waterways resulting in pollution.

11.2.3 Modification of Habitat for Soil Organisms

Soil is teeming with life, both macroscopic and microscopic. Soil organisms range in size from invisible microorganisms to easily visible insects, earthworms, and plant roots. Soil arguably harbors the greatest microbial diversity on Earth owing to the heterogeneity of the soil habitat and temporal variations in amounts and distribution of water and nutrients. A single gram of soil may contain 10^9–10^{10} microbial cells, including tens of thousands of different bacterial, archaeal, and fungal species (Torsvik and Øvreås, 2002). Soil solids are arranged in a very complex fashion, forming aggregates, and tortuous pore spaces where soil organisms live. Interfaces in soils (solid–liquid, solid–gas, and liquid–gas) also contribute to the formation of diverse microsites.

Soil aggregation influences microbial distribution and activity, as well as soil organic matter turnover. Soil pores inside aggregates provide refuge for soil microorganisms against predation and desiccation. Soil aggregates vary in size, as do the pores within. Aggregates smaller than 250 μm in diameter are considered microaggregates; larger ones are macroaggregates. Pore diameters in microaggregates range from 0.2 to 0.6 μm, while those in macroaggregates range from 25 to 100 μm (Paul and Clark, 1996). Pore size determines the occupants and pore neck size restricts entry into the pores. Bacteria usually live within microaggregates in pores with diameters ranging from 0.8 to 3 μm. Bacterial biomass increases as pore neck size increases from 0.2 to 1.2 μm (Hassink et al., 1993). Pores between microaggregates within macroaggregates are the main habitat for fungi. However, pores between macroaggregates and between microaggregates can be home for nematodes, protozoa, and fungi. Nematodes are usually found in pores of 30–90 μm in diameter. Protozoa and fungi often cannot enter pores with a neck size smaller than 6 μm. Therefore, most soil bacteria are physically separated from their predators, such as protozoa and nematodes. Soil mites are more abundant in macropores (Ducarme et al., 2004; Nielsen et al., 2008). Biological activities, such as root growth and earthworm movement, can form macropores (0.8–5 mm).

The primary effect of tillage is to physically disturb the habitat of living organisms. Different tillage systems disturb the soil matrix to different degrees and lead to changes in water and substrate distribution and spatial arrangement of the pore network (Young and Ritz, 2000). In general, intensive tillage breaks up large aggregates and thus aggregate sizes are smaller in tilled soil than in no-tillage soil. Drees et al. (1994) found that aggregate size in intensively tilled soil ranged from 0.25 to 0.39 mm in diameter and in no-tillage soil from 0.41 to 1.08 mm in a Maury silt loam from Kentucky, United States. Aggregates formed under conventional tillage are also less stable than those under conservation tillage practices (Pagliai et al., 2004; Grandy and Robertson, 2007). Kravchenko et al. (2011) examined intraaggregate pores in macroaggregates (4–6 mm) from a loamy Alfisol (0–20 cm) under 20 years of different soil management practices: conventional tillage, no-tillage, and natural succession vegetation. No-tillage and native succession treatments had been converted from an agricultural field under conventional tillage for over 100 years. Although distribution patterns of pores in the aggregates had similar trends among all three treatments, aggregates from conventional tillage had more pores in the size range of 15–60 μm (microcracks of nonbiological origin, e.g., wetting/drying and freezing/thawing) than from the other two treatments. Distribution of pore voxels (porosity based on pores >15 μm) belonging to large pores (>105 μm) was most heterogeneous in aggregates from the natural succession treatment, followed by no-tillage and conventional tillage. Visual observation of computer-assisted images confirmed the frequent presence of large pores of biological origin in the aggregate centers of no-tillage and natural succession aggregates. The authors speculated that macroaggregates break into microaggregates along pores >15 μm, especially along 15–60 μm microcracks. The abundance of 15–60 μm intraaggregate pores may be a cause of lower aggregate stability in conventionally tilled soil.

Both abiotic and biotic factors influence the size, shape, and continuity of the soil pore network. Tillage is one of the important abiotic factors that can be managed. Soil pores can be characterized indirectly by bulk density and porosity measurements, or directly by computer-assisted image analysis of soil thin sections (Young and Ritz, 2000). Here, we focus on results obtained from direct measurements. Drees et al. (1994) evaluated soil structural differences between conventionally tilled and no-tillage soils by characterizing the size, shape, and orientation of pores and aggregates in undisturbed samples from Kentucky, United States. Tillage treatments had been in place for 18 years under continuous corn (*Zea mays* L.) production with rye (*Secale cereale* L.) as the winter cover crop. Although no significant difference in pore area was found between tillage treatments, pores under conventional tillage were smaller. Average pore size ranged from 0.18 to 0.26 mm in diameter in the no-tillage soil and 0.12 to 0.17 mm in the conventionally tilled soil. Platy aggregates were found near the surface (1–5 cm) in no-tillage plots separated by interconnected planar pores, which formed an extensive network with other channels and chambers. In addition, earthworm burrows were abundant in the no-tillage soil at all sampling depths (down to 25 cm). In contrast, conventionally tilled soil had granular and fragmented structure in the surface layer; pores were complex without evidence of earthworm activity. In Italy, Pagliai and De Nobili (1993) found that total porosity was greater under conventional tillage than no-tillage in the A_p horizon after 12 years of tillage treatments in a clay loam soil (Vertic Xerofluvent) of a vineyard. Pore shape and size distribution were also different under both tillage treatments. Platy aggregates were present in the surface layer (0–6 cm) of conventionally tilled soil. In a more recent study, Pastorelli et al. (2013) showed that conventionally tilled soil had greater macroporosity, as well as total porosity than no-tillage soil in the surface layer (0–10 cm) after 12 years in a silty clay soil from central Italy. High macroporosity has been attributed to the presence of irregular and elongated pores larger than 500 μm (fissures) under conventional tillage in this Vertisol. The no-tillage treatment, however, had more macropores in the 50–200 μm size class (transmission pores) than the conventional tillage treatment (Pastorelli et al., 2013).

In general, soils under intensive tillage tend to have higher total porosity but smaller pores (except for Vertisols) than soils under no-tillage and also develop a plow pan over time. When soils under intensive tillage are converted to no-tillage, pores in the 100–500 μm size range typically increase. Fewer researchers have focused on the impact of tillage on micro- and mesopores (Kay and VandenBygaart, 2002). Intensive tillage also results in smaller and less stable aggregates. Tillage-induced habitat change typically depends on soil type and climatic conditions, as well as the time period between the last tillage and sampling time. The abundance and composition of soil biota are expected to change as their surrounding physical and chemical environments are altered by tillage system.

11.3 Residue Decomposition

Crop residue decomposition is the result of activities of heterotrophic organisms living in the soil. The majority of soil microorganisms are heterotrophs that use organic compounds as carbon (C) and energy sources. Soil microorganisms breakdown complex organic compounds in crop residues and use C, N, P, and other nutrients for their own growth while making nutrients available to plants. Residue decomposition studies have been primarily focused on nutrient release associated with mineralization of nutrients in crop residue, particularly N. Crop residue type, placement, level of incorporation, soil temperature, and water/aeration regimes affect microbial activity that governs N mineralization/immobilization of N that ultimately determines N release from crop residues (Aulakh et al., 1991). The C:N ratio is a common measurement used to determine N availability of crop residues and mineralization rate. Lignin and polyphenol contents of residues also affect mineralization (Palm and Sanchez, 1991; Vigil and Kissel, 1991); however, it is generally accepted that residues with C:N ratios <25 result in net N mineralization or N release, while C:N ratios >25 immobilize N (Jenkinson, 1981; Starovoytov et al., 2010). Subsequently, residues with low C:N ratios (high N contents) not only result in net N mineralization, but these residues tend to decompose faster (Mulvaney et al., 2010). As a result, crop residue persistence and synchronization of N released with N uptake of the following crop can be negatively impacted.

11.3.1 Fallow Systems

In a fallow system that includes intensive tillage, crop residue decomposition will occur faster due to incorporation of residue into the soil. A traditional fallow system is defined as leaving the land either uncropped and weed-free or with only volunteer vegetation during at least one period when a crop would traditionally be grown (SSSA, 1997). For the purpose of this discussion, fallow will only include residue from the previous crop with a weed-free period maintained with intensive tillage between crop sequences.

Faster decomposition of crop residue associated with intensive tillage results from complete exposure to the food web. Neher (1999) defined characteristics of a soil food web and that indicated it varies with geography and climate. However, Lachnicht et al. (2004) succinctly described the food web as the "decomposer community." In addition to geography and climate, this community can also change based on the tillage system. Frey et al. (1999) reported fungi dominate the microbial community in no-tillage systems, while bacteria dominate the community in conventional tillage systems. Obviously, residue remains on the surface in no-tillage systems, while residue is incorporated with conventional tillage. Faster decomposition of buried residue for conventional tillage systems is often related to higher water content and greater densities of decomposer organisms when compared to surface residue in no-tillage systems (Beare et al., 1992). Effects of tillage systems on soil microbial communities are typically manifested during the fallow period (see discussion in Section 11.5.4).

11.3.2 Multicrop Systems

In contrast to a fallow system, multicrop systems will increase crop residue amounts and subsequently also increase nutrient additions available to the decomposition process. Copeland and Crookston (1992) reported that crops grown in rotation produce more dry matter that is of higher quality compared to monoculture. Assuming the C:N ratio is <25, the organic N fraction will mineralize quickly, which is similar to residue from a fallow system using intensive tillage. As the mineralization process continues, this inorganic N becomes available for crop uptake as residual N, or susceptible for loss pathways that include leaching, denitrification, or runoff (Wood et al., 1990). However, another crop, which may include a cover crop or a cash crop in a double-crop situation, planted following this decomposition process can potentially use that N; thereby, reducing the N loss potential and improving N cycling. In contrast, Eck and Jones (1992) reported N loss increases during a fallow period because of continuous soil N mineralization, as a result of increased temperature and water content that coincided with the fallow period (Eck and Jones, 1992).

11.3.3 Nutrient Cycling

Nutrient cycling by a simple definition describes how nutrients in their organic form move into an inorganic form and back into the organic form. The N cycle is a famous example that is used to illustrate all transformation processes that organic and inorganic N go through within a given environment. An understanding of these processes helps in the development of management techniques that will minimize N losses and thereby increase efficiency of N use in the cropping system. Demand for this knowledge is driven by public concerns about sustainability of natural resources that was recognized during initial discussions about soil quality (Karlen et al., 1997).

Rapid decomposition of crop residues associated with intensive tillage may promote significant N leaching, particularly for fallow systems, which diminishes any advantage of the N contained within the crop residue for subsequent nutrient cycling into the following cropping system. For multicrop systems, this rapid decomposition and subsequent N availability could be perceived as an advantage, but this depends on how N within the system is regarded. For example, Doran and Werner (1990) compared an organic-based system with cover crops and legumes as green manure crops compared to a conventional system that included synthetic fertilizers. The organic system had lower levels of NO_3-N present in the early spring, which lowered the risk for potential N leaching, but this also resulted in lower available N, which was a disadvantage for the following corn crop (Doran and Werner, 1990).

Intensive tillage use may provide a management component to improve synchronization of nutrient release with nutrient uptake of a subsequent crop. As lag times increase between N release from decomposition of crop residues or specifically green manure crops and crop N demand, soils with inherent fertility and nutrient retention capability may offset extended lag times and improve synchronization between N release and crop N uptake (Cherr et al., 2006). N leaching losses can occur on sandy soils in warm, humid climates even with short lag times between N release and uptake (Nelson and King, 1996; Wyland et al., 1996; Cherr et al., 2006). Cherr et al. (2006) suggested that decomposition could be controlled to improve synchronization. Tillage was one management practice suggested, particularly in cooler environments, but negative environmental aspects previously described for intensive tillage compared to no-tillage may not justify tillage for improved N cycling.

11.4 Soil Carbon Under Intensive Tillage

Soil organic matter is a reservoir for nutrients that also interacts with soil particles to form soil aggregates (Ellert and Bettany, 1995). C is a major component of organic matter comprising of 48–58% (Nelson and Sommers, 1982). The global soil C pool is estimated at 2500 gigatons (Gt) that includes approximately 1550 Gt of soil organic C (SOC) and 950 Gt of soil inorganic C (Lal, 2004). Researchers have devoted significant attention to the SOC pool because this pool is sensitive to change and serves as a sink for atmospheric carbon dioxide (CO_2) (Baker et al., 2007).

Soil C sequestration is a process with potential to stabilize atmospheric greenhouse gas levels (Izaurralde et al., 2006). Converting to energy that does not include fossil fuels is an alternative for reducing emissions, but natural processes associated with the global C cycle must be fully understood (Causarano et al., 2006). Causarano et al. (2006) also referenced several textbooks that described how land-use and management affected C pools and fluxes. A common theme to increase SOC was the adoption of conservation management practices (Causarano et al., 2006). Conservation tillage combined with an environment that reduces decomposition rates can reduce C losses to the atmosphere (Follett, 2001).

The focus on sustainability, enhancing soil health, and mitigating greenhouse gas concentrations in the atmosphere does not warrant the use of intensive tillage. Tillage mixes the soil and breaks up aggregates, which allows decomposers greater access to further enhance SOC decomposition through CO_2 emissions (Reicosky et al., 1997; Post and Kwon, 2000; Quincke et al., 2007). Soil C at or near the soil surface is easily lost with increasing tillage intensity and soil erosion because this C is derived from plant material and/or animal excreta that enters the SOC pool and abrades into smaller sizes that are considered particulate organic matter (Follett, 2001).

11.4.1 Forms

Although tillage may strongly influence soil organic C decomposition, forms of C vary from easily decomposed to stable on a continuous basis (Quincke et al., 2007). Quincke et al. (2007) described two different labile organic C pools that consist of an active pool and slower reacting pool. The active pool consists of biologically active fractions that include microbial biomass (MBC) and potential C mineralization (PCM) (Sainju et al., 2008). Soil organic C changes may be better represented by these two fractions because they can change quickly and are sensitive to management that may include tillage, cropping system, cover crop, and/or N fertilization (Campbell et al., 1989; Sainju et al., 2006). The slower reacting C pool contains particulate organic C (POC) (>53 μm) that is considered a transitional fraction. Microbial decomposition may not be apparent following tillage, but land-use changes may affect POC before SOC (Quincke et al., 2007). The ability to identify and understand how these C fractions respond to management improves general knowledge of the global C cycle. These C fractions are important because examining only SOC, although an important component, typically changes slowly with time, represents a large C pool, and has inherent spatial variability (Franzluebbers et al., 1995; Sainju et al., 2008). However, Clapp et al. (2000) stated that long-term studies are the best way to predict how soil management changes affect soil C, a key indicator for soil health.

11.4.2 Distribution

Intensive tillage will move C fractions deeper into the soil profile and this disturbance associated with tillage is believed to be a major cause of historical C loss (Baker et al., 2007). De Jong and Kachanoski (1988) stated soil erosion can affect C redistribution, particularly on an ecosystem scale; however, tillage clearly incorporates C-containing residues into the soil profile that affects distribution. Environments can also affect general C contents. Typically, soil C increases with an increase in precipitation, a drop in mean annual temperature, and an increase in clay content (Jenny, 1980; Nichols, 1984; Burke et al., 1989; Franzluebbers, 2002).

Increased soil C at or near the soil surface in conservation tillage systems compared to conventional tillage systems is commonly associated with enhanced soil health, C sequestration, and subsequent crop production benefits (Lal et al., 1999; Follett, 2001; Causarano et al., 2006). Baker et al. (2007) cautioned that the propensity for SOC to accumulate at the soil surface in conservation tillage systems could be a sampling bias associated with sampling ≤30 cm and not a true increase of SOC using conservation tillage. The researchers argue that SOC is just distributed differently between tillage systems with no consistent difference observed between tillage systems. Baker et al. (2007) acknowledge benefits of conservation tillage that are advantageous compared to intensive tillage, but benefits associated with C sequestration should be closely scrutinized with respect to sampling depth. Franzluebbers (2002) recognized the stratification of soil organic C with depth and proposed a stratification ratio as an indicator of dynamic soil quality and as a method to detect management-induced changes in dynamic soil quality. This stratification ratio was designed to allow comparisons across ecoregions or landscape positions.

11.5 Soil Biota

The physical structure of soil plays a defining role in ecological processes (Ettema and Wardle, 2002). Tillage affects soil structure and hence modifies the habitat of soil biota. Different communities respond differently to tillage regimes. Frequent tillage is known to have a major effect on soil microbial communities and their function in nutrient cycling. Many of the tillage effects relate to composition of decomposer communities. In general, both the abundance and diversity of soil organisms increase with decreasing tillage intensity; however, responses to tillage are highly variable depending on climate, soil type, and cropping system (Roger-Estrade et al., 2010).

11.5.1 Soil Fauna

Although soil fauna comprise only a small fraction of soil biota biomass, they play a significant role in maintaining soil structure, as well as in accelerating organic matter decomposition and nutrient cycling. They interact with microorganisms by grazing on bacteria and fungi and regulate the size and composition of microbial communities. Because of their feeding, burrowing, and movement, soil fauna transport microorganisms and modify the soil structure and thus habitat of soil microorganisms. Their populations are controlled by both resources and predators (Fu et al., 2000).

Tillage accelerates decomposition of organic matter and thus affects the soil fauna community indirectly by reducing resources and subsequently decreasing fauna abundance and diversity (House and Parmelee, 1985). Tillage also has a direct effect on organisms themselves, i.e., killing or injuring macrofauna and exposing them to predators (Roger-Estrade et al., 2010). Total earthworm abundance, biomass, and species diversity increase with decreasing tillage intensity in various countries and under different climates (van Capelle et al., 2012). In general, plowing favors soil food webs comprising organisms with shorter generation times, smaller body size, rapid dispersal, and omnivorous feeding habits (Andren and Lagerlof, 1983; Steen, 1983). For example, a Polish study showed that tillage shifted the Annelida from earthworms to smaller and metabolically more active enchytraeids (potworms) (Golebiowska and Ryszkowski, 1977). At the Horseshoe Bend long-term experimental area in Georgia, United States, the abundance and biomass of earthworms were less and those of enchytraeids were greater in conventionally tilled soil than in no-tillage soil planted to sorghum (*Sorghum bicolor* L.) with winter cover crops (House and Parmelee, 1985). The type of winter cover crop influenced the magnitude of the difference. Evaluation of data from

German soils showed that enchytraeid abundance was numerically greater under various forms of conservation tillage than under conventional tillage and was the lowest under no-tillage (van Capelle et al., 2012). Enchytraeids appear to benefit from the mulch layer under conservation tillage, but need tillage to a certain extent. House and Parmelee (1985) found that all major microarthropods were lower under conventional tillage than under no-tillage at the Horseshoe Bend site (Table 11.1). The abundance of microarthropods in a winter wheat (*Triticum aestivum* L.) production system in the southwestern United States also showed the same trend except for mites in Prostigmata, which were more abundant under conventional tillage (Dubie et al., 2011). Two European studies found that collembolan (springtails) communities were not affected by tillage intensity (Sabatini et al., 1997; Petersen, 2002). In contrast, meta-analysis of German studies showed that the collembolan communities (both abundance and species diversity) flourished under conventional tillage and that the impact of tillage intensity depended on soil texture (van Capelle et al., 2012). Similarly, total individual numbers of mites decreased with reduced tillage intensity and were evenly distributed in the 0–30 cm depth under conventional tillage (van Capelle et al., 2012). It appears that mites and collembolans are less sensitive to mechanical injury and soil inversion exerted by tillage. There are conflicting results concerning tillage effects on nematodes (Roger-Estrade et al., 2010; van Capelle et al., 2012), with most studies focused on plant-parasitic nematodes. A wide range of nematode responses to different tillage regimes has been reported depending on nature, number, or timing of cultivation operations (Roger-Estrade et al., 2010). Moreover, different feeding types respond to tillage differently.

Tsiafouli et al. (2015) examined soil food webs from grasslands and extensive and intensive rotations in four agricultural regions across Europe (Sweden, the UK, the Czech Republic,

Table 11.1: Aggregated mean number per square meter of microarthropods over eight collection dates from conventional (CT) and no-tillage (NT) sorghum/crimson clover and sorghum/rye agroecosystems

	Sorghum/Clover		Sorghum/Rye	
	CT	NT	CT	NT
Acarina				
Mesostigmatids	2645	6799	2054	8639
Prostigmatids	25,978	63,859	23,318	32,523
Oribatids	5097	33,272	14,990	36,809
Astigmatids	3486	97	719	285
Collembola	7727	12,487	6244	14,684
Insects	1070	2594	2105	2548
Total	46,004	119,107	49,424	96,584

Source: Reproduced with permission of Elsevier from House, G.J., Parmelee, R.W., 1985. Comparison of soil arthropods and earthworms from conventional and no-tillage agroecosystems. Soil Tillage Res. 5, 351–360.

All tillage treatment pairs (except insects) significantly different ($p < 0.01$).

and Greece) and found consistent responses to increasing land-use intensity. In all four regions, species richness and taxonomic distinctness of earthworms, collembolans and oribatid mites were negatively affected by increased land-use intensity. They concluded that intensive agriculture reduces soil biodiversity, making soil food webs less diverse and composed of smaller organisms. Ultimately, these biological indicators are measures of the soil health environment and productivity.

11.5.2 Microbial Biomass

Microbial biomass represents a small fraction of the soil and has been used to estimate the soil biological status. Microbial biomass responds rapidly to conditions that eventually result in changes to soil organic matter content and has been used as a sensitive and early indicator of soil quality/health (Powlson et al., 1987; Sparling, 1997; Franzluebbers et al., 1999). In general, microbial biomass C increases with increasing soil organic matter content, comprising 1–4% of the total organic C (Rice et al., 1996). Many researchers have shown decreases in total microbial biomass in conventionally tilled soil (e.g., Doran, 1980, 1987; Feng et al., 2003). Recent studies using total phospholipid ester-linked fatty acids (PLFA) (Helgason et al., 2009; Mathew et al., 2012) or DNA (Kaurin et al., 2015) to estimate microbial biomass also corroborate this observation. In addition, the proportion of microbial biomass C in SOC is lower under conventional tillage than under no-tillage (Rice et al., 1996; Andrade et al., 2003). Bacteria, archaea, and fungi dominate soil microbial biomass. Their populations are controlled mostly by available resources rather than predators (Fu et al., 2000). It is well known that conservation tillage systems increase soil organic matter content in the surface soil, which stimulates soil biological activity.

11.5.3 Fungi

Fungal hyphae form an extensive network in soils. Tillage disrupts the extensive hyphal network, resulting in loss of cell contents and inducing stress conditions. Soil inversion by moldboard plowing deposits fungal propagules (spores, hyphae, and colonized root fragments) on the soil surface, thus affecting fungal populations due to higher temperatures, lower humidity, stronger UV radiation, and predator pressure. As a result, fungi are considered to be favored in no-tillage systems (Frey et al., 1999; Young and Ritz, 2000; Andrade et al., 2003).

Arbuscular mycorrhizal (AM) fungi are important for plant nutrient acquisition (particularly P), protection against root pathogens, drought resistance, and soil aggregate stability. Intensive tillage causes fragmentation of AM fungal hyphae, resulting in loss of cellular contents and thus loss in vitality as propagules. Consequently, the potential infectivity for subsequent crops is reduced. Fungal spores are often formed when growth becomes limiting (e.g., nutrient deprivation and environmental changes). Greater mycorrhizal sporulation has been found in some soils under conventional tillage compared to no-tillage, probably due to stress conditions induced by plowing (Andrade et al., 2003). On the other hand, other researchers (Galvez et al., 2001; Kabir, 2005) found that both spore populations and colonization of corn roots by AM fungi were lower in intensively tilled soil than in no-tillage soil. The magnitude of differences between tillage treatments fluctuated during the growing season and varied by depth (Fig. 11.3). The AM fungal spore abundance and species diversity under conventional tillage were also reduced in the surface soil (0–10cm) (Oehl et al., 2003). In a German study, the greatest number of AM fungal species and highest diversity were found at a soil depth of 20–35cm, just below the plow depth. However, in general, AM fungal spore abundance and species richness decreased with increasing soil depth (Oehl et al., 2005). Intensification of tillage practices,

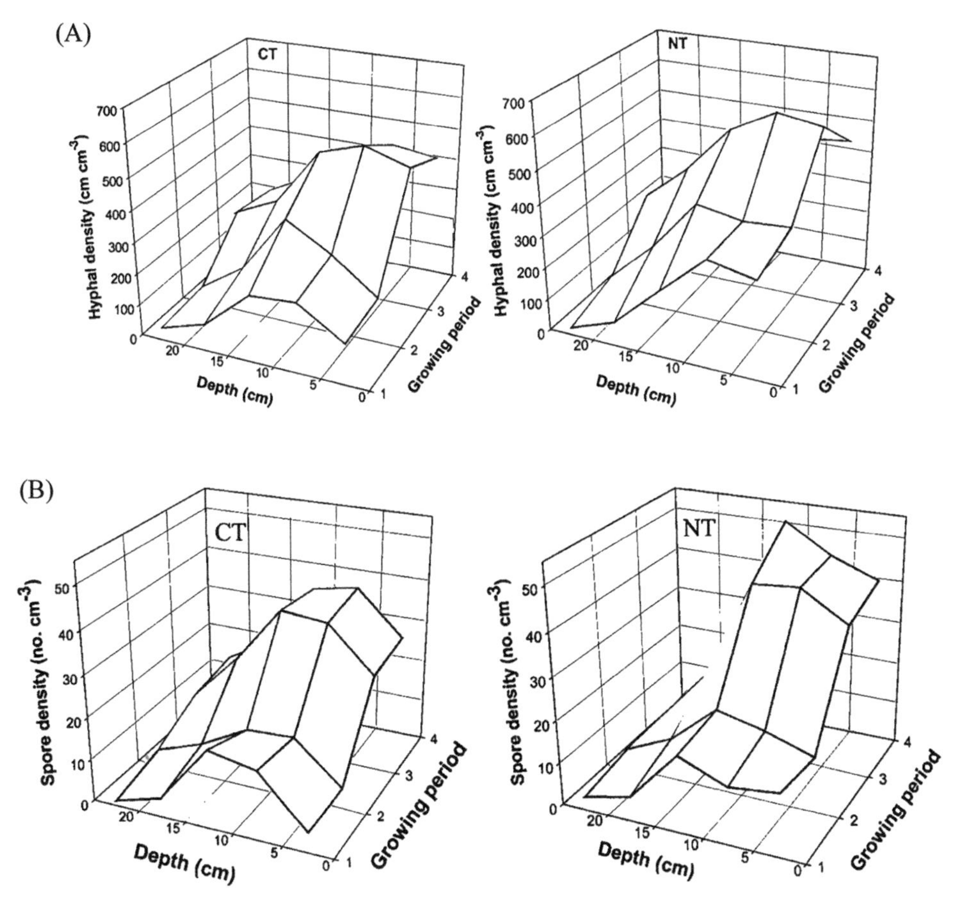

Figure 11.3

Seasonal and vertical distribution of mycorrhizal hyphae (A) and spores (B) in four corn-growing periods: five- to six-leaf stage (1), 10- to 12-leaf stage (2), silking stage (3), and mature stage (4) in conventional (CT) and no-tillage (NT) systems. *Reproduced with permission from Kabir, Z., 2005. Tillage or no-tillage: impact on mycorrhizae. Can. J. Plant Sci. 85, 23–29. © Canadian Science Publishing or its licensors.*

compared with low-intensity systems, appears to negatively affect AM fungi abundance and diversity, especially for species not belonging to the genus *Glomus* (Jansa et al., 2002; Oehl et al., 2005). Conventional tillage appears to favor proliferation of certain "generalist" species that colonize roots slowly but form spores rapidly, while no-tillage conditions favor the proliferation of other species (Oehl et al., 2003, 2005; Roger-Estrade et al., 2010). It is likely that tillage contributes to selection of AM fungal species that are better adapted to stress conditions; sporulation allows fungi to survive in disturbed soil, but not necessarily to maintain symbiotic efficiency (Andrade et al., 2003).

11.5.4 Microbial Communities

There are many studies reporting long-term and short-term effects of intensive tillage on gross microbial response (e.g., microbial biomass, respiration, and enzyme activity) and numbers of total or culturable soil microorganisms; however, fewer researchers have examined relationships between soil microbial community structure and tillage until recently. Characterization of soil microorganisms using culture-independent and community-level approaches, such as nucleic acid-based methods and PLFA analysis, are advantageous for "fingerprinting" the soil microbial community at a finer resolution to monitor changes in community structure resulting from different management practices. Recent advances in sequencing technology have made it possible to characterize the soil microbial community in greater depth, but reports on tillage effects are still limited (Sengupta and Dick, 2015). Table 11.2 summarizes soil microbial community studies discussed in this chapter.

Soil microbial communities under intensive tillage are significantly different from those under no-tillage or reduced tillage (Drijber et al., 2000; Feng et al., 2003; Peixoto et al., 2006; Mathew et al., 2012; Mbuthia et al., 2015). Since many environmental factors and management practices affect soil microorganisms, it is not always easy to discern the impact of tillage alone. Drijber et al. (2000) compared plots in mixed prairie sod and those cropped to wheat or left fallow under no-tillage, subtillage, or plow management in Nebraska, United States. Although PLFA profiles differentiated wheat and fallow systems by tillage, discrimination among tillage treatments was expressed most strongly during fallow. In Alabama, United States, Feng et al. (2003) found that the impact of tillage practices (conventional vs no-tillage for 12 years) on microbial communities as indicated by PLFAs was significant in February and May, but not in October in a continuous cotton (*Gossypium hirsutum* L.) production system. Results from both studies suggest that changes in microbial communities may be primarily determined by soil conditions responding to crop growth and environmental variables, such as moisture and temperature; during fallow or prior to crop establishment, community changes associated with tillage practices become more pronounced. This finding was verified in subsequent studies where soil samples from different tillage treatments were collected early in the growing season. Consequently, effects of tillage treatments on soil microbial communities were easily discerned

Table 11.2: Summary of studies on soil microbial community analysis

Site Location	Years of Field Experiment	Soil Texture	Cropping System	Sampling Time	Sampling Depth (cm)	Method of Analysis[a]	Target Organism	Reference
Brazil, central	4	Clay	Rice, soybean, sunhemp, millet	Jan (wet), Jun (dry)	0–5, 5–10	DGGE	Bacteria	Peixoto et al. (2006)
Canada	>24	Loam & clay loam	Wheat, lentil, pea, flax, triticale, canola	Two springs	0–15	PLFA	Bacteria, fungi, microfauna	Helgason et al. (2009)
Italy, central	13	Silty clay	Durum wheat, sunflower or corn	Jun, Nov	0–20	RT-PCR-DGGE	Bacteria	Pastorelli et al. (2013)
Mexico, central	>10	Clay loam	Corn, wheat	End of fallow period	0–15	Clone library	Bacteria	Ceja-Navarro et al. (2010)
Slovenia	12	Loam	Wheat, barley, oilseed rape, maize	Nov	0–10, 10–20	tRFLP	Archaea, bacteria, fungi	Kaurin et al. (2015)
USA, Midwest	>30	Silt loam	Wheat, soybean	May over 3 years	0–15	Pyrosequencing	Bacteria	Yin et al. (2010)
USA, Midwest	25	Loam	Wheat, fallow	Aug	0–15	PLFA	Bacteria, fungi, microfauna	Drijber et al. (2000)
USA, southeast	12	Silt loam	Cotton	Feb, May, Oct	0–3, 3–6, 6–12, 12–24	PLFA	Bacteria, fungi, microfauna	Feng et al. (2003)
USA, southeast	14	Silt loam	Corn	Apr	0–5, 5–15	PLFA & ARISA	Bacteria, fungi, microfauna	Mathew et al. (2012)
USA, southeast	31	Silt loam	Cotton	Jun	0–7.5	PLFA	Bacteria, fungi, microfauna	Mbuthia et al. (2015)

[a]*DGGE*, denaturing gradient gel electrophoresis; *PLFA*, phospholipid ester-linked fatty acids; *RT-PCR-DGGE*, reverse transcription-polymerase chain reaction-denaturing gradient gel electrophoresis; *tRFLP*, terminal restriction fragment polymorphism; *ARISA*, automated ribosomal intergenic spacer analysis.

(Mathew et al., 2012; Fig. 11.4). Fatty acids that are found primarily in particular microbial groups, the so-called PLFA markers, are often used in conjunction with multivariate analysis of PLFA profiles. The relative abundance of PLFA markers indicative of actinobacteria (10Me 16:0) and fungi (18:1ω9) appear to be important contributors to differences observed under conventional and no-tillage (Feng et al., 2003; Mathew et al., 2012). Soil organic C content was positively correlated with PLFA markers, except for bacterial PLFA (Mathew et al., 2012). Total PLFA in the surface layer (0–5 cm) was consistently greater in the no-tillage soil than in the conventionally tilled soil (Feng et al., 2003; Helgason et al., 2009; Mathew et al., 2012); bacterial and fungal PLFA followed the same trend (Drijber et al., 2000; Helgason et al., 2009; Mathew et al., 2012).

Mbuthia et al. (2015) investigated the impact of long-term (31 years) tillage (disk harrow vs no-tillage), cover crops, and N rates on soil microbial community structure and activity under continuous cotton production in West Tennessee, United States. Microbial communities, as indicated by PLFA, under conventional tillage were different from those under no-tillage and resulted in reduced microbial activities and C, N, and P levels. Fewer Gram-positive bacteria, mycorrhizae, and actinobacteria were found under conventional tillage compared with no-tillage (Mbuthia et al., 2015). Pastorelli et al. (2013) examined tillage effects and

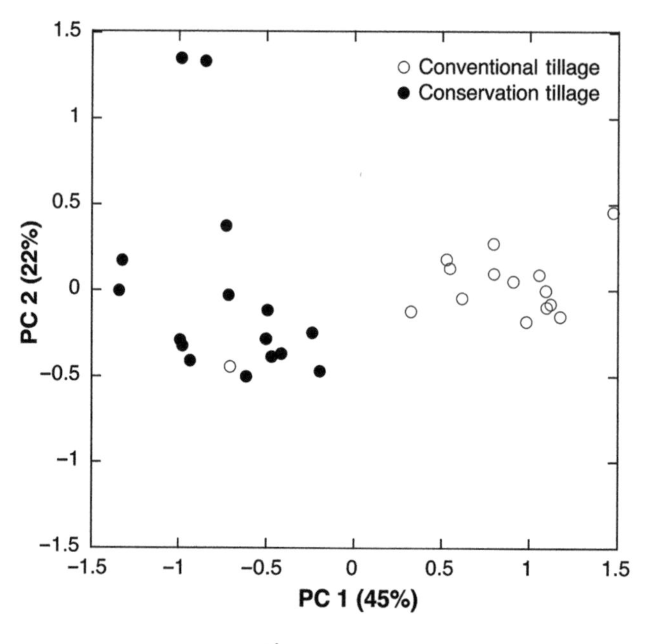

Figure 11.4

Plot of the first two principal component axes from the principal component analysis of PLFA profiles for conventional tillage and no-tillage treatments at 0–5 cm depth in Compass sandy loam located in central Alabama, United States. Soil samples were collected one month after cotton planting. *Unpublished data from Y. Feng.*

N fertilization on the active bacterial community in a clayey soil in central Italy using reverse transcription-polymerase chain reaction-denaturing gradient gel electrophoresis (RT-PCR-DGGE). Cluster analysis revealed differences between no-tillage and conventional tillage treatments; tillage appeared to be the main driver of differences in bacterial genetic diversity.

In a rice-soybean (*Oryza sativa* L./*Glycine max* L.) rotation study conducted in central Brazil, the impact of conventional and no-tillage with and without cover crops on soil bacterial community structure was determined using PCR-DGGE targeting 16S rRNA and *rpo*B (the gene encoding the RNA polymerase β subunit) genes (Peixoto et al., 2006). The field experiment had been in place for 4 years; responses of bacterial communities to cultivation, tillage, and soil depth, but not to cover cropping, were detected. Cluster analysis of both 16S rDNA and *rpo*B profiles showed that soil bacteria under conventional tillage formed a separate cluster from those under no-tillage and a forest control site. Bacterial community structure under no-tillage was more similar to that at the forest site. Ceja-Navarro et al. (2010) performed phylogenetic and multivariate analyses to determine the effects of tillage (conventional vs no-tillage) and residue management (removal vs retention) on soil bacterial communities in a long-term (>10 years) corn–wheat rotation experiment located in central Mexico. Based on analyses of 172 clones for each treatment, they found that bacterial communities under no-tillage with crop residue retention had the greatest diversity and richness. No-tillage had a positive effect on members of *Rhizobiales* and crop residue retention increased fluorescent *Pseudomonas* spp. and *Burkholderiales* groups. *Acidobacteriales* were the most abundant group found in all treatments with the highest proportion under conventional tillage with residue retention. Proportions of the *Sphingomonadales* group under conventional tillage were greater than those under no-tillage. Unfortunately, little information is available regarding functions of the members from these two bacterial orders for soil processes.

In Slovenia, Kaurin et al. (2015) examined soil bacterial, fungal, and archaeal communities using terminal restriction fragment polymorphism (tRFLP). They showed that microbial community composition changed 12 years after conversion from conventional tillage to minimum tillage in a crop rotation system consisting of winter wheat, winter barley (*Hordeum vulgare* L.), oilseed rape (*Brassica napus* L.), and corn. However, changes in microbial diversity as indicated by the number of terminal restriction fragments were not detected for any domain. The most pronounced shifts in composition were found for bacterial communities in the 10–20 cm layer, while composition of fungal communities changed slightly in the upper 0–10 cm of minimally tilled soil. Composition of archaeal communities was not affected by tillage or soil depth.

Analyzing soil microbial communities using high-throughput sequencing techniques provides taxonomic information at a higher resolution than PLFA, tRFLP, and DGGE. Yin et al. (2010) used pyrosequencing to study the effect of tillage (conventional vs no-tillage) and crop rotation (wheat-soybean vs continuous wheat) on total bacterial communities. Among the 299 phyla identified, Proteobacteria were the most abundant followed by Acidobacteria. Some members of Acidobacteria Group 2 were most frequently detected in the no-tillage treatment.

In spite of tremendous progress made in recent years, our understanding of soil microbial communities is still limited and the identities of most soil microorganisms remain largely unknown. Linking ecosystem functions to microbial communities remains the "Holy Grail" in microbial ecology. Our understanding of tillage impacts on soil microbial communities is based on what we can measure. Since each method has limitations, polyphasic approaches should be used when possible to better understand the influence of tillage-induced changes on soil microbial communities. The increasing interest in management of the biological component of soil to improve soil health requires better understanding on how management practices (e.g., tillage) and environmental conditions influence microbial communities. Intensive tillage more often than not simplifies the soil food web and leads to reduced soil biodiversity with potential consequences for loss of soil functions.

11.6 Conclusions

Intensive tillage often results in reduced organic matter content in the surface soil. Together with lack of ground cover, this leads to reduced infiltration and moisture retention. Tillage also accelerates crop residue decomposition by stimulating microbial activities. Therefore, mineralization rates in intensive tillage systems may exceed crop nutrient uptake rates at times. Tillage-induced microbial habitat changes typically depend on soil type, cropping systems, and climatic conditions, as well as time since the last tillage operation. Thus, effects on soil organisms are not always consistent. Nevertheless, physical disturbance of soil by tillage can have profound effects on soil biota (Table 11.3), most of which have

Table 11.3: Comparison of soil biological properties between intensive tillage (IT) and no-tillage (NT)[a]

Soil Biological Properties	IT Compared to NT
Soil Organic Matter in Surface Soil	↓ or No difference
Particulate or labile Organic Matter Fractions	↓
Soil Microbial Biomass	↓ or No difference
Fungal Populations	↓
Arbuscular Mycorrhizal Fungi	↓
Enzymatic Activity	↓
Beneficial Microorganisms	↓
Free-living (beneficial) Nematodes	↓
Plant-parasitic Nematodes	↑
Earthworms	↓
Arthropod Diversity	↓

[a]Modified after Palm et al. (2014).

negative consequences leading to loss of biodiversity and ultimately, loss of ecosystem services. Unfortunately, intensive tillage has been the traditional land management practice for many parts of the world. Only 35% of the world's arable area is under no-tillage (FAO, 2008). In the United States, 37.9% of tillable cropland is under conventional tillage (USDA, 2012). To meet increasing demands on food and fiber supplies, we need less costly, more sustainable approaches to improve productivity of a wide range of crops while minimizing the adverse environmental impact. Conservation tillage systems are the key to sustainable land management and soil biological diversity. Conserving or increasing biodiversity in arable soils will ensure sustained ecosystem services (e.g., residue decomposition, nutrient cycling, and maintenance of soil health) provided by soil biota.

References

Andrade, D.S., Colozzi-Filho, A., Giller, K.E., 2003. The soil microbial community and soil tillage. In: Titi, I.A.E. (Ed.), Soil Tillage in Agroecosystems. CRC Press LLC, Boca Raton, FL, pp. 51–81.

Andren, O., Lagerlof, J., 1983. Soil fauna (microarthropods, enchytraeids, nematodes) in Swedish agricultural cropping systems. Acta Agric. Scand. 33, 33–52.

Aulakh, M.S., Doran, J.W., Walters, D.T., Mosier, A.R., Francis, D.D., 1991. Crop residue type and placement effects on denitrification and mineralization. Soil Sci. Soc. Am. J. 55, 1020–1025.

Baker, J.M., Ochsner, T.E., Venterea, R.T., Griffis, T.J., 2007. Tillage and soil carbon sequestration—What do we really know? Agric. Ecosyst. Environ. 118, 1–5.

Balkcom, K.S., Terra, J.A., Shaw, J.N., Reeves, D.W., Raper, R.L., 2005. Soil management system and landscape position interactions on nutrient distribution in a Coastal Plain field. J. Soil Water Conserv. 60, 431–437.

Balkcom, K.S., Arriaga, F.J., van Santen, E., 2013. Conservation systems to enhance soil carbon sequestration in the Southeast U.S. Coastal Plain. Soil Sci. Soc. Am. J. 77, 1774–1783.

Balkcom, K.S., Duzy, L.M., Kornecki, T.S., Price, A.J., 2015. Timing of cover crop termination: management considerations for the Southeast. Crop, Forage Turfgrass Manage. 1, 1–7.

Baumhardt, R.L., Keeling, J.W., Wendt, C.W., 1993. Tillage and residue effects on infiltration into soils cropped to cotton. Agron. J. 85, 379–383.

Baumhardt, R.L., Unger, P.W., Dao, T.H., 2004. Seedbed surface geometry effects on soil crusting and seedling emergence. Agron. J. 96, 1112–1117.

Beare, M.H., Parmelee, R.W., Hendrix, P.F., Cheng, W., Coleman, D.C., Crossley, D.A., 1992. Microbial and faunal interactions and effects on litter nitrogen and decomposition in agroecosystems. Ecol. Monogr. 62, 569–591.

Blanco-Canqui, H., Lal, R., 2009. Crop residue removal impacts on soil productivity and environmental quality. CRC. Crit. Rev. Plant. Sci. 28, 139–163.

Blevins, R.L., Thomas, G.W., Cornelius, P.L., 1977. Influence of no-tillage and nitrogen fertilization on certain soil properties after 5 years of continuous corn. Agron. J. 69, 383–386.

Blevins, R.L., Thomas, G.W., Smith, M.S., Frye, W.W., Cornelius, P.L., 1983. Changes in soil properties after 10 years continuous non-tilled and conventionally tilled corn. Soil Tillage Res. 3, 135–146.

Burke, I.C., Yonker, C.M., Parton, W.J., Cole, C.V., Schimel, D.S., Flach, K., 1989. Texture, climate, and cultivation effects on soil organic matter content in U.S. grassland soils. Soil Sci. Soc. Am. J. 53, 800–805.

Campbell, C.A., Biederbeck, V.O., Schnitzer, M., Selles, F., Zentner, R.P., 1989. Effect of 6 years of zero tillage and N fertilizer management on changes in soil quality of an orthic Brown Chernozem in Southwestern Saskatchewan. Soil Tillage Res. 14, 39–52.

Causarano, H.J., Franzluebbers, A.J., Reeves, D.W., Shaw, J.N., 2006. Soil organic carbon sequestration in cotton production systems of the southeastern United States: a review. J. Environ. Qual. 35, 1374–1383.

Ceja-Navarro, J.A., Rivera-Orduna, F.N., Patino-Zuniga, L., Vila-Sanjurjo, A., Crossa, J., Govaerts, B., et al., 2010. Phylogenetic and multivariate analyses to determine the effect of different tillage and residue management practices on soil bacterial communities. Appl. Environ. Microbiol. 76, 3685–3691.

Cherr, C.M., Scholberg, J.M.S., McSorley, R., 2006. Green manure approaches to crop production: a synthesis. Agron. J. 98, 302–319.

Clapp, C.E., Allmaras, R.R., Layese, M.F., Linden, D.R., Dowdy, R.H., 2000. Soil organic carbon and ^{13}C abundance as related to tillage, crop residue, and nitrogen fertilization under continuous corn management in Minnesota. Soil Tillage Res. 55, 127–142.

Copeland, P.J., Crookston, R.K., 1992. Crop sequence affects nutrient composition of corn and soybean grown under high fertility. Agron. J. 84, 503–509.

Crozier, C.R., Naderman, G.C., Tucker, M.R., Sugg, R.E., 1999. Nutrient and pH stratification with conventional and no-till management. Commun. Soil Sci. Plant Anal. 30, 65–74.

De Jong, E., Kachanoski, R.G., 1988. The importance of erosion in the carbon balance of prairie soils. Can. J. Soil Sci. 68, 111–119.

Dick, W.A., 1983. Organic carbon, nitrogen, and phosphorus concentrations, and pH in soil profiles as affected by tillage intensity. Soil Sci. Soc. Am. J. 47, 102–107.

Doran, J.W., 1980. Soil microbial and biochemical changes associated with reduced tillage. Soil Sci. Soc. Am. J. 44, 765–771.

Doran, J.W., 1987. Microbial biomass and mineralizable nitrogen distributions in no-tillage and plowed soils. Biol. Fertil. Soils 5, 68–75.

Doran, J.W., Werner, M.R., 1990. Management and soil biology. In: Francis, C.A., Flora, C.B., King, L.D. (Eds.), Sustainable Agriculture in Temperate Zones. John Wiley & Sons, New York.

Drees, L.R., Wilding, L.P., Karathanasis, A.D., Blevins, R.L., 1994. Micromorphological characteristics of long-term no-till and conventionally tilled soils. Soil Sci. Soc. Am. J. 58, 508–517.

Drijber, R.A., Doran, J.W., Parkhurst, A.M., Lyon, D.J., 2000. Changes in soil microbial community structure with tillage under long-term wheat-fallow management. Soil. Biol. Biochem. 32, 1419–1430.

Dubie, T.R., Greenwood, C.M., Godsey, C., Payton, M.E., 2011. Effects of tillage on soil microarthropods in winter wheat. Southwest. Entomol. 36, 11–20.

Ducarme, X., André, H.M., Wauthy, G., Lebrun, P., 2004. Are there real endogeic species in temperate forest mites? Pedobiologia 48, 139–147.

Eck, H.V., Jones, O.R., 1992. Soil nitrogen status as affected by tillage, crops, and crop sequences. Agron. J. 84, 660–668.

Eckert, D.J., 1985. Effects of reduced tillage on the distribution of soil pH and nutrients in soil profiles. J. Fertil. Issues 2, 86–90.

Ellert, B.H., Bettany, J.R., 1995. Calculation of organic matter and nutrients stored in soils under contrasting management regimes. Can. J. Soil Sci. 75, 529–538.

Ettema, C.H., Wardle, D.A., 2002. Spatial soil ecology. Trends Ecol. Evol. 17, 177–183.

Feng, Y., Motta, A.C., Reeves, D.W., Burmester, C.H., van Santen, E., Osborne, J.A., 2003. Soil microbial communities under conventional-till and no-till continuous cotton systems. Soil. Biol. Biochem. 35, 1693–1703.

Follett, R.F., 2001. Soil management concepts and carbon sequestration in cropland soils. Soil Tillage Res. 61, 77–92.

Food and Agriculture Organization (FAO), 2008. Investing in Sustainable Agricultural Intensification: The Role of Conservation Agriculture – A Framework for Action, Report of the International Technical Workshop. FAO, Rome, Italy.

Franzluebbers, A.J., 2002. Soil organic matter stratification ratio as an indicator of soil quality. Soil Tillage Res. 66, 95–106.

Franzluebbers, A.J., Hons, F.M., 1996. Soil-profile distribution of primary and secondary plant-available nutrients under conventional and no tillage. Soil Tillage Res. 39, 229–239.

Franzluebbers, A.J., Hons, F.M., Zuberer, D.A., 1995. Soil organic carbon, microbial biomass, and mineralizable carbon and nitrogen in sorghum. Soil Sci. Soc. Am. J. 59, 460–466.
Franzluebbers, A.J., Langdale, G.W., Schomberg, H.H., 1999. Soil carbon, nitrogen, and aggregation in response to type and frequency of tillage. Soil Sci. Soc. Am. J. 63, 349–355.
Frey, S.D., Elliott, E.T., Paustian, K., 1999. Bacterial and fungal abundance and biomass in conventional and no-tillage agroecosystems along two climatic gradients. Soil. Biol. Biochem. 31, 573–585.
Fu, S., Cabrera, M.L., Coleman, D.C., Kisselle, K.W., Garrett, C.J., Hendrix, P.F., et al., 2000. Soil carbon dynamics of conventional tillage and no-till agroecosystems at Georgia Piedmont — HSB-C models. Ecol. Modell. 131, 229–248.
Galvez, L., Douds, D., Janke, R.R., Wagoner, P., 2001. Effect of tillage and farming system upon VAM fungus populations and mycorrhizas and nutrient uptake of maize. Plant. Soil. 228, 299–308.
Golebiowska, J., Ryszkowski, L., 1977. Energy and carbon fluxes in soil compartments of agroecosystems. Ecol. Bull. 25, 274–283.
Grandy, A.S., Robertson, G.P., 2007. Land-use intensity effects on soil organic carbon accumulation rates and mechanisms. Ecosystems 10, 59–73.
Hassink, J., Bouwman, L.A., Zwart, K.B., Brussaard, L., 1993. Relationships between habitable pore space, soil biota and mineralization rates in grassland soils. Soil. Biol. Biochem. 25, 47–55.
Helgason, B.L.W., Germida, F.L., James, J., 2009. Fungal and bacterial abundance in long-term no-till and intensive-till soils of the Northern Great Plains. Soil Sci. Soc. Am. J. 73, 120–127.
Hill, R.L., 1990. Long-term conventional and no-tillage effects on selected soil physical properties. Soil Sci. Soc. Am. J. 54, 161–166.
House, G.J., Parmelee, R.W., 1985. Comparison of soil arthropods and earthworms from conventional and no-tillage agroecosystems. Soil Tillage Res. 5, 351–360.
Howard, D.D., Essington, M.E., Tyler, D.D., 1999. Vertical phosphorus and potassium stratification in no-till cotton soils. Agron. J. 91, 266–269.
Izaurralde, R.C., Williams, J.R., McGill, W.B., Rosenberg, N.J., Jakas, M.C.Q., 2006. Simulating soil C dynamics with EPIC: model description and testing against long-term data. Ecol. Modell. 192, 362–384.
James, D.W., Wells, K.L., 1990. Soil sample collection and handling: technique based on source and degree of field variability. In: Westerman, R.L. (Ed.), Soil Testing and Plant Analysis. Soil Science Society of America, Madison, WI.
Jansa, J., Mozafar, A., Anken, T., Ruh, R., Sanders, I., Frossard, E., 2002. Diversity and structure of AMF communities as affected by tillage in a temperate soil. Mycorrhiza. 12, 225–234.
Jenkinson, D.S., 1981. The fate of plant and animal residues in soil. In: Greenland, D.J., Hayes, M.H.B. (Eds.), The Chemistry of Soil Processes. John Wiley & Sons, New York, pp. 505–561.
Jenny, H., 1980. The Soil Resource: Origin and Behaviour. Springer Verlag, New York.
Kabir, Z., 2005. Tillage or no-tillage: impact on mycorrhizae. Can. J. Plant Sci. 85, 23–29.
Karlen, D.L., Mausbach, M.J., Doran, J.W., Cline, R.G., Harris, R.F., Schuman, G.E., 1997. Soil quality: a concept, definition, and framework for evaluation (A Guest Editorial). Soil Sci. Soc. Am. J. 61, 4–10.
Kaurin, A., Mihelič, R., Kastelec, D., Schloter, M., Suhadolc, M., Grčman, H., 2015. Consequences of minimum soil tillage on abiotic soil properties and composition of microbial communities in a shallow Cambisol originated from fluvioglacial deposits. Biol. Fertil. Soils 51, 923–933.
Kay, B.D., VandenBygaart, A.J., 2002. Conservation tillage and depth stratification of porosity and soil organic matter. Soil Tillage Res. 66, 107–118.
Koller, K., 2003. Techniques of soil tillage. In: El Titi, A. (Ed.), Soil Tillage in Agroecosystems. CRC Press, Boca Raton, FL, pp. 1–25.
Kravchenko, A.N., Wang, A.N.W., Smucker, A.J.M., Rivers, M.L., 2011. Long-term differences in tillage and land use affect intra-aggregate pore heterogeneity. Soil Sci. Soc. Am. J. 75, 1658–1666.
Lachnicht, S.L., Hendrix, P.F., Potter, R.L., Coleman, D.C., Crossley Jr, D.A., 2004. Winter decomposition of transgenic cotton residue in conventional-till and no-till systems. Appl. Soil Ecol. 27, 135–142.

Lal, R., 2004. Soil carbon sequestration impacts on global climate change and food security. Science 304, 1623–1627.
Lal, R., Follett, R.F., Kimble, J., Cole, C.V., 1999. Managing U.S. cropland to sequester carbon in soil. J. Soil Water Conserv. 54, 374–381.
Langdale, G.W., Wilson, R.L., Bruce, R.R., 1990. Cropping frequencies to sustain long-term conservation tillage systems. Soil Sci. Soc. Am. J. 54, 193–198.
Mathew, R.P., Feng, Y., Githinji, L., Ankumah, R., Balkcom, K.S., 2012. Impact of no-tillage and conventional tillage systems on soil microbial communities. Appl. Environ. Soil Sci. http://dx.doi.org/10.1155/2012/548620.
Mbuthia, L.W., Acosta-Martínez, V., DeBruyn, J., Schaeffer, S., Tyler, D., Odoi, E., et al., 2015. Long term tillage, cover crop, and fertilization effects on microbial community structure, activity: Implications for soil quality. Soil Biol. Biochem. 89, 24–34.
Mulvaney, M.J., Wood, C.W., Balkcom, K.S., Shannon, D.A., Kemble, J.M., 2010. Carbon and nitrogen mineralization and persistence of organic residues under conservation and conventional tillage. Agron. J. 102, 1425–1433.
Neher, D.A., 1999. Soil community composition and ecosystem processes: comparing agricultural ecosystems with natural ecosystems. Agrofor. Syst. 45, 159–185.
Nelson, D.W., Sommers, L.E., 1982. Total carbon, organic carbon and organic matter. In: Page, A.L., Miller, R.H., Kenney, D.R. (Eds.), Methods of Soil Analysis Part 2: Chemical and Microbiological Properties. ASA and SSSA, Madison, WI.
Nelson, J.B., King, L.D., 1996. Green manure as a nitrogen source for wheat in the Southeastern United States. Am. J. Altern. Agric. 11, 182–189.
Nichols, J.D., 1984. Relation of organic carbon to soil properties and climate in the Southern Great Plains. Soil Sci. Soc. Am. J. 48, 1382–1384.
Nielsen, U.N., Osler, G.H.R., van der Wal, R., Campbell, C.A., Burslem, D.F.R.P., 2008. Soil pore volume and the abundance of soil mites in two contrasting habitats. Soil. Biol. Biochem. 40, 1538–1541.
Novak, J.M., Frederick, J.R., Bauer, P.J., Watts, D.W., 2009. Rebuilding organic carbon contents in Coastal Plain soils using conservation tillage systems. Soil Sci. Soc. Am. J. 73, 622–629.
Oehl, F., Sieverding, E., Ineichen, K., Mader, P., Dubois, D., BoUer, T., et al., 2003. Impact of long-term conventional and organic farming on the diversity of arbuscular mycorrhizal fungi. Oecologia. 138, 574–583.
Oehl, F., Sieverding, E., Ineichen, K., Ris, E., Boller, T., Wiemken, A., 2005. Community structure of arbuscular mycorrhizal fungi at different soil depths in extensively and intensively managed agroecosystems. New Phytol. 165, 273–283.
Pagliai, M., De Nobili, M., 1993. International workshop on methods of research on soil structure/soil biota interrelationships: relationships between soil porosity, root development and soil enzyme activity in cultivated soils. Geoderma 56, 243–256.
Pagliai, M., Vignozzi, N., Pellegrini, S., 2004. Soil structure and the effect of management practices. Soil Tillage Res. 79, 131–143.
Palm, C.A., Sanchez, P.A., 1991. Nitrogen release from the leaves of some tropical legumes as affected by their lignin and polyphenolic contents. Soil. Biol. Biochem. 23, 83–88.
Palm, C., Blanco-Canqui, H., DeClerck, F., Gatere, L., Grace, P., 2014. Conservation agriculture and ecosystem services: an overview. Agr. Ecosyst. Environ. 187, 87–105.
Pastorelli, R., Vignozzi, N., Landi, S., Piccolo, R., Orsini, R., Seddaiu, G., et al., 2013. Consequences on macroporosity and bacterial diversity of adopting a no-tillage farming system in a clayish soil of Central Italy. Soil. Biol. Biochem. 66, 78–93.
Paul, E.A., Clark, F.E., 1996. Soil Microbiology and Biochemistry, 2nd ed. Academic Press, San Diego, CA.
Peixoto, R.S., Coutinho, H.L.C., Madari, B., Machado, P.L.O.A., Rumjanek, N.G., Van Elsas, J.D., et al., 2006. Soil aggregation and bacterial community structure as affected by tillage and cover cropping in the Brazilian Cerrados. Soil Tillage Res. 90, 16–28.
Petersen, H., 2002. Effects of non-inverting deep tillage vs. conventional ploughing on collembolan populations in an organic wheat field. Eur. J. Soil Biol. 38, 177–180.
Post, W.M., Kwon, K.C., 2000. Soil carbon sequestration and land-use change: processes and potential. Global Change Biol. 6, 317–327.

Powlson, D.S., Brookes, P.C., Christensen, B.T., 1987. Measurement of soil microbial biomass provides an early indication of changes in total soil organic matter due to straw incorporation. Soil. Biol. Biochem. 19, 159–164.

Quincke, J.A., Wortmann, C.S., Mamo, M., Franti, T., Drijber, R.A., 2007. Occasional tillage of no-till systems: carbon dioxide flux and changes in total and labile soil organic carbon. Agron. J. 99, 1158–1168.

Reicosky, D.C., Dugas, W.A., Torbert, H.A., 1997. Tillage-induced soil carbon dioxide loss from different cropping systems. Soil Tillage Res. 41, 105–118.

Rice, C.W., Moorman, T.B., Beare, M., 1996. Role of microbial biomass carbon and nitrogen in soil quality. In: Doran, J.W., Jones, A.J. (Eds.), Methods for Assessing Soil Quality. Soil Science Society of America, Inc., Madison, WI.

Roger-Estrade, J., Anger, C., Bertrand, M., Richard, G., 2010. Tillage and soil ecology: partners for sustainable agriculture. Soil Tillage Res. 111, 33–40.

Sabatini, M.A., Rebecchi, L., Cappi, C., Bertolani, R., Fratello, B., 1997. Long-term effects of three different continuous tillage practices on Collembola populations. Pedobiologia, 185–193.

Sainju, U.M., Singh, B.P., Whitehead, W.F., Wang, S., 2006. Carbon supply and storage in tilled and nontilled soils as influenced by cover crops and nitrogen fertilization. J. Environ. Qual. 35, 1507–1517.

Sainju, U.M., Senwo, Z.N., Nyakatawa, E.Z., Tazisong, I.A., Reddy, K.C., 2008. Tillage, cropping systems, and nitrogen fertilizer source effects on soil carbon sequestration and fractions. J. Environ. Qual. 37, 880–888.

Sengupta, A., Dick, W.A., 2015. Bacterial community diversity in soil under two tillage practices as determined by pyrosequencing. Microb. Ecol. 70, 853–859.

So, H.B., Grabski, A., Desborough, P., 2009. The impact of 14 years of conventional and no-till cultivation on the physical properties and crop yields of a loam soil at Grafton NSW, Australia. Soil Tillage Res. 104, 180–184.

Soil Science Society of America, 1997. Glossary of Soil Science Terms 1996. SSSA, Madison, WI.

Sparling, G.P., 1997. Soil microbial biomass, activity and nutrient cycling as indicators of soil health. In: Pankhurst, C., Doube, B.M., Gupta, V.V.S.R. (Eds.), Biological indicators of soil health. CAB International, New York, pp. 97–119.

Starovoytov, A., Gallagher, R.S., Jacobsen, K.L., Kaye, J.P., Bradley, B., 2010. Management of small grain residues to retain legume-derived nitrogen in corn cropping systems. Agron. J. 102, 895–903.

Steen, E., 1983. Soil animals in relation to agricultural practices and soil productivity. Swedish J. Agric. Res. 13, 157–165.

Taylor, H.M., Gardner, H.R., 1963. Penetration of cotton seedling taproots as influenced by bulk density, moisture content, and strength of soil. Soil. Sci. 96, 153–156.

Torsvik, V., Øvreås, L., 2002. Microbial diversity and function in soil: from genes to ecosystems. Curr. Opin. Microbiol. 5, 240–245.

Touchton, J.T., Sims, J.T., 1987. Tillage systems and nutrient management in the east and southeast. In: Boersma, L. (Ed.), Future Developments in Soil Science Research. Soil Science Society of America, Madison, WI.

Tsiafouli, M.A., Thebault, E., Sgardelis, S.P., de Ruiter, P.C., van der Putten, W.H., Birkhofer, K., et al., 2015. Intensive agriculture reduces soil biodiversity across Europe. Global Change Biol. 21, 973–985.

USDA, 2012. US Census of Agriculture. <http://www.agcensus.usda.gov/Publications/2012/>.

van Capelle, C., Schrader, S., Brunotte, J., 2012. Tillage-induced changes in the functional diversity of soil biota - A review with a focus on German data. Eur. J. Soil Biol. 50, 165–181.

Vigil, M.F., Kissel, D.E., 1991. Equations for estimating the amount of nitrogen mineralized from crop residues. Soil Sci. Soc. Am. J. 55, 757–761.

Wood, C.W., Westfall, D.G., Peterson, G.A., Burke, I.C., 1990. Impacts of cropping intensity on carbon and nitrogen mineralization under no-till dryland agroecosystems. Agron. J. 82, 1115–1120.

Wright, A.L., Hons, F.M., Lemon, R.G., McFarland, M.L., Nichols, R.L., 2007. Stratification of nutrients in soil for different tillage regimes and cotton rotations. Soil Tillage Res. 96, 19–27.

Wyland, L.J., Jackson, L.E., Chaney, W.E., Klonsky, K., Koike, S.T., Kimple, B., 1996. Winter cover crops in a vegetable cropping system: impacts on nitrate leaching, soil water, crop yield, pests and management costs. Agric. Ecosyst. Environ. 59, 1–17.

Yin, C., Jones, K.L., Peterson, D.E., Garrett, K.A., Hulbert, S.H., Paulitz, T.C., 2010. Members of soil bacterial communities sensitive to tillage and crop rotation. Soil. Biol. Biochem. 42, 2111–2118.

Young, I.M., Ritz, K., 2000. Tillage, habitat space and function of soil microbes. Soil Tillage Res. 53, 201–213.

CHAPTER 12

Row-Crop Production Practices Effects on Greenhouse Gas Emissions

Jane M.-F. Johnson[1], Virginia L. Jin[2], Caroline Colnenne-David[3], Catherine E. Stewart[4], Claudia Pozzi Jantalia[5] and Zhengqin Xiong[6]

[1]USDA-ARS, Morris, MN, United States [2]USDA-ARS, Lincoln, NE, United States [3]INRA Research Center at Versailles, Versailles, France [4]USDA-ARS, Fort Collins, CO, United States [5]Embrapa Agrobiology, Rio de Janeiro, Brazil [6]Nanjing Agricultural University, Nanjing, China

12.1 Introduction

With the global population expected to exceed 9.7 billion by 2050 and food demand projected to double by 2050 compared to 2005 (Alexandratos and Bruinsma, 2012; UNEP, 2015), food security is of tantamount importance worldwide. Indeed, one of the grand challenges facing humankind is global food security in a world undergoing global climate change (IPCC, 2014). Productive healthy soils are crucial to meeting global needs. Agricultural extensification (i.e., land use change) and intensification are expected to help meet food demands (Angelsen, 2010; Pfaff and Walker, 2010), which could have significant, negative impacts on soil health and environmental quality, including agricultural contributions to atmospheric greenhouse gas (GHG) levels (Garnett et al., 2013; Godfray and Garnett, 2014). The implementation of socioeconomically feasible agronomic management practices, improvements in high-yielding technologies, and the transfer of both practices and technologies to low-yielding developing nations are needed to meet demands and significantly reduce GHG emissions at the global scale. China, for example, is using an integrated soil-crop system management as a food security approach (Chen et al., 2011). This model-driven approach is used to identify the best combination of practices to optimize productivity and identify the most effective N management to support yield with minimal N lost to the environment (Chen et al., 2011). In France, a similar effort has been launched to compare mitigation or abatement potential among conservation practices (Pellerin et al., 2013). Realizing sustainable intensification includes reducing the demand for food with a high environmental footprint, reducing food waste, and other social-economic aspects, but this chapter is focused on agronomic

Soil Health and Intensification of Agroecosystems.
DOI: http://dx.doi.org/10.1016/B978-0-12-805317-1.00012-9

management to reduce GHG emissions, increase soil organic carbon (SOC) sequestration, and the subsequent improvement of soil health.

Globally, GHGs emitted into the atmosphere in 2010 were ~20% more than in 2000, and 14% greater than the predicted median of 2020 emissions level necessary to have a likely chance of meeting the 2°C target (UNEP, 2015). Agricultural CO_2 emissions are small compared to the energy, industry, and transportation sectors (IPCC, 2013; UNEP, 2015). The agricultural sector, however, is considered a global source of methane (CH_4) and nitrous oxide (N_2O) (IPCC, 2013; Tubiello et al., 2015). The IPCC – Fifth Assessment Report (IPCC, 2014) based on a 100-year time horizon sets the global warming potential (GWP) of CH_4 at 28 and N_2O at 265 relative to CO_2. The relative contribution of agriculture to global emissions decreased from 12.3± 0.6% in the 1990s to 11.2± 0.4% in 2010, but actual emission rates increased annually by ~1% due to greater land application of synthetic fertilizer and animal manures (IPCC, 2013; FAO, 2014; Tubiello et al., 2015). Several extensive reviews have addressed cropland GHG emissions for multiple countries and systems. For example, Eve et al. (2014), Johnson et al. (2005), and Liebig et al. (2012) summarized emissions in the United States, Zhang et al. (2014b) in China, Pellerin et al. (2013) in France, and Aguilera et al. (2013) in Mediterranean climates, and Linquist et al. (2012a) provided a global meta-analysis. Agriculture has potential to offset GHG emissions by the sequestration of CO_2 via photosynthetic C fixation in crop biomass and subsequent transfer of fixed C into soil organic matter (SOM). In general, mitigation or exacerbation of atmospheric GHG levels in row-crop production systems depends upon the balance among SOC storage, CH_4 (emission or consumption), and soil N_2O emissions. In upland crops, CH_4 is negligible, but it is a major GHG source from soil under paddy rice production (Nayak et al., 2015). This chapter focuses on agricultural row-crop production of maize (*Zea mays* L.), wheat (*Triticum aestivum* L.), and paddy rice (*Oriza sativa* L.) production systems within the context of sustaining yield and avoiding or mitigating GHG emissions. Section 12.2 addresses direct management of N-inputs (rate, timing, placement, and enhanced efficiency fertilizers). Section 12.3 focuses on conservation practices (e.g., tillage and residue management) and related soil health implications. Section 12.4 discusses GWP and yield-scaled emissions related to agricultural intensification.

12.2 Nitrogen Management

Historically, agricultural intensification was achieved by increasing fertilizer and other chemical inputs, enhancing genetics, and implementing mechanical advances to increase crop yield (Ruttan, 1982). Nitrogenous inputs are necessary for crop production, but can have direct impacts on soil emissions of N_2O. The relationship between crop yield and fertilizer application is nonlinear, exhibiting sharp declines in crop fertilizer (i.e., N) use efficiency at higher application rates (Cassman et al., 2003). The complexity of N_2O formation as part of the N cycle and the multiple environmental and management interactions make it challenging to find generalizable mitigation/abatement options (Venterea et al., 2012). The application of N fertilizer is assumed to be directly proportional to N_2O production and is estimated to be 1% of the N fertilizer applied under IPCC Tier 1 guidelines (IPCC, 2013). Measured and modeled estimates of N_2O as a function of N applied tend to converge on this 1% prediction when values are aggregated over large land areas and/or over time (Del Grosso et al., 2008; Jin et al., 2014). However, the IPCC Tier 1 guidelines approach does not address nonlinear relationships between N application rate and N_2O emissions, which require more complex IPCC Tier 2 or Tier 3 modeling approaches (Ogle et al., 2014). Emissions have been measured and modeled to be relatively stable when N fertilizer rates approximate crop N needs, followed by significant increases in N_2O emissions once fertilizer rates exceed crop N demand (e.g., McSwiney and Robertson, 2005; van Groenigen et al., 2010). Thus, excess fertilizer regardless of the source increases the risk of negative environmental impacts (e.g., eutrophication, reduced water quality, direct and indirect N_2O emissions). Crop selection in a crop rotation can affect N_2O emissions (Cavigelli and Parkin, 2012) insofar as crops vary in their N use efficiencies and N fertilizer needs. Thus, strategies for mitigating N_2O strive to minimize the availability of N-substrate for conversion to N_2O by manipulating N-rate, timing, placement, formulation, and through various N-cycle inhibitors, while optimizing crop N use efficiency.

12.2.1 Mitigation Potential for N Conservation Practices

Reducing N-fertilizer rate to mitigate or avoid N_2O emissions must avoid yield loss, which has been demonstrated in multiple countries (Linquist et al., 2012a; Pellerin et al., 2013; Nayak et al., 2015). In the United States, many Midwestern states have reduced recommended fertilizer rates for maize (Ogle et al., 2014), and in the Western states, reduced rates did not affect net profitability (Archer and Halvorson, 2010). Determining optimum N-fertilizer rates can be complicated because management practices that produce high N use efficiency can also result in high N_2O emissions (Fujinuma et al., 2011; Gagnon et al., 2011). Further, "over-application" of fertilizer may be perceived as necessary to account for N-losses in a favorable weather year (Venterea et al., 2012). Nonetheless, reducing N-rates without sacrificing yield has been clearly demonstrated. Nayak et al. (2015) reported that, in China, reducing very high N-fertilizer rates by 10–60% avoided overapplication without yield penalties and could reduce N_2O emissions by 8–49% in rice, 11–38% in wheat, and 17–49% in maize compared to current national application rates. Linquist et al. (2012b) suggested that optimal N for rice yield without increasing N_2O emission would reduce yield-scaled CH_4, but these authors did not offer a specific rate of reduction. In simulating GHG emissions and SOC sequestration across China, Cheng et al. (2014) found that combining reduced fertilizer application and intermittent flooding had the highest potential to reduce N_2O and CH_4 emissions in paddy rice systems, estimating mitigation potential ranging from 0.08 to 0.83 Mg CO_2 equivalent Mg^{-1}

rice grain. In France, adjusting mineral fertilizer rates to realistic yield targets was identified as a cost-effective GHG abatement method (Pellerin et al., 2013).

Matching crop N demand with the timing of fertilizer application (i.e., fall to a spring application, or providing multiple) can decrease N loss potential (Pellerin et al., 2013; Decock, 2014). Many farmers may apply N fertilizer before planting because of convenience, cost-effectiveness, or necessity. For example, in the Midwest region of the United States, even though it is not considered a best management practice for environmental reasons, N-fertilizer is frequently applied after fall harvest as anhydrous ammonia when the soil temperatures are below 10°C. As a consequence, there are several months between N-application and planting of the next crop, during which N may be lost as gaseous N forms (e.g., N_2O) or by leaching. Switching from fall to spring application (Hao et al., 2001; Burton et al., 2008), or from pre- to post-plant N applications (Matson et al., 1998; Phillips et al., 2009) has shown inconsistent results, but more consistent reductions in N_2O were reported through multiple applications (Decock, 2014). Johnson et al. (2010a) found that only 6% of annual N_2O emission was associated with a split-application of anhydrous ammonia compared to 65% related to spring-thaw events, which they attributed to utilizing best management practices for N-fertilizer timing.

Fertilizer placement affects its susceptibility to leaching and gaseous losses, depending on climatic and soil variables. Important factors for N fertilizer placement include: (1) broadcasting versus banding; (2) application depth; and (3) uniform versus a spatially variable rate (precision application). Banding N application can place fertilizer closer to plant roots and facilitate plant N uptake, thereby reducing N_2O emissions compared to broadcasting (Malhi and Nyborg, 1985). Banding, however, can also increase N_2O emissions compared to broadcasting due to the highly concentrated zones of N within the band (Engel et al., 2010). A meta-analysis of maize-based systems found no difference between broadcast and banded N fertilizer placement (Decock, 2014). Another meta-analysis found N_2O emission response to N application depth was inconsistent among the reported results (Ogle et al., 2014). Precision N application is a technique that can spatially adjusts N rate based upon soil properties, topography, and expected yield potential across a field. Although this technique has been shown to increase N use efficiency (Mamo et al., 2003; Scharf et al., 2005), few scientists have assessed this approach for N_2O mitigation potential (Sehy et al., 2003). Therefore, additional research on the efficacy of precision N-application to reduce N_2O emissions is needed.

The chemical form of N fertilizer (e.g., anhydrous ammonia, urea-ammonium nitrate, ammonium nitrate, and urea) may impact N_2O emissions. For example there have been reports of injected anhydrous ammonia having 200–400% greater N_2O emissions compared to broadcast urea (e.g., Fujinuma et al., 2011; Venterea et al., 2011). In rice systems, using ammonium sulfate instead of urea reduced CH_4 emission by >30%, but N_2O reduction was equivocal (Linquist et al., 2012b). Decock (2014) found numerical but nonsignificant increases in N_2O emissions for anhydrous ammonia (12 observations) and nitrate-based fertilizers (66 observations) compared to urea (79 observations), suggesting that the high variability in N_2O emissions could be attributed to other management or climatic events and interactions with soil properties. Soil properties (e.g., bulk density, texture, SOM) alter aeration, porosity, and the likelihood of denitrifying conditions, which can increase N_2O formation. In a recent global review of long-term synthetic fertilizer effects on soil microorganisms, Geisseler and Scow (2014) found that soil microbial biomass was 15% greater in fertilized compared to unfertilized soils, especially when soil pH>5. The authors attributed this increase in soil microbial biomass to an increase in SOC in fertilized soils; however, they also noted that specific changes in soil microbial community compositions and how that could affect soil functioning were highly variable and dependent on environmental and crop management-related factors (Geisseler and Scow, 2014).

12.2.2 Mitigation Potential from Fertilizer Formulated for Enhanced Efficiency

Agricultural N_2O emissions can also be reduced by using fertilizers formulated to enhance nitrogen (N) use efficiency, including those with nitrification inhibitors (NIs), urease inhibitors, or controlled-release fertilizers (polymer-coated urea or PCU). NIs can be synthetic (e.g., dicyandiamide, nitrapyrin) or naturally occurring [e.g., neem (Majumdar et al., 2000) or brachialactone (Subbarao et al., 2007, 2009)]. In general, NIs work by blocking the action of the enzyme ammonia monooxygenase, disrupting the enzymatic pathway to NO_3 and thereby limiting the substrate pools available for N_2O production (McCarty, 1999; Di et al., 2009). In addition, by blocking ammonia monooxygenase, NI also can reduce CH_4 oxidation by methanotrophic microbes (Bronson and Mosier, 1994; McCarty, 1999), which can be important in paddy rice systems. Unlike NIs, urease inhibitors (e.g., N-(*n*-butyl) thiophosphoric triamide) block the hydrolysis of urea. Lastly, polymer-coating in PCU can be used to physically or chemically modify fertilizer products to reduce the rate of urea dissolution.

Multiple studies have been conducted producing results to show that N_2O can be reduced by NIs, with variable efficacy between rainfed and irrigated systems, and among tillage management practices (Table 12.1). Consistent reductions (20–47% for maize, rice, and wheat) were observed in irrigated systems (37 studies, Table 12.1), similar to meta-analyses that report a 35–44% reduction in N_2O emissions using various NIs and fertilizers (Akiyama et al., 2010; Decock, 2014; Nayak et al., 2015; Ruser and Schulz, 2015; Gilsanz et al., 2016). Meta-analysis of paddy rice systems by Nayak et al. (2015) and Linquist et al. (2012b) found that NIs decreased N_2O emissions >20%, and reduced CH_4 emissions by 11–28%. However, the effect of NIs on N_2O in rainfed maize systems appears inconsistent (Table 12.1). This could be due in part to the short-term study durations, as N_2O emissions associated with an NI can simply shift emission timing rather than result in actual emissions reduction as

Table 12.1: Average effects of nitrification inhibitor (NI) or polymer-coated urea (PCU) on N_2O emission expressed as % reduction compared to control mean (standard deviation) calculated among cited studies

Crop[a]	System[b]	Tillage[c]	N_2O Reduced by NI		% N_2O Reduced by PCU		Country (State, Region, or Province)	References[d]
			#[e]	%	#	%		
M	Rf	CT	8	27.9 (35.3)	7	16.2 (41.0)	Germany; Indonesia; USA (Iowa, Indiana, Minnesota)	1
		ST	3	7.5 (9.8)	1	0 (–)	USA (Iowa)	2
		NT	9	0.2 (39.4)	8	−37.1 (39.0)	USA (Indiana, Kentucky, Minnesota)	3
		N/A	1	43.3 (–)	1	9.1 (–)	Indonesia	4
	I	CT	8	49.9 (10)	2	−21.7 (30.2)	USA (Colorado); China (Hebei, Shanxi, North China Plain)	5
		ST	4	48.2 (22.8)	1	45.9 (–)	USA (Colorado)	6
		NT	5	44.1 (12.4)	2	44.7 (5.6)	USA (Colorado); China (Henan)	7
		N/A	3	44.7 (21.1)	-	-	Australia; Spain	8
W	Rf	RT	–	-	1	22.1 (–)	Canada (Manitoba)	9
	I	CT	7	24.1 (14.5)	1	33.0 (–)	China (Hebei, Shanxi, North China Plain); India	10
		NT	3	17.0 (2.0)	-	-	India	11
		N/A	13	20.4 (9.5)	-	-	Australia; India	12
Ra/W	N	CT	1	−4 (–)	1	48.4 (–)	Canada (Manitoba)	13
R	F	N/A	22	23.8 (14.9)	-	-	China; India	14

[a]M, maize; W, wheat; Ra, rapeseed; R, rice.
[b]I, irrigated; Rf, rainfed; N, not reported; F, flooded or paddy.
[c]CT, conventional tillage; N, not reported; NT, no tillage; ST, strip tillage.
[d]1= [(Bremner et al., 1981; Burzaco et al., 2013; Hadi et al., 2008; Omonode and Vyn, 2013; Parkin and Hatfield, 2010; Venterea et al., 2011)] 2= [(Parkin and Hatfield, 2014)]; 3= [(Dell et al., 2014; Omonode and Vyn, 2013; Sistani et al., 2011; Venterea et al., 2011)]; 4= [(Jumadi et al., 2008)]; 5= [(Bronson et al., 1992; Halvorson et al., 2016; Hu et al., 2013; Ju et al., 2011; Liu et al., 2013)]; 6= [(Halvorson et al., 2014)]; 7= [(Ding et al., 2010; Halvorson et al., 2014)]; 8= [(De Antoni Migliorati et al., 2014; Sanz-Cobena et al., 2012)]; 9= [(Burton et al., 2008)]; 10= [(Bhatia et al., 2010; Hu et al., 2013; Ju et al., 2011; Liu et al., 2013)]; 11= [(Bhatia et al., 2010)]; 12= [(De Antoni Migliorati et al., 2014; Majumdar et al., 2002; Pathak et al., 2002)]; 13= [(Asgedom et al., 2014)]; 14= [(Cheng et al., 2014; Ghosh et al., 2003; Kumar et al., 2000; Li et al., 2009; Majumdar et al., 2000; Malla et al., 2005; Pathak et al., 2002)].
[e]#, number of observations.

cautioned by Parkin and Hatfield (2010). Many of the rainfed studies were conducted on soils high in organic matter, which has been found to reduce the effectiveness of NIs (Jacinthe and Pichtel, 1992). Further, Halvorson et al. (2016) found that ammonia losses can reduce available N and consequently N_2O emissions compared to NI treatments, reducing the apparent effectiveness of NIs.

In major cereal crops under varying agricultural management practices, the effectiveness of PCUs at reducing N_2O emissions was highly variable between rainfed or irrigated systems and tillage practices (Table 12.1). The lack of consistent effectiveness has been attributed to asynchronization of N availability and plant demand (Decock, 2014) as well as due to interactions with land-use, soil type, site, and regional characteristics (Akiyama et al., 2010; Decock, 2014). Continued research is needed to delineate which, when, and how enhanced efficiency fertilizers can be used to reduce N_2O emissions among a wide range of soil, climate, and management practices.

12.2.3 Dryland and Irrigation in Cropping Systems

Semiarid rainfed row-crop systems typically have low N_2O flux compared to humid rainfed systems (Liebig et al., 2012; Aguilera et al., 2013; Trost et al., 2013; Eve et al., 2014). Emission factors for irrigated systems were similar to IPCC Tier 1 estimate of 1% (0.98%) though rainfed emission factors in Mediterranean climates systems were significantly less (0.08%) (Aguilera et al., 2013). Irrigation events mimic resource pulses in rainfed systems, with greater N_2O efflux occurring after irrigation events (López-Fernández et al., 2007; Trost et al., 2013). Flood or furrow irrigation and associated increases in N inputs can lead to greater potential N_2O losses; however, N_2O emissions can be reduced using drip-irrigation, which avoids extended periods of saturated or near-saturated conditions compared to furrow or flood-irrigated soils (Aguilera et al., 2013). Irrigation may increase N_2O emission but also has been demonstrated to increase SOC because of the increased plant productivity (Cochran et al., 2007). In a global meta-analysis of irrigated systems, Trost et al. (2013) found that SOC increases were dependent on climate and initial SOC level, with greatest SOC gains in irrigated desert soils (90–500% increase), moderate gains in irrigated semiarid soils (11–35% increase), and highly variable SOC responses to irrigation in humid regions. Because SOC is tightly linked to soil health benefits (see Section 12.3), the effect of irrigation on soil health will likely be mediated through the relative change in SOC depending on climate, initial soil conditions, and other management practices. Although numerous researchers have measured N_2O and/or SOC responses in semiarid rainfed, humid rainfed, and irrigated systems separately, a comprehensive meta-analysis among systems that explicitly incorporates soil health is still needed, but is beyond the scope of this chapter. Other aspects of irrigation, such as the environmental impacts of using geologic water (i.e., Ogallala aquafer, United States) or diversion of surface waters, were not included in this chapter but are important aspects of sustainable intensification.

12.3 Conservation Management and Soil Health Implications

Carbon (C) and N cycles are closely coupled. SOM is directly proportional to SOC, so SOM or SOC frequently is used as a primary indicator for various soil health functions because of its links to numerous soil biological, chemical, and physical properties (Lehman et al., 2015a, b). Soil health properties that are directly and/or indirect linked to SOM or SOC include aeration, permeability, water-holding capacity, resilience to soil compaction, nutrient-holding capacity, and nutrient cycling as reviewed by Doran and Parkin (1994) and Johnson et al. (2010b). Conservation practices to increase SOC (e.g., tillage, biomass management, cover crops) can also impact GHG emission directly or indirectly by altering C and/or N substrate, water-filled pore space, microclimate, or other soil properties.

12.3.1 Mitigation Potential of Conservation Tillage

Conservation tillage decreases soil disturbance and decomposition, leaving surface residue cover that can increase water retention, soil C and N, and potentially crop yield. Although no- or reduced tillage can decrease GHG emissions and sequester soil C, these benefits can be region-specific and require careful accounting. Changing tillage impacts residue placement and microclimate, interacts with soil properties, and in turn can impact N_2O emission. No-tillage management can increase (Rochette, 2008), decrease (Mosier et al., 2006), or have no effect (Grandy et al., 2006; Decock, 2014) on soil N_2O emissions. Increased N_2O emissions under no-tillage occurred on soils with greater clay content (Burford et al., 1981; Rochette, 2008). A meta-analysis by van Kessel et al. (2013) found no overall reduction in yield-scaled N_2O emissions in no-tillage compared to tilled systems (239 comparisons). However, they also reported that the impact of no-tillage varied regionally, with long-term experiments in dry climates showing that no-tillage did reduce N_2O emissions compared to tilled systems. van Kessel et al. (2013) emphasized the importance of fertilizer placement on N_2O emissions, particularly in wet humid climates when fertilizer was placed deeper than 5 cm in no-tillage or ridge-tillage systems (rather than surface broadcast). They also noted a significant study duration effect, where shorter-term studies (<10 years) had greater N_2O emissions. A meta-analysis from China (240 publications) reported that although converting to conservation tillage in upland crops (maize, wheat) increased N_2O emission by 46%, the rate of SOC accrual successfully offset the N_2O emissions, such that conservation tillage was considered a mitigation option (Nayak et al., 2015).

Conservation practices that retain soil on the landscape are expected to increase SOC and mitigate GHG emissions. Early meta-analysis (West and Post, 2002) suggested that converting from moldboard plowing to no-tillage management can sequester SOC. Adoption of no-tillage management had a SOC sequestration rate of 0.40 ± 0.61 Mg C ha^{-1} $year^{-1}$ (n=44, average depth 30 cm) in the central United States (Johnson et al., 2005) and 0.58 ± 0.71 Mg

C ha^{-1} $year^{-1}$ (n=37, average depth 21 cm) in the eastern region of the United States (Dell and Novak, 2012). In a meta-analysis of 69 studies, Luo et al. (2010) reported SOC sequestration following conversion to no-tillage varied by depth, with sequestration occurring in the surface soils (0–10 cm) and declines or no difference in SOC occurring at greater depths. Noteworthy among the reviews and meta-analysis is the huge variability ranging from reports of SOC loss, no SOC difference, and substantial SOC sequestration in response to converting to no-tillage management.

Documenting SOC sequestration as a consequence of adopting no-tillage management is fraught with challenges, as noted in reviews by Yang et al. (2013) and by Olson (2013). Briefly, issues raised included challenges related to making an equitable comparison based on equivalent soil mass, identifying redistribution of SOC instead of actual accrual, and resolving challenges to these comparisons that are exacerbated by shallow sampling, lack of information on baseline SOC, lack of accounting of impact of soil erosion between tillage treatments, and variability masking treatment effects. In recent meta-analyses of 74 studies covering 14 countries (Ogle et al., 2012) and of 41 studies (239 comparisons) (van Kessel et al., 2013), researchers found that no-tillage and even reduced tillage reduced crop yield compared to conventional tillage. Reduced crop production and decreases in associated C inputs could reduce SOC accrual or result in SOC losses if the rate of C mineralization exceeds the rate of humification (Johnson et al., 2010b). However, because the mean residence time of C inputs appeared to increase about 15% in soils without tillage compared to those with tillage, SOC accrual was predicted unless yield was suppressed more than 15% (Ogle et al., 2012). Results of these studies emphasize the importance of regionally analyzing mitigation strategies and the need for high-quality SOC data to address the issues noted by Yang et al. (2013). Conservation tillage practices were developed for erosion control, and serve a critical function in protecting soil and water resources. While converting to no-tillage or reduced tillage is not a panacea nor a guaranteed method for SOC sequestration, the evidence generally supports the potential for SOC sequestration and associated benefits of higher SOC on soil health functions under conservation tillage practices.

12.3.2 Mitigation Potential of Biomass Management

Residue harvest or planting cover crops can alter the quantity and quality of biomass inputs. A multiyear, multisite study reported that harvesting maize residue tended to decreased N_2O emissions, as a result of changes in C and N inputs, and microclimatic conditions associated with changes in soil cover (Jin et al., 2014). Chen et al. (2013) found that N_2O emissions were positively correlated with applying low C:N crop residues, but negatively correlated when C:N were >100. Since maize residue has a C:N ratio >50 (Johnson et al., 2010c), moderate levels of maize residue harvest may reduce N_2O emissions, but excessive harvest is likely to reduce soil quality and SOC (Johnson et al., 2006, 2014). The amount of biomass needed to avoid losses in

SOC is influenced by climate, soil, and sampling depth (Johnson et al., 2014). High biomass-yielding crops, perennial crops, grass cover (Kong et al., 2005; Ogle et al., 2005), reduction of fallow (Sherrod et al., 2003, 2005), or adding a cover crop can increase biomass inputs and increase SOC. A recent meta-analysis ($n = 139$, 37 sites) suggested that the adoption of cover crops could provide an additional and sustained input of 0.32 ± 0.08 Mg C ha^{-1} $year^{-1}$ for an estimated 155 years before achieving steady state (Poeplau and Don, 2015). Cover crops could reduce N_2O emissions by acquiring soil N otherwise lost through gaseous forms or leaching, and possibly offset fertilization rates. However, there are no studies demonstrating that N inputs from cover crops instead of mineral fertilizer will reduce direct N_2O emissions (Ogle et al., 2014). In paddy rice, adding a legume cover crop increased N_2O and CH_4 emissions by increasing N substrate (Pramanik et al., 2013). A meta-analysis of 26 studies (106 observations) found that the impact of cover crops on N_2O emission was highly equivocal (Basche et al., 2014). A literature review by Justes et al. (2012) reported N_2O emissions increased by 0.11 ± 1.12 kg N ha^{-1} $year^{-1}$ in the year following cover crop use, based on French agricultural practices. Cover crops, however, have a high potential for GHG abatement when SOC sequestration was included (Pellerin et al., 2013), but the high cost to French farmers caused by additional cultivation operations may limit adoption. Environmental restraints [i.e., northern climates (Wilson et al., 2014) or semiarid climates] can also hinder cover crop adoption or cause reductions in cash crop yield (Nielsen et al., 2016). Nonetheless, other benefits of cover crop use could encourage adoption, including the protection of SOC-rich topsoil from erosion; provision of plant biomass inputs that in turn support the soil ecosystem needed to build SOC; and enhancement of above- and belowground biodiversity that provide ecosystem services associated with healthy soils. Thus, utilizing cover crops where feasible is an important tool to achieve sustainable intensification.

12.4 Global Warming Potential and Sustainable Intensification

Previous sections either focused on the mitigation of N_2O emissions or SOC sequestration. Accounting for increasing SOC storage and CH_4 and N_2O emissions among management practices or systems can provide a more complete assessment of their global footprint. Likewise, assessing greenhouse gas intensity (GHGI = GWP per unit yield) provides an index to assess whether the dual goals of increasing yield and reducing C-footprint have been achieved. Several examples of GWP and GHGI results are presented in Table 12.2.

Multiple approaches have been used to estimate the CO_2 contributions of various agricultural sources when calculating GWP. The direction and magnitude of GHGI for various management practices can change depending upon the approach chosen for calculating GHGI. One common approach is to estimate no-tillage or reduced tillage effects on SOC relative to conventional tillage by assuming that both tillage practices had the same initial SOC stock; the validity of this type of approach has been questioned (Olson, 2013; Yang et al.,

2013). A second approach estimates net ecosystem exchange using small chambers, sometimes called a respiration approach (Liu et al., 2012; Sainju et al., 2014). Cavigelli and Parkin (2012) cautioned that chamber-based CO_2 measurements should not be used to calculate GWP; rather, net ecosystem exchange should be evaluated in the near-term using micrometeorological approaches (Baker and Griffis, 2005; Glenn et al., 2010) or inferred over the long term by taking measurements of SOC changes over time. A third approach calculates a change in C stocks at a sample time compared to the baseline, which ideally is based on three or more time points (Huang et al., 2013; Sainju et al., 2014). The selection of small chamber-based respiration approach or the SOC change over time approach to calculate GWP and GHGI can impact the treatment effect magnitude, as well as whether a system is a GHG source or sink (Table 12.2) (Mosier et al., 2006; Sainju et al., 2014). While Zhang et al. (2014a) and Ma et al. (2013) used several difference approaches (methods four, five, or six; Table 12.2) to estimate CO_2/SOC sequestration impacts on GWP and GHGI. They found all accounting methods had inherent biases that were significant enough to change the apparent ranking among treatments (Ma et al., 2013; Zhang et al., 2014a). Making comparisons among management treatments even within a study and method may be insightful, but may not accurately reflect whether a given practice is truly achieving the desired effect of sustainable intensification. The lack of uniformity in how the CO_2 contribution is addressed makes it very difficult to compare among studies. This has important policy and management implications when using GHGI as a criterion for sustainable intensification. Other criteria for sustainable intensification are also needed; as noted by Lal and Stewart (2015), achieving the *sustainable* intensification of arable landscape will require balancing rather than tradeoffs to feed the world and protect the environment. For example, successful and widespread adoption of conservation management practices could lead to additional SOC accrual and associated improvements in soil health. Soil health improvements, in turn, could maintain or increase yield with lower inputs of water, agrochemicals, and energy. While soil health benefits may take some years to realize following a change in practice, finding socioeconomically feasible ways to implement practical and site-specific conservation practices in the present day could have significant positive economic and environmental outcomes in the long term.

12.5 Conclusions

Row-crop production is a crucial component of meeting the demands of the expanding human population. Although the optimum N-rate may vary, avoiding excessive N fertilizer is a vital strategy for sustainable intensification. Reducing or eliminating tillage may not always result in clear GHG mitigation, but preventing soil erosion is critical—without topsoil there can be no food security. Providing plant biomass input from crop residues and cover crops can promote the sequestration of SOC and provide the synergistic benefits of improving soil quality and soil health. Individual studies, reviews, and meta-analyses covered in this chapter present both hypothesized and demonstrated benefits of conservation management

Table 12.2: Examples of global warming potential (GWP) and greenhouse gas intensity (GHGI) values reported in the literature based on several approaches

Country (State/Region or Province)	Crop[a]	Treatment (kg N ha^{-1})	GWP Mg CO_2 eq ha^{-1} yr^{-1}	Yield unit	GHGI kg CO_2 eq kg^{-1} yr^{-1}	Approach[b] (Depth cm)	References
Brazil (Parana)	M	CT[c]	2.9	Silage	0.17	One (100)	Piva et al. (2012)
		NT	−0.55		−0.03		
China (Jiangsu)	R	No Fe(III)	−24.3	Grain	−2.02	Two	Liu et al. (2012)
		Moderate Fe(III)	−32.9		−2.62		
		High Fe(III)	−35.7		−2.72		
USA (Colorado)	M	CT (0)	1.95	Grain	0.09		Mosier et al. (2006)
		CT (202)	−1.74		−0.15		
		NT (0)	−0.83		−0.16		
		NT (202)	−4.30		−0.39	Two	
		CT (0)	0.08		0.013	Three (7.5)	
		CT (202)	0.13		0.13		
		NT (0)	−0.23		−0.04		
		NT (202)	−0.09		−0.08		
USA (Montana)	Irr. B	CT (67–134)	−8.0	Grain	−2.0	Two	Sainju et al. (2014)
		NT (67–134)	−8.0		−2.0		
		CT (67–134)	1.5		0.50	Three (10)	
		NT (67–134)	1.1		0.35		

China (Northern China Plain)	M/W	No straw (0)	4.06	Grain	0.562	Three (20)	Huang et al. (2013)
		Straw (0)	1.35		0.181		
		No straw (Opt)	2.88		0.258		
		Straw (Opt)	2.83		0.252		
		No straw (Con)	4.68		0.393		
		Straw (Con)	3.93		0.310		
China (Jiangu)	R/W	0	7.87	Grain	1.038	Three (20)	Zhang et al. (2015)
		Farmer 300/180	11.5		0.810		
		270/162	10.4		0.664		
China (Jiangu)	R/W	0	5.7	Grain	0.74	Four	Ma et al. (2013)
		Farmer 300/180	8.36		0.58		
		270/162	6.66		0.41		
		0	4.72		0.61	Five	
		Farmer 300/180	8.96		0.62		
		270/162	8.19		0.52		
		0	9.3		1.21	Six	
		Farmer 300/180	5.58		0.39		
		270/162	1.0		0.06		

[a]Crop: A, alfalfa; Irr. B, irrigated barley; M, maize; R, paddy rice; S, soybean; W, wheat (drained wheat in locations in China).
[b]Approaches used: One—SOC in the no-tillage system was estimated relative to conventional, assumes both tillage practices had the same initial C stock (0–100 cm); Two—estimates CO_2 from net ecosystem exchange using small chamber or sometimes called a respiration approach. Three—calculates a change in C stocks at a sample time compared to the baseline, which ideally is based on three or more time points. Four—derived SOC changes assuming it was 0.213 of net ecosystem C balance. Five—excluded SOC change. Six—derived SOC change assuming it was equal to net ecosystem C balance.
[c]CT, conventional tillage; RT, reduced tillage; NT, no-tillage; No straw, straw was harvested; Straw, straw was returned to field; Opt, optimizing N rate; Con, conventional N treatment.

practices that could contribute to enhancing soil resilience to climate change and supporting agroecosystem services such as food production, air quality, and water quality. A systems approach that considers GHG production and soil C sequestration, as well as other measures of environmental and soil health, will be critical for accurately evaluating agricultural management effects.

References

Aguilera, E., Lassaletta, L., Sanz-Cobena, A., Garnier, J., Vallejo, A., 2013. The potential of organic fertilizers and water management to reduce N_2O emissions in Mediterranean climate cropping systems. A review. Agric. Ecosyst. Environ. 164, 32–52.

Akiyama, H., Yan, X., Yagi, K., 2010. Evaluation of effectiveness of enhanced-efficiency fertilizers as mitigation options for N_2O and NO emissions from agricultural soils: Meta-analysis. Glob. Change Biol. 16, 1837–1846.

Alexandratos, N., Bruinsma, J., 2012. World Agriculture Towards 2030/2050: The 2012 Revision. ESA Working Paper No. 12-03., Agricultural Development Economics Division, Food and Agriculture Organization of the United Nations. UN FAO, Rome, Italy.

Angelsen, A., 2010. Policies for reduced deforestation and their impact on agricultural production. Proc. Natl. Acad. Sci. 107, 19639–19644.

Archer, D.W., Halvorson, A.D., 2010. Greenhouse gas mitigation economics for irrigated cropping systems in northeastern Colorado. Soil Sci. Soc. Am. J. 74, 446–452.

Asgedom, H., Tenuta, M., Flaten, D.N., Gao, X., Kebreab, E., 2014. Nitrous oxide emissions from a clay soil receiving granular urea formulations and dairy manure. Agron. J. 106, 732–744.

Baker, J.M., Griffis, T.J., 2005. Examining strategies to improve the carbon balance of corn/bean agriculture using eddy covariance and mass balance techniques. Agric. Forest Meteorol. 128, 163–177.

Basche, A.D., Miguez, F.E., Kaspar, T.C., Castellano, M.J., 2014. Do cover crops increase or decrease nitrous oxide emissions? A meta-analysis. J. Soil Water Conserv. 69, 471–482.

Bhatia, A., Sasmal, S., Jain, N., Pathak, H., Kumar, R., Singh, A., 2010. Mitigating nitrous oxide emission from soil under conventional and no-tillage in wheat using nitrification inhibitors. Agric. Ecosyst. Environ. 136, 247.

Bremner, J.M., Breitenbeck, G.A., Blackmer, A.M., 1981. Effect of nitrapyrin on emission of nitrous oxide from soil fertilized with anhydrous ammonia. Geophys. Res. Lett. 8, 353–356.

Bronson, K.F., Mosier, A.R., 1994. Suppression of methane oxidation in aerobic soil by nitrogen fertilizers, nitrification inhibitors, and urease inhibitors. Biol. Fertil. Soils 17, 263–268.

Bronson, K.F., Mosier, A.R., Bishnoi, S.R., 1992. Nitrous oxide emissions in irrigated corn as affected by nitrification inhibitors. Soil Sci. Soc. Am. J. 56, 161–165.

Burford, J.R., Dowdell, R.J., Crees, R., 1981. Emission of nitrous oxide to the atmosphere from direct-drilled and ploughed clay soils. J. Sci. Food. Agric. 32, 219–223.

Burton, D.L., Li, X., Grant, C.A., 2008. Influence of fertilizer nitrogen source and management practice on N_2O emissions from two Black Chernozemic soils. Can. J. Soil Sci. 88, 219–227.

Burzaco, J.P., Smith, D.R., Vyn, T.J., 2013. Nitrous oxide emissions in Midwest US maize production vary widely with band-injected N fertilizer rates, timing and nitrapyrin presence. Environ. Res. Lett. 8, 1–11.

Cassman, K.G., Dobermann, A., Walters, D.T., Yang, H., 2003. Meeting cereal demand while protecting natural resources and improving environmental quality. Ann. Rev. Environ. Resour. 28, 315–358.

Cavigelli, M.A., Parkin, T., 2012. Cropland management contributions to greenhouse gas flux. In: Liebig, M., Franzluebbers, A., Follett, R.F. (Eds.), Managing Agricultural Greenhouse Gases: Coordinated Agricultural Research Through GRACEnet to Address Our Changing Climate. Academic Press an imprint of Elsevier, Waltham, MA.

Chen, H.H., Li, X.C., Hu, F., Shi, W., 2013. Soil nitrous oxide emissions following crop residue addition: a meta-analysis. Glob. Change Biol. 19, 2956–2964.

Chen, X.-P., Cui, Z.-L., Vitousek, P.M., Cassman, K.G., Matson, P.A., Bai, J.-S., et al., 2011. Integrated soil–crop system management for food security. Proc. Natl. Acad. Sci. 108, 6399–6404.

Cheng, K., Ogle, S.M., Parton, W.J., Pan, G., 2014. Simulating greenhouse gas mitigation potentials for Chinese Croplands using the DAYCENT ecosystem model. Glob. Change Biol. 20, 948–962.

Cochran, R.L., Collins, H.P., Kennedy, A., Bezdicek, D.F., 2007. Soil carbon pools and fluxes after land conversion in a semiarid shrub-steppe ecosystem. Biol. Fert. Soils 43, 479–489.

De Antoni Migliorati, M., Scheer, C., Grace, P.R., Rowlings, D.W., Bell, M., McGree, J., 2014. Influence of different nitrogen rates and DMPP nitrification inhibitor on annual N_2O emissions from a subtropical wheat–maize cropping system. Agric. Ecosyst. Environ. 186, 33–43.

Decock, C., 2014. Mitigating nitrous oxide emissions from corn cropping systems in the Midwestern U.S.: Potential and data gaps. Environ. Sci. Technol. 48, 4247–4256.

Del Grosso, S.J., Wirth, T., Ogle, S.M., Parton, W.J., 2008. Estimating agricultural nitrous oxide emissions. Eos, Trans. Am. Geophys. Union 89 529–529.

Dell, C., Novak, J.M., 2012. Cropland management in the Eastern United States for improved soil organic carbon sequestration. In: Liebig, M.A., Franzleubbers, A.J., Follett, R.F. (Eds.), Managing Agricultural Greenhouse Gases: Coordinated Agricultural Research Through GRACEnet to Address Our Changing Climate. Academic Press, New York, NY, pp. 23–40.

Dell, C.J., Han, K., Bryant, R.B., Schmidt, J.P., 2014. Nitrous oxide emissions with enhanced efficiency nitrogen fertilizers in a rainfed system. Agron. J. 106, 723–731.

Di, H.J., Cameron, K.C., Shen, J.P., Winefield, C.S., O'Callaghan, M., Bowatte, S., et al., 2009. Nitrification driven by bacteria and not archaea in nitrogen-rich grassland soils. Nat. Geosci. 2, 621–624.

Ding, Y., Liu, Y.-X., Wu, W.-X., Shi, D.-Z., Yang, M., Zhong, Z.-K., 2010. Evaluation of biochar effects on nitrogen retention and leaching in multi-layered soil columns. Water Air. Soil. Pollut. 213, 47–55.

Doran, J.W., Parkin, T.B., 1994. Defining and assessing soil quality. In: Doran, J.W., Coleman, D.C., Bezdicek, D.F., Stewart, B.A. (Eds.), Defining Soil Quality for a Sustainable Environment. SSSA, ASA, Madison, WI, pp. 3–21. SSSA Spec. Publ. 35.

Engel, R., Liang, D.L., Wallander, R., Bembenek, A., 2010. Influence of urea fertilizer placement on nitrous oxide production from a silt loam soil. J. Environ. Qual. 39, 115–125.

Eve, M., Pape, D., Flugge, M., Steele, R., Man, D., Riley-Gilbert, M., et al., 2014. Quantifying Greenhouse Gas Fluxes in Agriculture and Forestry: Methods From Entity-Scale Inventory. Office of the Chief Economist, U.S. Department of Agriculture, Washington, DC606., Technical Bulletin Number 1939.

FAO, 2014. Agriculture, Forestry and Other Land Use Emissions by Sources and Removals by Sinks: 1990–2011 Analysis. FAO, Rome, Italy, FAO Statistics Division Working Paper Series.

Fujinuma, R., Venterea, R.T., Rosen, C., 2011. Broadcast urea reduces N_2O but increases NO emissions compared with conventional and shallow-applied anhydrous ammonia in a coarse-textured soil. J. Environ. Qual. 40, 1806–1815.

Gagnon, B., Ziadi, N., Rochette, P., Chantigny, M.H., Angers, D.A., 2011. Fertilizer source influenced nitrous oxide emissions from a clay soil under corn. Soil Sci. Soc. Am. J. 75, 595–604.

Garnett, T., Appleby, M.C., Balmford, A., Bateman, I.J., Benton, T.G., Bloomer, P., et al., 2013. Sustainable intensification in agriculture: premises and policies. Science 341, 33–34.

Geisseler, D., Scow, K.M., 2014. Long-term effects of mineral fertilizers on soil microorganisms – a review. Soil Biol. Biochem. 75, 54–63.

Ghosh, S., Majumdar, D., Jain, M.C., 2003. Methane and nitrous oxide emissions from an irrigated rice of North India. Chemosphere 51, 181–195.

Gilsanz, C., Báez, D., Misselbrook, T.H., Dhanoa, M.S., Cárdenas, L.M., 2016. Development of emission factors and efficiency of two nitrification inhibitors, DCD and DMPP. Agric. Ecosyst. Environ. 216, 1–8.

Glenn, A.J., Amiro, B.D., Tenuta, M., Stewart, S.E., Wagner-Riddle, C., 2010. Carbon dioxide exchange in a northern Prairie cropping system over three years. Agric. Forest Meteorol. 150, 908–918.

Godfray, H.C.J., Garnett, T., 2014. Food security and sustainable intensification. Philos. Trans. R. Soc. London B Biol. Sci. 369.

Grandy, A.S., Loecke, T.D., Parr, S., Robertson, G.P., 2006. Long-term trends in nitrous oxide emissions, soil nitrogen, and crop yields of till and no-till cropping systems. J. Environ. Qual. 35, 1487–1495.
Hadi, A., Jumadi, O., Inubushi, K., Yagi, K., 2008. Mitigation options for N_2O emission from a corn field in Kalimantan, Indonesia. Soil Sci. Plant Nutr. 54, 644–649.
Halvorson, A.D., Snyder, C.S., Blaylock, A.D., Del Grosso, S.J., 2014. Enhanced-efficiency nitrogen fertilizers: Potential role in nitrous oxide emission mitigation. Agron. J. 106, 715–722.
Halvorson, A.D., Del Grosso, S.J., Stewart, C.E., 2016. Manure and inorganic N affect trace gas emissions under semi-arid irrigated corn. J. Environ. Qual. (in-press).
Hao, X., Chang, C., Carefoot, J.M., Janzen, H.H., Ellert, B.H., 2001. Nitrous oxide emissions from an irrigated soil as affected by fertilizer and straw management. Nutr. Cycl. Agroecosyst. 60, 1–8.
Hu, X.-K., Su, F., Ju, X.-T., Gao, B., Oenema, O., Christie, P., et al., 2013. Greenhouse gas emissions from a wheat–maize double cropping system with different nitrogen fertilization regimes. Environ. Pollut. 176, 198–207.
Huang, T., Gao, B., Christie, P., Ju, X., 2013. Net global warming potential and greenhouse gas intensity in a double-cropping cereal rotation as affected by nitrogen and straw management. Biogeosciences 10, 7897–7911.
IPCC, 2013. Climate change 2013: the physical science basis. In: Stocker, T., Qin, D., Plattner, G., Tignor, M., Allen, S., Boschung, J., Nauels, A., Xia, Y., Bex, V., Midglev, P. (Eds.), Contribution of Working Group I to the Fifth Assessment Report of the Intergovernmental Panel on Climate Change. Cambridge University Press, Cambridge and New York.
IPCC, 2014. In: Pachauri, R.K., Meyers, L.A. (Eds.), Climate Change 2014: Synthesis Report of Working Groups I, II, and III to the Fifth Assessment Report of the Intergovernmental Panel on Climate Change. IPCC, Geneva, Switzerland, pp. 151.
Jacinthe, P.A., Pichtel, J.R., 1992. Interaction of nitrapyrin and dicyandiamide with soil humic compounds. Soil Sci. Soc. Am. J. 56, 465–470.
Jin, V.L., Baker, J.M., Johnson, J.M.F., Karlen, D.L., Lehman, R.M., Osborne, S.L., et al., 2014. Soil greenhouse gas emissions in response to corn stover removal and tillage management across the US Corn Belt. BioEnergy Res., 1–11.
Johnson, J.M.F., Reicosky, D.C., Allmaras, R.R., Venterea, R.T., Sauer, T.J., Dell, C.J., 2005. Greenhouse gas contributions and mitigation potential of agriculture in the central USA. Soil Till. Res. 83, 73–94.
Johnson, J.M.F., Allmaras, R.R., Reicosky, D.C., 2006. Estimating source carbon from crop residues, roots and rhizodeposits using the national grain-yield database. Agron. J. 98, 622–636.
Johnson, J.M.F., Archer, D.W., Barbour, N.W., 2010a. Greenhouse gas emission from contrasting management scenarios in the Northern Corn Belt. Soil Sci. Soc. Am. J. 74, 396–406.
Johnson, J.M.F., Papiernik, S.K., Mikha, M.M., Spokas, K.A., Tomer, M.D., Weyers, S.L., 2010b. Soil processes and residue harvest management. In: Lal, R., Stewart, B.A. (Eds.), Carbon Management, Fuels, and Soil Quality. Taylor and Francis, LLC, New York, NY, pp. 1–44.
Johnson, J.M.F., Wilhelm, W.W., Karlen, D.L., Archer, D.W., Wienhold, B., Lightle, D., et al., 2010c. Nutrient removal as a function of corn stover cutting height and cob harvest. BioEnergy Res. 3, 342–352.
Johnson, J.M.F., Novak, J.M., Varvel, G.E., Stott, D.E., Osborne, S.L., Karlen, D.L., et al., 2014. Crop residue mass needed to maintain soil organic carbon levels: can it be determined? BioEnergy Res. 7, 481–490.
Ju, X., Lu, X., Gao, Z., Chen, X., Su, F., Kogge, M., et al., 2011. Processes and factors controlling N_2O production in an intensively managed low carbon calcareous soil under sub-humid monsoon conditions. Environ. Pollut. 159, 1007–1016.
Jumadi, O., Hala, Y., Muis, A., Ali, A., Palennari, M., Yagi, K., et al., 2008. Influences of chemical fertilizers and a nitrification inhibitor on greenhouse gas fluxes in a corn (*Zea mays* L.) field in Indonesia. Microbes Environ. 2, 29–34.
Justes, E., Beaudoin, N., Bertuzzi, P., Charles, R., Constantin, J., Dürr, C., et al., 2012. Cover crops to reduce nitrate leaching. Effect on water and nitrogen balance and other ecosystem services. Proceedings - International Fertiliser Society.

Kong, A.Y.Y., Six, J., Bryant, D.C., Denison, R.F., van Kessel, C., 2005. The relationship between carbon input, aggregation, and soil organic carbon stabilization in sustainable cropping systems. Soil Sci. Soc. Am. J. 69, 1078–1085.

Kumar, U., Jain, C.M., Pathak, H., Kumar, S., Majumdar, D., 2000. Nitrous oxide emission from different fertilizers and its mitigation by nitrification inhibitors in irrigated rice. Biol. Fertil. Soils 32, 474–478.

Lal, R., Stewart, B.B., 2015. Sustainable intensification of smallholder agriculture. In: Lal, R., Stewart, B. (Eds.), Soil Management of Smallholder Agriculture. CRC Press, Boca Raton, FL, pp. 385–392.

Lehman, R.M., Acosta-Martinez, V., Buyer, J.S., Cambardella, C.A., Collins, H.P., Ducey, T.F., et al., 2015a. Soil biology for resilient, healthy soil. J. Soil Water Conserv. 70, 12A–18A.

Lehman, R.M., Cambardella, C., Stott, D., Acosta-Martinez, V., Manter, D., Buyer, J., et al., 2015b. Understanding and enhancing soil biological health: the solution for reversing soil degradation. Sustainability 7, 988.

Li, X., Zhang, X., Xu, H., Cai, Z., Yagi, K., 2009. Methane and nitrous oxide emissions from rice paddy soil as influenced by timing of application of hydroquinone and dicyandiamide. Nutr. Cycl. Agroecosyst. 85, 31–40.

Liebig, M.A., Franzluebbers, A.J., Follett, R.F., 2012. Managing agricultural Greenhouse Gases. Coordinated Agricultural Research Through GRACEnet to Address Our Changing Climate. Elselvier, Waltham, MA p. 547.

Linquist, B., van Groenigen, K.J., Adviento-Borbe, M.A., Pittelkow, C., van Kessel, C., 2012a. An agronomic assessment of greenhouse gas emissions from major cereal crops. Glob. Change Biol. 18, 194–209.

Linquist, B.A., Adviento-Borbe, M.A., Pittelkow, C.M., van Kessel, C., van Groenigen, K.J., 2012b. Fertilizer management practices and greenhouse gas emissions from rice systems: a quantitative review and analysis. Field Crops Res. 135, 10–21.

Liu, C., Wang, K., Zheng, X., 2013. Effects of nitrification inhibitors (DCD and DMPP) on nitrous oxide emission, crop yield and nitrogen uptake in a wheat–maize cropping system. Biogeosciences 10, 2427–2437.

Liu, S., Zhang, L., Liu, Q., Zou, J., 2012. Fe(III) fertilization mitigating net global warming potential and greenhouse gas intensity in paddy rice-wheat rotation systems in China. Environ. Pollut. 164, 73–80.

López-Fernández, S., Díez, J.A., Hernáiz, P., Arce, A., García-Torres, L., Vallejo, A., 2007. Effects of fertiliser type and the presence or absence of plants on nitrous oxide emissions from irrigated soils. Nutr. Cycl. Agroecosyst. 78, 279–289.

Luo, Z., Wang, E., Sun, O.J., 2010. Can no-tillage stimulate carbon sequestration in agricultural soils? A meta-analysis of paired experiments. Agric. Ecosyst. Environ. 139, 224–231.

Ma, Y.C., Kong, X.W., Yang, B., Zhang, X.L., Yan, X.Y., Yang, J.C., et al., 2013. Net global warming potential and greenhouse gas intensity of annual rice–wheat rotations with integrated soil–crop system management. Agric. Ecosyst. Environ. 164, 209–219.

Majumdar, D., Kumar, S., Pathak, H., Jain, M.C., Kumar, U., 2000. Reducing nitrous oxide emission from an irrigated rice field of North India with nitrification inhibitors. Agric. Ecosyst. Environ. 81, 163–169.

Majumdar, D., Pathak, H., Kumar, S., Jain, M.C., 2002. Nitrous oxide emission from a sandy loam Inceptisol under irrigated wheat in India as influenced by different nitrification inhibitors. Agric. Ecosyst. Environ. 91, 283–293.

Malhi, S.S., Nyborg, M., 1985. Methods of placement for increasing the efficiency of N fertilizers applied in the fall1. Agron. J. 77, 27–32.

Malla, G., Bhatia, A., Pathak, H., Prasad, S., Jain, N., Singh, J., 2005. Mitigating nitrous oxide and methane emissions from soil in rice–wheat system of the Indo-Gangetic plain with nitrification and urease inhibitors. Chemosphere 58, 141–147.

Mamo, M., Malzer, G.L., Mulla, D.J., Huggins, D.R., Strock, J., 2003. Spatial and temporal variation in economically optimum nitrogen rate for corn. Agron. J. 95, 958–964.

Matson, P.A., Naylor, R., Ortiz-Monasterio, I., 1998. Integration of environmental, agronomic, and economic aspects of fertilizer management. Science 280, 112–115.

McCarty, W.G., 1999. Modes of action of nitrification inhibitors. Biol. Fertil. Soils 29, 1–9.

McSwiney, C.P., Robertson, G.P., 2005. Nonlinear response of N_2O flux to incremental fertilizer addition in a continuous maize (*Zea mays* L.) cropping system. Glob. Change Biol. 11, 1712–1719.

Mosier, A.R., Halvorson, A.D., Reule, C.A., Liu, X.J., 2006. Net global warming potential and greenhouse gas intensity in irrigated cropping systems in Northeastern Colorado. J. Environ. Qual. 35, 1584–1598.
Nayak, D., Saetnan, E., Cheng, K., Wang, W., Koslowski, F., Cheng, Y., et al., 2015. Management opportunities to mitigate greenhouse gas emissions from Chinese agriculture. Agric. Ecosyst. Environ. 209, 108–124.
Nielsen, D.C., Lyon, D.J., Higgins, R.K., Hergert, G.W., Holman, J.D., Vigil, M.F., 2016. Cover crop effect on subsequent wheat yield in the central Great Plains. Agron. J. 108
Ogle, S.M., Breidt, F.J., Paustian, K., 2005. Agricultural management impacts on soil organic carbon storage under moist and dry climatic conditions of temperate and tropical regions. Biogeochemistry 72, 87–121.
Ogle, S.M., Swan, A., Paustian, K., 2012. No-till management impacts on crop productivity, carbon input and soil carbon sequestration. Agric. Ecosyst. Environ. 149, 37–49.
Ogle, S.M., Adler, P.R., Breidt, F.J., Del Grosso, S.J., Derner, J.D., Franzluebbers, A.J., et al., 2014. Chapter 3: quantifing greenhouse gas sources and sinks in cropland and grazing land systems. In: Eve, M., Pape, D., Flugge, M., Steele, R., Man, D., Riley-Gilbert, M., Bigger, S. (Eds.), Quantifying Greenhouse Gas Fluxes in Agriculture and Forestery: Methods for Entity-Scale Inventory. Office of the Chief Economist, Washington DC, pp. 606.
Olson, K.R., 2013. Soil organic carbon sequestration, storage, retention and loss in U.S. croplands: issues paper for protocol development. Geoderma 195–196, 201–206.
Omonode, R.A., Vyn, T.J., 2013. Nitrification kinetics and nitrous oxide emissions when nitrapyrin is coapplied with urea–ammonium nitrate. Agron. J. 105, 1475–1486.
Parkin, T.B., Hatfield, J.L., 2010. Influence of nitrapyrin on N_2O losses from soil receiving fall-applied anhydrous ammonia. Agric. Ecosyst. Environ. 136, 81–86.
Parkin, T.B., Hatfield, J.L., 2014. Enhanced efficiency fertilizers: effect on nitrous oxide emissions in Iowa. Agron. J. 106, 694–702.
Pathak, H., Bhatia, A., Prasad, S., Singh, S., Kumar, S., Jain, M.C., et al., 2002. Emission of nitrous oxide from rice-wheat systems of Indo-Gangetic plains of India. Environ. Monit. Assess. 77, 163–178.
Pellerin, S., Bamiere, L., Anger, D., Beline, F., Benoit, M., Butault, J.P., et al., 2013. How can French agriculture contribute to reducing greenhouse gas emissions? Abatement potential and cost of ten technical measures. Summary of the study report. INRA, France, p. 92.
Pfaff, A., Walker, R., 2010. Regional interdependence and forest "transitions": substitute deforestation limits the relevance of local reversals. Land Use Policy 27, 119–129.
Phillips, R.L., Tanaka, D.L., Archer, D.W., Hanson, J.D., 2009. Fertilizer application timing influences greenhouse gas fluxes over a growing season. J. Environ. Qual. 38, 1569–1579.
Piva, J.T., Dieckow, J., Bayer, C., Zanatta, J.A., de Moraes, A., Pauletti, V., et al., 2012. No-till reduces global warming potential in a subtropical Ferralsol. Plant Soil 361, 359–373.
Poeplau, C., Don, A., 2015. Carbon sequestration in agricultural soils via cultivation of cover crops - a meta-analysis. Agric. Ecosyst. Environ. 200, 33–41.
Pramanik, P., Haque, M.M., Kim, P.J., 2013. Effect of nodule formation in roots of hairy vetch (*Vicia villosa*) on methane and nitrous oxide emissions during succeeding rice cultivation. Agric. Ecosyst. Environ. 178, 51–56.
Rochette, P., 2008. No-till only increases N_2O emissions in poorly-aerated soils. Soil Till. Res. 101, 97–100.
Ruser, R., Schulz, R., 2015. The effect of nitrification inhibitors on the nitrous oxide (N_2O) release from agricultural soils—a review. J. Plant Nutr. Soil Sci. 178, 171–188.
Ruttan, V.W., 1982. Agricultural Research Policy. University of Minnesota Press, Minneapolis, MN.
Sainju, U.M., Barsotti, J.L., Wang, J., 2014. Net global warming potential and greenhouse gas intensity affected by cropping sequence and nitrogen fertilization. Soil Sci. Soc. Am. J. 78, 248–261.
Sanz-Cobena, A., Sánchez-Martín, L., García-Torres, L., Vallejo, A., 2012. Gaseous emissions of N_2O and NO and NO^{-}_3 leaching from urea applied with urease and nitrification inhibitors to a maize (*Zea mays*) crop. Agric. Ecosyst. Environ. 149, 64–73.
Scharf, P.C., Kitchen, N.R., Sudduth, K.A., Davis, J.G., Hubbard, V.C., Lory, J.A., 2005. Field-scale variability in optimal nitrogen fertilizer rate for corn. Agron. J. 97, 452–461.

Sehy, U., Ruser, R., Munch, J.C., 2003. Nitrous oxide fluxes from maize fields: relationship to yield, site-specific fertilization, and soil conditions. Agric. Ecosyst. Environ. 99, 97–111.

Sherrod, L.A., Peterson, G.A., Westfall, D.G., Ahuja, L.R., 2003. Cropping intensity enhances soil organic carbon and nitrogen in a no-till agroecosystem. Soil Sci. Soc. Am. J. 67, 1533–1543.

Sherrod, L.A., Peterson, G.A., Westfall, D.G., Ahuja, L.R., 2005. Soil organic carbon pools after 12 years in no-till dryland agroecosystems. Soil Sci. Soc. Am. J. 69, 1600–1608.

Sistani, K.R., Jn-Baptiste, M., Lovanh, N., Cook, K.L., 2011. Atmospheric emissions of nitrous oxide, methane, and carbon dioxide from different nitrogen fertilizers. J. Environ. Qual. 40, 1797–1805.

Subbarao, G.V., Rondon, M., Ito, O., Ishikawa, T., Rao, I.M., Nakahara, K., et al., 2007. Biological nitrification inhibition (BNI) - is it a widespread phenomenon? Plant Soil. 294, 5–18.

Subbarao, G.V., Nakahara, K., Hurtado, M.P., Ono, H., Moreta, D.E., Salcedo, A.F., et al., 2009. Evidence for biological nitrification inhibition in Brachiaria pastures. Proc. Natl. Acad. Sci. 106, 17302–17307.

Trost, B., Prochnow, A., Drastic, D., Meyer-Aurich, A., Ellmer, F., Baumecker, M., 2013. Irrigation, soil organic carbon and N_2O emissions. A review. Agron. Sustain. Dev. 33, 733–749.

Tubiello, F.N., Salvatore, M., Ferrara, A.F., House, J., Federici, S., Rossi, S., et al., 2015. The contribution of agriculture, forestry and other land use activities to global warming, 1990–2012. Glob. Change Biol. 21, 2655–2660.

UNEP, 2015. United Nations, Department of Economic and Social Affairs, Population Division, World Population Prospects: The 2015 Revision, Key Findings and Advance Tables.

van Groenigen, J.W., Velthof, G.L., Oenema, O., Van Groenigen, K.J., Van Kessel, C., 2010. Towards an agronomic assessment of N_2O emissions: a case study for arable crops. Eur. J. Soil Sci. 61, 903–913.

van Kessel, C., Venterea, R., Six, J., Adviento-Borbe, M.A., Linquist, B., van Groenigen, K.J., 2013. Climate, duration, and N placement determine N_2O emissions in reduced tillage systems: a meta-analysis. Glob. Change Biol. 19, 33–44.

Venterea, R.T., Bijesh, M., Dolan, M.S., 2011. Fertilizer source and tillage effects on yield-scaled nitrous oxide emissions in a corn cropping system. J. Environ. Qual. 40, 1521–1531.

Venterea, R.T., Halvorson, A., Kitchen, N., Liebig, M., Cavigelli, M., Grosso, S.D., et al., 2012. Technical challenges and opportunities for mitigating nitrous oxide emissions in fertilized cropping systems. Front. Ecol. 10, 562–570.

West, T.O., Post, W.M., 2002. Soil organic carbon sequestration rates by tillage and crop rotation: a global data analysis. Soil Sci. Soc. Am. J. 66, 1930–1946.

Wilson, M.L., Allan, D.L., Baker, J.M., 2014. Aerially seeding cover crops in the northern US Corn Belt: Limitations, future research needs, and alternative practices. J. Soil Water Conserv. 69, 67A–72A.

Yang, X., Drury, C.F., Wander, M.M., 2013. Wide view of no-tillage practices and soil organic carbon sequestration. Acta Agric. Scand. 63, 523–530.

Zhang, X., Zhou, Z., Liu, Y., Xu, X., Wang, J., Zhang, H., et al., 2015. Net global warming potential and greenhouse gas intensity in rice agriculture driven by high yields and nitrogen use eficiency: a 5 year field study. Biogeosci. Discuss. 12, 18883–18911.

Zhang, X.X., Fan, C.H., Ma, Y.C., Liu, Y.L., Li, L., Zhou, Q.S., et al., 2014a. Two approaches for net ecosystem carbon budgets and soil carbon sequestration in a rice-wheat rotation system in China. Nutr. Cycl. Agroecosyst. 100, 301–313.

Zhang, Y., Mu, Y., Zhou, Y., Liu, J., Zhang, C., 2014b. Nitrous oxide emissions from maize–wheat field during 4 successive years in the North China Plain. Biogeosciences 11, 1717–1726.

CHAPTER 13

Low-Input and Intensified Crop Production Systems Effects on Soil Health and Environment

Sharon L. Weyers[1] and Greta Gramig[2]

[1]USDA - ARS, North Central Soil Conservation Research Lab, Morris, MN, United States
[2]North Dakota State University, Fargo, ND, United States

13.1 Introduction

"Low-input" is a term that came into use in the early 1990s. As suggested in two publications released by the Soil and Water Conservation Society (Edwards et al., 1990; Schnepf, 1990), "low-input agriculture" is an concept that stems from Rachel Carson's 1962 publication of Silent Spring, the 1970s energy crisis, or as counter to the "Green Revolution." All three events have imbued the public with a sense of responsibility for their health and wellbeing. In the United States, this decade also coincided with new agricultural policies, and the development of the Low-Input Sustainable Agriculture Program, "LISA," (O'Connell, 1990), which is now the Sustainable Agriculture Research and Education program, "SARE." According to Ikerd (1990), "low-input" has "no universally accepted definition," though the term invokes an implied meaning of sustainable, which is the interlinking of environmentally sound, socially acceptable, and economically viable production. As an ideal, "low-input agriculture" embraces a wide range of practices that produce marketable farm products while employing a reduced level of external inputs. As a management system, "low-input agriculture" likely requires definition at the farm level.

"Intensified agriculture" stands in contrast to low-input. "Intensified agriculture," however, evokes more of a process, rather than an explicit method of production. Over millennia a variety of technological advances in agricultural practices have led to intensification of agriculture. In this sense, intensification of agriculture can be defined as "increasing productivity on a set area of land." Intensification is distinguished from "extensification," which can be defined as "increasing productivity by increasing land area under production." With intensification, genetic as well as chemical advances in agricultural technologies have led to both stabilization and destabilization of the biological and chemical nature of soils

Soil Health and Intensification of Agroecosystems.
DOI: http://dx.doi.org/10.1016/B978-0-12-805317-1.00013-0

(Matson et al., 1997). With extensification, conversion of natural areas, which continues to be pervasive, particularly in the tropics and developing nations, has more recently led to environmental degradation (Lambin and Meyfroidt, 2011; Garnett et al., 2013).

Society's repudiation of "extensification," as well as potential destabilization resulting from intensification, is driving research and policy efforts increasingly toward development of "sustainable intensification" practices (Tilman et al., 2011; Garnett et al., 2013). As with low-input, no explicit defining practice for sustainable intensification exists, thus allowing room for the development of "ecological intensification" theory (Peterson and Snapp, 2015). As the implicit goal of intensification is to increase productivity, primarily through increasing yields (Garnett et al., 2013), the implication of sustainable intensification is to do so, with environmentally sound, socially acceptable, and economically viable practices. Regardless of using "sustainable" or "ecological" terminology, the approach clearly embraces concepts of agroecological theory (Dore et al., 2011).

Agroecology emerged as a scientific discipline in the 1930s. With growing understanding of the agricultural system as an ecosystem, agroecological theory evolved into a wide assemblage of management approaches through the 1970s and 1980s (Wezel et al., 2009, 2014). Agroecological management developed concurrently with organic agriculture, which originated but eventually branched out from "biodynamic agriculture" (Harwood, 1990). Proponents of the agroecological approach to managed production have outlined numerous guiding principles for achieving sustainability (Gliessman, 1998). To this day, however, only organic agriculture has developed into well-formulated and highly regulated practices that can be certified to obtain price premiums.

The terms "sustainable," "conservation," "agroecosystem," "eco-agriculture," "low-input," and so forth have apparently been used interchangeably to describe management systems that use components derived from agroecological theory, or the so-called guiding principles. The many forms or practical applications of sustainable, agroecological practices have one clear defining goal, which is to develop a system capable of supporting the internal cycling and supply of nutrients, providing pest and disease control, reducing fuel and labor costs, and maintaining the economic viability of the farmstead by supporting the export of a farm commodity. However, Rosset and Altieri (1997) argue against the sustainability of approaches that use "input-substitution" rather than true agroecology to manage production systems that are simply a succession of monocultures still dependent to an extent on commercially available inputs, such as organic agriculture.

Prior to modern agricultural technology and inputs, traditional agricultural systems were often marked with a long track record of providing for local needs when operating independently from external inputs by using locally available and renewable resources, e.g., livestock manures, and emphasizing nutrient recycling (Gliessman, 1998). Other internal resources come from the "natural capital," which includes naturally captured inputs such as N fixation and intrinsic properties of a natural system such as biological control. To increase yields premodern systems also relied on extensification, i.e., increasing land base. On a stable land base, premodern systems may, in theory, be seen as evolving in two opposing directions to meet increased yield demands, either toward greater use of external inputs or greater use of captured inputs (Fig. 13.1). In one direction are production systems that exhibit increasing intensification by increasing reliance on and use of external inputs (Fig. 13.1A). However, these systems might tend to degrade the soil resource base (Matson et al., 1997). In the opposing direction are production systems that exhibit increasing biological efficiency by enhancing their natural capital (Fig. 13.1B). Natural capital is gained by building a strong soil resource base, such as increasing soil organic matter (SOM), enhancing biological diversity, generating internal capacity to fix N, and increasing nutrient cycling efficiency.

The objective of this chapter is to use this theoretical framework to characterize and contrast multiple agricultural production systems evaluated through research or practical case studies. These systems will be evaluated on their ability to support increased yields and benefit soil health and ecological services. Research case studies presented provide comparisons along gradients of low to high use of external inputs with practices labeled, e.g., as low-input, organic, and conventional. On-farm or practical case studies demonstrate the practical application and benefits of adaptive approaches. Using this theoretical framework, conclusions will be drawn on the practicality of developing "sustainable and intensified" production systems that can meet increased yield demand and still support soil health.

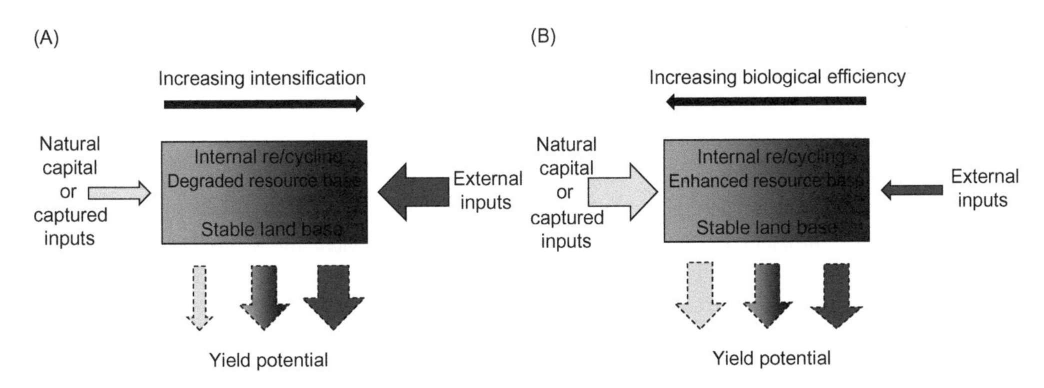

Figure 13.1

Theoretical framework for the two directions that modern agriculture has evolved toward. (A) Agricultural system moving toward intensification through the increased use of external inputs. (B) Agricultural system moving toward biological efficiency by limiting use of external inputs, generating natural capital, and benefiting from captured inputs. Yield potential reflects the expectations along an input-gradient from captured to external inputs.

13.2 Research Case Studies Addressing Impact of Low-Input Systems on Soil Health and Crop Yields

This section details a variety of long-term experiments with 10 or more years of data on crop yields and soil health measures (Table 13.1). These experiments were chosen based on publicly accessible databases, availability of detailed production and management inputs and outcomes, and within-trial comparisons of highly contrasting approaches. They were also chosen to

Table 13.1: Brief overview of long-term research trial case studies using input gradients in production system comparisons, giving location, soil types, annual rainfall and temperature, and production systems under evaluation

Case Study	Soils, Rainfall, Annual Temperature	Production System Designations	Notes and Other Information
Broadbalk winter wheat trial 51°48′ N, 0°22′ W	Aquic Paleudalf 700 mm 10°C	Nil (no amendments) FYM[a] FYM + Nx Nx + PK ± Mg where x = 0, 45, 96, 192, 288 kg N ha^{-1}	The main treatment structure is based on N application; details on the sectioning of fertility treatments are given in the text. Crop is mainly continuous wheat; wheat is also in rotation with corn, potato, and beans http://www.rothamsted.ac.uk/
Biodynamic-Organic-Conventional (DOK) trial 47°30′ N, 7°33′ E	Typic Hapludalf 785 mm 9.5°C	CNVMIN[b] CNVFYM BIOORG BIODYN	The CNVMIN is split into a no-fertilizer and high-fertilizer treatment; other treatments are split into low N and high N fertility based on livestock units of 0.7 and 1.4, respectively. Rotation is 7 years and includes grain, forage, and root crops as detailed in the text. http://www.fibl.org/en/switzerland/research/soil-sciences/bw-projekte/dok-trial.html
Sustainable agricultural farming systems (SAFS) 38°32′ N, 121°47′ W	Coarse-loamy, mixed, nonacid, thermic Mollic Xerofluvents 470 mm 15.6°C	Conv-2 Conv-4 Low-input Organic	Mineral fertilizers are used in the Conv-2, Conv-4 and to a limited extent the low-input; manure, compost, and green covers are used in the low-input and organic systems. Rotations include a vegetable and grain; Conv-2 is a 2-year rotation; the others are 4-year rotations. All treatments are irrigated. http://169.237.124.147:8081/ASI/programs/rr

[a]FYM, indicates farmyard manure.
[b]CNV MIN, conventional with mineral N, CNVFYM, conventional with farmyard manure, BIOORG, bioorganic, BIODYN, biodynamic.

cover a diversity of approaches used to implement low-input production, the diversity of commodities produced, and global range. A brief overview of the management structure of the experiment is provided for each case study. The impact of this management structure on yield and soil health is then discussed. Soil health metrics of focus in all studies were SOM, soil organic C (SOC), soil nitrogen (N), and microbial biomass C (MBC) (Table 13.1). Additional aspects involving soil health or environmental benefits are addressed where data are available.

13.2.1 The Broadbalk Trial, Rothamsted Research, Harpenden, UK

The Rothamsted's Broadbalk winter wheat (*Triticum aestivum* L.) trial is one of the oldest continuously monitored experimental agricultural systems in the world, with scientific studies established and soil samples archived since the early 1840s (Rothamsted Research, 2012). The Broadbalk was initiated in a continuous wheat production system with treatment strips, 6m × 320m, initially varying by use of farmyard manure (FYM), and combinations of inorganic N, P, potassium (K), sodium (Na), and magnesium (Mg) (Fig. 13.2). Inorganic N has been applied at rates of 0, 48, 96, 144, 192, and 288kg N ha^{-1} $year^{-1}$. The FYM has been applied and incorporated with tillage each fall at a consistent application rate of 35t ha^{-1}, typically supplying 248kg N ha^{-1} and 35kgP ha^{-1} (Ogilvie et al., 2008). In 1926, treatment strips were separated into five sections to evaluate a 5-year fallow cycle on weed control. In 1968, sections were further subdivided to compare continuous wheat (sections 0, 1, and 9), continuous wheat without herbicides (section 8), wheat in rotation with spring beans (*Phaseolus vulgaris* L.) and potato (*Solanum tuberosum* L.) (sections 2, 4, and 7), and 2 years of wheat with 1 year of fallow (sections 3, 5, and 6) (Dyke et al., 1983). More recently forage corn (*Zea mays* L.) and oats (*Avena sativa* L.) have replaced beans and potato in rotation; fallow is used occasionally, fungicide use is restricted in section 6; and wheat straw is retained (incorporated) in section 0, but harvested elsewhere (Fig. 13.2; Watts et al., 2006; Rothamsted Research, 2012). Uncultivated pathways separate the 10 sections, and individual plots, 6m × 23m, within these sections lay side-by-side.

Since inception, wheat yields in the Broadbalk trial in all but the unfertilized treatment have increased over the years, clearly demonstrating the impact of agricultural intensification from technological advances (Fig. 13.3). The greatest gains in wheat yields were attributed to breeding improvements, with use of modern cultivars beginning in 1967. Average grain yields of the Cappelle wheat cultivar grown 1969–78 were greater when grown in rotation and using only 96kg N ha^{-1} than when grown continuously, even with using 144kg N ha^{-1}, and were 6t ha^{-1} and 5t ha^{-1}, respectively (Johnston, 1997). These findings were similar for other cultivars, Flander grown from 1978–84 (Johnston, 1997) and Hereward grown between 1996 and 2005 (Rothamsted Research, 2012). Introduction of herbicides in 1964 and fungicides in 1978 to control weeds and diseases, also had positive impacts on yields (Johnston, 1997). Crop rotation management, instituted in 1968, was another aspect of technological changes that improved wheat yields in the Broadbalk trial by moderating both weed and disease pressure (Johnston, 1997; Fig. 13.3.)

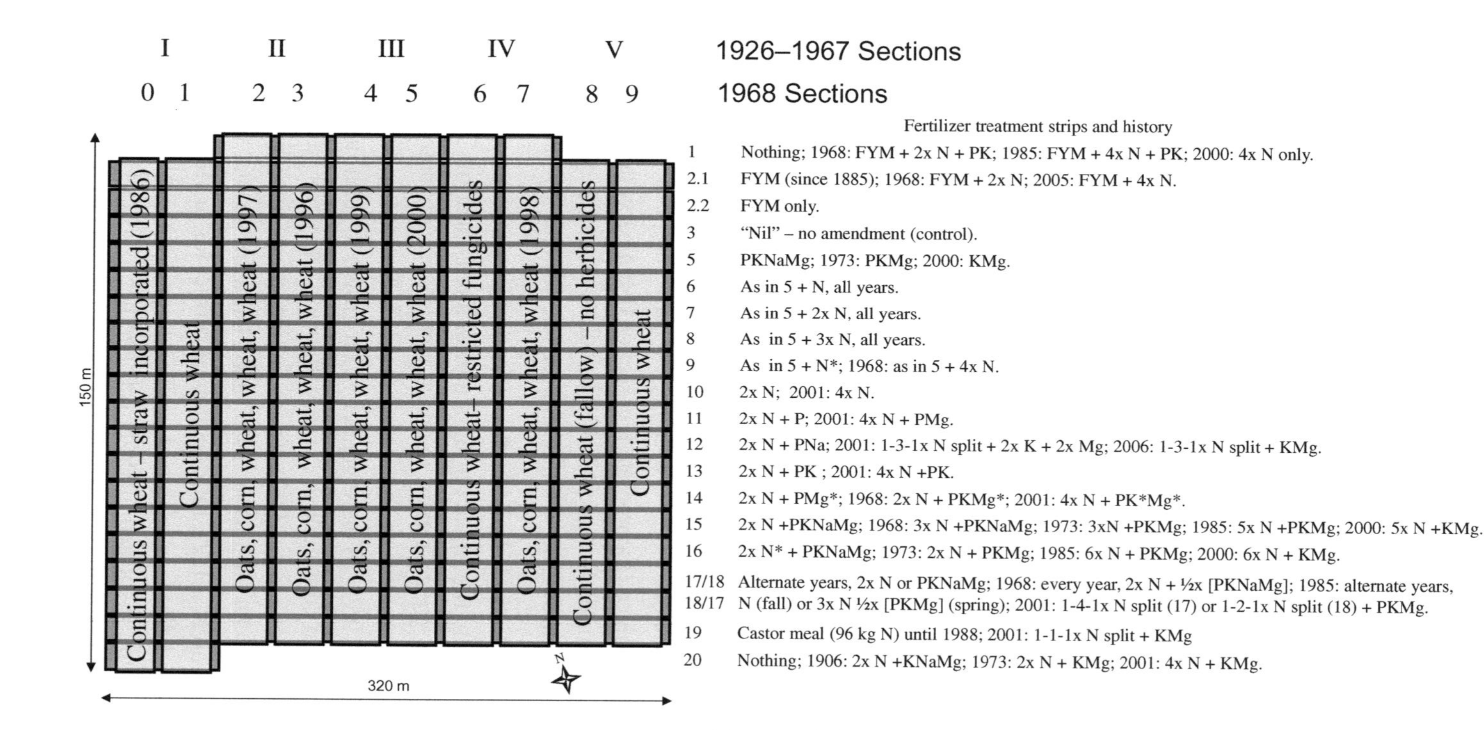

Figure 13.2

Treatment structure of the Broadbalk winter wheat experiment, describing the section separations (north to south) made in 1926 and again in 1968, and a simplified treatment history of the 20 fertility treatment strips (east to west), since 1852 with changes noted by year. *Treatments*: FYM = 35 t farmyard manure ha^{-1}. N = 48 kg ammonium sulfate-N ha^{-1}, or 2, 3, 4, 5 or 6 × N; N* = 48 kg N ha^{-1} as sodium nitrate, as calcium ammonium nitrate (1968), then as ammonium nitrate (1985); N split = mid-March, mid-April and mid-May application. P = 35 kg triple superphosphate-P ha^{-1}. K = 90 kg potassium sulfate-K ha^{-1}, or 2 × K (2001–2005). K* = 90 kg potassium chloride-K ha^{-1}. Mg = 11 kg magnesium sulfate-Mg ha^{-1} or 2 × Mg, then 12 kg Kieserite-Mg ha^{-1} (1972), with 35 kg Kieserite-Mg ha^{-1} every third year (from 1974–2000). Mg* = 31 kg Kieserite-Mg ha^{-1}, then 30 kg Kieserite-Mg ha^{-1} (1974–2000). Na = 16 kg sodium sulfate-Na ha^{-1}; for strip 12, Na = 57 kg ha^{-1}, then 55 kg ha^{-1} (1973–2000). Note: Details combined for strips 17 and 18; treatments applied in alternate years, starting even years with 2 × N in strip 17 and PKNaMg in strip 18. *Diagram and information adapted from Rothamsted Research, 2012. Guide to the Classical and Other Long-Term Experiments, Datasets and Sample Archive. Lawes Agricultural Trust Co. Ltd. Online at: <http:www.rothamsted.ac.uk> and Watts, C.W., Clark, L.J., Poulton, P.R., Powlson, D.S., Whitmore, A.P., 2006. The role of clay, organic carbon and long-term management on mouldboard plough draught measured on the Broadbalk wheat experiment at Rothamsted. Soil Use Manag. 22, 334–341.*

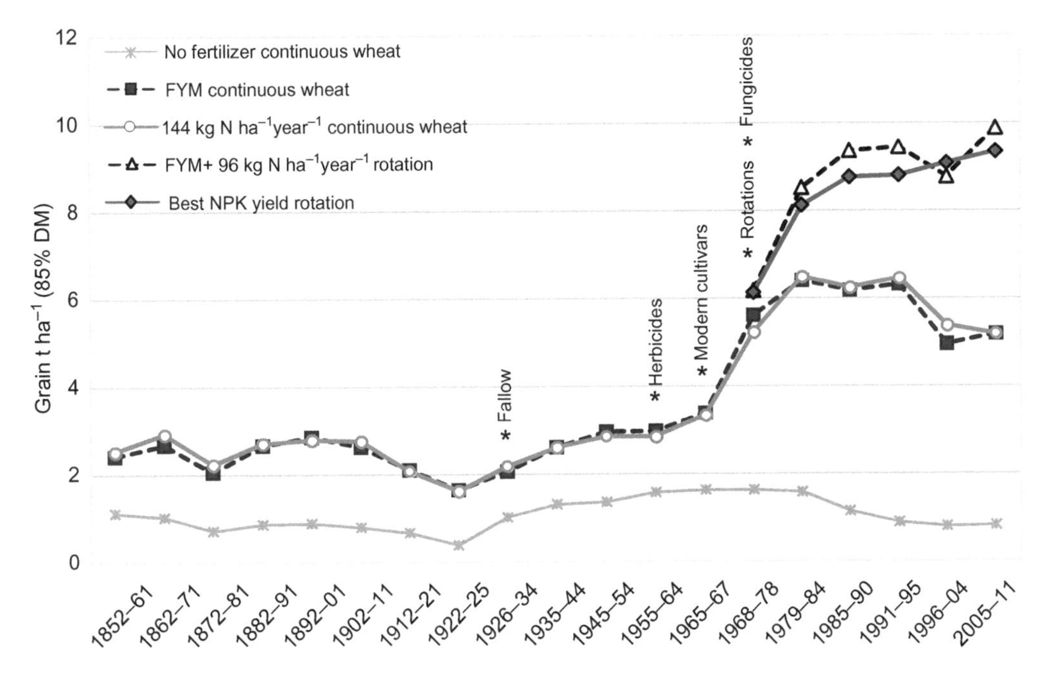

Figure 13.3

Average wheat grain yields over the long-term Broadbalk Trial from selected fertilizer treatments in continuous and rotation wheat. Asterisks (*) indicate periods when new management practices were instituted (see text). *Modified from Rothamsted Research, 2012. Guide to the Classical and Other Long-Term Experiments, Datasets and Sample Archive. Lawes Agricultural Trust Co. Ltd. Online at: <http:www.rothamsted.ac.uk>.*

In recent years, yields achieved with FYM were surpassed by yields obtained with FYM plus an additional 96 kg N ha^{-1} (increased in 2005 to 144 kg N ha^{-1}), as well as yields obtained using higher levels of inorganic N, ranging 96 to 288 kg N ha^{-1} depending on cultivar (Fig. 13.3). However, from 1985 to 1990 wheat grain yields for inorganic and FYM treatments were different by only 5%, respectively averaging 8.3 and 7.9 t ha^{-1} (Edmeades, 2003). In this study, Edmeades (2003) found comparable yield gaps between inorganically fertilized or manure-fertilized treatments that were deemed insignificant. Despite this 5% yield gap, using FYM had a substantial positive impact on SOM content in the surface 0–23 cm (Watts et al., 2006; Powlson et al., 2012). Further, combining the use of fertilizer with FYM seems to have had an even greater impact on SOC in comparison to just FYM (Table 13.2). However, examining the relative gain in SOC between the 1893 baseline and the 2010 data, FYM-only treatments had a substantially greater net increase in SOC, increasing from 1.55% to 2.89%, than FYM with added fertilizer, increasing from 2.15% to 2.97% (Table 13.2).

Total soil N showed similar patterns and changes to SOC, increasing over time with greater increase in the FYM treatments (Table 13.2). Labeled ^{15}N experiments (Shen et al., 1989)

Table 13.2: Soil organic C (SOC), total soil N (N), microbial biomass C (MBC), and microbial biomass P (MBP) for select dates and treatments from the Broadbalk winter wheat trial at Rothamsted

		Treatment or Inorganic N Rate kg N ha^{-1} (Treatment Number)						
Parameter	**Year**	**Control (0)**	**FYM (2.1)**	**FYM + 96/144[a] (2.2)**	**144 (8)**	**240 (15)**	**288 (16)**	**Citation and Notes on Sampling Location**
SOC %	1893	0.78	1.55	2.15	1.07			Rothamsted Research (2012); except 1893, values are means of samples from sections 1, 6, and 9
	1966	0.87	2.28	2.61	1.08			
	1992	0.77	2.67	2.68	1.07			
	2000	0.84	2.72	2.80	1.08	1.14	1.19	Watts et al. (2006)
	2010	0.85	2.89	2.97	1.04			Rothamsted Research (2012); means sections 1, 6, and 9
N %	1893			0.222	0.119			Glendining et al. (1996); Glendining and Powlson (1990); means all sections
	1966			0.254	0.121			Ibid.; means all sections
	1987			0.270	0.126			Ibid.; means section 1 and 9
	1998	0.098	0.246	0.234	0.114	0.117	0.114	Blair et al. (2006); means sections 0 and 1
MBC μg g^{-1}	1976	102 ± 7.6		236 ± 7.6	111 ± 7.6			Jenkinson and Powlson (1976)
	1982	158		342	190			Brooks et al. (1984)
	2006				174 ± 8	264 ± 17		De Nobili et al. (2006); plot 8, section 9; plot 15, section 0, top,
						184 ± 11		and plot 15, section 1 bottom
MBP μg g^{-1}	1982	6.0		28.9	5.3			Brooks et al. (1982, 1984)

[a]Treatment 2.2 received FYM and 96 kg N ha^{-1} through 2004, thereafter received FYM and 144 kg N ha^{-1}.

indicated that the N gains in inorganically fertilized treatments were derived directly from the fertilizer. Increased soil N contributed to greater N mineralization rates in inorganically fertilized treatments, but greater soil mineral N content at harvest in FYM treatments (Glendining and Powlson, 1990; Glendining et al., 1996). Blair et al. (2006) correlated improvements in soil aggregation and unsaturated hydraulic activity with increasing soil content of both C and N. However, increased N mineralization and availability resulted in substantial leaching losses of nitrate (Goulding et al., 2000). Application of phosphorus (P) fertilizers also contributed to an increase in P leaching (Heckrath et al., 1995). Though autumn FYM application led to greater N loss, P loss was low (Heckrath et al., 1995; Goulding et al., 2000).

Microbial biomass, activity, and diversity have been measured in several of the Broadbalk management systems (Table 13.2). Jenkinson and Powlson (1976) reported a significantly greater microbial biomass in continuous wheat, section 1, in the FYM treatment compared to the nonfertilized and 144 kg N ha^{-1} treatments. These samples were taken shortly after harvest, and were only slightly greater within respective treatments than biomass measured in section 3 of continuous wheat during a fallow stage. Six years later, Brookes et al. (1982, 1984) reported a similar finding with MBC as well as microbial biomass P in these same treatments and sections. De Nobili et al. (2006) evaluated microbial biomass in the 240 kg N ha^{-1} treatment, finding that straw incorporation in section 0 substantially increased biomass compared to straw removal in section 1.

Functional gene diversity, investigated by Ogilvie et al. (2008), was qualitatively and quantitatively similar among the nonfertilized, FYM, and 240 kg N ha^{-1} treatments. Clark et al. (2012) performed a more detailed analysis on denitrification processing genes in the FYM and 0, 144, and 288 kg N ha^{-1} treatments. Though Clark et al. (2012) did not find differences in total gene abundance or diversity among these treatments, they did find qualitative differences in community structure of genes. The *nirS* genes, coding for nitrate reductase, were greater in plots with the 0 or 144 kg N ha^{-1}, whereas *nirK*, an alternate nitrate reductase gene, was greater in the 288 kg N ha^{-1} and FYM treatments. The *nosZ* gene, coding for nitrous oxide reductase which converts N_2O into N_2, had the highest, but not different, number of gene copies in the 288 kg N ha^{-1}, followed by the FYM. In addition, NO_3-substrate-induced N_2O emission was five times greater in the FYM than the other N treatments. The greater flux of N_2O observed in the FYM was positively correlated with increased *nirK* gene copies, but decreased *nosZ* gene copies, as well as decreased bulk density and increased soil C and soil N that might have enhanced the water-holding capacity in this treatment.

13.2.2 The "DOK" Trial, Agroscope Reckenholz-Tänikon Research Station and the Research Institute of Organic Agriculture (FiBL), Therwil, Switzerland

The DOK trial was established in 1978 to compare three main types of management systems, biodynamic (D), organic (O), and conventional (K, for "konventionell"), with conventional divided into one manure-based (CONFYM) and one mineral-based (CONMIN) fertilization system. Each of these four management systems was also split into a reduced and a normal fertility level. In the CONMIN system reduced fertility was at zero (NOFERT). However the other systems had a reduced fertility level at an N equivalent of 0.7 livestock units, and normal fertility level at an N equivalent of 1.4 livestock units, with average inputs for organic C, total N, soluble N, P, and K varying by the fertility source (Table 13.3). Plot size was 5m by 20m.

Seven-year rotations were used, with some shifts in crops each full rotation phase (Table 13.3). Phase one, 1978–84, was potato, winter wheat, cabbage (*Brassica oleracea* L.), winter wheat, winter barley (*Hordeum vulgare* L.) and 2 years of grass-clover-ley (*Trifolium pretense* L., *T. repens* L., *Dactylis glomerata* L., *Festuca rubra* L., *Phleum pratense* L., and *Lolium perennne* L.); green manure and fodder crops were grown after harvest of potato and first winter wheat (Fließbach et al., 2007). Phase two, 1985–91, replaced cabbage with beetroot (*Beta vulgaris*). Phase three, 1991–98, introduced a sunflower/vetch (*Helianthus* sp./*Vici* sp.) catch crop in place of fodder after first winter wheat, removed barley for a 3-year cycle of grass-clover-ley, and added *Poa pratensis* L. to the grass-clover mix (Fließbach et al., 2007). Phase four began in 1999 with potato, followed by soybean (*Glycine max*), silage corn, winter wheat, and grass-clover-ley (Espershültz et al., 2007; Hildermann et al., 2010). Generally, plant extracts, bio-controls, and mechanical controls were used for weed, disease, and insect pest control in BIODYN and BIOORG, and NOFERT, and mechanical approaches in addition to chemical pesticides and plant growth regulators were used in CONFYM and CONMIN. The BIODYN system was differentiated from the BIOORG by the use of specialized biodynamic preparations.

By the third phase in rotation, crop yield in biodynamic and organic management systems were relatively comparable for wheat and grass-clover production (Table 13.3). The occurrence of *Phytophthora infestans* compromised potato yields in the BIOORG (Mäder et al., 2002). Despite the slight reduction, overall energy use was lower, 12.8 and 13.3 GJ ha^{-1} $year^{-1}$, in BIODYN and BIOORG, respectively, compared to 20.9 and 24.1 GJ ha^{-1} $year^{-1}$ for CONFYM and CONMIN, respectively (Mäder et al., 2002). Mäder et al. (2002) suggested this might lead to greater profitability of lower-input systems, but clarified organic farms in Europe typically generate the same profits as conventional farms. Food quality assessment of the winter wheat indicated that nutritional value did not differ among the different farming systems (Mäder et al., 2007). Thus, lower energy use and higher nutrient use efficiency found in the organic and biodynamic systems provided the same quality of wheat achieved with the other conventional approaches.

Soil health aspects were quite remarkably better under the alternative than the conventional production systems (Mäder et al., 2002; Birkhofer et al., 2008). The BIODYN and BIOORG systems had better SOC and MBC (Table 13.4), as well as aggregate stability, soil percolation, enzyme activity, and faunal abundance (not shown) than in the CONFYM, which

Table 13.3: Treatment structure used on the DOK experiment by system, fertility levels, yearly average amendment rates from 1978 to 2005 for the High-fertility level treatments, and the crops used in each 7-year rotation phase

	BIODYN[a]		BIOORG		CNVFYM		NOFERT	CNVMIN
Fertility levels livestock unit (l.u.) equivalent								
	0.7	1.4	0.7	1.4	0.7	1.4	0	1.4
Source	Compost FYM + slurry		Rotted FYM + slurry		Stacked FYM + slurry		-	Mineral[b]
Amendment rate[c] (1.4 l.u.) kg ha^{-1} $year^{-1}$								
Organic C	1818		2272		2272		-	-
Total N	99		102		157		-	122
Mineral N	31		35		101		-	122
P	23		27		41		-	40
K	165		157		258		-	250

Phases	Crop Rotations						
	Year 1	Year 2	Year 3	Year 4	Year 5	Year 6	Year 7
1978–1984	Potato, green manure	Winter wheat, fodder crop	White cabbage	Winter wheat	Winter barley	Grass-clover-ley	Grass-clover-ley
1985–1991	Potato, green manure	Winter wheat, fodder crop	Beetroot	Winter wheat	Winter barley	Grass-clover-ley	Grass-clover-ley
1992–1998	Potato	Winter wheat, sunflower/vetch	Beetroot	Winter wheat	Grass-clover-ley	Grass-clover-ley	Grass-clover-ley
1999–2005	Potato	Soybean	Silage corn	Winter wheat	Grass-clover-ley	Grass-clover-ley	Grass-clover-ley

[a]BIODYN, biodynamic; BIOORG, bioorganic; CNVFYM, conventional management with farmyard manure; NOFERT, no fertility treatment as part of the CNVMIN, conventional management with mineral fertilizers.
[b]No fertilizers were applied from 1978 to 1984.
[c]Data from Birkhofer et al. (2008).

Table 13.4: Approximate yields in 1998, average change in Soil C 1992–98 compared to 1977 baseline, and microbial biomass C (MBC) in 1998 for the DOK trial

	Yields[a]			Soil Parameters[b]		
	Wheat	Potato	Grass-Clover	Change C[b]	MBC μg g^{-1}	
	Grain t ha^{-1}	Tuber t ha^{-1}	t dry matter ha^{-1}	%	1998[b]	2005[c]
BIODYN	4.1	32	12	1	360	460
BIOORG	4.1	32	11	−9	313	430
CNVFYM	4.6	51	15	−7	267	305
CNVMIN	4.6	48	12.8	−15	218	230
NOFERT	–[d]	–	–	−22	235	–

[a]Mäder et al. (2002).
[b]Fließbach et al. (2007).
[c]Birkhofer et al. (2008).
[d]Data not available.

was better in these regards than the CONMIN. Assessment of the microbial community with phospholipid fatty acid (PLFA) analysis indicated the diversity of PLFA was higher in the BIOORG and CONFYM than the BIODYN system, but the community was the same in the BIOORG and BIODYN (Espershültz et al., 2007). Arbuscular mycorrhizal root colonization of bioassay plants was better in soils taken from the BIODYN and BIOORG systems than CONFYM and CONMIN (Mäder et al., 2000; Hildermann et al., 2010).

13.2.3 Sustainable Agriculture Farming Systems, SAFS, Russell Ranch, UC Davis, CA, United States

The SAFS was established in 1988 nearby the University of California, Davis campus, primarily as a means to explore the effects of transitioning from conventional into low-input or organic management practices (Temple et al., 1994a; Poudel et al., 2001a). The traditional management practice in the region at the time was compared to three alternative management systems. The traditional, "conventionally managed" system was a 2-year rotation of processing tomato (*Lycopersicon esculentum* L.) and wheat (Conv-2). An alternative conventional system used a 4-year rotation, which included canning tomato, safflower (*Carthamus tinctorius* L.), corn, and winter wheat double cropped with beans (*Phaseolus vulgaris* L.) (Conv-4). The third and fourth systems were managed as "organic" and as "low-input." The organic and low-input management systems also had a 4-year rotation of tomato, safflower, corn, and bean, which included various summer and winter cover crops and mixtures of sorghum sudan (hybrid *Sorghum bicolor* (L.) Moench/*Sorghum sudanense*), cowpea (*Vigna unguiculata* L.), lablab (*Dolichos lablab* L.), oats (*Avena sativa* L.), vetch (*Vicia* spp.), pea (*Pisum* sp.), and lupin (*Lupinus* spp.).

Conventional systems used inorganic fertilizers, chemical weed control, and integrated pest management for insect and disease control. Organic and low-input systems used cover crops with N-fixing legumes, and occasional use of composted manure for fertility. Organic management also included mechanical weed control and infrequent applications of organically approved pesticides. Low-input management was characterized by supplemental use of manures or inorganic fertilizers, replacement of long-lived preemergence herbicides with shorter-lived postemergence herbicides, and sparing use of other chemicals when productivity might be compromised. Tillage, typically disk tillage, was used in all systems, though tillage was reduced when possible in the low-input system (Temple et al., 1994b). Individual experimental plots were 0.135 ha in size and were irrigated.

In the early years of transition (1989–92), tomato and corn yields were lowest in the organic system (Table 13.5). However, in the latter years, the low-input system outperformed both the organic and conv-4. Dry beans, however, performed best under organic or low-input management (Table 13.5). During the transition period, yield deficits were generally significant with regard to the organic system (Temple et al., 1994b). In later years, significant

Table 13.5: Average yields of three major crops and microbial biomass C (MBC) over time in the Sustainable Agriculture Farming Systems (SAFS) trial, University of California, Davis.

		System				
Crop	Years	Organic	Low-Input	Con-4	Con-2	Citations and Notes
Tomato Mg ha^{-1}	1989–1992	71	81	92	85	Temple et al. (1994b); Poudel et al. (2001b, 2002); Clark et al. (1999b)
	1994–1998	67	71	75	72	
	1989–1999	67	75	77	73	
Corn Mg ha^{-1}	1989–1992	10.3	11.3	11.1		
	1994–1998	11.2	12.4	11.3		
	1989–1999	10.5	11.9	11.2		
Dry beans Mg ha^{-1}	1989–1992	2.5	2.4	2.1		
	1989–1999	1.9	1.9	1.7		
SOC %	1996	1.08a[a]	1.03b	0.90c	0.92c	Doane and Horwath (2004); 0–15 cm depth; averages across all crops/all plots within a treatment
	2000	1.13a	1.04b	0.84c	0.88c	
% Increase in SOC	1988–2000	36a	26b	12b	7b	
Total N %	1998	0.146a	0.126b	0.113bc	0.11c	Poudel et al. (2001b)
MBC μg g^{-1} soil	1990	68	68	65	58	Scow et al. (1994); Gunapala and Scow (1998); September samples 0–15 cm depth
	1992	93	106	55	47	
	1993	90	85	65	67	

[a]lowercase letters within a row indicate reported significant differences.

differences all but disappeared (Poudel et al., 2002). Over the entire course of the experiment, 1989–99, crop yields were comparable among production systems, with a maximum tomato yield gap of only 13% in the organic system, and less than 3% in the low-input system, compared to Conv-4. Poudel et al. (2001a) attributed this deficit to an N limitation in the transition years that was eventually surmounted with long-term management.

Initial SOC in 1988, across all experimental plots, was 0.79% for the 0–30 cm depth (Scow et al., 1994). This initial value was recalculated as 0.83% for a 0–15 cm depth to compare to SOC to measurements made in 1996 and 2000 (Doane and Horwath, 2004). The SOC increased in all treatments over time, eventually reaching a significantly greater increase in the organic system, with no difference in the percentage increase among the low-input, Conv-4, or the Conv-2 across all crops (Doane and Horwath, 2004; Table 13.5). Dynamics of these changes, investigated with ^{13}C, were highly dependent on the cropping system. For example, corn inputs seasonally increased SOC by as much as 36%, but in the next year loss could be more than the gain, particularly under conventional production (Doane and Horwath, 2004). These soils apparently had a very low capacity for building soil C, or at the very least, had an extremely slow rate for SOC buildup and retention, which may have been due to the use of tillage.

As noted above, soil N dynamics played a crucial role in the productivity of these four management systems. Total soil N in the 0–15 cm depth, measured in 1998, followed a similar trend to SOC, being significantly greater in the organic versus the other production systems (Table 13.5). This did not translate to the lower 15–30 cm depth, where no difference occurred in total N, which ranged 0.09–0.10%, however, total soil N storage capacity for the 0–30 cm depth was substantially greater in the organic system, followed by the low-input (Poudel et al., 2001b). The conv-4 and conv-2 systems were also prone to greater N loss. Soil N availability exhibited seasonal and yearly differences that may have masked management system-level differences; however, mineralization potential was consistently greater in the organic system, followed by the low-input and conventional systems (Clark et al., 1999a; Poudel et al., 2002). Where mineralization potential was high the rate of turnover was low, providing a reason for the accumulation of N in organic and low-input systems, as well as explaining the potential for nitrate leaching in the conventional systems (Poudel et al., 2002).

Changes in soil MBC over time did support the increases observed in later years in SOC (Table 13.5) and were perhaps a more sensitive indicator of soil resource changes. In 1990, soil MBC was generally too variable over the growing season to be significantly different among cropping systems; however, significant peaks in extractable mass occurred with April incorporation of cover crops in the organic and low-input systems (Scow et al., 1994). By 1992, MBC was significantly greater, by an estimated 30–60%, in the organic and low-input over the conventional systems throughout the growing season (Scow et al., 1994). This significantly higher MBC was maintained through the 1993 growing season

(Gunapala and Scow, 1998; Table 13.5). Laboratory investigations on soil samples taken in 1996 continued to indicate that MBC was greater in the organic and low-input systems by nearly 40% compared to Conv-4 and Conv-2 (Lundquist et al., 1999).

Poudel et al. (2001a) reviewed aspects of weed dynamics, disease issues, and biodiversity. Weed pressures were clearly greatest in the organic system, but also a problem in the low-input system, resulting in a high cost for weed management (Clark et al., 1999b; Poudel et al., 2002). Though weed pressure was high, Clark et al. (1999a) determined that it did not interfere with crop N uptake. On the other hand, plant diseases, particularly of tomato, were more of an issue in the conv-2 system, likely because the short rotation back to tomato was insufficient to deter reoccurrence (Poudel et al., 2001a). The organic and low-input systems promoted higher soil biodiversity, including nematode food-web structure (Berkelmans et al., 2003) and ground-dwelling beetles (Clark, 1999). The healthier well-regulated nematode community of the organic and low-input systems was a stark contrast to the depauperate community prone to high population numbers of plant parasitic nematodes of the conv-4 system (Berkelmans et al., 2003).

13.3 Farm Case Studies Evaluating the Implementation of Adaptive, Low-Input Production Practices

Three case studies were selected to represent examples of low-input agriculture practiced on small, medium, and national scales. For each case study, the extent of the farming practice is addressed in relation to the fiscal and intellection inputs driving their development. The first case study discusses a small-scale producer faced with the need to improve productivity on a degraded landscape. The second case study examines a producer's management practices with the goal to increase production system diversity as a means to build resilience to market and economic pressures. The final case study addresses the Cuban national response to a political and economic embargo on technological advances, input commodities, and national food security.

13.3.1 Seven Pines Farm, Wadena County, Minnesota, United States

Seven Pines is a small, 35-ha, diversified farm raising pasture grazed dairy cattle, laying hens, and hogs operated by Kent and Linda Solberg near Verndale, MN, with a major focus on soil health. The farmland, acquired in 2003, had highly degraded, loamy sand soils with only 1% SOM. Annual rainfall averages 771 mm and mean temperatures range from a July high of 20.6°C to a January low of −12.8°C. Building and maintaining fertility of these sandy soils was the main challenge, and initially required more external inputs compared to other more productive soils, to achieve sufficient productivity. Consequently, the Seven Pines Farm operation continues to employ limited purchased inputs, including lime to raise the soil pH and some purchased manure for fertility, when affordable. Even though the climate is quite dry, there is no irrigation due to economic constraints; instead, the farm management focuses on building soil water-holding capacity by improving soil health. The only other purchased consumable inputs include seeds and a limited amount of herbicide needed to combat invasive weeds.

To rebuild soil health, Seven Pines management uses a combination of cultural soil-building practices. Most of the farm is planted to perennials, with living roots established in the soil year-round. A diverse mixture of grazed crops and cover crops grows on all of the acres. Perennial pastures include intermediate wheat grass (*Thinopyrum intermedium* (Host) Barkworth and D.R. Dewey), orchard grass (*Dactylis glomerata* L.), tall fescue (*Festuca arundinacea* Schreb.), alfalfa (*Medicago sativa* L.), yellow sweet clover (*Melilotus officinalis* (L.) Pall.), red clover (*Trifolium pratense* L.), or birds foot trefoil (*Lotus corniculatus* L.), and chicory (*Cichorium intybus* L.). High stock density and mob grazing are used for managing dairy cattle. Two to three paddock shifts per day are required, which keeps dry matter intake high, facilitates trampling of uneaten forages, and keeps manure distribution even.

On this farm, other livestock used to help manage the pasture include chickens and hogs. During the grazing season, laying hens are transported in moveable trailers into recently grazed pasture. Foraging chickens break up dung pats to feed on insect larvae, they feed on seeds that may include weed species, and they add fertility with their manure (Entz and Theissen Martens, 2009; Hilimire, 2011). Hogs are used to renovate pastures. Their rooting activity serves as a disturbance method that eliminates the need for tillage (Andresen et al., 2001). Once the paddocks are prepared by the "hog-tillage," they are then seeded to a multispecies blend of cover crops to maintain plant diversity. Warm season grasses and broadleaf plants complement naturally occurring cool season grasses and broadleaves. By increasing pasture diversity, this management practice is enhancing soil health by fostering the production of a diverse microbial community (Zak et al., 2003). The hogs also provide natural weed control by rooting, removing, and eating rhizomes of weedy species, including perennial grasses such as quackgrass (*Elymus repens* (L.) Gould) (Bond et al., 2001).

Cattle are "outwintered" during the winter, meaning they are kept in a paddock, and fed bales of self-grown hay and other forage supplements such as kelp. Cattle have access to water, a windbreak, and a straw bedding pack. Winter forage is laid out in strips to assist in spreading manure. The outwintering and building of the manure pack helps manage soil fertility across the farm. Outwintering dramatically reduced manure hauling chores, which are still necessary for the chickens and hogs kept in winter pens.

Approximately half of the acreage is used to produce winter forage, something that was not possible the first 8 years on the farm. All acreage is divided into paddocks that are rotated on a 5–6-year basis between annual crops, which may include early harvest millet (*Pennisetum glaucum* (L.) R.Br.), followed by triticale (×*Triticosecale Wittm.* ex *A. Camus.*), and forage turnip (*Brassica rapa* var. *rapa* L.) that are fall grazed into October or November. Fall forage is undersown with a pasture mix (see above) and kept in this new pasture for 4–5 years before rotating back to millet. Out wintering precedes as well as follows the annual crops. Over the years, this process had increased the productivity of the farm, with forage production doubling to tripling in paddocks that had been rotated compared to those that had not.

The farm is not currently certified organic, but plans are to become certified and to move toward organic milk production. Milk, pork, feeder pigs, and some beef are sold as farm products, but eggs are for family use only. All farm income is invested and the farm is debt-free. The farm has sufficient capacity to employ two part-time farm hands and Mr. Solberg enjoys a thriving second career as an agricultural educator and speaker.

13.3.2 Brown's Ranch, Burleigh County, North Dakota, United States

Brown's Ranch is an operation of approximately 2000 ha of owned and leased land managed by Gabe Brown and family in Burleigh County, near Bismarck, ND. Soils on the farm include loam, silt loam, and silty clay loam; average annual rainfall is 453 mm, and mean temperatures range from a July high of 21.7°C to a January low of –10.7°C. Since 1993, this farm has been managed using no-tillage coupled with diverse crop and forage rotations, along with carefully managed grazing-based livestock. The Brown's Ranch goal with this highly integrated crop and livestock production system is to build soil tilth, fertility, and to enhance beneficial soil biota. Over time, the management system has evolved to the extent that use of synthetic fertilizers, fungicides, and insecticides has been eliminated. Herbicides are used infrequently and judiciously, with a future goal of eventually eliminating this input.

The crop and forage production systems include a diverse assemblage of warm and cool season grasses and broadleaf species of cash crops, grown with companion crops, or followed by warm season cover crop mixtures (Table 13.6). Growing diverse rotations provides many benefits for agroecosystems, including maximizing efficiency of nutrient cycling and disrupting regeneration niches of pests (Bezdicek and Granatstein, 1989; Kremen and Miles, 2012). These diverse cover crop mixtures increase SOM and also provide a way to nutritionally optimize the forage intake, as they can be grazed anytime between October and January. Companion planting has been explored recently as a means to further increase diversity and to provide direct benefits to an accompanying cash crop. For example, intercropping hairy vetch, subterranean clover, or sweet clover with corn has been observed to provide benefits to the corn crop by reducing soil surface evaporation and enhancing soil mycorrhizal communities (Dabney et al., 2001). These companion crops are also grazed by livestock after the cash crop is harvested. Yield comparisons on Brown's Ranch have shown that cover crop polycultures can have up to three times the productivity of monocultures.

Livestock have been integrated into the annual crop production system in innovative ways that maximize efficiency and productivity by improving soil resources. Brown's Ranch runs about 350 cow–calf pairs specifically choosing breeds that reduce inputs and costs of production

Table 13.6: Examples of the cash crops, companion crops, and cover crops used on Brown's Ranch

Cash Crops
Alfalfa; corn (*Zea mays* subsp. *Mays* L.); grain and forage pea (*Pisum sativum* L.); hairy vetch (*Vicia villosa* Roth) oat (*Avena sativa* L.); spring wheat; sunflower (*Helianthus annuus* L.), winter triticale (× *Triticosecale Wittm.* ex *A. Camus.*)
Example Companion Crop Mixtures
Winter triticale/hairy vetch Corn/hairy vetch Corn/subterranean clover (*Trifolium subterraneum* L.)/ sweetclover (*Melilotus officinalis* (L.) Lam.)
Example Cover Crop Mixture
Cowpea (*Vigna unguiculata* (L.) Walp.), hybrid pearl millet (*Pennisetum glaucum* (L.) R.Br.), radishes (*Raphanus sativus* L.), sorghum/sudangrass (*Sorghum×drummondi* (Nees ex. Steud.)), soybean (*Glycine max* (L.) Merr.), sunflower, sunn hemp (*Crotalaria juncea* L.), turnips (*Brassica rapa* var. *rapa* L.)
Other Cover Crops
Canola (*Brassica rapa* L.), buckwheat (*Fagopyrum esculentum* Moench), kale (*Brassica oleracea* L.) lentil (*Lens culinaris* Medikus), mung bean (*Vigna radiata* (L.) R. Wilczek), pasja (*Brassica Rapa*, syn. *B. campestris*), phacelia (*Phacelia tanacetifolia* Benth.), proso millet (*Panicum miliaceum* L.), rape (*Brassica napus* L.), red clover, ryegrass (*Lolium multiflorum* Lam.), sugarbeet (*Beta vulgaris* L.), sweetclover

by being low-maintenance and smaller framed for efficient biomass gain, and thriving on a pure forage diet. Even in the harsh northern winter climate of ND, the cattle are typically grazed at least 9 months of the year. Efficiency is gained by using the cattle to add and spread their manure. The addition of manure and urine helps to cycle and recycle nutrients through the macro- and micro-biomass that influences the supply of available nutrients needed for subsequent crops (Hilimire, 2011). Other livestock, including chickens and Katahdin hair sheep influence the systems in similar ways, but they also provide diverse impacts on forage. Free-ranging chickens, transported with mobile houses, feed on insects and plant seeds, and are used after the cattle grazing to break up the manure pats. The sheep not only feed on different forages, which benefits nutrient turnover, but they also provide cattle disease protection as they are a dead-end host for some parasites (Esmall, 1991). Pasture hogs will be added soon to turn compost piles, keeping the compost active, thus providing energy savings over using fossil fuels to accomplish this task.

Brown's Ranch farming system continues to evolve over time, with ongoing experimentation key to developing innovative new techniques and strategies that capitalize on the integrated efficiencies of a diverse and multifunctional system. Step by step, this farming family is learning how to assemble a farm that looks and functions more like a natural ecosystem than a typical input-intensive cropping system. The fundamental strategy is to substitute integrated functionality of crops and livestock to replace inputs of fossil fuels, synthetic chemical inputs, and human labor. Not only has this approach resulted in high production efficiency, it has maximized the number of calories produced per hectare because the same unit of land supplies the necessary inputs for multiple enterprises—diverse cash crops and diverse livestock.

The success of the Brown's Ranch approach can be evaluated using various metrics. Corn yields on this farm have been as high as 7965 kg ha^{-1}, far exceeding the Burleigh County average for dryland corn yield of less than 6272 kg ha^{-1}. Furthermore, the total cost of corn production was calculated at \$3.50 ha^{-1}, which enhances profitability even when prices are low. In comparison, 2013–14 projected operating costs (including seed, fertilizer, fuel, and irrigation) for conventional corn production in the Northern Great Plains averaged \$804 ha^{-1} (ERS, 2015). Despite not using fertilizers, corn leaf tissue analysis revealed nutrient levels that were sufficient or more than sufficient for production. When the ranch was established in 1991, SOM ranged from 1.7 to 1.9% and the water infiltration rate was only 12.7 mm h^{-1}. After more than 20 years of practicing no-tillage and cover cropping, the SOM currently ranges from 5.3 to 6.1% and the water infiltration rate is about 203 mm h^{-1}. Mr. Brown credits the power of living roots and undisturbed soil over sustained periods of time for these improvements in soil properties. Livestock grazing improved soil P content; soil from fields grazed 2 out of 6 years contained 267 kg P ha^{-1}, whereas fields that were nongrazed had only 67 kg P ha^{-1}. These metrics demonstrate the success of a large-scale low-input approach to agricultural production in terms of numerous outcomes—both those associated with short-term profitability and those associated with long-term sustainability.

13.3.3 Campesino-a-Campesino Agriculture, Republic of Cuba

The transformation of the agricultural production system in Cuba is probably the largest-scale example of successful conversion from input-intensive to low-input production. Prior to 1989, Cuba's agriculture system was artificially propped up with support from the Soviet Union, which provided fuel, pesticides, fertilizers, machinery, and a market for Cuban agricultural products. The resulting agricultural production system was a model of Green Revolution-era success, producing the second-highest grain yields in Latin America. As much as 30% of the land base was devoted to production of export crops, and thus food imports accounted for 57% of the caloric intake of the Cuban people (Rosset and Benjamin, 1994).

With the collapse of the Soviet Union in 1989, supports ceased and Cuba's agricultural sector began to collapse, seriously jeopardizing the nation's food security (Perfecto, 1994; Oppenheim, 2001). In response, Cuba's farmers, scientists, and planners gathered together with full government support to re-imagine and re-engineer the food production system of this small island country. Since most high-energy inputs were no longer available, farmers needed to develop low-input production systems. Principles from agroecology were combined with a grassroots framework for testing and disseminating new ideas called "Campesino-a-Campesino," or "farmer-to-farmer" methodology (CAC). This methodology is based on the idea that farmers learn more readily and are more likely to buy into ideas from other farmers demonstrating success. The CAC extension method proved much more successful than conventional extension activities for spreading knowledge about agroecological approaches to low-input sustainable agriculture (Rosset et al., 2011).

A detailed account of the agroecological transformation of Cuban agriculture is beyond the scope of this chapter, but the basic approaches are summarized briefly (for further information see Rosset and Benjamin, 1994; Funes et al., 2002). Organic sources of soil fertility and soil conservation measures to improve soil were heavily emphasized, including the use of manures, sugarcane filter-cake mud, compost, earthworm humus (vermicompost), mulches, soil microorganism inoculants, and cover crops (Gersper et al., 1983). Ecological management of pests and diseases without synthetic chemicals was also a considerable challenge and focus. Biological control agents consisting of insect predators, insect pathogens, and disease antagonists were employed as substitutes for conventional pesticides (Perfecto, 1994; Oppenheim, 2001).

Other approaches, such as crop rotations, crop polycultures, or intercrops, provided the means to suppress pests, increase yields, and improve soil condition. Measures were taken to return to locally adapted breeds of animals, and to integrate livestock grazing with other farming activities such as crop production. Conservation tillage approaches, contour plowing, living barriers, ground covers including locally adapted pasture plants, and cover crops were put in place to combat degraded soils. Although pasture was shown to be ideal for soil health (e.g., SOC, soil bulk density, aggregation, enzyme activity, microbial biomass and activity), comparative studies with forage and polycrop systems in Northwest Cuba showed these other methods of production helped maintain resilience of the soil through the dry season (Izquierdo et al., 2003).

The transformation of Cuba's agricultural system has been credited, in part, with the country's recovery from economic collapse during the early 1990s. The Cuban government has supported the agricultural transformation with efforts such as the land reform of 2008 and the Lineamientos of 2011 (Leitgeb et al., 2015). For example, in response to a food crisis in 2007–08, these reforms aided the transformation of urban landscapes into food production systems. City residents used organic techniques and agroecological principles to supply basic nutrition for their families and neighbors (Leitgeb et al., 2015). The success of these urban gardens greatly increased food security and access to fresh produce for Cuba's urban residents (Funes et al., 2002). Recently, expanded export markets and subsidized oil from Venezuela have further stoked Cuba's economic engine. Nevertheless, Cuba is still heavily dependent on foreign imports, with the United States supplying at least 30% of food and agricultural imports (USDA, 2008).

13.4 Potential for Sustainable Intensification of Production

The research case studies presented in this chapter placed evaluated input intensity with measures of yield and soil health as metrics. These practices can be contrasted along an input gradient between external and captured inputs and the impact on soil health indicators and yield gap or gains evaluated (Table 13.7). For the Broadbalk trial, FYM is conceivably a captured input, assuming it would be self-produced on the farm. Use of FYM, with or without additional inorganic fertilization increased soil organic C, microbial biomass, and other measures of soil health (Tables 13.2 and 13.7). In comparison to a similar level of inorganic fertilization in continuous or rotated wheat, FYM use increased yields. Of course, benefits to soil health and substantial yield gains were obtained just by using rotations (which likely benefitted the soils' natural capital) versus continuous production of wheat, regardless of the amount of fertilizer applied.

The DOK trial in Switzerland contrasts greatly with the Broadbalk trial, in that all systems investigated used a highly, temporally diverse rotation. The systems investigated contrasted not necessarily in the form of N applied, as FYM compost was used in most of them, but in the use of chemical amendments versus substituted inputs such as plant extracts and biodynamic preparations (see Section 13.2.2). Though the BIODYN had perceptively better soil health than the BIOORG, the two systems did not differ in yield output (Table 13.7). They did differ substantially from the CNVFYM system in a positive way by supporting soil health parameters, and in a negative way by providing substantially reduced yields, with an 11% and 37% yield gap for wheat and potato production, respectively. The BIODYN and BIOORG systems lie on the less energy-intensive side of the input gradient, using up to 11.3 GJ ha^{-1} $year^{-1}$ less than the conventional systems (Mäder et al., 2002).

The SAFS trial generated a contrast between a reduced input intensity, with the low-input system, and an input-substitution system, with the organic system. Both systems, as well as the Conv-4, embraced practices conforming to sustainable ideals, such as employing diverse crop rotations, though only the low-input and organic systems used cover crops (see Section 13.2.3). The organic system appeared to be superior to the low-input system only because of higher values of the soil health measures, as the organic system had a substantial yield gap for both corn and tomato, averaging 12.5% (Table 13.7). The low-input system did come out as superior to the Conv-4, as both soil health and yield gains were observed, and perhaps better than the organic system in comparison to the Conv-4, as both soil health and yields were on the positive side. Considering just the economics, however, the low-input system had the lowest profits of all production systems, primarily as a result of higher operating costs in comparison to the conventional systems, and lower gross returns in comparison to the organic system, which received organic price premiums (Clark et al., 1999b).

Yield gains over a century of production in the Broadbalk trial clearly demonstrate the outcome of the technological achievements that have driven the global intensification of agriculture. Matson et al. (1997) has argued that this intensification has led to declines in agricultural diversity and other environmental detriments. The research case studies presented in this chapter demonstrate that intensification and the continuation of high-input practices to obtain increased yields do contribute to declines in crop diversity and soil health.

Table 13.7: Summary of impacts on soil health indicators and consequential yield gap from select comparisons of agricultural practices from lower to higher input in research case studies

	Comparisons[a]		Indicator[b]	Crop and % Yield Gap[c]	
Case Study[a]	**Lower Input**	**Higher Input**	**↑ Increase/ ↓ Decrease**	**↑ Increase/↓ Decrease**	
Broadbalk	FYM	Inorganic N 144	↑ SOC, ↑ SN, ↑ MBC ↑ N loss ↓ P loss	Continuous wheat	0%
	FYM + 96 kg N ha^{-1}	Inorganic N	↑ SOC ↑ N loss ↓ P loss	Rotated wheat	↑ 6%
	Rotated wheat	Continuous wheat	↑ SOC	Wheat	↑ 86%
DOK	BIODYN	BIOORG	↑ MBC, ↑ SOC	Wheat	0%
				Potato	0%
	BIOORG	CNVFYM	↑ MBC, ↑ SOC	Wheat	↓ 11%
				Potato	↓ 37%
	CNVFYM	CNVMIN	↑ MBC, ↑ SOC	Wheat	0%
				Potato	↑ 6%
SAFS	Low-input	Organic	↓ MBC, ↓ SOC ↓ SN ↓ PMN	Tomato	↑ 12%
				Corn	↑ 13%
				Beans	0%
	Low-input	Conv-4	↑ MBC, ↑ SOC ↑ SN ↓ N loss	Tomato	↓ 3%
				Corn	↑ 10%
				Beans	↑ 12%
	Organic	Conv-4	↑ MBC, ↑ SOC ↑ SN ↓ N loss	Tomato	↓ 13%
				Corn	↓ 6%
				Beans	↑ 12%
	Conv-4	Conv-2	↑ MBC	Tomato	↑ 5%

[a]Practice and case study notations are given as in Table 13.1.
[b]Indicators are: SOC, soil organic C; SN, soil N; MBC, microbial biomass C; MBP, microbial biomass P; PMN, potentially mineralizable N.
[c]Yield gaps are calculated using yield from the last year or last interval reported as ((lower input yield − higher input yield)/ higher input yield)*100.

Koh et al. (2013) echoed the decline in diversity when they theorized that global food demands could be met though optimizing agricultural land use by selecting and only growing the single best producing crop for a given location. However, intensification through high-input systems is likely to continue to meet the demands of a growing global population.

A number of models and metrics have been developed to evaluate the sustainability of food production. For industrialized countries, nationwide assessments using life cycle analysis have been proposed. Heller and Keoleian (2003) evaluated the sustainability of the US food system by evaluating a variety of economic, social, and environmental indicators in the life cycle of the entire supply chain from grower, to processor, to consumer. They concluded the current trends threatening the sustainability of the US food system might be reversed by influencing consumption behavior. Primarily, their assessment suggests that a change in behavior to reduce consumption and food wastage can improve human and environmental health, and result in a sevenfold reduction in energy use for each calorie saved.

The on-farm and Cuban national case studies presented in this chapter offer an alternative point of view. These case studies show that sustainable and agroecology practices can be successfully adopted at the farm level, and even the national scale, in ways that maintain the economic viability of the producer, as seen with Seven Pines and Brown's Ranch, but also the production needs, i.e., food security, of an entire nation. This suggests that behaviors can be changed, and these changes go beyond individual consumption and waste. Most importantly, Brown's Ranch, by demonstrating their ability to reduce or eliminate commercial chemical inputs, epitomizes the ability to adopt sustainable, agroecological practices, and drive the system toward increased biological efficiency to generate greater natural capital for increasing availability and use of capture inputs (see Fig. 13.1). Collectively, the case studies presented in this chapter have demonstrated the enhancement of ecological services such as carbon (C) sequestration, enhanced water quality, and flow regulation (water use), and enhanced biodiversity that will have positive benefits for the environment (Foley et al., 2005).

13.5 Conclusions

Hanspach et al. (2013) wrote in a letter to *Science* that the sustainability of intensified production remains unknown, and that sustainable intensification as a goal has to be applied on an individual basis based on the benefits that are being sought (cf. food production vs food security). On-farm case studies presented in this chapter indicate that "sustainable intensification" is achievable by approaching needs on a case-by-case basis. This aspect complicates the ability to derive the relevant quantitative data needed to develop widely adoptable, sustainable, yet intensified management approaches. Thus, scaling up from a research study to a farm-level practice does need more research and development.

Regardless of the lack of consensus on approach or definition, "sustainable intensification" seems to encompass strategies consistent with agroecological theory. From the case studies presented here, low-input production systems are clearly viable means to achieve similar or greater yields than comparable conventional practices with the added benefits to soil health and an enhanced soil resource base. Although the low-input and organic strategies are similar in their positive impact on biodiversity and soil health, they will differ in the energy intensiveness of the system, potentially driven higher under organic approaches through input-substitution, and reliance on and use of commercialized inputs, rather than input-reduction. Individual producers have the potential to improve the sustainability of the entire farming system by generating their own intellectual capital from on-farm trials and embracing agroecological ideals.

References

Andresen, N., Ciszuk, P., Ohlander, L., 2001. Pigs on grassland: animal growth rate, tillage work and effects in the following winter wheat crop. Biol. Agric. Hortic. 18, 327–343.

Berkelmans, R., Ferris, H., Tenuta, M., van Bruggen, A.H.C., 2003. Effects of long-term crop management on nematode trophic levels other than plant feeders disappear after 1 year of disruptive soil management. Appl. Soil Ecol. 23, 223–235.

Bezdicek, D.F., Granatstein, D., 1989. Crop rotation efficiencies and biological diversity in farming systems. Am. J. Altern. Agric. 4, 111–119.

Birkhofer, K., Bezemer, T.M., Bloem, J., Bonkowski, M., Christensen, S., Dubois, D., et al., 2008. Long-term organic farming fosters below and aboveground biota: implications for soil quality, biological control and productivity. Soil Biol. Biochem. 40, 2297–2308.

Blair, N., Faulkner, R.D., Till, A.R., Poulton, P.R., 2006. Long-term management impacts on soil C, N and physical fertility: Part I: Broadbalk experiment. Soil Till. Res. 91, 30–38.

Bond, W., Davies, G., Turner, R.J., 2001. The biology and non-chemical control of common couch (*Elytrigia repens* (L.) Nevski). Available online at <http://www.gardenorganic.org.uk/weeds/common-couch>.

Brookes, P.C., Powlson, D.S., Jenkinson, D.S., 1982. Measurement of microbial biomass phosphorus in soil. Soil Biol. Biochem. 14, 319–329.

Brookes, P.C., Powlson, D.S., Jenkinson, D.S., 1984. Phosphorus in the soil microbial biomass. Soil Biol. Biochem. 16, 169–175.

Clark, I.M., Buchkina, N., Jhurreea, D., Goulding, K.W.T., Hirsch, P.R., 2012. Impacts of nitrogen application rates on the activity and diversity of denitrifying bacteria in the Broadbalk Wheat Experiment. Philos. Trans. R. Soc. B 367, 1235–1244.

Clark, M.S., 1999. Ground beetle abundance and community composition in conventional and organic tomato systems of California's Central Valley. Appl. Soil Ecol. 11, 199–206.

Clark, M.S., Horwath, W.R., Shennan, C., Scow, K.M., Lantni, W.T., Ferris, H., 1999a. Nitrogen, weeds and water as yield-limiting factors in conventional, low-input, and organic tomato systems. Agric. Ecosyst. Environ. 73, 257–270.

Clark, S., Klonsky, K., Livingston, P., Temple, S., 1999b. Crop-yield and economic comparisons of organic, low-input and conventional farming systems in California's Sacramento Valley. Am. J. Altern. Agric. 14, 109–121.

Dabney, S.M., Delgado, J.A., Reeves, D.W., 2001. Using winter cover crops to improve soil and water quality. Commun. Soil Sci. Plant Anal. 32, 1221–1250.

De Nobili, M., Contin, M., Brookes, P.C., 2006. Microbial biomass dynamics in recently air-dried and rewetted soils compared to others stored air-dry for up to 103 years. Soil Biol. Biochem. 38, 2871–2881.

Doane, T.A., Horwath, W.R., 2004. Annual dynamics of soil organic matter in the context of long-term trends. Global Biogeochem. Cycl. 18, GB3008. http://dx.doi.org/10.1029/2004GB002252.

Doré, T., Makowski, D., Malézieux, E., Munier-Jolain, N., Tchamitchian, M., Tittonell, P., 2011. Facing up to the paradigm of ecological intensification in agronomy: revisiting methods, concepts and knowledge. Eur. J. Agron. 34, 197–210.

Dyke, G.V., George, B.J., Johnston, A.E., Poulton, P.R., Todd, A.D., 1983. The Broadbalk Wheat experiment 1968-78: yields and plant nutrients in crops grown continuously and in rotation. Rothamsted Exp. Station Rep. 1982 (Part 2), 5–44. GRID: ResReport1982 p2-5-44.

Edmeades, D.C., 2003. The long-term effects of manures and fertilisers on soil productivity and quality: a review. Nutr. Cycl. Agroecosyst. 66, 165–180.

Edwards, C.A., Lal, R., Madden, P., Miller, R.H., House, G., 1990. Sustainable Agricultural Systems. Soil and Water Conservation Society, Iowa.

Entz, M.H., Theissen Martens, J.R., 2009. Organic crop-livestock systems. In: Francis, C. (Ed.), Organic Farming: The Ecological System. Agronomy Monograph 54. American Society of Agronomy, Madison, pp. 69–84.

ERS, 2015. Corn Production Costs and Returns per Planted Acre, Excluding Government Payments, 2013–2014. US Department of Agriculture, Economic Research Service., <http://www.ers.usda.gov/data-products/commodity-costs-and-returns.aspx> (accessed 15.04.16.).

Esmall, S.H.M., 1991. Multispecies grazing by cattle and sheep. Rangelands 13, 35–37.

Espershültz, J., Gattinger, A., Mäder, P., Schloter, M., Fließbach, A., 2007. Response of soil microbial biomass and community structures to conventional and organic farming systems under identical crop rotations. FEMS Microbiol. Ecol. 61, 26–37.

Fließbach, A., Oberholzer, H.R., Gunst, L., Mäder, P., 2007. Soil organic matter and biological soil quality indicators after 21 years of organic and conventional farming. Agric. Ecosyst. Environ. 118, 273–284.

Foley, J.A., DeFries, R., Asner, G.P., Barford, C., Bonan, G., Carpenter, S.R., et al., 2005. Global consequences of land use. Science 309, 570–574.

Funes F, García, L., Bourque, M., Pérez, N., Rosset, P., 2002. Sustainable Agriculture and Resistance: Transforming Food Production in Cuba. Food First, Oakland.

Garnett, T., Appleby, M.C., Balmford, A., Bateman, I.J., Benton, T.G., Bloomer, P., et al., 2013. Sustainable Intensification in Agriculture: Premises and Policies. Science 341, 33–34.

Gersper, P.L., Rodríguez-Barbosa, S., Orlando, L.F., 1983. Soil conservation in Cuba: a key to the new model for agriculture. Agric. Hum. Values 10, 16–23.

Glendining, M.J., Powlson, D.S., 1990. 130 Years of inorganic nitrogen fertilizer applications to the Broadbalk wheat experiment: the effect on soil organic matter. Trans. 14th International Congress of Soil Science, Kyoto, Japan, August 1990, vol. IV, pp. 9–13.

Glendining, M.J., Powlson, D.S., Poulton, P.R., Bradbury, N.J., Palazzo, D., Li, X., 1996. The effects of long-term applications of inorganic nitrogen fertilizer on soil nitrogen in the Broadbalk Wheat Experiment. J. Agric. Sci. 127, 347–363.

Gliessman, S.R., 1998. Agroecology: Ecological Processes in Sustainable Agriculture. CRC Press, Boca Raton FL.

Goulding, K.W.T., Poulton, P.R., Webster, C.P., Howe, M.T., 2000. Nitrate leaching from the Broadbalk Wheat Experiment, Rothamsted, UK, as influenced by fertilizer and manure inputs and the weather. Soil Use Manag. 16, 244–250.

Gunapala, N., Scow, K.M., 1998. Dynamics of soil microbial biomass and activity in conventional and organic farming systems. Soil Biol. Biochem. 30, 805–816.

Hanspach, J., Abson, D.J., Loos, J., Tichit, M., Chappell, M.J., Fischer, J., 2013. Develop, then intensify. Science 341, 713.

Harwood, R.R., 1990. A history of sustainable agriculture. In: Edwards, C.A., Lal, R., Madden, P., Miller, R.H., House, G. (Eds.), Sustainable Agricultural Systems. Soil and Water Conservation Society, St. Lucis Press, Florida, pp. 3–19.

Heckrath, G., Brookes, P.C., Poulton, P.R., Goulding, K.W.T., 1995. Phosphorus leaching from soils containing different phosphorus concentrations in the Broadbalk Experiment. J. Environ. Qual. 24, 904–910.

Heller, M.C., Keoleian, G.A., 2003. Assessing the sustainability of the US food system: a life cycle perspective. Agric. Syst. 76, 1007–1041.

Hildermann, I., Messmer, M., Dubois, D., Boller, T., Wiemken, A., Mäder, P., 2010. Nutrient use efficiency and arbuscular mycorrhizal root colonisation of winter wheat cultivars in different farming systems of the DOK long-term trial. J. Sci. Food Agric. 90, 2027–2038.

Hilimire, K., 2011. Integrated crop/livestock agriculture in the United States: a review. J. Sustain. Agric. 35, 376–393.

Ikerd, J.E., 1990. Agriculture's search for sustainability and profitability. J. Soil Water Conserv. 45, 8–23.

Izquierdo, I., Caravaca, F., Alguacil, M.M., Roldán, A., 2003. Changes in physical and biological soil quality indicators in a tropical crop system (Havana, Cuba) in response to different agroecological management practices. Environ. Manag. 32, 639–645.

Jenkinson, D.S., Powlson, D.S., 1976. The effects of biocidal treatments on metabolism in soil V. A method for measuring soil biomass. Soil Biol. Biochem. 8, 209–213.

Johnston, A.E., 1997. The value of long-term field experiments in agricultural, ecological and environmental research. Adv. Agron. 59, 291–333.

Koh, L.P., Koellner, T., Ghazoul, J., 2013. Transformative optimisation of agricultural land use to meet future food demands. PeerJ 1, e188. http://dx.doi.org/10.7717/peerj.188.

Kremen, C., Miles, A., 2012. Ecosystem services in biologically diversified versus conventional farming systems: benefits, externalities, and trade-offs. Ecol. Soc. 17, 40. <http://www.ecologyandsociety.org/vol17/iss4/art40/>.

Lambin, E.F., Meyfroidt, P., 2011. Global land use change, economic globalization, and the looming land scarcity. Proc. Nat. Acad. Sci. 108, 3465–3472.

Leitgeb, F., Schneider, S., Vogl, C.R., 2015. Increasing food sovereignty with urban agriculture in Cuba. Agric. Hum. Values, 1–12. http://dx.doi.org/10.1007/s10460-015-9616-9.

Lundquist, E.J., Jackson, L.E., Scow, K.M., Hsu, C., 1999. Changes in microbial biomass and community composition, and soil carbon and nitrogen pools after incorporation of rye into three California agricultural soils. Soil Biol. Biochem. 31, 221–236.

Mäder, P., Edenhofer, S., Boller, T., Wiemken, A., Niggli, U., 2000. Arbuscular mycorrhizae in a long-term field trial comparing low-input (organic, biological) and high-input (conventional) farming systems in a crop rotation. Biol. Fertil. Soils 31, 150–156.

Mäder, P., Fließbach, A., Dubois, D., Gunst, L., Fried, P., Niggli, U., 2002. Soil fertility and biodiversity in organic farming. Science 296, 1694–1697.

Mäder, P., Hahn, D., Dubois, D., Gunst, L., Alföldi, T., Bergmann, H., et al., 2007. Wheat quality in organic and conventional farming: results of a 21 year field experiment. J. Sci. Food Agric. 87, 1826–1835.

Matson, P.A., Parton, W.J., Power, A.G., Swift, M.J., 1997. Agricultural intensification and ecosystem properties. Science 277, 504–509.

O'Connell, P.F., 1990. Policy development for the low-input sustainable agriculture program. In: Edwards, C.A., Lal, R., Madden, P., Miller, R.H., House, G. (Eds.), Sustainable Agricultural Systems. Soil and Water Conservation Society, St. Lucis Press, Florida, pp. 453–458.

Ogilvie, L., Hirsch, P.R., Johnston, A.W.B., 2008. Bacterial diversity of the Broadbalk "Classical" winter wheat experiment in relation to long-term fertilizer inputs. Microb. Ecol. 56, 525–537.

Oppenheim, S., 2001. Alternative agriculture in Cuba. Am. Entomol. 47, 216–227.

Perfecto, I., 1994. The transformation of Cuban agriculture after the cold war. Am. J. Altern. Agric. 9, 98–108.

Petersen, B., Snapp, S., 2015. What is sustainable intensification? Views from experts. Land Use Policy 46, 1–10.

Poudel, D.D., Ferris, H., Klonsky, K., Horwath, W.R., Scow, K.M., van Bruggen, A.H.C., et al., 2001a. The sustainable agriculture farming system project in California's Sacramento Valley. Outlook Agric. 30, 109–116.

Poudel, D.D., Horwath, W.R., Mitchell, J.P., Temple, S.R., 2001b. Impacts of cropping systems on soil nitrogen storage and loss. Agric. Syst. 68, 253–268.

Poudel, D.D., Horwath, W.R., Lanini, W.T., Temple, S.R., van Bruggen, A.H.C., 2002. Comparison of soil N availability and leaching potential, crop yields and weeds in organic, low-input and conventional farming systems in northern California. Agric. Ecosyst. Environ. 90, 125–137.

Powlson, D.S., Bhogal, A., Chambers, B.J., Coleman, K., Macdonald, A.J., Goulding, K.W.T., et al., 2012. The potential to increase soil carbon stocks through reduced tillage or organic material additions in England and Wales: a case study. Agric. Ecosyst. Environ. 146, 232–233.

Rosset, P., Benjamin, M., 1994. The Greening of the Revolution: Cuba's Experiment with Organic Agriculture. Ocean Press, Melbourne.

Rosset, P.M., Altieri, M.A., 1997. Agroecology versus input substitution: a fundamental contradiction of sustainable agriculture. Soc. Nat. Res. 10, 283–295.

Rosset, P.M., Sosa, B.M., Jaime, A.M.R., Lozano, D.R.A., 2011. The Campesino-to-Campesino agroecology movement of ANAP in Cuba: social process methodology in the construction of sustainable peasant agriculture and food sovereignty. J. Peasant Stud. 38, 161–191.

Rothamsted Research, 2012. Guide to the Classical and Other Long-Term Experiments, Datasets and Sample Archive. Lawes Agricultural Trust Co. Ltd.., Online at: <http:www.rothamsted.ac.uk >.

Schnepf, M., 1990. The promise of low-input agriculture: a search for sustainability and profitability. J. Soil Water Conserv. 45, 1–159.

Scow, K.M., Somasco, O., Gunapala, N., Lau, S., Venette, R., Howard, R., et al., 1994. Transition from conventional to low-input agriculture changes soil fertility and biology. California Agric. 48, 20–26.

Shen, S.M., Hart, P.B.S., Powlson, D.S., Jenkinson, D.S., 1989. The nitrogen cycle in the Broadbalk wheat experiment: ^{15}N-labelled fertilizer residues in the soil and in the soil microbial biomass. Soil Biol. Biochem. 21, 529–533.

Temple, S.R., Friedman, D.B., Somasco, O., Ferris, H., Scow, K., Klonsky, K., 1994a. An interdisciplinary, experiment station-based participatory comparison of alternative crop management systems for California's Sacramento Valley'. Am. J. Altern. Agric. 9, 65–72.

Temple, S.R., Somasco, O., Kirk, M., Friedman, D., 1994b. Conventional, low-input and organic farming systems compared. California Agric. 48, 14–19.

Tilman, D., Balzer, C., Hill, J., Befort, B.L., 2011. Global food demand and the sustainable intensification of agriculture. Proc. Nat. Acad. Sci. 108, 20260–20264.

USDA, 2008. Cuba's food and agriculture situation report. Office of Global Analysis, FAS, USDA. Available online at <https://www.ilfb.org/media/546435/fas_report_on_cuba.pdf > .

Watts, C.W., Clark, L.J., Poulton, P.R., Powlson, D.S., Whitmore, A.P., 2006. The role of clay, organic carbon and long-term management on mouldboard plough draught measured on the Broadbalk wheat experiment at Rothamsted. Soil Use Manag. 22, 334–341.

Wezel, A., Bellon, S., Doré, T., Francis, C., Vallod, D., David, C., 2009. Agroecology as a science, a movement or a practice. A review. Agron. Sustain. Dev. 29, 503–515.

Wezel, A., Casagrande, M., Celette, F., Vian, J.F., Ferrer, A., Peigné, J., 2014. Agroecological practices for sustainable agriculture. A review. Agron. Sustain. Dev. 34, 1–20.

Zak, D.R., Holmes, W.E., White, D.C., Peacock, A.D., Tilman, D., 2003. Plant diversity, soil microbial communities, and ecosystem function: are there any links? Ecology 84, 2042–2050.

CHAPTER 14

Agroforestry Practices and Soil Ecosystem Services

Ranjith P. Udawatta, Clark J. Gantzer and Shibu Jose
University of Missouri, Columbia, MO, United States

14.1 Introduction

According to the 2005 Millennium Ecosystem Assessment (MEA; MEA, 2005), humans are depleting the Earth's natural resources and living capital, decreasing biodiversity, and putting the well-being of current and future humanity at risk. Soil degradation, including soil erosion and loss of soil fertility, is a serious issue that has been affecting human welfare for millennia. The Food and Agricultural Organization (FAO; FAO and ITPS, 2015) has identified 13 causes of soil degradation. Furthermore, soil degradation has intensified in recent decades due to anthropogenic pressure. Continuous soil degradation historically has caused the disappearance of numerous civilizations (Montgomery, 2007). Land stewardship determines how the land will sustainably provide into the future.

Agricultural production requires a fertile productive soil. As petroleum fuels become increasingly scarce or their use unacceptable in this century, the use of nitrogen (N) fertilizers to improve soil fertility will become more expensive. Phosphorus (P), an "essential nutrient," is becoming unavailable and expensive, without major change in its use it will become unrecoverable within a few centuries (USGS, 2016). Increasing demand for food and decreasing land resources have called attention to the importance of soil conservation and soil health (Lal, 2010). While investments in soil conservation have been slowly increasing, the projected population of nine billion people by 2050 raises serious concerns about the sustainability of our soil resources for providing food and fiber (FAO and ITPS, 2015). Current agricultural practices cause great stress on soil, water, and biodiversity. These stresses are not sustainable; thus, highly productive farming systems that have a lower environmental impact are urgently needed (Ponisio et al., 2015).

Increased environmental concern led to a draft international assessment: "Protecting Our Planet, Securing Our Future: Linkages Among Global Environmental Issues and Human Needs" (UNEP, NASA, WB, 1998). This publication called for "a more integrative

Soil Health and Intensification of Agroecosystems.
DOI: http://dx.doi.org/10.1016/B978-0-12-805317-1.00014-2

assessment process for selected scientific issues, a process that can highlight the linkages between questions relevant to climate, biodiversity, desertification, and forest issues" (MEA, 2005). It focused on environmental service functions of land use, and the increased importance of an ecosystem approach for resource management (Palm et al., 2007). This emphasis complements agroforestry's perennial multispecies integration of trees and/or shrubs with crops and livestock.

14.2 Agroforestry

Agroforestry (AGF) can meet human needs for food, fiber, and fuel as well as restore degraded soils and contribute to biodiversity while providing beneficial ecosystem services (FAO World Forestry Congress, 2005; MEA, 2005; Jose, 2009). Agroforestry is an "intensive land-use management that optimizes the benefits (physical, biological, ecological, economic, and social) from biophysical interactions created when trees and/or shrubs are intentionally integrated with crops and/or livestock to achieve economic, conservation, and economical goals" (Gold and Garrett, 2009). While agroforestry has historic roots, the formal study, promotion, and use of agroforestry only began in the 1970s (Steppler and Nair, 1987). Early studies conducted in the tropics evaluated socioeconomic factors. Studies conducted during the last 25 years have examined component interactions, production, environmental services, and the socioeconomics of AGF (Jose et al., 2004).

In North America and other temperate zones, riparian buffers, alley cropping, windbreaks, silvopasture, and forest farming are the major AGF practices. Riparian areas (a term relating to wetlands adjacent to rivers and streams) and riparian buffers are different. Riparian buffers are designed plantings located between streams, adjacent wetlands, and upland areas. Riparian buffers along small-order streams are effective in improving and maintaining water and soil quality (Fig. 14.1) (Naiman et al., 2005). These buffers serve as a final defense against sediment and nutrient loss from watersheds while protecting streams from sedimentation.

In alley cropping, agronomic or horticultural crops or pastures are grown between widely spaced single or multispecies trees, shrubs, and grass planted in multiple rows. Thus, spatial heterogeneity is produced by using an agroforestry system that creates differences within a landscape and in above- and below-ground production.

The use of windbreaks for protection of structures, crops, and livestock from wind and blowing snow is a practice that also provides beneficial wildlife habitat (Brandle et al., 2004, 2009). Generally windbreaks are designed with one or more rows of trees or shrubs to reduce wind speed that causes soil erosion and to produce a sheltered local microclimate. Since windbreaks reduce evaporative loss of water from soil and plants, crops grown beyond the fetch of windbreaks may have improved growing conditions (Brandle et al., 2009; Rivest and Vézina, 2015).

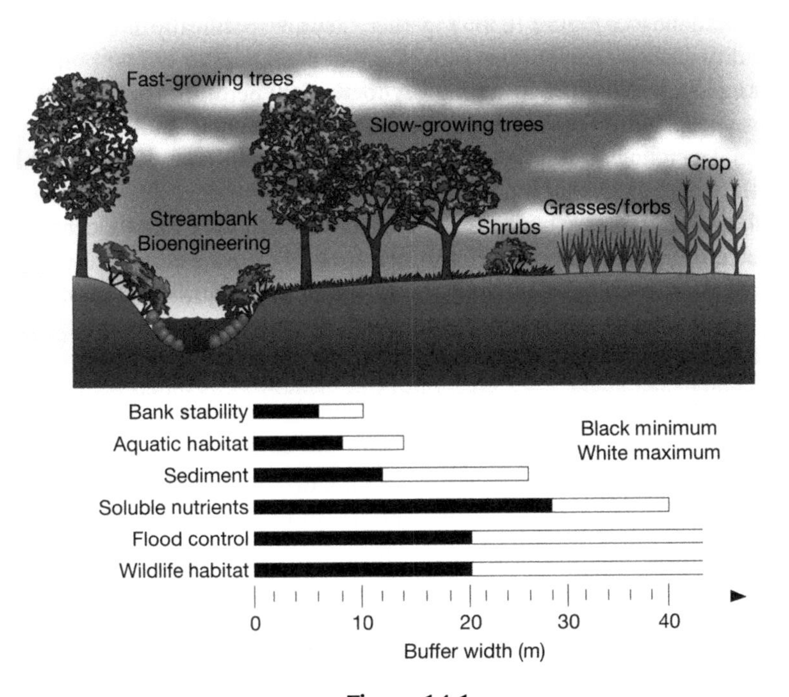

Figure 14.1

Four-zone riparian buffer and environmental benefits and associated buffer widths. *Naiman, R.J., Decamps, H., McClain, M.E., 2005. Conservation restoration. In: Naiman, R.J. et al. (Eds.), Riparia: Ecology, Conservation, and Management of Streamside Communities. Elsevier, New York, pp. 269–325.*

Silvopasture integrates trees, forage crops, and livestock into a functional structure to optimize benefits of improved biophysical interactions among soil, water, nutrients, biology, and microclimate. It is the most common form of AGF in North America (Clason and Sharrow, 2000; Nair et al., 2008; Sharrow et al., 2009). Silvopasture requires regular, consistent management of trees, forages, and livestock. To be successful, a producer must understand management of all three components and their interactions. These systems produce a variety of products including nontimber forest products such as wild game, mushrooms, honey, cork, and firewood. An open oak parkland, known as "dehesa," a type of savannah, is a multifunctional silvopasture system practiced over thousands of years in Spain and Portugal (Linares, 2007).

In forest farming, high-value specialty crops are cultivated under protection of a forest canopy for shade and microclimate control. Crops like ginseng, mushrooms, and ferns are grown under forests and are sold for medicinal, culinary, and ornamental uses.

In tropical regions with small landholding, these practices are seldom segregated, and two or three practices are integrated within an area. In most of these farms, trees and shrubs may be found with crops and livestock with different canopy levels. Thus, a combination of terms such as agrosilvopasture and agrisilviculture are used to describe agroforestry practices in the tropical regions. In these systems, it is difficult to estimate the beneficial soil ecosystem services from a particular AGF practice.

The selection of perennial species for agroforestry practices including trees, shrubs, and grasses is determined by landowner objectives, suitability and availability of plant species, income potential, and environmental and wildlife concerns. Within each AGF practice, there is a continuum of options available to landowners depending on their goals to design the landscape (e.g., whether to maximize the production of interplanted crops, animal forage, or trees). Thus, each AGF practice for each farm increases landscape diversity. The plant, soil microbial, insect, and microclimate differences within these systems stimulate significant above- and below-ground processes leading to healthier ecosystems (Stamps and Linit, 1998; Jose, 2009; Udawatta et al., 2009; Jose et al., 2012). Currently, landowners and farmers are increasingly realizing agroforestry's benefits, and are implementing agroforestry practices for production, conservation, social, and economic benefits (Jose et al., 2012).

Soil ecosystem services of AGF include soil conservation; soil quality or health; carbon sequestration (CS); nutrient cycling and enrichment; sequestration, transformation, and detoxification of chemicals to nontoxic forms. These benefits are difficult to measure independently, because they are interlinked. For example, increased CS increases soil nutrient status by increasing cation exchange capacity, soil stability, and soil quality, and reduces soil erosion.

14.3 Soil Conservation

Ancient civilizations began cultivating valleys containing fertile soils and then began moving up the slopes as the need for food increased. Soil degradation from erosion caused land abandonment that in turn led to clearing of additional forests and grassland for cultivation. After years of fallowing the abandoned land is often returned to agriculture production. During the fallow years, vegetation was reestablished, soil erosion decreased, and soil health improved and thus helped to restore land productivity.

"Land degradation is a long-term loss of ecosystem function and services, caused by disturbances from which the system cannot recover unaided" (Dent, 2007). The FAO and ITPS (2015) have identified 13 forms of soil degradation causing land degradation. Soil degradation caused by poor land use has been identified as a major cause of C loss from soils and increased carbon dioxide (CO_2) emission to the atmosphere (Albrecht and Kandji, 2003). Land degradation from erosion of conventional agricultural systems is three times greater than that in conservation agriculture, and over 75 times greater compared to erosion under native vegetation (Montgomery, 2007). The FAO and ITPS (2015) estimated global

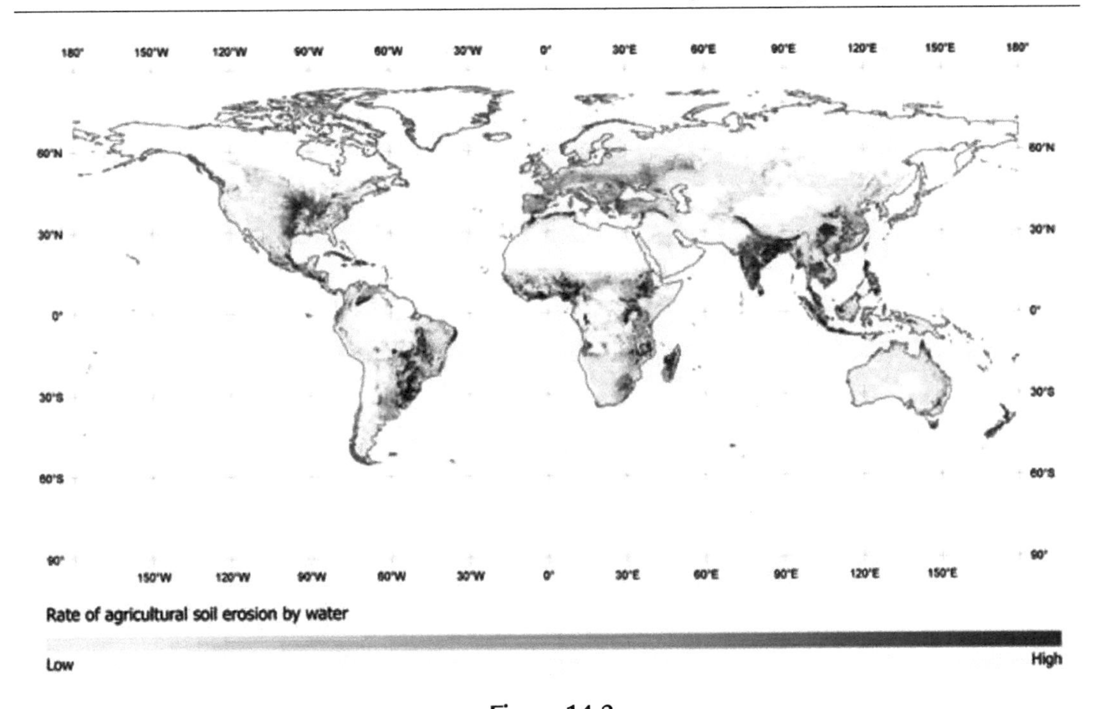

Figure 14.2

Global soil erosion by water on agricultural lands. *FAO and ITPS, 2015. Status of the World's Soil Resources (SWSR) – Main Report. Food and Agriculture Organization of the United Nations and Intergovernmental Technical Panel on Soils, Rome, Italy.*

soil erosion by water is 20–30 Gt year^{-1}, and erosion caused by tillage may amount to ~5 Gt year^{-1} (Fig. 14.2). In Mediterranean regions, Spain has the greatest estimated soil loss of ~200 t ha^{-1} year^{-1}, while France had ~15 t ha^{-1} year^{-1} loss (European Environment Agency, 2003; Correal et al., 2009). In the United States, sheet and rill soil erosion on cropland estimated with the Revised Universal Soil Loss Equation (RUSLE) decreased from 1.7×10^9 t year^{-1} in 1982 to 960×10^6 t year^{-1} in 2007, a 43% reduction (USDA, 2011).

Soil degradation damages soil properties and degrades soil functions, reducing ecosystem services of soil and land productivity (Fig. 14.3A) (Montgomery, 2007; Al-Kaisi, 2008). Soil erosion results in a shrinkage of agricultural land as degraded land (land degraded caused by erosion) is abandoned (Lal, 1996; Bakker et al., 2005; Boardman, 2006). In many regions, soil erosion greatly exceeds the rate of soil formation (Fig. 14.3B). Soil erosion may account for a 10% increase in the total US agriculture energy usage to offset losses of nutrients, water, and soil productivity (Pimentel et al., 1995). The loss of soil productivity may be reduced by increased use of pesticides, fertilizers, and irrigation, but these practices often contribute to off-farm pollution and hazards associated with water pollution (Pimentel et al., 1995).

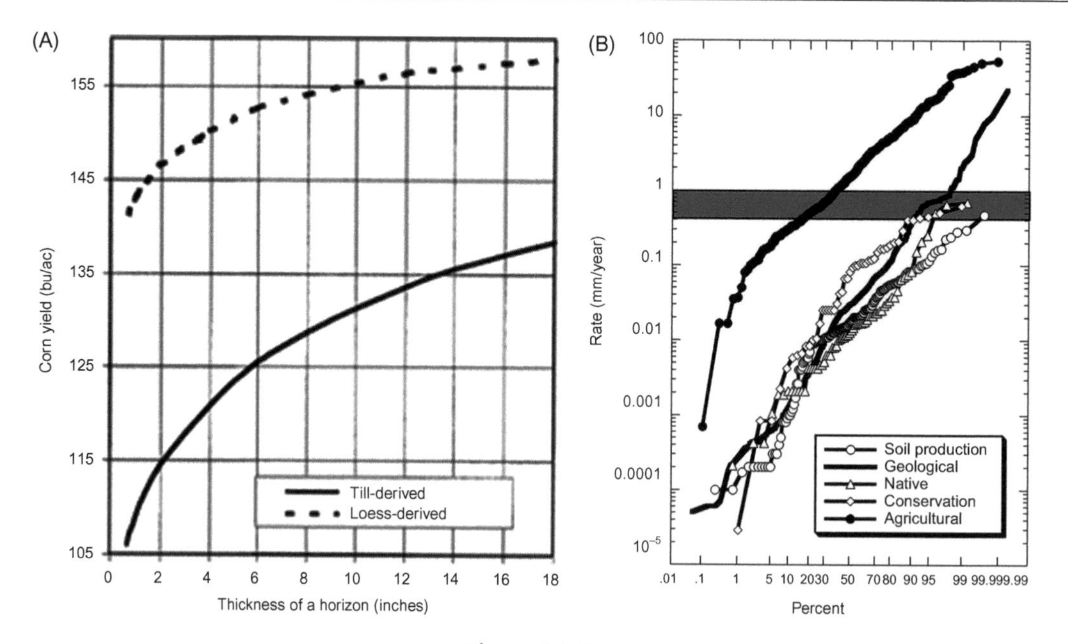

Figure 14.3
Top soil loss and reduction in crop yields (A) and soil production and loss rates for geologic, native, conservation, and agricultural practices (B). *(A) Kazemi, M.L., Dumenil, L.C., Fenton, T.E., 1990. Effects of accelerated erosion on corn yields of loess derived and till-derived soils in Iowa. Unpublished Technical Report. Department of Agronomy, Iowa State University, pp. 1–102 (Kazemi et al., 1990); (B) Montgomery, D.R., 2007. Soil erosion and agricultural sustainability. Proc. Natl. Acad. Sci. USA. 104, 13268–13272.*

The role of AGF in reducing soil degradation has been reviewed by Young (1997) and Buresh and Tian (1998) among others. The simplest explanation is that trees and shrubs provide plant litter that protects soil from raindrop impact that dislodges soil. A single large raindrop falling at terminal velocity has kinetic energy of 2.01 mJ that dislodges soil to produce splash erosion (Gantzer et al., 1987). AGF also may slow runoff by obstructing the flow pathway and thus reduce its capacity to transport sediment.

Rainfall erosivity, slope degree, slope length, and protective vegetative cover determine the susceptibility of soil to water erosion (Wischmeier and Smith, 1978; Renard et al., 1997; FAO and ITPS, 2015). Increased runoff is a main reason for increased sediment transport to surface waters causing nonpoint source pollution (NPSP; Seobi et al., 2005). Upland agroforestry buffers planted on the contour have shown significant reductions in sediment loss in long-term corn (*Zea mays* L.)-soybean (*Glycine max* (L.) Merr.) rotational paired watershed study in the Central Claypan Areas Illinois and Missouri, United States (Major Land Resource Area [MLRA] 113) (Fig. 14.4) (USDA-NRCS, 2006; Udawatta et al., 2011a,b). Claypan soils have a thin silt loam topsoil but are underlain with a high clay content subsoil having an extremely low hydraulic conductivity, which restricts downward water movement

Figure 14.4
Upland pin, bur, swamp white oak with grass buffers on a corn-soybean alley cropping agroforestry practice in Missouri, United States.

(Seobi et al., 2005). Thus, these soils have a potential for high runoff and enhanced sediment and nutrient loss (Zhu et al., 1989; Blanco-Canqui et al., 2004). Seobi et al. (2005) showed there was 33% greater porosity in the root zone with AGF compared to cropped areas, improving soil hydraulic function. Upland contour buffers also reduce slope length and slope degree, further reducing the soil erosion.

In the US Midwest, researchers have shown improvements in infiltration, water-holding capacity, bulk density, and porosity parameters when AGF was integrated in row crop and grazing management practices (Fig. 14.5) (Bharati et al., 2002; Seobi et al., 2005; Udawatta et al., 2006; Kumar et al., 2008). Infiltration rates were five times greater within AGF buffer soils compared to soils cultivated with corn, soybean, and pasture (Bharati et al., 2002). Infiltration was improved by 80 times in the soils under buffers as compared to grazing areas in loess soils (Kumar et al., 2008). In a corn-soybean rotation, AGF buffer soils had 40 times greater infiltration compared to crop claypan soils (Seobi et al., 2005).

Living and dead plant material on the soil surface reduce the flow velocity by intercepting and dissipating energy and enhance in-field sedimentation, reducing soil, nutrient, and chemical loss to surface water (Fig. 14.6) (Zhu et al., 1989). Dense, deep roots enhance water infiltration and improve water redistribution within soils (Udawatta et al., 2011a). Studies in temperate North America and South America have shown that AGF buffers maintained lower water contents, increasing air-filled porosity, creating potential soil water storage for future rainfall that increased infiltrability, and thus reducing runoff (Balandier et al., 2008; Udawatta et al., 2011a).

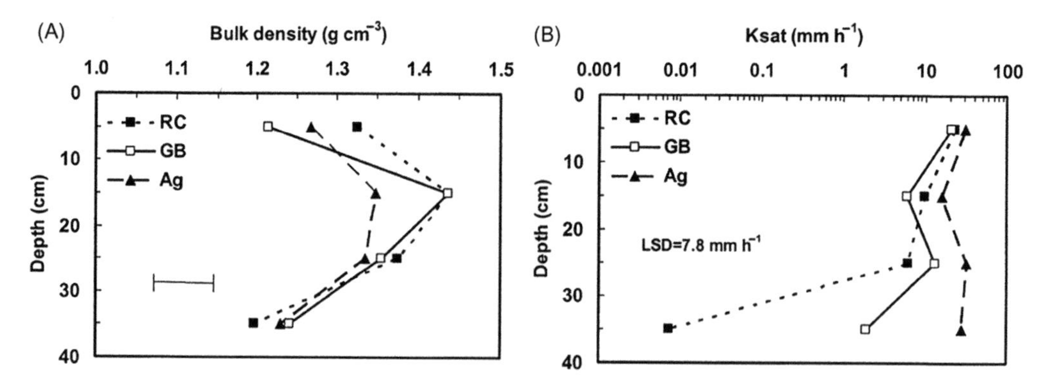

Figure 14.5

Bulk density (A) and saturated hydraulic conductivity (Ksat) (B) by depth for row crop (RC), grass buffer (GB), and agroforestry (Ag) buffer at the paired watershed study at Greenley Center, University of Missouri, United States. The bar and the value indicate LSD at 0.05 for bulk density and Ksat. *Seobi, T., Anderson, S.H., Udawatta, R.P., Gantzer, C.J., 2005. Influences of grass and agroforestry buffer strips on soil hydraulic properties. Soil Sci. Soc. Am. J. 69, 893–901.*

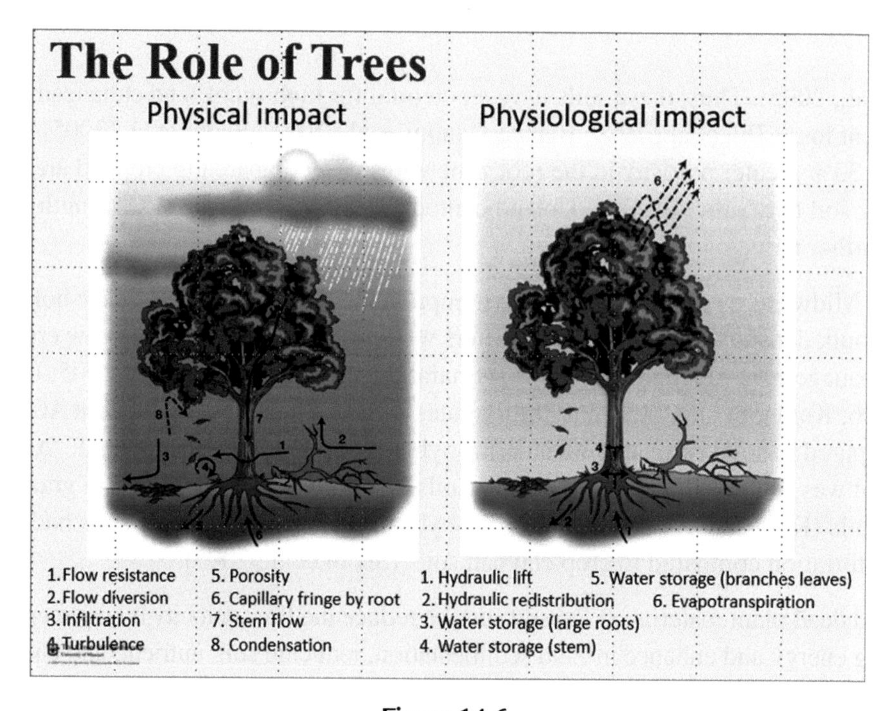

Figure 14.6

Physical and physiological impact of trees helping reduce soil erosion and nonpoint source pollution. *Naiman, R.J., Decamps, H., McClain, M.E., 2005. Conservation restoration. In: Naiman, R.J. et al. (Eds.), Riparia: Ecology, Conservation, and Management of Streamside Communities. Elsevier, New York, pp. 269–325.*

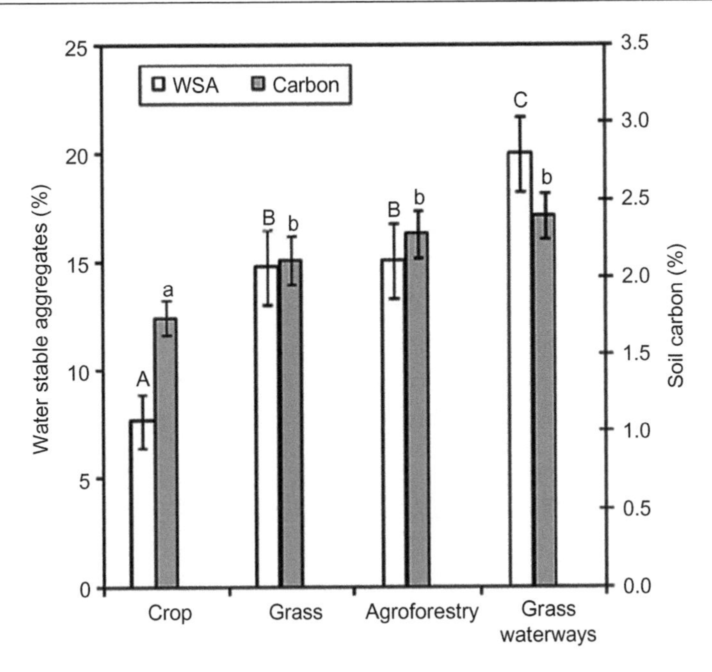

Figure 14.7
Water-stable aggregates and soil carbon for crop, grass, agroforestry, and grass waterway of a corn-soybean alleycropping watershed with agroforestry (tree + grass) and grass buffers at Greenley Center, University of Missouri, United States. *Udawatta, R.P., Kremer, R.J., Garrett, H.E., Anderson, S.H., 2009. Soil enzyme activities and physical properties in a watershed managed under agroforestry and row-crop system. Agric. Ecosyst. Environ. 131, 98–104.*

Studies in some tropical areas and in Europe also have shown significant reduction in soil erosion, improved infiltration, and reduced bulk density as influenced by AGF. Studies throughout the world have proven the benefit of AGF for reducing soil erosion and degradation because trees, shrubs, and other perennial vegetation help soil conservation and improve soil health (Figs. 14.5–14.7) (Naiman et al., 2005; Hatfield and Sauer, 2011).

Agroforestry areas of perennial vegetation are usually managed with little soil disturbance; thus, soil aggregate stability and CS increase in AGF areas (Fig. 14.7) (Udawatta et al., 2008; Gupta et al., 2009; Paudel et al., 2011). Greater accumulation of organic matter, nutrients in biomass, and root density reduce runoff from cropland when AGF is integrated into fields compared to soil with only cropland. As soil organic matter (SOM) increases, a concomitant increase in aggregate stability has been found, reducing soil erodibility (Al-Kaisi et al., 2014). Generally, multispecies systems provide greater organic root exudates and a more favorable microenvironment to support microbial and soil faunal communities. Root exudates, faunal excretion, and detritus add to SOM and improve soil aggregate stability (Kremer and Kussman, 2011).

Riparian areas are ecosystems within and bordering aquatic ecosystems such as streams, rivers, lakes, estuaries, and bays, and their adjacent side channels, flood plains, and wetlands (Schultz et al., 2009). The USDA Forest Service (Welsch, 1991) defines these as: "the aquatic ecosystem and the portions of the adjacent terrestrial ecosystem that directly affect or are affected by the aquatic environment." Since these areas are located at the edge of the water, they serve as the final defense against sediment and nutrient loss from fields to the water. In an ideal landscape, the goal is to keep soil and materials close to their point of origin. The US Dept. of Agriculture Practice codes No 332, 393, 412, and 601 (http://www.nrcs.usda.gov/wps/portal/nrcs/detailfull/national/technical/cp/ncps/?cid=nrcs143_026849#A) have defined contour buffer strips, filter strips, grassed waterways, and vegetative barriers, respectively, for upland areas to increase sediment and nutrient retention in the areas of origin.

Riparian buffers can filter nutrients and sediment and reduce water erosion (Fig. 14.1), (Broadmeadow and Nisbet, 2004; Nair and Graetz, 2004; Schultz et al., 2004; Udawatta et al., 2002, 2011b). Sediment filtering efficiencies as high as 98% have been observed with various buffer widths (Dillaha et al., 1989; Robinson et al., 1996; Patty et al., 1997; Rhoades et al., 1998a; Schmitt et al., 1999; Lee et al., 2003). Research also has shown that most of the soil sediments within the first 4–7.5 m of the buffer strip and sediment deposition through the buffer diminishes exponentially with increasing width (Foster, 1982; Schmitt et al., 1999; Blanco-Canqui et al., 2004, 2006). In a metadata analysis, Liu et al. (2008) showed increasing sedimentation with buffer width (Fig. 14.8). In their study, sedimentation decreased with increasing buffer widths greater than 8 m and this agreed with findings of other studies. The width of the riparian and other types of buffers determines the efficiency of each buffer

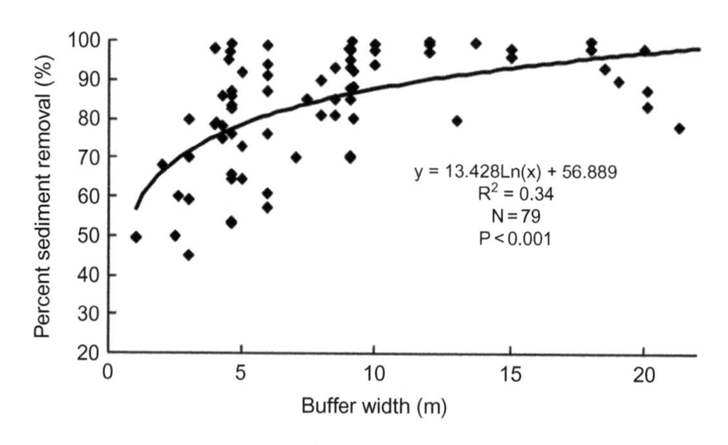

Figure 14.8

Sediment trapping efficiency and buffer width. *Liu, X., Zhang, X., Zhangm M., 2008. Major factors influencing the efficacy of vegetated buffers on sediment trapping: a review and analysis. J. Environ. Qual. 37, 1667–1674.*

type. The selection of width is dependent on landowner objectives, landscape, weather, and management practices. Wider buffers are more effective for dissolved nutrient retention, while wider grass and shrub zones are effective in controlling sediment loss from watersheds (Fig. 14.1).

14.4 Carbon Sequestration

In December 2015, leaders from 195 countries met in Paris to negotiate a landmark accord on climate change (FCCC, 2015). The agreement called for all nations to pledge to limit carbon (C) emissions enough to keep average temperature to below 2°C above the preindustrial levels. The governments also agreed to meet every 5 years (starting in 2020) with updated plans on cutting CO_2 emissions as well as publicly reporting their progress every 5 years (starting in 2023). Secretary-General Ban Ki-Moon said, "For the first time, we have a truly universal agreement on climate change, one of the most crucial problems on earth." Although an ambitious plan to combat climate change, the agreement will not begin until the 55 countries that produce at least 55% of global greenhouse gas (GHG) emissions ratify the agreement. This agreement is historic in recognizing the importance of "reducing GHG emissions from deforestation and forest degradation, and the role of conservation, sustainable management of forests and enhancement of forest C stocks in developing countries, …[and] the integral and sustainable management of forests," explicitly in the text.

Success in reducing C emissions depends on global peer-pressure and the actions of future governments. Countries have to adopt measures to reduce C emissions, and capture more C from the atmosphere (store in the biomass/soil/deep sea ….) for the indefinite future. CS is removing atmospheric CO_2, by the uptake of CO_2 during photosynthesis process of carbohydrate accumulation and subsequently storing it in soil and biomass for an extended time (UNFCCC, 2007). The world needs CS techniques that provide social, environmental, and economic benefits while reducing atmospheric CO_2 concentrations.

Agroforestry practices, as a strategy for soil CS, were approved by both the afforestation and reforestation programs and under the Clean Development Mechanisms of the Kyoto Protocol (IPCC, 2007; Smith et al., 2007; Watson et al., 2000). It was recognized that perennial vegetation in AGF has many advantages over agricultural monoculture and its applicability in agricultural lands and reforestation programs for storing C in above- and belowground biomass, soil, living and dead organisms, and by root exudates (Cairns and Meganck, 1994; Albrecht and Kandji, 2003; Pinho et al., 2012). Perennial vegetation (trees, shrubs, and grasses) in AGF is more efficient in CS than annual monoculture crops because both forest and grassland sequestration and storage patterns are active and allocate a higher percentage of C to belowground biomass through an extended growing season (Figs. 14.9 and 14.10) (Schroeder, 1994; Kort and Turnock, 1999; Sharrow and Ismail, 2004; Morgan et al., 2010).

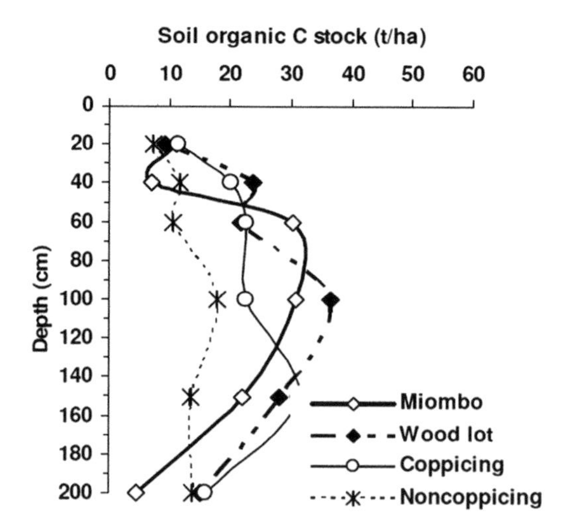

Figure 14.9

Soil carbon storage belowground Miombo, woodlots, coppicing, and noncoppicing fallow in Zambia. *Sileshi, G., Akinnifesi, F.K., Ajayi, O.C., Chakeredza, S., Kaonga, M., Matakala, P.W., 2007. Contributions of agroforestry to ecosystem services in the miombo eco-region of eastern and southern Africa. Afr. J. Environ. Sci. Technol. 1, 68–80.*

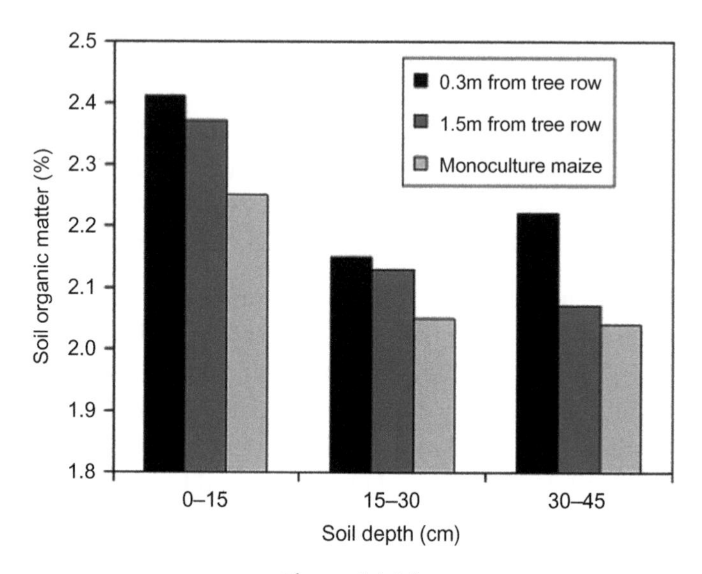

Figure 14.10

Soil organic matter percentage decreased with increasing distance and depth from tree rows for a 4-year-old red alder-corn alley cropping system in western Oregon, United States. *Adapted from Seiter, S., William, R.D., Hibbs, D.E., 1999. Crop yield and tree-leaf production in three planting patterns of temperate-zone alley cropping in Oregon, USA. Agroforest. Syst. 46 273–288 (Seiter et al., 1999).*

The enhanced CS concept is based on the efficient use of resources by plant communities in AGF that are structurally and functionally more diverse and complex than monoculture or simple grass systems (Sanchez, 2000; Sharrow and Ismail, 2004; Thevathasan and Gordon, 2004; Steinbeiss et al., 2008). For example, light energy use at different canopy levels of a pine-pasture alley cropping system is more efficient than in a monocropping system. Trees, grasses, and shrubs in a corn-soybean alley cropping practice begin photosynthesis before the cash crop is established and continue after the cash crop is harvested. Physiological resources needed by plants vary in amounts and availability during the year by species, and ecosystems possess different structural and functional means to obtain these needs (Jose et al., 2004). For example, if one species differs in use of even one of the resources such as light saturation of C3 versus C4 plants, CS will increase. Multiple vegetative species promote larger and more diverse microbial and fauna communities that could contribute to additional CS (Udawatta et al., 2009; Paudel et al., 2011). The sequestration potential varies by ecosystem region, soil, climate, management practices, species composition, and ages of individual species (Jose, 2009; Pinho et al., 2012). CS does take time and in some alley cropping practices soil C may decrease with time as the system matures (Thevathasan and Gordon, 2004; Oelbermann et al., 2006a, b; Bambrick et al., 2010).

While soil C is beneficial for agricultural productivity, it is 40–70 times more valuable for improving environmental services (Sparling et al., 2006). C can be stored in trees in situ in biomass and soil, and ex situ in products (Montagnini and Nair, 2004). C in soils, deep roots, or as wood-products can stay in these pools of C for extended periods of time, and if the systems are managed sustainably, C can be retained for centuries (Dixon, 1995). For example, restoring degraded cropland and pasturelands is an effective means for storing C in soil (Albrecht and Kandji, 2003). Riparian buffers, being well-watered, have an enormous potential for CS while improving other ecosystem services (Udawatta and Jose, 2012).

Indirect benefits of AGF for CS, such as improved nutrient enrichment, have been reported (Chirwa, 2006; Oelbermann et al., 2006b). Deep roots bring mineral nutrients to the topsoil and provide them for growth of cash crops and AGF vegetation (Chirwa, 2006). The quantity and quality of residue (pruned material, natural leaf fall, and litter) supplied by trees, shrubs, and grass in AGF systems increase soil C as it is incorporated into soil during breakdown (Oelbermann et al., 2006b). Decomposition of AGF plant materials and legumes enriches soil nutrient status and enhances the CS potential. Repeated applications of AGF residue improve soil quality and provide improved conditions for soil organisms and plants. Fertilizers and manures for cash crops or animals also provide nutrients for AGF's perennial vegetation for optimum growth. Because of the network of nutrient cycling, greater soil C, microbial diversity, and soil N have been observed in alley cropping practices, which in turn enhance CS (Chirwa et al., 2007; Udawatta et al., 2009, 2014; Weerasekera et al., 2016). Improved soil moisture conditions have also been observed in alley cropping and silvopasture systems that also help improve plant growth (Balandier et al., 2008; Udawatta et al., 2011a).

Agroforestry can help reduce pressure on forests, which are the largest sink of terrestrial C (Montagnini and Nair, 2004). Researchers in Kenya, Madagascar, and Sumatra have shown positive effects of AGF and reduced pressure on forests (Young, 1997; Sileshi et al., 2007). Microclimate induced shade, improved soil conditions, and reduced stress on trees help enhance CS (Sauer et al., 2007; Udawatta et al., 2009). AGF also can decrease the C footprint of agriculture by reducing the need for synthetic fertilizers and by substituting forest material and energy that is produced from CO_2. Generally, AGF systems improve soil aggregate stability and likely influence more C storage in finer soil aggregates (Fig. 14.7). In a 30-year mature parkland study, more C was found in finer soil aggregates ($<53\,\mu m$), a fraction which is more stable than C in larger aggregates (Takimoto et al., 2008). Others have shown more C in larger soil aggregates of mixed (8–55-year-old) and 8-year-old systems with trees (Haile et al., 2008; Takimoto et al., 2008). This also implies the importance of long-term maintenance for C integration into more stable soil aggregates.

For AGF small land holdings in the tropics, the CS ranges from 1.5 to 3.5 t C ha^{-1} $year^{-1}$ (Montagnini and Nair, 2004). In Zimbabwe, Kaonga (2005) reported 74 and 140 t ha^{-1} in aboveground and soil, in 2–12-year-old *Leucaena* spp. woodlots. In these AGF systems, coppicing stored more C in aboveground material and soil than noncoppiced systems (Fig. 14.9), (Sileshi et al., 2007). CS in agroforestry systems of West African Sahel was investigated by Takimoto et al. (2009) by comparing two traditional parkland systems, two improved agroforestry systems (live fence and fodder bank), and abandoned land. The authors concluded that improved AGF practices, such as living fences and fodder banks, sequestered more C, especially in soils, than traditional parklands. Kaonga and Bailey-Smith (2009) quantified tree and soil C stocks and their response to different tree species and soil clay contents in 2-, 4-, and 10-year-old improved fallows in Zambia and concluded that C stored in trees and soil of fallow land could be increased by planting selected tree species on soils with high clay content. In a poplar-based AGF system in northwest India, soil organic C concentration and pools were greater in soils under AGF and increased with tree age (Fig. 14.11) (Gupta et al., 2009). Saha et al. (2009) examined how tree density and plant-stand characteristics of Indian home gardens influenced soil CS. Results showed that the potential to sequester soil C increased with species richness and tree density. All these case studies add to the body of literature indicating that AGF systems can sequester greater amounts of above- and belowground C compared to traditional farming systems. These authors concluded that, generally, AGF on arid, semiarid, and degraded soils had a lower CS potential than those on fertile humid sites; and temperate AGF systems had lower rates compared to tropical systems.

Well-managed silvopastoral systems also improve CS and land productivity. Total C in silvopastoral systems in Latin America varied between 68 and 204 t ha^{-1}, with most C stored in the soil, while annual C increments varied between 1.8 and 5.2 t ha^{-1} (López et al., 1999). On the slopes of the Ecuadorian Andes, total soil C increased from 7.9% under open *Setaria sphacelata* pasture to 11.4% beneath the canopies of *Inga* sp., but no differences were

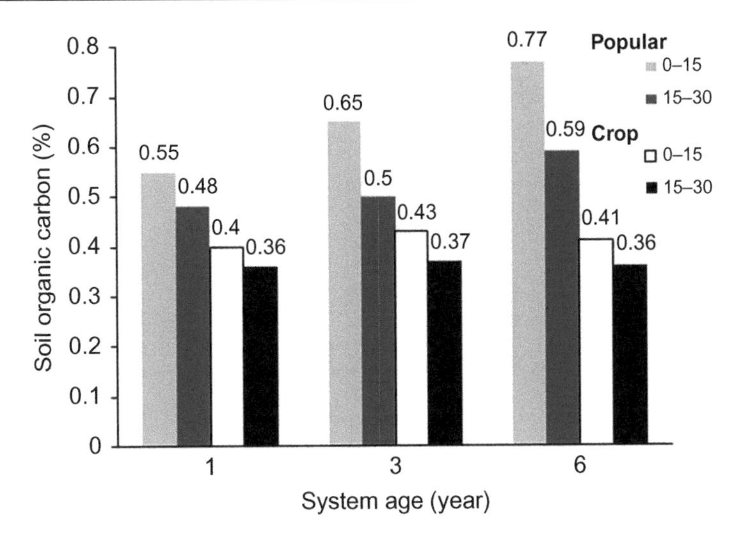

Figure 14.11

Soil organic carbon in 0–15 and 15–30 cm depths for poplar-based agroforestry system and crop area in India. *Gupta, N., Kukal, S.S., Bawa, S.S., Dhaliwal, G.S., 2009. Soil organic carbon and aggregation under poplar based agroforestry system in relation to tree age and soil type. Agroforest. Syst. 76, 27–35.*

observed under *Psidium guajava*. Soils under *Inga* sp. contained an additional 20 Mg C ha^{-1} in the upper 15 cm compared to open pasture (Rhoades et al., 1998b).

Udawatta and Jose (2012) have estimated that riparian buffers, alley cropping, silvopasture, windbreaks, and forest farming could contribute an additional 548.4 Tg C $year^{-1}$. The analysis used peer-reviewed data and land areas under each management and potential lands that could be maintained under these practices. For example, CS of riparian buffers (4.7 Tg C $year^{-1}$) was estimated by considering a 60-m wide buffer on 5% of the total river length multiplied by a conservative CS value of 2.6 Mg C ha^{-1} $year^{-1}$ accrual rate. Fig. 14.12 also compares the forest, pasture, and cropland C sequestration with other published data. The total potential CS by AGF in the United States could offset the current US CO_2 emissions (1.6 Pg C $year^{-1}$ from burning fossil fuel such as coal, oil, and gas) by 34%.

The CS potential of agroforestry for global, regional, and continental scales has been documented (Dixon, 1995; IPCC, 2000; Nair et al., 2009b; Udawatta and Jose, 2012). Dixon (1995) estimated a global sequestration potential of 1.1–2.2 Pg of C over 50 years on 585–1215 million ha of land across Africa, Asia, and the Americas. The global CS potential is 1.9 Pg C over 50 years on 1023 million ha under AGF with median sequestration of 94 Mg ha^{-1} (Dixon, 1995; Jose, 2009; Nair et al., 2009b). IPCC (2000) estimated 12 Gg C ha^{-1} $year^{-1}$ by 2010 and 17 Gg C ha^{-1} $year^{-1}$ by 2040 CS by improving current management practices on existing AGF practices. Converting 630 million ha of unproductive crop and pasturelands to AGF could sequester an additional 391 Pg C $year^{-1}$ by 2010 and 586 Pg C $year^{-1}$ by 2140 (IPCC, 2000).

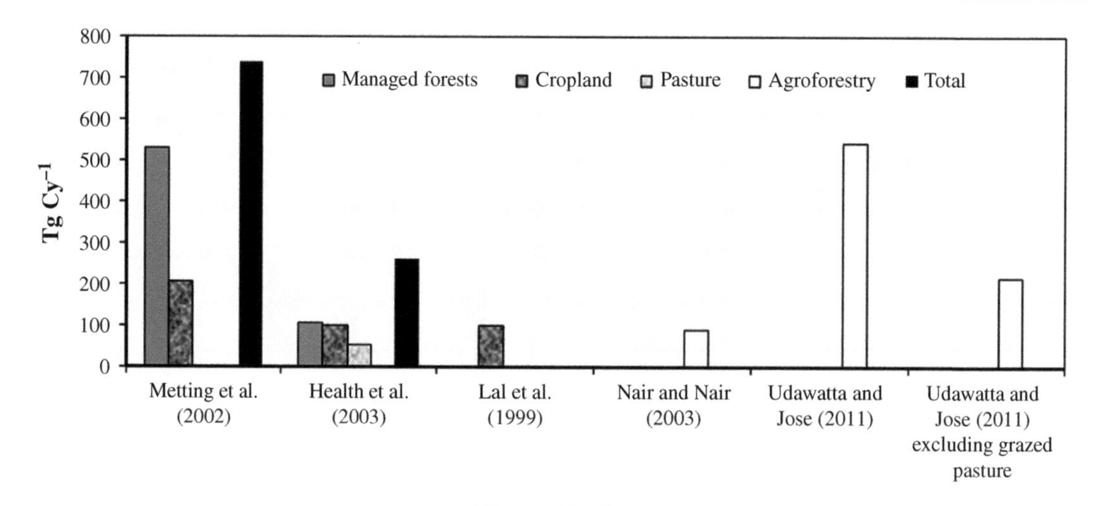

Figure 14.12

Carbon sequestration potential for various management systems in the United States. *Udawatta, R.P., Jose, S., 2012. Agroforestry strategies to sequester carbon in temperate North America. Agroforest. Syst. doi:10.1007/s10457-012-9561-1.*

Results of many studies recently reported the CS potential of AGF by management practice, country, region, continental, and global scales. However, our understanding of C storage and dynamics in AFS is still limited (Nair et al., 2009a). The inherent variability in C storage associated with soil and climate, and the lack of standard methodologies, long-term studies, and estimates for all regions and soils limit comparisons (Nair, 2012; Udawatta and Jose, 2012). Agroforestry is complex with multiple species, diverse soil, climate, landscape, and management practices and therefore, it is necessary to establish regional long-term studies. A comprehensive data set on C distributions in AFS for many countries is lacking in the literature; therefore, commitments and investments are needed to collect such data. Such information will help design better AGF systems for agricultural landscape to enhance CS (Udawatta and Godsey, 2010).

14.5 Nutrient Cycling, Sequestration, and Enrichment

In the United States, average farm size has almost doubled between 1982 (238 ha) and 2007 (448 ha; MacDonald et al., 2013). Grass strips, riparian buffers, filter strips, and other perennial vegetative conservation practices were removed to increase the farm size. Nutrient input can be reduced by reducing the cropland area; however, it would increase the grain price, land value, and the cost of conservation programs (Wu and Lin, 2010; Whittaker et al., 2015). The public demands better nutrient management plans that will improve water quality in all water bodies including rivers, lakes, estuaries, ground water, and hypoxia zones.

Figure 14.13
Improved crop performance around agroforestry practice due to improved soil and water status. *Dent, D., et al., 2007. Chapter 3: land. In: Mnatsakanian, R. (Ed.), Global Environment Outlook 4: Environment for Development. United Nations Environment Programme; Stationery Office distributor, pp. 81–114.*

World Agroforestry Center (ICRAF) reported that 43% of the planet's agricultural land has more than 10% tree cover and 160 million ha of agricultural land has more than 50% tree cover (Zomer et al., 2009). The ICRAF-sponsored project used 1-km resolution geospatial remote sensing data and reported the area under AGF to be about one billion ha of agriculture lands worldwide. Agroforestry's role in maintaining long-term soil productivity, enhancing soil quality, and improving land sustainability has been well documented (Jose, 2009; Udawatta et al., 2009, 2011b; Zomer et al., 2009). Agroforestry research in Africa first examined ways of improving and maintaining soil fertility in annual cropping systems. These studies were conducted to evaluate soil nutrient improvements in parklands and alley cropping by legumes (Charreau and Vidal, 1965; Kang and Reynolds, 1989).

Initial individual tree studies have shown radial patterns in soil nutrient status (Zinke, 1962). These microenvironment islands influenced soil conditions and plant productivity (Pinho et al., 2012). For example, soil rehabilitation and stress reduction improve plant growth and further enhance land productivity (Fig. 14.13) (Fassbender et al., 1991; Beer et al., 1998). The authors have shown improved soil nutrient status because of improvements in soil physical, chemical, and biological status.

Establishment of perennial vegetative buffers along water bodies and upland areas has reduced nutrient losses from row crop and grazing practices and stored those nutrients in the crop areas (Schultz et al., 2009; Udawatta et al., 2011b). Schultz et al. (2009) have proposed a four-zone buffer system consisting of fast-growing trees next to the waterbody, slow-growing trees away from the waterbody in the second zone, shrubs in the third zone and grasses in the fourth zone (Fig. 14.1). Grasses and shrubs help reduce concentration flow and enhance sedimentation. Sediment-bound nutrients are deposited on the soil surface while the roots of the perennial vegetation uses nutrients from the subsurface horizons. Because of its position, riparian zones effectively remove nutrients from the surface and subsurface flow before surface, subsurface, and ground water enters water bodies. Riparian vegetation intercepts soil solution as it passes through the rooting zone prior to entering the water bodies. The AGF buffers enrich soil nutrient status and improve soil properties in these buffers and adjacent areas, and thereby enhance soil nutrient status and land productivity (Jose et al., 2004; Udawatta et al., 2006; Schultz et al., 2009).

Deep roots of perennial vegetation can capture nutrients that were lost from the crop root zone and from the weathering zones and store those nutrients in roots, stems, branches, and leaves (van Noordwijk et al., 1996; Jose et al., 2000; Allen et al., 2004). Subsequently, these nutrients are released to the soil surface when plant material decays. In Kenya, sesbania has reduced subsoil nitrate through plant uptake and reduced the leaching losses (Hartemink et al., 1996; Mekonnen et al., 1997). The "safety net" and "nutrient pump mechanisms" can be improved by selecting species with proportionately greater vertical rooting, thus reducing competition for resources near the crop–tree interface, allowing for improved nutrient uptake for crops (Udawatta et al., 2006, 2014).

Legumes are used in AGF for N-fixation, soil enrichment, production of green manure, fuel wood, and fodder (Rao et al., 1998; Jose, 2009; Nygren and Leblanc, 2009). Increased soil and microbial C, residual N after legume-enriched fallow, long-term accumulation of C and N in soil and increased N mineralization and nitrification rates are other benefits of AGF (Mazzarino et al., 1993; Sierra et al., 2002; Ståhl et al., 2002; Dulormne et al., 2003; Sierra and Nygren, 2005). In Oregon, United States, using an ^{15}N injection technique, Seiter et al. (1995) showed that 32–58% of the total N for corn was obtained from N-fixed by red alder (*Alnus rubra*). The N transfer decreased with distance from the tree. Another ^{15}N study in Costa Rica showed that cocoa (*Theobroma cacao*) displayed some N transfer from legumes (Nygren and Leblanc, 2009).

Initial AGF research (especially in Africa) focused on maintaining soil fertility in annual cropping systems by using leguminous shrub species; e.g., in parkland (Charreau and Vidal, 1965) and in alley cropping (Kang and Reynolds, 1989). Those studies displayed beneficial effects of leguminous species compared to the controls. Additionally, leguminous species provide more nutrients, such as P, potassium (K), calcium (Ca), and magnesium (Mg) than other herbaceous vegetation provides (Szott and Palm, 1996). The authors suggest selection of fast-growing leguminous species for faster restoration of soil nutrients. Various alley cropping systems developed in Europe also demonstrated the beneficial effects of leguminous species in AGF systems.

In the tropics, because of higher temperature and moisture conditions, soils are highly weathered and hold less nutrients; these conditions are conducive to the decomposition of organic matter. In some regions, weathering has resulted in formation of aluminum and iron oxides and these particles do not hold cations required for crop and pasture growth (Pinho et al., 2012). These nutrient's poor soils can be improved by integration of AGF (Pinho et al., 2012). The importance of litter material supplied by AGF for nutrient enrichment and plant growth has been documented for several tropical regions (Uphoff et al., 2006; Sileshi et al., 2009). Additionally, litter and other materials provide cover and food for micro-, meso-, macro-fauna to carryout organic matter decomposition and nutrient cycling. Greater enzyme activity and diversity within AGF promote efficient nutrient cycling (Fig. 14.14) (Udawatta et al., 2009; Paudel et al., 2011). Examining new, established, and old home gardens and a savannah in northern Brazil, Pinho et al. (2011) found progressively increased P, K, Ca, Mg, and organic matter contents and pH in the old home garden (Table 14.1). Trees and shrubs in AGF systems provide organic matter and litter material that are major sources of nutrients.

The quantity and quality of the material, types of nutrients, and release dynamics are determined by the agroforestry vegetation, soil, and climate conditions. Significant increases in basic cations were observed in southern Cameron for Senna (*Senna spectabilis* [DC]) and Flemingia on Ultisols (Hulugalle and Ndi, 1994). Agroforestry in general also provides more efficient cycling of nutrients, improved release and availability of nutrients, a suitable soil environment, and more increased activity of beneficial microorganisms in the rooting zone (Lee and Jose, 2003).

Perennial vegetation in AGF systems provides plant nutrients; however, as the system matures the perennial vegetation competes for resources (Jose et al., 2000; Senaviratne et al., 2012; Udawatta et al., 2014, 2016). Changing the cropping system, selecting suitable species, and employing a correct land management practice may help reduce competition for resources while improving land productivity (Jose et al., 2000; Udawatta et al., 2014, 2016). In mature systems, techniques such as root pruning, root barriers, branch pruning, thinning, and selecting less competitive species can help reduce competition for resources.

14.6 Soil Biological Services

Numerous researchers have reported improvements in physical and chemical soil parameters by AGF in silvopasture, alley cropping, windbreaks, and riparian areas (Figs. 14.5, 14.7, 14.9–14.12) (Seobi et al., 2005; Kumar et al., 2008, 2010; Sauer and Hernandez-Ramirez, 2011). Agroforestry also improves biological soil parameters (Udawatta et al., 2009; Paudel et al., 2011). Soil enzyme activity, microbial diversity, and macrofauna activity

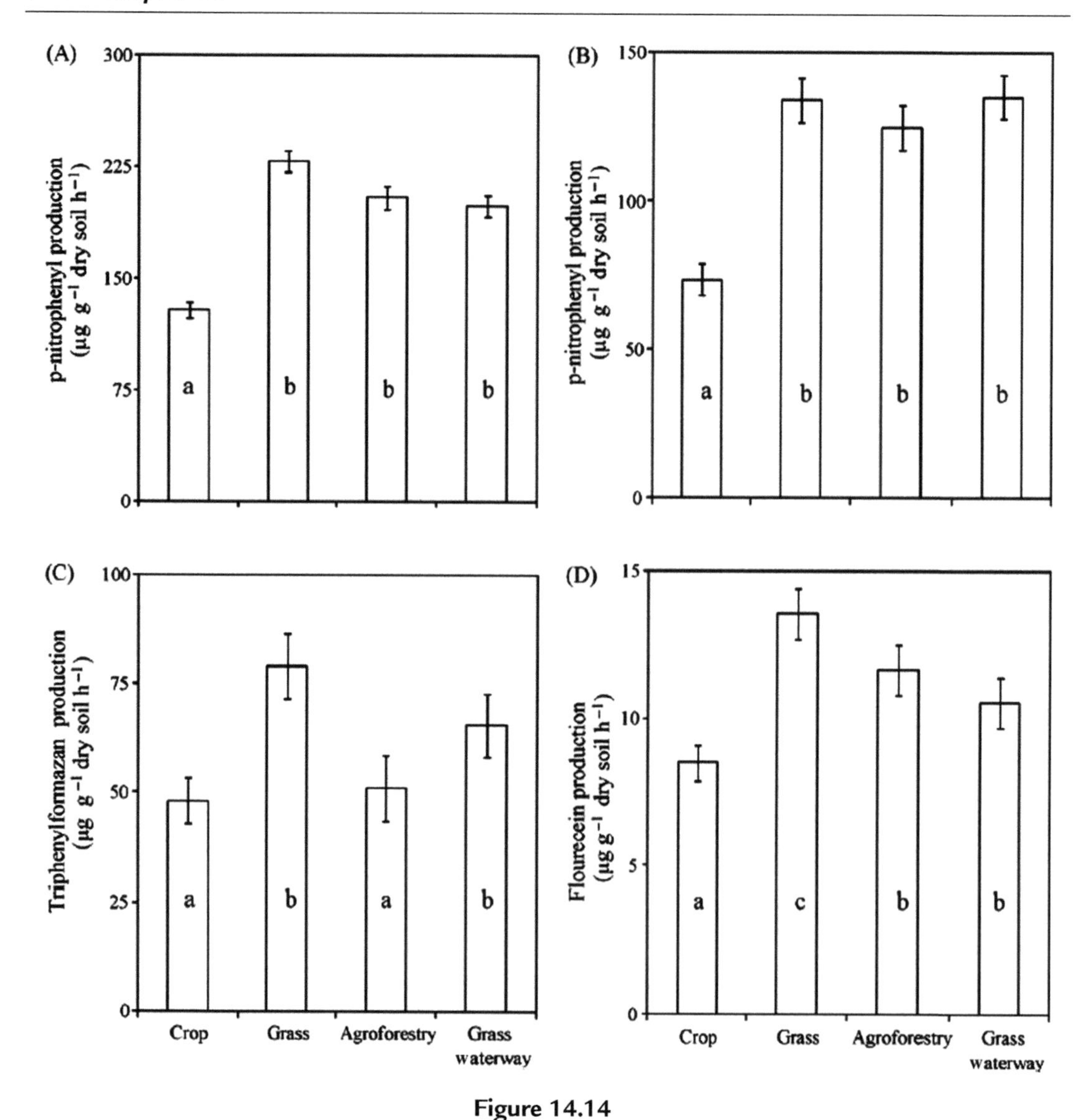

Figure 14.14

β-Glucosidase (A), glucosaminidase (B), dehydrogenase (C), and FDA (D) activities for crop, grass, agroforestry, and grass waterway of a corn-soybean alleycropping watershed with agroforestry (tree + grass) and grass buffers at Greenley Center, University of Missouri, United States. *Udawatta, R.P., Kremer, R.J., Garrett, H.E., Anderson, S.H., 2009. Soil enzyme activities and physical properties in a watershed managed under agroforestry and row-crop system. Agric. Ecosyst. Environ. 131, 98–104.*

were greater within AGF as compared to grazing and row crop management (Fig. 14.14) (Mungai et al., 2005; Sauer and Hernandez-Ramirez, 2011). In West Africa, Buresh and Tian (1998) observed 2–3 times greater earthworm populations under improved fallow as compared to corn. Similar reports are available from studies conducted in Central America,

Table 14.1: Soil nutrient enrichment and pH from savannah to new, established, and old homegarden agroforestry systems in Brazil

	Soil Depth	Treatment			
			Homegarden Age (Year)		
Parameter	cm	Savannah	0–10	15–35	>40
P (mg kg^{-1})	0–10	2.1	14.3	26.5	43.7
	10–20	1.1	8.5	12.2	37.9
	20–30	0.6	7.5	7.5	31.2
K (cmolc kg^{-1})	0–10	0.07	1.08	0.11	0.18
	10–20	0.06	0.08	0.10	0.15
	20–30	0.06	0.07	0.10	0.13
Ca (cmolc kg^{-1})	0–10	0.32	0.66	1.48	1.93
	10–20	0.22	0.24	1.02	1.41
	20–30	0.17	0.18	0.63	1.03
Mg (cmolc kg^{-1})	0–10	0.16	0.18	0.30	0.48
	10–20	0.09	0.10	0.24	0.33
	20–30	0.06	0.07	0.18	0.28
OM (g kg^{-1})	0–10	12.9	12.2	20.6	27.6
	10–20	10.0	7.9	14.6	19.5
	20–30	8.3	6.9	10.4	14.3
pH	0–10	5.34	5.39	5.48	5.45
	10–20	5.16	5.28	5.39	5.49
	20–30	5.13	5.29	5.30	5.32

Source: Pinho, R.C., Miller, R.P., Alfaia, S.S., 2012. Review Article. Agroforestry and the improvement of soil fertility: a view from Amazonia. App. Environ. Soil Sci. 2012, ID 616383, 11 pages. doi:10.1155/2012/616383.

Nigeria, Zambia, and other tropical countries (Budowski and Russo, 1997; Rao et al., 1998; Adejuyigbe et al., 1999). Many of these benefits were explained in previous sections.

Researchers also have shown other soil ecosystem benefits of AGF (Jose, 2009). Biodegradation of antibiotics in runoff from grazing watersheds with AGF practices were observed by Chu et al. (2010). Antibiotics were strongly bonded in soils under AGF areas as compared to grazing and crop soils, thus reducing antibiotic losses in surface and subsurface waters. They attributed these benefits to diverse microbial communities, exudates, and chemistry in the soils. Grass buffers (Chu et al., 2010; Lin et al., 2010) can remove about 80% of antibiotics.

Studies by Lin et al. (2004, 2007, 2010, 2011) reported shorter half-life and faster degradation of herbicides, antibiotics, and degradation products in multispecies AGF buffers (Fig. 14.15) (Yang et al., 2011, 2012). Enzymatic activities, root exudates, and physical parameters have been the major reason for these ecosystem benefits of AGF. Infiltration through the soil profile because of improved soil physical parameters was the initial mechanism of reduction of losses of antibiotics to water (Krutz et al., 2004; Lin et al., 2004). Diverse microbial

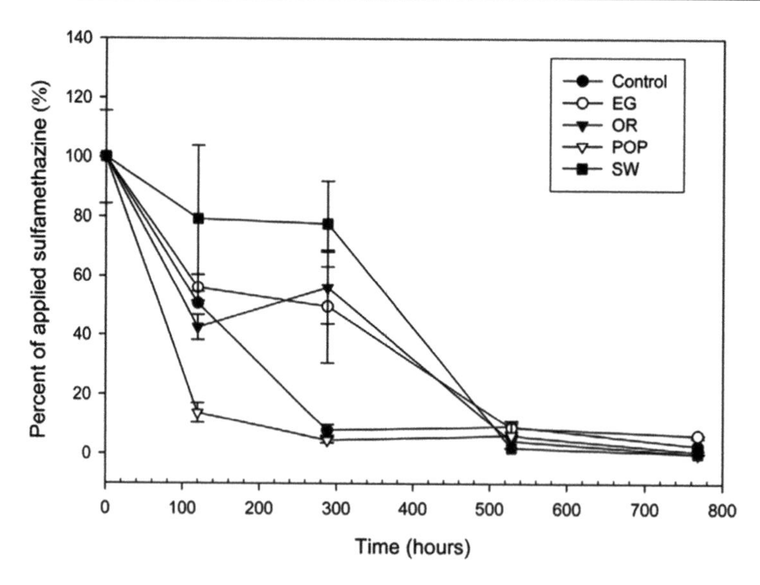

Figure 14.15
^{3}H-sulfamethazine in root zones of control, eastern gamma grass (EG), orchard grass (OG), poplar (POP), and switch grass (SG) with time. *Lin, C.H., Goyne, K.W., Kremer, R.J., Lerch, R.N., Garrett, H.E., 2010. Dissipation of sulfamethazine and tetracycline in the root zone of grass and tree species. J. Environ. Qual. 39, 1269–1278.*

communities growing in the root zone metabolize pollutants by numerous biochemical processes and produce less harmful compounds and CO_2 (Lin et al., 2004, 2010, 2011; Mandelbaum et al., 1993). In the Lin et al. studies the half-life was 2.1 and 7.6 days in poplar and control treatments, respectively, for sulfamethazine.

14.7 Conclusions

Modern agriculture most often uses continuous monocropping on a farm or a Major Land Resource Area using large machinery and inputs including fertilizers and agrochemicals. These practices often degrade soil health and ecosystem services over time. Long-term damage to soil will increase the cost of food production and incur restoration activities to improve degraded land. Agroforestry is a possible solution to restore some of the damages from farming to improve the ecosystem services associated with soil and water.

The array of conservation practices within agroforestry can be integrated into agricultural management practices to conserve and enhance the quality of soil, water, and the environment, and help sustain food and fiber production. Intentional integration of native species in strategically placed locations within farms can further enhance ecosystem functions and reduce the production cost. Researches from the tropical and temperate regions have shown enhanced soil conservation, CS, nutrient enrichment, and biological services due to adoption of AGF. Landowners and farmers who adopt these practices provide multiple services and benefits to society that are enjoyed at local, regional, continental, and ecoregional global scales. Some of these benefits are invaluable and may surpass just soil ecosystem services.

Rewards or economic incentives are limited to those who sacrifice their resources for establishment and maintenance of AGF, and thus further adoption is not encouraged. Proper guidelines, incentives, policies, and site- and system-specific research studies are needed to promote wide-scale adoption of AGF.

References

Adejuyigbe, C.O., Tian, G., Adeoye, G.O., 1999. Soil microarthropod populations under natural and planted fallows in southwestern Nigeria. Agroforest. Syst. 47, 263–272.

Albrecht, A., Kandji, S.T., 2003. Carbon sequestration in tropical agroforestry systems. Agric. Ecosyst. Environ. 99, 15–27.

Al-Kaisi, M., 2008. Soil Erosion, Crop Productivity and Cultural Practices. University Extension, Iowa University Press., 4p.

Al-Kaisi, M.M., Douelle, A., Kwaw-Mensah, D., 2014. Soil microaggregate and macroaggregate decay over time and soil carbon change as influenced by different tillage systems. J. Soil Water Conserv. 69, 574–580.

Allen, S.C., Jose, S., Nair, P.K.R., Brecke, B.J., Nkedi-Kizza, P., Ramsey, C.L., 2004. Safety-net role of tree roots: evidence from a pecan (Carya illinoensis K. Koch)–cotton (*Gossypium hirsutum* L.) alley cropping system in the southern United States. For. Ecol. Manage. 192, 395–407.

Bakker, M.M., Govers, G., Kosmas, C., Vanacker, V., van Oost, K., Rounsevell, M., 2005. Soil erosion as a driver of land-use change. Agric. Ecosyst. Environ. 105, 467–481.

Balandier, P., de Montard, F., Curt, T., 2008. Root competition for water between trees and grass in a silvopastoral plots of 10 year old *Prunus avium*. In: Batish, D.R., Kohli, R.K., Jose, S., Singh, H.P. (Eds.), Ecological Basis of Agroforestry. CRC Press, New York, pp. 253–270.

Bambrick, A.D., Whalen, J.K., Bradley, R.L., Cogliastro, A., Gordon, A.M., Olivier, A., et al., 2010. Spatial heterogeneity of soil organic carbon in tree-based intercropping systems in Quebec and Ontario, Canada. Agroforest. Syst. 79, 343–353.

Beer, J., Harvey, C.A., Ibrahim, M., Harmand, J.M., Somarriba, E., Jiménez, M., 2003. Service functions of agroforestry systems. Congress Proc, pp. 417–424. Available at: <http://www.fao.org/docrep/article/wfc/xii/ms20-e.htm>.

Bharati, L., Lee, K.H., Isenhart, T.M., Schultz, R.C., 2002. Soil-water infiltration under crops, pasture, and established riparian buffer in Midwestern USA. Agroforest. Syst. 56, 249–257.

Blanco-Canqui, H., Gantzer, C.J., Anderson, S.H., Alberts, E.E., Thompson, A.L., 2004. Grass barrier and vegetative filter strip effectiveness in reducing runoff, sediment, nitrogen, and phosphorus loss. Soil Sci. Soc. Am. J. 68, 1670–1678.

Blanco-Canqui, H., Gantzer, C.J., Anderson, S.H., 2006. Performance of grass barriers and filter strips under interrill and concentrated flow. J. Environ. Qual. 35, 1969–1974.

Boardman, J., 2006. Soil erosion science: reflections on the limitations of current approaches. Catena 68, 73–86.

Brandle, J.R., Hodges, L., Zhou, X.H., 2004. Windbreaks in North American agricultural systems. Agroforest. Syst. 61, 65–78.

Brandle, J.R., Hodges, L., Tyndall, J., Sudmeyer, R.A., 2009. Windbreak practices. In: Garrett, H.E. (Ed.), North American Agroforestry, an Integrated Science and Practice, second ed. American Society of Agronomy, Madison, WI, pp. 75–104.

Broadmeadow, S., Nisbet, T.R., 2004. The effects of riparian forest management on the freshwater environment: a literature review of best management practice. Hydrol. Earth Syst. Sci. 8, 286–305.

Budowski, G., Russo, R.O., 1997. Nitrogen-fixing trees and nitrogen fixation in sustainable agriculture: research and challenges. Soil Biol. Biochem. 29, 767–770.
Buresh, R.J., Tian, G., 1998. Soil improvement by trees in sub-Saharan Africa. Agroforest. Syst. 38, 51–76.
Cairns, M.A., Meganck, R.A., 1994. Carbon sequestration, biological diversity, and sustainable development: integrated forest management. Environ. Manage. 18, 13–22.
Charreau, C., Vidal, P., 1965. Influence de l'*Acacia albida* Del. sur le sol, nutrition minérale et rendements des mils Pennisetum au Sénégal. Agron. Trop. 20, 600–626.
Chirwa, P.W., 2006. Nitrogen dynamics in cropping systems in Southern Malawi containing *Gliricidia sepium*, pigionpea and maize. Agroforest. Syst. 67, 93–106.
Chirwa, P.W., Ong, C.K., Maghembe, J., Black, C.R., 2007. Soil water dynamics in intercropping systems containing *Gliricidia sepium*, pigeon pea and maize in southern Malawi. Agroforest. Syst. 69, 29–43.
Chu, B., Goyne, K.W., Anderson, S.H., Lin, C.H., Udawatta, R.P., 2010. Veterinary antibiotic sorption to agroforestry buffer, grass buffer and cropland soils. Agroforest. Syst. 79, 67–80.
Clason, T.R., Sharrow, S.H., 2000. Silvopastural practices. In: Garrett, H.E., Rietveld, W.J., Fisher, R.F. (Eds.), North American Agroforestry: An integrated Science and Practice. American Society of Agronomy., Madison, WI, pp. 119–147.
Correal, E., Erena, M., Ríos, S., Robledo, A., Vicente, M., 2009. Agroforestry systems in southeastern Spain. In: Rigueiro-Rodriguez, A. (Ed.), Agroforestry in Europe. Springer, Dordrecht, The Netherlands, pp. 183–210.
Dent, D., 2007. Chapter 3: land. In: Mnatsakanian, R. (Ed.), Global Environment Outlook 4: Environment for Development. United Nations Environment Programme; Stationery Office distributor, pp. 81–114.
Dillaha, T.A., Reneau, R.B., Mostaghimi, S., Lee, D., 1989. Vegetative filter strips for agricultural nonpoint-source pollution control. Trans. ASAE 32, 513–519.
Dixon, R.K., 1995. Agroforestry systems: sources or sinks of greenhouse gases? Agroforest. Syst. 31, 99–116.
Dulormne, M., Sierra, J., Nygren, P., Cruz, P., 2003. Nitrogen fixation dynamics in a cut-and-carry silvopastoral system in the subhumid conditions of Guadeloupe, French Antilles. Agroforest. Syst. 59, 121–129.
EEA, 2003. (European Environmental Agency) Europe's environment: the third assessment. Environmental Assessment report # 10. European Environment Agency Copenhagen.
FAO and ITPS, 2015. Status of the World's Soil Resources (SWSR) – Main Report. Food and Agriculture Organization of the United Nations and Intergovernmental Technical Panel on Soils, Rome, Italy.
FAO World Forestry Congress, 2005. Foresters Call for Action: Future land management needs better integration of sectors. Recommendations from the XIV World Forestry Congress. XIV World Forestry Congress, September 7–11, 2005, Durban, South Africa.
Fassbender, H.W., Beer, J., Heuveldop, J., Imbach, A., Enriquez, G., Bonnemann, A., 1991. Ten year balances of organic matter and nutrients in agroforestry systems at CATIE, Costa Rica. For. Ecol. Manage. 45, 173–183.
FCCC, 2015. (Framework Convention on Climate Change, FCCC) Adoption of the Paris agreement. Available at: <http://unfccc.int/resource/docs/2015/cop21/eng/l09r01.pdf > (accessed 15.12.15.).
Foster, G.R., 1982. Modeling the erosion process. In: Haan, C.T., Johnson, H.P., Brakensiek, D.L. (Eds.), Hydrologic Modeling of Small Watersheds. ASAE Monograph Number 5, St. Joseph, Michigan, pp. 297–375.
Gantzer, C.J., Buyanovsky, G.A., Alberts, E.E., Remley, P.A., 1987. Effects of soybean and corn residue decomposition on soil strength and splash detachment. Soil Sci. Soc. Am. J. 51, 202–206.
Gold, M.A., Garrett, H.E., 2009. Agroforestry nomenclature, concepts, and practices. In: Garrett, H.E. (Ed.), North American Agroforestry, an Integrated Science and Practice, second ed. American Society of Agronomy, Madison, WI, pp. 45–55.
Gupta, N., Kukal, S.S., Bawa, S.S., Dhaliwal, G.S., 2009. Soil organic carbon and aggregation under poplar based agroforestry system in relation to tree age and soil type. Agroforest. Syst. 76, 27–35.
Haile, S.G., Nair, P.K.R., Nair, V.D., 2008. Carbon storage of different soil-size fractions in Florida silvopastoral systems. J. Environ. Qual. 37, 1789–1797.
Hartemink, A.E., Janssen, B.H., Buresh, R.J., Jama, B., 1996. Soil nitrate and water dynamics in sesbania fallows, weed fallows, and maize. Soil Sci. Soc. Am. J. 60, 568–574.

Hatfield, J.L., Sauer, T.J., 2011. Soil Management. Am. Soc. Agronomy and Soil Sci. Soc. Am., Madison, WI, USA.

Hulugalle, N.R., Ndi, J.N., 1994. Changes of soil properties in a newly cleared Ultisol due to establishment of hedgerow species in alleycropping systems. J. Agric. Sci. 222, 435–443.

IPCC, 2000. In: Watson, R.T., Noble, I.R., Bolin, B., Ravindranath, N.H., Verardo, D.J., Dokken, D.J. (Eds.), Land-use, land-use change, and forestry. A special report of the IPCC. Cambridge University Press, Cambridge, UK, pp. 375.

IPCC, 2007. Intergovernmental Panel on Climate Change 2007. Synthesis Report. <http://www.ipcc.ch/pdf/assessment-report/ar4/syr/ar4_syr.pdf > (accessed 01.01.16.).

Jose, S., 2009. Agroforestry for ecosystem services and environmental benefits: an overview. Agroforest. Syst. 76, 1–10.

Jose, S., Gillespie, A.R., Seifert, J.R., Mengel, D.B., Pope, P.E., 2000. Defining competition vector in a temperate alley cropping system in Midwestern USA. 2. Competition for water. Agroforest. Syst. 48, 41–59.

Jose, S., Gillespie, A.R., Pallardy, S.G., 2004. Interspecific interactions in temperate agroforestry. Agroforest. Syst. 61, 237–255.

Jose, S., Gold, M.A., Garrett, H.E., 2012. The future of temperate agroforestry in the United States. Pp 2-7-245. In: Nair, P.K.R., Garrity, D. (Eds.), Agroforestry - The Future of Global 217 Land Use, Advances in Agroforestry 9. <http://dx.doi.org/10.1007/978-94-007-4676-3_14 > , © Springer Science+ Business Media Dordrecht 2012.

Kang, B.T., Reynolds, L. (Eds.), 1989. *A*lley farming in the humid and sub humid Tropics. Proceedings. Ottawa, Canada, IDRC, 251 pp.

Kaonga, M.L., 2005. Understanding carbon dynamics in agroforestry systems in eastern Zambia. Phd Dissertation Fitzwilliam College, University of Cambridge, UK.

Kaonga, M.L., Bailey-Smith, T.P., 2009. Carbon pools in tree biomass and the soil in improved fallows in eastern Zambia. Agroforest. Syst. 76, 37–51.

Kazemi, M.L., Dumenil, L.C., Fenton, T.E., 1990. Effects of Accelerated Erosion on Corn Yields of Loess Derived and Till-Derived Soils in Iowa. Unpublished Technical Report. Department of Agronomy, Iowa State University, pp. 1–102.

Kort, J., Turnock, R., 1999. Carbon reservoir and biomass in Canadian Prairie Shelterbelts. Agroforest. Syst. 44, 175–186.

Kremer, R.J., Kussman, R.D., 2011. Soil quality in a pecan - kura clover alley cropping system in the Midwestern USA. Agroforest. Syst. 83, 213–223.

Krutz, L.J., Senseman, S.A., Dozier, M.C., Hoffman, D.W., Tierney, D.P., 2004. Infiltration and adsorption of dissolved metolachlor, metolachlor oxanilic acid, and metolachlor ethanesulfonic acid by buff alograss (*Buchloe dactyloides*) filter strips. Weed Sci. 52, 166–171.

Kumar, S., Anderson, S.H., Bricknell, L.G., Udawatta, R.P., 2008. Soil hydraulic properties influenced by agroforestry and grass buffers for grazed pasture systems. J. Soil Water Conserv. 63, 224–232.

Kumar, S., Udawatta, R.P., Anderson, S.H., 2010. Root length density and carbon content of agroforestry and grass buffers under grazed pasture systems in a Hapludalf. Agroforest. Syst. 80, 85–96.

Lal, R., 1996. Deforestation and land-use effects on soil degradation and rehabilitation in western Nigeria. III. Runoff, soil erosion and nutrient loss. Land Degrad. Dev. 7, 87–98.

Lal, R., 2010. Managing soil and ecosystems for mitigating anthropogenic carbon emissions and advancing global food security. Bioscience 60, 708–721.

Lee, K.H., Jose, S., 2003. Soil respiration and microbial biomass in soils under alley cropping and monoculture cropping systems in southern USA. Agroforest. Syst. 58, 45–54.

Lee, K.H., Isenhart, T.M., Schultz, R.C., 2003. Sediment and nutrient removal in an established multi-species riparian buffer. J. Soil Water Conserv. 58, 1–8.

Lin, C.H., Lerch, R.N., Garrett, H.E., George, M.F., 2004. Incorporating forage grasses in riparian buffers for bioremediation of atrazine, isoxaflutole and nitrate in Missouri. Agroforest. Syst. 63, 91–99.

Lin, C.H., Lerch, R.N., Garrett, H.E., Li, Y.Z., George, M.F., 2007. An improved HPLC-MS/MS method for determination of isoxaflutole (Balance™) and its metabolites in soils and forage plants. J. Agric. Food Chem. 55, 3805–3815.

Lin, C.H., Goyne, K.W., Kremer, R.J., Lerch, R.N., Garrett, H.E., 2010. Dissipation of sulfamethazine and tetracycline in the root zone of grass and tree species. J. Environ. Qual. 39, 1269–1278.
Lin, C.-H., Lerch, R.N., Goyne, K.W., Garrett, H.E., 2011. Reducing herbicides and veterinary antibiotic losses from agroecosystems using vegetative buffers. J. Environ. Qual. 40, 791–799.
Linares, A.M., 2007. Forest planning and traditional knowledge in collective woodlands of Spain: the dehesa system. For. Ecol. Manage. 249, 71–79.
Liu, X., Zhang, X., Zhangm, M., 2008. Major factors influencing the efficacy of vegetated buff ers on sediment trapping: a review and analysis. J. Environ. Qual. 37, 1667–1674.
López, M., Schlönvoigt, A.A., Ibrahim, M., Kleinn, C., Kanninen, M., 1999. Cuantificación del carbono almacenado en el suelo de un sistema silvopastoril en la zona Atlántica de Costa Rica. Agroforestería en las Américas 6 (23), 51–53.
MacDonald, J.M., Korb, P., Hoppe, R.A., 2013. Farm Size and the Organization of U.S. Crop Farming. USDA, Economic Research Service, Economic Research Report Number 152, 61 p.
Mandelbaum, R.T., Wackett, L.P., Allan, D.L., 1993. Rapid hydrolysis of atrazine to hydroxyatrazine by soil bacteria. Environ. Sci. Technol. 27, 1943–1944.
Mazzarino, M.J., Szott, L., Jimenez, M., 1993. Dynamics of soil total C and N, microbial biomass, and water-soluble C in tropical agroecosystems. Soil Biol. Biochem. 25, 205–214.
MEA (Millennium Ecosystem Assessment), 2005. Ecosystems and Human Well-Being: Synthesis. A Framework for Assessment. Island Press, Washington, DC.
Mekonnen, K., Buresh, R.J., Jama, B., 1997. Root and inorganic nitrogen distributions in *Sesbania* fallow, natural fallow and maize fields. Plant Soil 188, 319–327.
Montagnini, F., Nair, P.K.R., 2004. Carbon sequestration: an underexploited environmental benefit of agroforestry systems. Agroforest. Syst. 61, 281–295.
Montgomery, D.R., 2007. Soil erosion and agricultural sustainability. Proc. Natl. Acad. Sci. USA. 104, 13268–13272.
Morgan, J.A., Follett, R.F., Allen, L.H., Grosso, S.D., Derner, J.D., Dijkstra, F., et al., 2010. Carbon sequestration in agricultural land of the United States. J. Soil Water Conserv. 65, 6A–13A.
Mungai, N.W., Motavalli, P.P., Kremer, R.J., Nelson, K.A., 2005. Spatial variation of soil enzyme activities and microbial functional diversity in temperate alley cropping systems. Biol. Fert. Soils 42, 129–136.
Naiman, R.J., Decamps, H., McClain, M.E., 2005. Conservation restoration. In: Naiman, R.J. (Ed.), Riparia: Ecology, Conservation, and Management of Streamside Communities. Elsevier, New York, pp. 269–325.
Nair, P.K.R., 2012. Carbon sequestration studies in agroforestry systems: a reality check. Agroforest. Syst. 86, 243–253.
Nair, P.K.R., Gordon, A.M., Mosquera-Losada, M.R., 2008. Agroforestry. In: Jorgenson, S.E., Faith, B.D. (Eds.), Encyclopedia of Ecology. Elsevier, Oxford, UK, pp. 101–110.
Nair, P.K.R., Kumar, B.M., Nair, V.D., 2009a. Agroforestry as a strategy for carbon sequestration. J. Plant Nutr. Soil Sci. 172, 10–23.
Nair, P.K.R., Nair, V.D., Gama-Rodriguez, E.F., Garciad, R., Haile, S.G., Howlett, D.S., et al., 2009b. Soil carbon in agroforestry systems: an unexplored treasure? Nature Precedings: < hdl:10101/npre.2009.4061.1> .
Nair, V.D., Graetz, D.A., 2004. Agroforestry as an approach to minimizing nutrient loss from heavily fertilized soils: the Florida experience. Agroforest. Syst. 61, 269–279.
Nygren, P., Leblanc, H.A., 2009. Natural abundance of 15N in two cacao plantations with legume and non-legume shade trees. Agroforest. Syst. 76, 303–315.
Oelbermann, M., Voroney, R.P., Kass, D.C.L., Schlönvoigt, A.M., 2006a. Soil carbon and nitrogen dynamics using stable isotopes in 19- and 10-yr old tropical agroforestry systems. Geoderma 130, 356–367.
Oelbermann, M., Voroney, R.P., Thevathasan, N.V., Gordon, A.M., Kass, D.C.L., Schlönvoigt, A.M., 2006b. Soil carbon dynamics and residue stabilization in a Costa Rican and southern Canadian alley cropping system. Agroforest. Syst. 68, 27–36.
Palm, C., Sanchez, P., Ahamed, S., Awiti, A., 2007. Soils: a contemporary perspective. Annu. Rev. Environ. Resou. 32, 99–129.

Patty, L., Real, B., Gril, J.J., 1997. The use of grassed buffer strips to remove pesticides, nitrate and soluble phosphorus compounds from runoff water. Pestic. Sci. 49, 243–251.
Paudel, B.R., Udawatta, R.P., Anderson, S.H., 2011. Agroforestry and grass buffer effects on soil quality parameters for grazed pasture and row-crop systems. App. Soil Ecol. 48, 125–132.
Pimentel, D., Harvey, C., Resosudarmo, P., Sinclair, K., Kurz, D., McNair, M., et al., 1995. Environmental and economic costs of soil erosion and conservation benefits. Science 267, 1117–1123.
Pinho, R.C., Alfaia, S.S., Miller, R.P., Uguen, K., Magalhães, L.D., Ayres, M., et al., 2011. Islands of fertility: soil improvement under indigenous homegardens in the savannas of Roraima, Brazil. Agroforest. Syst. 81, 235–247.
Pinho, R.C., Miller, R.P., Alfaia, S.S., 2012. Review article. Agroforestry and the improvement of soil fertility: a view from Amazonia. App. Environ. Soil Sci. 2012 ID 616383, 11 pages. doi:10.1155/2012/616383.
Ponisio, L.C., M'Gonigle, L.K., Mace, K.C., Palomino, J., de Valpine, P., Kremen, C., 2015. Diversification practices reduce organic to conventional yield gap. Proc. R. Soc. B 282 20141396.
Rao, M.R., Nair, P.K.R., Ong, C.K., 1998. Biophysical interactions in tropical agroforestry systems. Agroforest. Syst. 38, 3–50.
Renard, K.G., Foster, G.R., Weesies, G.A., McCool, D.K., Yoder, D.C., 1997. Predicting Soil Erosion by Water: A Guide to Conservation Planning With the Revised Universal Soil Loss Equation (RUSLE), U.S. Department of Agriculture Agriculture Handbook No. 703, 404 pp.
Rhoades, C.C., Eckert, G.E., Coleman, D.C., 1998a. Effect of pasture trees on soil nitrogen and organic matter: implications for tropical montane forest restoration. Resto. Ecol. 6, 262–270.
Rhoades, C.C., Nissen, T.M., Kettler, J.S., 1998b. Soil nitrogen dynamics in alley cropping and no-till systems on ultisols of Georgia Piedmont, USA. Agroforest. Syst. 39, 31–44.
Rivest, D., Vézina, A., 2015. Maize yield patterns on the leeward side of tree windbreaks are site-specific and depend on rainfall conditions in eastern Canada. Agroforest. Syst. 89, 237–246.
Robinson, C.A., Ghaffarzadeh, M., Cruse, R.M., 1996. Vegetative filter strip effects on sediment concentration in cropland runoff. J. Soil Water Conserv. 50, 227–230.
Saha, S.K., Nair, P.K.R., Nair, V.D., Kumar, B.M., 2009. Soil carbon stock in relation to plant diversity in Kerala, India. Agroforest. Syst. 76, 53–65.
Sanchez, P.A., 2000. Linking climate change research with food security and poverty reduction in the tropics. Agric. Ecosyst. Environ. 82, 71–383.
Sauer, T.J., Hernandez-Ramirez, G., 2011. Agroforestry. In: Hatfield, J.L., Sauer, T.J. (Eds.), Soil Management: Building A Stable Base for Agriculture. American Society of Agronomy and Soil Science of America, Madison, WI, pp. 351–370.
Sauer, T.J., Cambardella, C.A., Brandle, J.R., 2007. Soil carbon and litter dynamics in a red cedar-scotch pine shelterbelt. Agroforest. Syst. 71, 163–174.
Schmitt, T.J., Dosskey, M.G., Hoagland, K.D., 1999. Filter strip performance and processes for different widths and contaminants. J. Environ. Qual. 28, 1479–1489.
Schroeder, P., 1994. Carbon storage benefits of agroforestry systems. Agroforest. Syst. 27, 89–97.
Schultz, R.C., Isenhart, T.M., Simpkins, W.W., Colletti, J.C., 2004. Riparian forest buffers in agroecosystems—Lessons learned from the Bear Creek Watershed, central Iowa, U.S.A. Agroforest. Syst. 61, 35–50.
Schultz, R.C., Isenhart, T.M., Colletti, J.P., Simpkins, W.W., Udawatta, R.P., Schultz, P.L., 2009. Riparian and upland buffer practices. In: Garrett, H.E. (Ed.), North American Agroforestry, an Integrated Science and Practice, second ed. American Society of Agronomy, Madison, WI, pp. 163–218.
Seiter, S., Ingham, E.R., William, R.D., Hibbs, D.E., 1995. Increase in soil microbial biomass and transfer of nitrogen from alder to sweet corn in an alley cropping system. In: Ehrenreich, J.H., Ehrenreich, D.L., Lee, H.W. (Eds.), Growing a Sustainable Future. University of Idaho, Boise, ID, pp. 56–158.
Seiter, S., William, R.D., Hibbs, D.E., 1999. Crop yield and tree-leaf production in three planting patterns of temperate-zone alley cropping in Oregon, USA. Agroforest. Syst. 46, 273–288.
Senaviratne, G.M.M.M.A., Udawatta, R.P., Nelson, K.A., Shannon, K., Jose, S., 2012. Temporal and spatial influence of perennial upland buffers on corn and soybean yields. Agron. J. 104, 1356–1362.

Seobi, T., Anderson, S.H., Udawatta, R.P., Gantzer, C.J., 2005. Influences of grass and agroforestry buffer strips on soil hydraulic properties. Soil Sci. Soc. Am. J. 69, 893–901.

Sharrow, S.H., Ismail, S., 2004. Carbon and nitrogen storage in agroforests, tree plantations, and pastures in western Oregon, USA. Agroforest. Syst. 60, 123–130.

Sharrow, S.H., Brauer, D., Clason, T., 2009. Silvopastoral practices. In: Garrett, H.E. (Ed.), North American Agroforestry: An Integrated Science and Practice, second ed. American Society of Agronomy, Madison, WI, pp. 105–132.

Sierra, J., Nygren, P., 2005. Role of root inputs from a dinitrogen-fixing tree in soil carbon and nitrogen sequestration in a tropical silvopastoral system. Aust. J. Soil Res. 43, 667–675.

Sierra, J., Dulormne, M., Desfontaines, L., 2002. Soil nitrogen as affected by Gliricidia sepium in a silvopastoral system in Guadeloupe, French Antilles. Agroforest. Syst. 54, 87–97.

Sileshi, G., Akinnifesi, F.K., Ajayi, O.C., Chakeredza, S., Kaonga, M., Matakala, P.W., 2007. Contributions of agroforestry to ecosystem services in the miombo eco-region of eastern and southern Africa. Afr. J. Environ. Sci. Technol. 1, 68–80.

Sileshi, G., Akinnifesi, F.K., Ajayi, O.C., Place, F., 2009. Evidence for impact of green fertilizers on maize production in sub-Saharan Africa: a meta-analysis, ICRAF Occasional Paper No. 10, World Agroforestry Centre, Nairobi, Kenya.

Smith, P., Martono, D., Cai, Z., Gwary, D., Janzen, H., Kumar, P., et al., 2007. Agriculture. In: Metz, B. (Ed.), Climate Change 2007: Mitigation. Cambridge Univ. Press, Cambridge, UK, pp. 497–540.

Sparling, G.P., Wheeler, D., Vesley, E.T., Schipper, L.A., 2006. What is soil organic matter worth? J. Environ. Qual. 35, 548–557.

Ståhl, L., Nyberg, G., Högberg, P., Buresh, R.L., 2002. Effects of planted tree fallows on soil nitrogen dynamics above ground and root biomass, N2-fixation and subsequent maize crop productivity in Kenya. Plant Soil 243, 103–117.

Stamps, W.T., Linit, M.J., 1998. Plant diversity and arthropod communities: implications for temperate agroforestry. Agroforest. Syst. 39, 73–89.

Steinbeiss, S., Beβler, H., Engels, C., Temperton, V.M., Buchmanns, N., Roscher, C., et al., 2008. Plant diversity positively affects short-term soil carbon storage in experimental grasslands. Glob. Clim. Biol. 14, 2937–2949.

Steppler, H.A., Nair, P.K.R. (Eds.), 1987. Agroforestry: A Decade of Development. ICRAF, Nairobi.

Szott, L.T., Palm, C.A., 1996. Nutrient stocks in managed and natural humid tropical fallows. Plant Soil 186, 293–309.

Takimoto, A., Nair, P.K.R., Nair, V.D., 2008. Carbon stock and sequestration potential of traditional and improved agroforestry systems in the West African Sahel. Agric. Ecosyst. Environ. 125, 159–166.

Takimoto, A., Nair, V.D., Nair, P.K.R., 2009. Contribution of trees to soil carbon sequestration under agroforestry systems in the West African Sahel. Agroforest. Syst. 76, 11–25.

Thevathasan, N.V., Gordon, A.M., 2004. Ecology of tree intercropping systems in the north temperate region: experiences from southern Ontario, Canada. Agroforest. Syst. 61, 257–268.

Udawatta, R.P., Godsey, L.D., 2010. Agroforestry comes of age: putting science into practice. Agroforest. Syst. 79, 1–4.

Udawatta, R.P., Jose, S., 2012. Agroforestry strategies to sequester carbon in temperate North America. Agroforest. Syst. http://dx.doi.org/10.1007/s10457-012-9561-1.

Udawatta, R.P., Krstansky, J.J., Henderson, G.S., Garrett, H.E., 2002. Agroforestry practices, runoff, and nutrient loss: a paired watershed comparison. J. Environ. Qual. 31, 1214–1225.

Udawatta, R.P., Anderson, S.H., Gantzer, C.J., Garrett, H.E., 2006. Agroforestry and grass buffer influence on macropore characteristics: a computed tomography analysis. Soil Sci. Soc. Am. J. 70, 1763–1773.

Udawatta, R.P., Kremer, R.J., Adamson, B.W., Anderson, S.H., 2008. Variations in soil aggregate stability and enzyme activities in a temperate agroforestry practice. Appl. Soil Ecol. 39, 153–160. http://dx.doi.org/10.1016/j.apsoil.2007.12.002.

Udawatta, R.P., Kremer, R.J., Garrett, H.E., Anderson, S.H., 2009. Soil enzyme activities and physical properties in a watershed managed under agroforestry and row-crop system. Agric. Ecosyst. Environ. 131, 98–104.

Udawatta, R.P., Anderson, S.H., Motavalli, P.P., Garrett, H.E., 2011a. Clay and temperature influences on sensor measured volumetric soil water content. Agroforest. Syst. 82, 61–75.
Udawatta, R.P., Garrett, H.E., Kallenbach, R.L., 2011b. Agroforestry buffers for non point source pollution reductions from agricultural watersheds. J. Environ. Qual. 40, 800–806.
Udawatta, R.P., Nelson, K.A., Jose, S., Motavalli, P.P., 2014. Temporal and spatial differences in crop yields of a mature silver maple alley cropping system. Agron. J. 106, 407–415.
Udawatta, R.P., Gantzer, C.J., Reinbott, T.M., Wright, R.L., Pierce, R.A., 2016. Temporal and spatial yield differences influenced by riparian buffers and CRP. Agron. J. 108, 647–655.
UNEP, NASA, WB, 1998. (United Nations Environmental Program, National Aeronautics and Space Administration, The World bank) Protecting Our Planet Securing Our Future, 95p.
UNFCCC, 2007. Report of the conference of parties on its thirteenth session. Bali, Indonesia. In: United Nations framework convention on climate change, Geneva, Switzerland.
Uphoff, N., Ball, A.S., Fernandes, E. (Eds.), 2006. Biological Approaches to Sustainable Soil Systems. CRC Press, Boca Raton, FL UN. 2016, The 2015 Revision of World Population Prospects World Population Prospects Population Division, United Nations, New York, NY Available at <http://esa.un.org/unpd/wpp/unpp/panel_population.htm > (accessed 6.01.16.).
USDA-NRCS, 2006. Land Resource Regions and Major Land Resource Areas of the United States, the Caribbean, and the Pacific Basin. US Department of Agriculture Handbook 296.
USDA (U.S. Department of Agriculture), 2011. RCA Appraisal 2011 Soil and Water Resources Conservation Act. [Online] USDA-NRCS, Washington, DC. Available at: <www.nrcs.usda.gov/Internet/FSE_DOCUMENTS/stelprdb1044939.pdf > (accessed 4.03.16.).
USGS (U.S. Geological Survey), 2016. Phosphate Rock Statistics and Information. USGS Northeast Region, Reston, VA [Online]. Available at: <http://minerals.usgs.gov/minerals/pubs/commodity/phosphate_rock/index.html > (accessed 6.01.16).
van Noordwijk, M., Lawson, G., Soumare, A., Groot, J.J.R., Hairiah, K., 1996. Root distribution of trees and crops: competition and/or complementarity. In: Ong, C.K., Huxley, P. (Eds.), Tree-Crop Interactions. CAB International, Wallingford, UK, pp. 319–364.
Watson, R.T., Noble, I.R., Bolin, B., Ravindranathan, N.R., Verardo, D.J., Dokken, D.J., 2000. IPCC special report on land use, land use change, and forestry. <http://www.rida.no/climate/ipcc/land_use/ > .
Weerasekara, C., Udawatta, R.P., Jose, S., Kremer, R.J., Weerasekara, C., 2016. Soil quality differences in a row-crop watershed with agroforestry and grass buffers. Agroforest. Syst. http://dx.doi.org/10.1007/s10457-016-9903-5.
Welsh, D.J., 1991. Riparian Forest Buffers: Function and Design for Protection and Enhancement of Water Resources. NA-PR-07-91. USDA Forest Service, Randor, PA.
Whittaker, G., Barnhart, B.L., Srinivasan, R., Arnold, J.G., 2015. Cost of areal reduction of gulf hypoxia through agricultural practice. Sci. Total Environ. 505, 149–153.
Wischmeier, W.H., Smith, D.D., 1978. Predicting Rainfall Erosion Losses—A Guide to Conservation Planning. Agriculture Handbook No. 537. USDA, Washington, DC.
Wu, J., Lin, H., 2010. The effect of the conservation reserve program on land values. Land Econ. 86, 1–21.
Yang, C., Wang, M., Chen, H., Li, J., 2011. Responses of butachlor degradation and microbial properties in a riparian soil to the cultivation of three different plants. J. Environ. Sci. 23, 1437–1444.
Yang, C., Wang, M., Cai, W., Li, J., 2012. Bensulfuron-methyl biodegradation and microbial parameters in a riparian soil as affected by simulated saltwater incursion. Clean–Soil Air Water 40, 348–355.
Young, A., 1997. Agroforestry for Soil Conservation. CAB International, Wallingford, UK.297.
Zhu, J.C., Gantzer, C.J., Anderson, S.H., Alberts, E.E., Beuselinck, P.R., 1989. Runoff, soil, and dissolved nutrient losses from no-till soybean with winter cover crops. Soil Sci. Soc. Am. J. 53, 1210–1214.
Zinke, P.J., 1962. The pattern of influence of individual forest trees on soil properties. Ecology 43, 130–133.
Zomer, R.J., Trabucco, A., Coe, R., Place, F., 2009. Trees on Farm: Analysis of Global Extent and Geographical Patterns of Agroforestry, ICRAF Working Paper no. 89, World Agroforestry Centre, Nairobi, Kenya.

CHAPTER 15

Targeted Use of Perennial Grass Biomass Crops in and Around Annual Crop Production Fields to Improve Soil Health

Randall D. Jackson
University of Wisconsin-Madison, Madison, WI, United States

15.1 Introduction

One form of sustainable intensification of agroecosystems is the addition of perennial plants into areas dominated by annual cropping systems in ways that can improve ecosystem services including economic gain, clean water, climate stabilization, flood reduction, and wildlife habitat (Heaton et al., 2013; Asbjornsen et al., 2014). Typically, this means increasing the diversity of agricultural areas in one of two general ways. The first is planting perennial grasses, shrubs, and/or trees along field margins of annual crops, and in key areas within these fields, and harvesting them as biomass feedstock for bioenergy, bioproducts, or livestock feed. The second is planting or harvesting herbaceous forage or biomass crops in the understory of fruit and nut tree crops—so-called agroforestry (see chapter: Agroforestry Practices and Soil Ecosystem Services). Both of these are examples of increasing patch-level diversity of farmed landscapes with perennial vegetation. The focus of this chapter is on perennial grasses sown for biomass production in and around annual cropping systems. Recent advances in our understanding about how perennial grasses affect ecosystem stocks and processes that emerge from improved soil health are explored while addressing how the planting, management, and placement of these perennial crops for biomass influences their ability to improve soil health.

Perennial biomass crops can be used to improve many of the ecosystem services derived from agricultural landscapes (Schulte et al., 2006; Jordan and Warner, 2010; Dale et al., 2014; Holland et al., 2015). A biofuels and bioproducts industry that draws on a range of perennial biomass crops can enable production systems that better optimize the supply of food, fiber, feed, and fuel and increase the total productivity and efficiency of land, water, and nutrient use (Gopalakrishnan et al., 2009; Spiertz and Ewert, 2009). Such systems can also improve conservation of biodiversity, wildlife habitat, and water quality (Robertson and Swinton, 2005; Robertson et al., 2008; Tilman et al., 2009; Jordan and Warner, 2010),

Soil Health and Intensification of Agroecosystems.
DOI: http://dx.doi.org/10.1016/B978-0-12-805317-1.00015-4

while improving profitability of the farming enterprise and creating new revenue streams for landowners (Robertson and Swinton, 2005; Jordan et al., 2007; Swinton et al., 2007). However, a significant barrier to development of perennial grass-based biomass production systems is inadequate understanding of their agronomic and environmental performance, especially under alternative management strategies (Robertson et al., 2008; Fargione et al., 2009; Webster et al., 2010; Searle and Malins, 2014). Primary management factors are (1) choice of crop, (2) placement of crop on the landscape, (3) establishment approach including tillage and weed management, (4) soil fertility amendments, and (5) harvest frequency and intensity (Ventura et al., 2012).

15.2 Perennial Grass Biomass Crops to Improve Soil Health

Perennial grasses can influence many components of soil health, primarily by stabilizing soils physically, increasing nutrient and water-holding capacity, and promoting beneficial microbial communities relative to annual cropping systems. In all of these categories the implication of better soil health is improved agricultural yields and reduced environmental pollution loads. The mechanisms by which perennial grasses lead to these benefits mainly stem from their highly productive, dense, and fibrous roots, combined with the relative lack of soil disturbance associated with their agronomic production (Dietzel et al., 2015; Ferchaud et al., 2015; Gauder et al., 2016). These phenomena are addressed here under the subheadings: (1) soil building and conservation, (2) soil nutrient and water retention, and (3) beneficial soil biology.

15.2.1 Soil Building and Conservation

The accumulation and retention of soils is key to all ecosystem functions, human and nonhuman processes alike, and as such is considered a *supporting* ecosystem service (Dominati et al., 2010). Soil building and retention rely significantly on soil particle aggregation and soil organic carbon (SOC) stabilization, which both lead to a *regulating* service—climate stabilization—as well as the *provisioning* service of crop-yield increase and stability via improved nutrient and water retention (Smith et al., 2008; Dominati et al., 2010).

Soil aggregation occurs when soil particles of various sizes clump together because of differential affinities among these particles resulting from combinations of moisture and aeration conditions (i.e., not too dry and not too wet). These affinities are facilitated and reinforced by microbes, plants, and their exudates and residues acting as binding and flocculation agents for soil particles (Tisdall and Oades, 1982; Blankinship et al., 2016) (Fig. 15.1).

Soil disturbance undermines aggregation because particles are physically dispersed, allowing organic exudates and residues coating particle surfaces, sometimes called biofilms, to dry

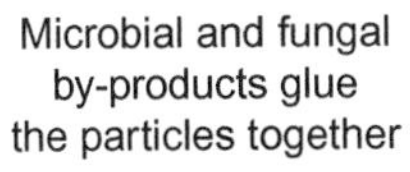
Microbial and fungal
by-products glue
the particles together

Dispersed state

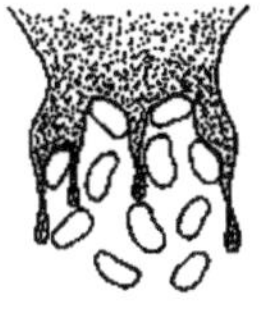

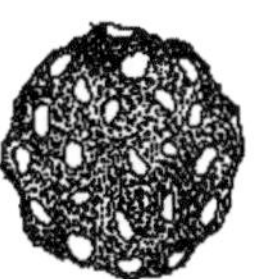
Aggregated state

Figure 15.1
Simplified schematic of soil particles in a dispersed state and an aggregated state indicating the importance of microbial residues as binding agents. *From http://soilandhealth.org.*

out. Also, fungal hyphae that play a significant role in helping to hold together soil particles are severed, allowing organic matter otherwise protected from microbial decomposition to be respired to the atmosphere as CO_2. Because of their prolific, fungi-promoting root systems, perennial grasses lead to increased soil aggregation relative to soils under annual crops that are disturbed annually for planting, but often more frequently for fertility and weed management (Blanco-Canqui et al., 2014; Cates et al., 2016).

Partly because of greater aggregate formation and stability, perennial grasses result in more SOC accumulating in soils than in annual cropping systems (Kong et al., 2005; Sanford, 2014; Agostini et al., 2015). Warm-season (C4 photosynthesis) prairie grasses of North America, with their deeper root systems, appear to do this more than cool-season (C3 photosynthesis) grasses that tend to have shallower, but more fibrous root systems (Fornara and Tilman, 2008). Sanford (2014) found significant SOC accumulation across the 1-m soil horizon in a Wisconsin Mollisol under 10-year-old warm-season grasslands, while nearby cool-season pastures accumulated SOC only in the surface 15-cm of soil but lost C from depths >30 cm (Sanford et al., 2012). Research in riparian buffers planted in Illinois and Nebraska, United States demonstrated the relative abilities of grasses to accumulate organic matter, with warm-season grasses doing so to a higher degree (Tufekcioglu et al., 1998, 2003; Blanco-Canqui et al., 2014).

When considering the effects of harvesting perennial grasses on soil C, it is important to note the findings of Harris et al. (2015) that conversion from arable land to perennial grass resulted in significant gains in soil C (~25%). However, converting extant unmanaged grasslands, such as those enrolled in set-aside agricultural programs, to ostensibly more productive dedicated bioenergy grasslands resulted in a 10% loss of soil C, on average. These results align with those of Gelfand et al. (2011), who tracked C balance during conversion of set-aside grassland in southern Michigan, United States, to an annually harvested warm-season switchgrass (*Panicum virgatum*). The attendant "payback" times for soil C loss during conversion were estimated to be 25 years if the perennial grass stand was established with tillage, but 0 years if established without tillage. However, while no-tillage establishment techniques for warm-season perennial grass stands can be effective, they are well known for resulting in slower-establishing, sometimes weed-compromised stands, which indicates room for significant agronomic improvement (Sanford et al., 2016). For example, Miesel et al. (2012) showed that weed control with herbicides at planting was key for quick establishment of switchgrass and diverse prairie in southern Wisconsin, United States, which put these perennial grasses ahead of weed growth and ensured high cover and biomass yields. However, Smith et al. (2015) showed that soil characteristics were more important determinants of stand success in more southern climates of the United States.

15.2.2 Soil Nutrient and Water Retention

In many parts of the world, P availability in soils is limiting to crop production (Roberts and Johnston, 2015). Because the source of soil P is mineral weathering, the planting of perennial vegetation has little effect on its availability except where it serves to slow the loss of existing P. This is the main benefit of perennial grasses in regions where soil P is excessive—usually the result of decades of manure spreading—and therefore prone to loss via runoff. Phosphorous (P) is not particularly mobile within the soil matrix because it strongly adsorbs to soil particles, so it tends to be lost from agroecosystems attached to soil particles that are eroding from the system (Soranno et al., 1996; Roberts and Johnston, 2015). Hence, P retention is accomplished via the same mechanisms described above for soil conservation—increased rainfall interception, infiltration, and plant uptake—but also physical impedance of displaced soil particles from moving further downslope.

Nitrogen (N) can be lost from ecosystems via aqueous-phase leaching and gaseous-phase denitrification, nitrification, and volatilization, which makes it a difficult nutrient to retain in agricultural systems (Robertson and Vitousek, 2009). Fortunately, perennial grasses, especially diverse assemblages of prairie plants, are superior to annual systems at N uptake and conservation (Dietzel et al., 2015), and when diverse assemblages include a significant legume component, N replenishment of the soil can occur (Tilman et al., 2006). Perennial grasses effectively decelerate the N cycle relative to annual cropping systems in ways that reduce losses of N to aquatic and atmospheric sinks (Hudiburg et al., 2015; Oates et al., 2015). Retaining N in the agroecosystems is beneficial not only to downstream ecosystems, but also to the profitability of the agricultural enterprise because fertilizer N is one of the most expensive production costs (de Vries and Bardgett, 2012).

In a central Iowa, United States study, the presence of perennial grasses nested within the annual crop matrix of a corn (*Zea mays*)-soybean (*Glycine max*) rotation resulted in less soil and P loss to downstream aquatic ecosystems compared to fields of the same crops without the perennial grasses (Gutierrez-Lopez et al., 2014). Zhou et al. (2014) showed that the presence of perennial vegetation, either within the annual crop field or at the toe-slope at the bottom of the catchment, resulted in annual losses of nitrate, total N, and total P that were 67%, 84%, and 90%, respectively, of fields without perennials. While the amount and distribution of perennial grass strips had no significant impact on runoff and nutrient yields in this study, other work by Perez-Suarez et al. (2014) showed that N could be better distributed in uplands with strips of perennial vegetation distributed upslope, away from the bottomlands of watersheds. However, N transformations and losses are highly variable and context-dependent. In a Denmark study, N leaching was very different under perennial crops from year to year, while differences between annual and perennial crops were highly significant (Pugesgaard et al., 2015). Working in the southeastern United States, Sarkar and Miller (2014) found a lesser degree of N reduction than many studies from perennial grass buffers along the edges of annually cropped cotton (*Gossypium* spp.) fields.

Gaseous loss of N as the potent greenhouse gas N_2O is of primary concern in agricultural systems. Oates et al. (2015) showed >twofold reductions in N_2O emissions over a 3-year perennial grass establishment period in southern Wisconsin, United States. However, Walter et al. (2015) showed that N_2O emissions from German fields growing unfertilized perennial grasses with clover were not different than fertilized corn and also found clear site differences related to N availability in soils linked to past land use. Duran et al. (2016) showed that higher plant diversity in warm-season perennial grasslands of southern Wisconsin, United States, harvested for biomass can reduce N_2O emissions relative to switchgrass monocultures, which were similar to the findings of Epie et al. (2015) in European cool-season grass stands of reed canarygrass (*Elymus elytrigia*).

While sediment, P, and N are clearly retained to a greater degree in agricultural watersheds containing perennials, the overall water balance of these types of systems in the North Central United States does not appear to be much affected. Gutierrez-Lopez et al. (2014) found that watersheds with perennials had greater soil water storage in 1 of 4 years in central Iowa, United States. Hamilton et al. (2015) found no significant differences in water use, and therefore soil water storage, in a comparison of corn and various stands of perennial warm-season grasses—switchgrass, mixed grasses, and diverse prairie. These authors point to similarities in rooting depth and growing season for these results and caution against extrapolating their results to drier areas where evapotranspiration differences among crops have been observed.

15.2.3 Beneficial Soil Biology

The role of soil biota in shaping soil health cannot be overstated. Microbes (i.e., body sizes <100 μm; mainly archaea, bacteria, and fungi) transform N between organic and inorganic forms and between aqueous and gaseous states, which can have subtle to profound influences on plant uptake, production, and composition (van der Heijden et al., 2008). These effects on plants exert feedbacks to ecosystem-level C dynamics, which are also partially driven by microbial processing of organic matter. Mesofauna (i.e., body sizes 100 μm to ~4 mm) are key drivers of ecosystem C and N cycling, playing critical roles in organic matter decomposition, soil aeration, and soil turnover (Lavelle et al., 2006; Culliney, 2013).

N fertilizer experiments in warm-season perennial grass stands sometimes show positive yield response to N addition (e.g., Hoagland et al., 2013), but with almost equal frequency show no response (Parrish and Fike, 2005; Jach-Smith and Jackson, 2015). Many posit that these variable results must be attributable to variations in soil biology (Baer et al., 2003; van der Heijden et al., 2008). Irrespective of the underlying causes, there is significant uncertainty about how environmental context and genotypic variation in plants modify N limitation of warm-season perennial grasses (see Craine and Jackson, 2009), and as a result, how they should be managed to minimize N input costs as well as N pollution. An abundance of fungi in the microbial community of agroecosystems indicates tighter N cycling, less N loss, and therefore healthier soils (de Vries and Bardgett, 2012). It is believed that a higher degree of arbuscular mychorrizal (AM) fungi can provide N to crop plants at agronomically relevant levels, i.e., replacing inputs of highly labile inorganic N, which is more readily lost to downstream aquatic ecosystems and the atmosphere. However, evidence for this level of N supply via AM fungi is scant (Correa et al., 2015).

Planting perennial grasses to soils previously in annual crops has been shown to rather quickly (2 years after sowing) restore AM fungi to levels found in sites with long-term perennial grasses (Herzberger et al., 2014) (Fig. 15.2). This quick response of the soil microbial community offers hope that land managers may ultimately have some ability to purposefully manipulate them in ways that benefit crop production, nutrient retention, and (C) sequestration.

Orr et al. (2015) compared microbial communities in corn, mixed prairie dominated by the warm-season perennial grass big bluestem (*Andropogon gerardii*), switchgrass, miscanthus (*Miscanthus* × *giganteus*), and sorghum (*Sorghum bicolor*). Bacteria dominated all systems except mixed prairie, which was dominated by fungi. After 3 years of establishment, the soil microbial community under switchgrass was beginning to resemble mixed prairie. These findings were similar to results in southern Wisconsin where switchgrass and mixed prairie soils were enriched in AM fungi biomarkers, but these markers were not more abundant when these perennial grass crops were amended with N at levels targeted to replace N removed with harvest (~65 kg N ha^{-1} $year^{-1}$ Oates et al. 2016). Working in northern Italy, Cattaneo et al. (2014) concluded that the rhizomatous perennial grasses miscanthus and giant reed (*Arundo donax*) increased soil biogeochemical activity and microbial diversity relative to annual cropping systems.

Perennial grasses have the capacity to improve a range of ecosystem stocks and processes that underpin important ecosystem services supplied to humans. This is especially true when perennial grasses are replacing highly disturbed, input-intensive annual crops. Many of these stocks and processes are key drivers of improved soil health, namely, increased aggregate

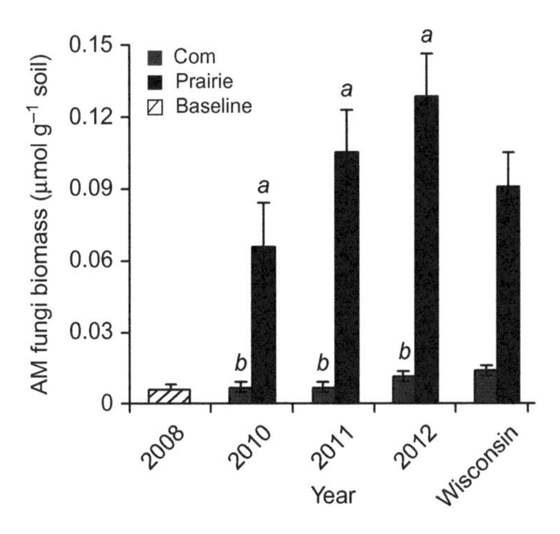

Figure 15.2

Absolute abundance of arbuscular mycorrhizal fungi lipid biomarkers. Baseline data from 2008 pretreatment plots compared to corn and prairie established on same plots in 2010, 2011, and 2012 and "Wisconsin" plots, which were sampled in 2008 on corn and long-term prairie sites across south central Wisconsin. Bars are ±1 standard error. Groups sharing a letter are not significantly different at $P < 0.05$. *Modified from Herzberger, A.J., Duncan, D.S., Jackson, R.D., 2014. Bouncing back: plant-associated soil microbes respond rapidly to prairie establishment. PLoS ONE, e115775.*

formation and stability, C accumulation and nutrient retention, and fungal abundance and dominance. Whether these perennial biomass crops are likely to be adopted widely to convey these services is the topic of the next section.

15.3 Incentivizing Adoption of Perennial Grasses

Synergies may be realized where ecosystem services positively covary, or when they are provided to an even greater degree, as a result of sustainable intensification via landscape diversification (Robertson et al., 2008, 2014; Werling et al., 2014; Liere et al., 2015). However, finding and managing these synergies can be elusive because of the agronomic momentum provided by economic policies driving highly specialized production and supply chains that incentivize low-diversity, high-input cropping systems in most industrialized societies. Synergies derived from landscape diversification with perennial grasses mainly arise from the potential for overall improvement of agronomic yields while simultaneously realizing gains from harvest and sale of perennial biomass for cellulosic bioenergy or forage. But, it is clear that configuration of fields and landscapes—to displace only low-yielding food crops or to leverage particularly good nutrient retention areas (i.e., *hotspots*)—are a critical component of minimizing the downside of tradeoffs that will make adoption of perennials unattractive (Thelemann et al., 2010; Gelfand et al., 2013).

Much of the economic research into biomass production under the current US policy environment indicates that adoption of perennial grass crops is unattractive and unlikely. These studies indicate that corn stover would be the initially preferred cellulosic feedstock (especially in the North Central United States), because it is jointly produced with (and a residue of) already profitable corn grain (Jiang and Swinton, 2009). In one scenario for southern Michigan, United States, to cover average costs and current earnings from corn production, a typical Michigan farmer was found to require \$50 to 100 Mg^{-1} biomass to plant perennial grasses instead of corn (James et al., 2010). However, this type of breakeven analysis ignores the higher risk and "sunk costs" of perennial crops associated with reduced ability to follow market trends by planting different crops each year; including these more than double the estimated breakeven price (Song et al., 2011). These costs do not include logistical challenges of storage capacity and quality, transportation, or processing and conversion bottlenecks that currently impede the nascent bioeconomy (but see Eranki et al., 2013; Dale and Ong, 2014). Moreover, as bioenergy crops are grown on larger areas of US cropland, they are expected to drive food prices higher (Babcock, 2008), even as they increase total crop acreage (Feng and Babcock, 2010), so market feedbacks favor food crops over dedicated energy crops on current cropland under the current US policy landscape (Barham et al., 2016).

Currently, grasses are being grown as energy crops to a higher degree in Europe, particularly in Germany where about 16% of arable land was used for bioenergy feedstocks in 2011 (Gissén et al., 2014). A significant fraction of this biomass was to feed biogas production, which has been incentivized with government policies that insure "buyback" rates of electricity generated at the farm level that compensates farmers for capital investment in infrastructure (Bilek et al., 2011). It is clear that US policies will need to change to incentivize perennial grass crop production. Such change could take many forms, but would need to reward land managers for providing society more than food and feed. They will need to reward stewardship that promotes C sequestration, nutrient retention, and biodiversity. However, to effectively manage landscapes for these phenomena sophisticated, but easy-to-use, tools must continue to evolve and be available (i.e., web-based and easy to use) to policymakers, land managers, and citizens at large. These types of decision support systems are most useful in the context of *stakeholder engagement*, where scenarios of plausible landscapes are imagined and interrogated in an effort to build consensus about how to manage land use and land cover configurations (Jordan et al., 2011).

15.4 Integrating Perennial Grass Biomass Crops Into Landscapes

When adding perennial biomass crops to landscapes dominated by annual cropping systems, the spatial configuration of the crops within the landscape is an important consideration. There are two general approaches relevant here, either adding the perennials to the field edges or adding the perennials within the field matrix. It is not clear how best to decide between these two general configurations, partly because it is not hard to imagine lower yields resulting from both. In general, reduced annual crop yields are more likely on the field margins/edges where equipment turnaround and travel are more frequent. On the other hand, locating perennial grasses on sensitive resource areas within fields—i.e., boggy, dry, shallow, or rocky soils—may allow the farmer to remove low-productivity land from annual crop production, with its attendant annual soil disturbance, in a way that ultimately increases the overall farm profit and reduces soil losses. This is the essence of sustainable intensification (Heaton et al., 2013). These so-called marginal lands, where the price fetched for the agronomic product is not covered by the cost of production, are the ideal areas for perennial biomass production and provision of ecological services that were outlined previously. This has been explored in the context of the highly productive US corn-belt (Gelfand et al., 2013; Brandes et al., 2016), as well as the drier Mediterranean region (Fernando et al., 2015; Schmidt et al., 2015).

15.4.1 Planting Perennial Grasses at Field Edges to Improve Soil Health

When considering perennials on field margins, most researchers have focused on so-called buffer strips, which typically are *unharvested* perennials (although they are sometimes mowed for weed control) planted along drainage and riparian areas to physically and biogeochemically impede movement of soils and nutrients with the flow of water (Hoffmann et al., 2009). In areas of high rainfall and steep topography, lands in annual crops can lose as much as 10 Mg soil ha^{-1} $year^{-1}$ via runoff (Helmers et al., 2012). This is not only a critical resource loss to the farmer, but also compromises downstream aquatic ecosystems where nutrients attached to soil particles degrade water quality and clarity, fisheries, and human health (Rabalais et al., 2002; Turner and Rabalais, 2003). Results from hundreds of studies have shown that buffer strips are effective at slowing runoff, trapping sediments, and immobilizing nutrients and, as a result, buffer strips have become a key part of federal and state policy in the United States. However, little improvement in downstream water quality has been observed as a result of millions of hectares of buffer strips being planted (Dodd and Sharpley, 2015). Reasons for this lack of observed/documented efficacy are likely related to (1) the importance of the continuity and contiguity of buffer strips in time and space, (2) uncertainty about how best to manage buffer strips for consistent efficacy, and (3) effective placement of buffer strips within a field or landscape to maximize its efficiency in interception of runoff (Reed and Carpenter, 2002). The first two of these are addressed here and the third is treated in Section 15.4.3.

Buffer strip experiments consistently show efficacy at the experimental level (Al-wadaey et al., 2012), where buffers are placed directly in the path of soils and nutrients moving downslope and these resources are being measured on the upslope and downslope sides of the buffer. However, when placed in a real-world context where the vagaries of weather, vegetation, and disturbance can exert their influence, many buffers are compromised at key places as a result of mass loss, ruts from field equipment, and other preferential flow pathways. These compromised buffers can then act as input pathways for runoff from fields to streams (Helmers et al., 2005). Hence, at relatively small spatial scales the buffers are "working," but at the integrated landscape level, losses of soil and nutrients can still be quite high. A similar situation can occur in time, where during certain periods of the year, the plants growing as buffer strips are not actively taking up nutrients and the plant stems from the previous year are senesced, so not physically impeding overland runoff. Therefore, the ability of the buffer to slow water flow and take up nutrients, perhaps at the time when it is most critical (e.g., spring thaw), can be low to nonexistent.

Perhaps most relevant to the topic at hand, the best practices for managing buffer strips to maintain their efficacy have not received much attention. Simply planting perennial grasses and never disturbing them is unlikely to maintain their long-term productivity and hence their ability to trap soil and take up nutrients. The analogy of a sponge is useful here. The buffer strip might be effective at P and N uptake for some period of time, but eventually those nutrients will be returned to the soil within the buffer and made available for future plant uptake (Jarvie et al., 2013). Additional nutrients entering the system will have a lower and lower likelihood of being taken up by plants. Worse, when an extreme weather event occurs, those nutrients that have accumulated within the buffer strip are ready to be flushed in a pulse into downstream ecosystems. Hence, it would seem important that these nutrients that are taken up in buffer strip plants be harvested periodically, and either removed from the system as a product, say a bioenergy feedstock or as livestock feed. Ideally, some grazing animal would harvest these nutrients and then return most of them upslope to the annual cropping system fields, which would serve to maintain the capacity of the buffer while improving soil health of the matrix cropping system via nutrient and C return. It is not clear how repeated annual mechanical harvest of perennial grasses planted as riparian buffers would affect their efficacy, but harvest and removal of buffer vegetation in Finland was found to reduce potential for N and P loss to streams (Räty et al., 2009). In this study, harvesting was compared to grazing, but the exposure of bare soil made grazing less effective at potential nutrient retention because of the risk of soil erosion. Tufekcioglu et al. (2003) and Blanco-Canqui et al. (2014) showed soil C should accumulate to a greater degree within the soils of the buffer area as a result of the dense, fibrous root system maintained by perennials and the undisturbed soil. Less clear is how the existence of buffers being harvested for biomass would improve soil health in adjacent annual cropping systems, though transfer of sediment and nutrients from buffers to annual crop fields by organisms such as birds, arthropods, and microbes (known in the ecological literature as allochthony) is a possibility that has not been explored.

15.4.2 Planting Perennial Grasses Within Fields to Improve Soil Health

The best and perhaps only example of planting perennial grasses within fields to improve soil-based ecosystem services on the landscape comes from the Science-based Trials of Row-crops Integrated with Prairies (STRIPs) project in Iowa, United States (Liebman et al., 2013). Results from this watershed-level experiment showed that incorporating some perennial grass cover into corn and soybean fields, as little as 10% of the area, can have a disproportionately large positive effect on sediment and nutrient loss reduction (Helmers et al., 2012; Hernandez-Santana et al., 2013). The researchers conducting this experiment compared 3 configurations of prairie plantings and a control in 12 small catchments ($n = 4$ watersheds/treatment) dominated by corn and soybean production (Fig. 15.3).

Treatments were 10% prairie in strips along contours, 10% prairie at the toeslope of the watershed, and 20% prairie at the toeslope of the watershed. While there were sometimes differences among these configurations, depending on the response variable and the year in which they were monitoring, they observed 70 to 90% reductions in nutrient and sediment loss with any prairie in the catchment compared to the annual crop-only controls.

15.4.3 Locating "Hotspots" on the Landscape for Perennial Grass Production

The land area sown to perennial cover relative to the area in annual crops must be considered in addition to locating key spots on the landscape where placement of perennials is likely to result in the greatest degree of ecosystem services (Qiu and Turner, 2013). Understanding where so-called hotspots on the landscape are located is a difficult enough challenge for any land manager, but further understanding of the degree to which a particular spot sown to perennials will simultaneously provide nutrient retention, C sequestration, wildlife habitat,

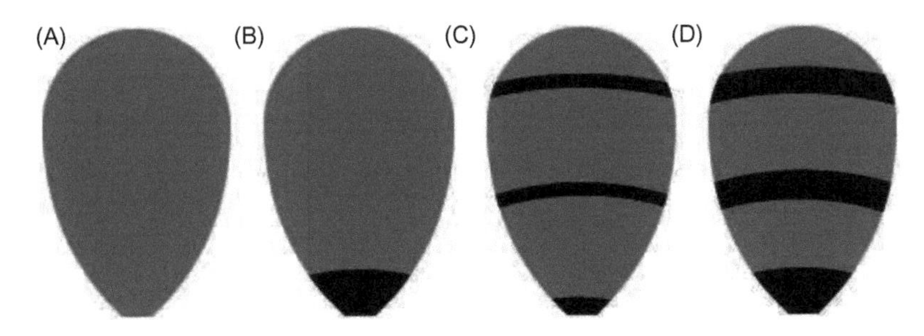

Figure 15.3

The original STRIPS project includes from left to right: (A) 100% row-crop, (B) 90% row-crop with 10% prairie at the bottom of the watershed, (C) 90% row-crop with 10% prairie integrated along strips, and (D) 80% row-crop with 20% prairie integrated along strips. *Taken from http://www.nrem.iastate.edu/research/STRIPs/content/research-overview.*

and other ecosystem services is exceedingly complex. It is this complexity in optimization that makes decision support systems critical to successful integration of perennial biomass crops into agricultural landscapes. While many of these tools are emerging, an example of this type of decision support system is SmartScape™ (Tayyebi et al., 2016), which allows users to transform land use and land cover configurations and visualize multiple ecosystem service responses (Fig. 15.4).

Historically, focus has been on provisioning services, such as grain yields, when evaluating agroecosystems because markets were explicitly developed to value these commodities and products. More recently, regulating, supporting, and cultural ecosystem services are being evaluated simultaneously with considerable emphasis on methodologies for integrating and comparing numerous metrics of sustainability that emerge from these services, but have very different currencies or denominators. Meehan et al. (2013) provided an example of such an analysis. Using SmartScape™ to focus on streams and creeks in the Yahara River watershed of south central Wisconsin, United States, these authors found that converting continuous corn areas within 100 m of streams to perennial grasses harvested for biomass, decreased annual income of farmers by 75% (assuming \$100 Mg^{-1} dry grass biomass), while decreasing annual P loading to surface waters by 29%, increasing C sequestration by 30%, and decreasing annual N_2O emissions by 84%. They expressed tradeoffs between income and other ecosystem services as benefit–cost ratios, which averaged 0.84 kg of avoided P pollution, 18.97 Mg of sequestered C, and 1.99 kg of avoided N_2O emissions for every \$1000 reduction in income for the entire watershed. More important than the actual estimated values of these ecosystem

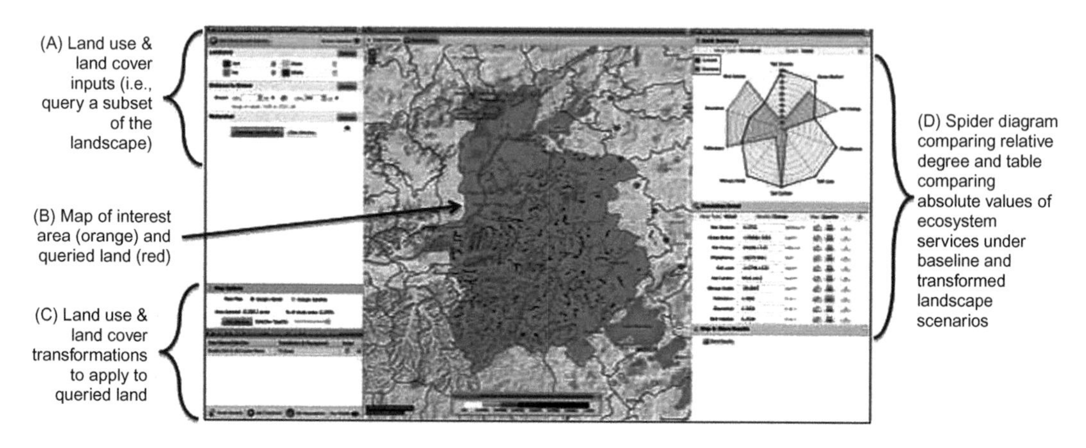

Figure 15.4

Annotated screenshot of SmartScape™, a web-based decision support system that allows users to select land for transformation (A) and visualize these selections on the landscape (B) based on specific criteria, transform the land to alternative land use and land cover with alternative management options (C), and to rapidly observe and compare the consequences of these changes on a variety of ecosystem and economic performance measures (D).

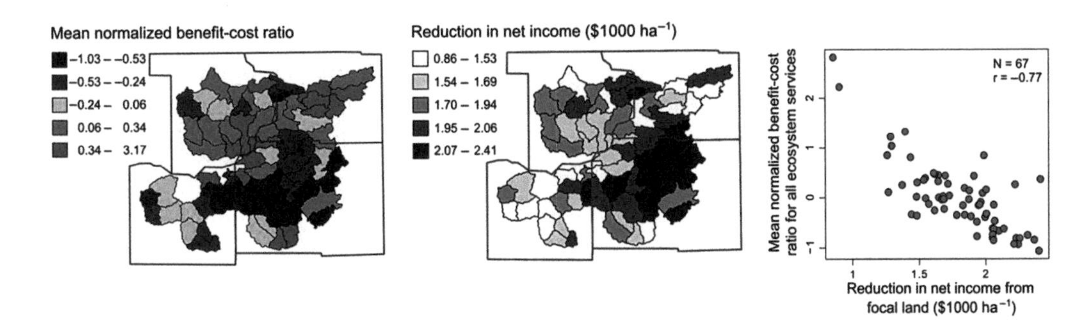

Figure 15.5

Relationship between generalized benefit-cost ratios and opportunity costs. Spatial (maps) and bivariate (scatterplot, r = Pearson's correlation coefficient) relationships between average normalized benefit-cost ratios and opportunity costs of the perennial-grass scenario (difference in net income between continuous-corn and perennial-grass scenarios) for 67 study watersheds in southern Wisconsin, United States. Average normalized ratios illustrate hotspots—watersheds where switching from continuous corn to perennial-grass brought relatively high environmental benefits for relatively low opportunity cost. *Taken from Meehan et al. (2013).*

services was the ability of the decision support system (SmartScape™) to identify where these land use and land cover conversions would result in the greatest gains in ecosystem services per unit of lost income to farmers watershed-wide, i.e., the "hotspots" (Fig. 15.5).

What is clear from the maps in Fig. 15.5 is that the cost of improving soil health with this approach would not be borne equally by those who enjoy the benefits, i.e., all stakeholders within the watershed and even the greater region and the globe. Hence, the need to employ this type of decision support system as part of a collaborative process of stakeholder landscape design (Slotterback et al., 2015). In these stakeholder sessions, users can "paint the landscape" into alternative land use and land cover mosaics and visualize tradeoffs and synergies in services. Moreover, they can identify economic "winners" and "losers" resulting from these scenarios and begin to identify policy mechanisms to make up for lost revenue to particular landowners and ecosystem service gains to society. This is the essence of a process known as *collaborative geodesign*, as coined by Slotterback et al. (2015). From these processes, employed at the watershed to regional levels of spatial organization, should emerge more sustainable landscape designs and implementations (Jordan and Warner, 2010; Jordan et al., 2011, 2013).

15.5 Conclusions

Perennial grass biomass crops can improve many soil-based ecosystem resources and processes—namely, soil organic matter, nutrient retention, and beneficial soil biota—leading to improved soil health, but management is critical. Gains in soil health can be quickly and easily undone with soil disturbance and excessive nutrient inputs, which are both very likely because of the high inherent variability of perennial grass productivity. Moreover, yield variability is likely to be even higher when perennial grasses are grown on field margins and other less productive parts of the landscape to avoid displacing food and feed crops. However, planting perennial grasses in these so-called marginal lands is a form of sustainable intensification that has the potential to improve the profitability of a farm enterprise. Because the economics of perennial grass production are not particularly favorable under current policies that favor continued fossil fuel use and annual crop production, new mechanisms are needed to incentivize perennial grass cropping systems. These mechanisms should reward farmers for the ecosystem services they provide, especially those leading to improved soil health. However, to effectively manage integrated landscape mosaics that include perennial grass crops, decision support systems are critical tools for helping land owners and other stakeholders with their placement and management—understanding tradeoffs and synergies of alternative landscape configurations—which should lead to higher levels of societal acceptance, and therefore more sustainable agroecosystems.

References

Agostini, F., Gregory, A.S., Richter, G.M., 2015. Carbon sequestration by perennial energy crops: is the jury still out? BioEnergy Res. 8, 1057–1080.

Al-wadaey, A., Wortmann, C.S., Franti, T.G., Shapiro, C.A., Eisenhauer, D.E., 2012. Effectiveness of grass filters in reducing phosphorus and sediment runoff. Water Air Soil Pollut. 223, 5865–5875.

Asbjornsen, H., Hernandez-Santana, V., Liebman, M., Bayala, J., Chen, J., Helmers, M., et al., 2014. Targeting perennial vegetation in agricultural landscapes for enhancing ecosystem services. Renew. Agric. Food Syst. 29, 101–125.

Babcock, B.A., 2008. Breaking the Link between Food and Biofuels, CARD Briefing Paper 08-BP 53. Iowa State University, Ames, IA.

Baer, S.G., Blair, J.M., Collins, S.L., Knapp, A.K., 2003. Soil resources regulate productivity and diversity in newly established tallgrass prairie. Ecology 84, 724–735.

Barham, B.L., Mooney, D., Swinton, S., 2016. Inconvenient truths about landowner (un)willingness to grow dedicated bioenergy crops. Choices 31, 1–8.

Bilek, A., McCord, A., Plachinski, S., Radloff, G., Starke, J.A., 2011. German biogas experience and lessons for wisconsin. Biocycle 52, 48–51.

Blanco-Canqui, H., Gilley, J.E., Eisenhauer, D.E., Jasa, P.J., Boldt, A., 2014. Soil carbon accumulation under switchgrass barriers. Agron. J. 106, 2185–2192.

Blankinship, J.C., Fonte, S.J., Six, J., Schimel, J.P., 2016. Plant versus microbial controls on soil aggregate stability in a seasonally dry ecosystem. Geoderma 272, 39–50.

Brandes, E., McNunn, G., Schulte, L., Bonner, I., Muth, D., Babcock, B., et al., 2016. Subfield profitability analysis reveals an economic case for cropland diversification. Environ. Res. Lett. 11, 014009.

Cates, A.M., Ruark, M.D., Hedtcke, J.L., Posner, J.L., 2016. Long-term tillage, rotation and perennialization effects on particulate and aggregate soil organic matter. Soil Tillage Res. 155, 371–380.

Cattaneo, F., Di Gennaro, P., Barbanti, L., Giovannini, C., Labra, M., Moreno, B., et al., 2014. Perennial energy cropping systems affect soil enzyme activities and bacterial community structure in a South European agricultural area. Appl. Soil Ecol. 84, 213–222.

Correa, A., Cruz, C., Ferrol, N., 2015. Nitrogen and carbon/nitrogen dynamics in arbuscular mycorrhiza: the great unknown. Mycorrhiza 25, 499–515.

Craine, J.M., Jackson, R.D., 2009. Plant nitrogen and phosphorus limitation in 98 North American grassland soils. Plant Soil 334, 73–84.

Culliney, T., 2013. Role of arthropods in maintaining soil fertility. Agriculture 3, 629–659.

Dale, B.E., Ong, R.G., 2014. Design, implementation, and evaluation of sustainable bioenergy production systems. Biofuels Bioprod. Biorefin.-Biofpr. 8, 487–503.

Dale, B.E., Anderson, J.E., Brown, R.C., Csonka, S., Dale, V.H., Herwick, G., et al., 2014. Take a closer look: biofuels can support environmental, economic and social goals. Environ. Sci. Technol. 48, 7200–7203.

de Vries, F.T., Bardgett, R.D., 2012. Plant–microbial linkages and ecosystem nitrogen retention: lessons for sustainable agriculture. Front. Ecol. Environ. 10, 425–432.

Dietzel, R., Jarchow, M.E., Liebman, M., 2015. Above- and belowground growth, biomass, and nitrogen use in maize and reconstructed prairie cropping systems. Crop. Sci. 55, 910–923.

Dodd, R.J., Sharpley, A.N., 2015. Conservation practice effectiveness and adoption: unintended consequences and implications for sustainable phosphorus management. Nutr. Cycl. Agroecosyst. 104, 373–392.

Dominati, E., Patterson, M., Mackay, A., 2010. A framework for classifying and quantifying the natural capital and ecosystem services of soils. Ecol. Econ. 69, 1858–1868.

Duran, B.E., Duncan, D.S., Oates, L.G., Kucharik, C.J., Jackson, R.D., 2016. Nitrogen fertilization effects on productivity and nitrogen loss in three grass-based perennial bioenergy cropping systems. PLoS ONE 11, e0151919.

Epie, K.E., Saikkonen, L., Santanen, A., Jaakkola, S., Makela, P., Simojoki, A., et al., 2015. Nitrous oxide emissions from perennial grass-legume intercrop for bioenergy use. Nutr. Cycl. Agroecosyst. 101, 211–222.

Eranki, P.L., Manowitz, D.H., Bals, B.D., Izaurralde, R.C., Kim, S., Dale, B.E., 2013. The watershed-scale optimized and rearranged landscape design (WORLD) model and local biomass processing depots for sustainable biofuel production: Integrated life cycle assessments. Biofuels Bioprod. Biorefin.-Biofpr. 7, 537–550.

Fargione, J.E., Cooper, T.R., Flaspohler, D.J., Hill, J., Lehman, C., McCoy, T., et al., 2009. Bioenergy and wildlife: threats and opportunities for grassland conservation. Bioscience 59, 767–777.

Feng, H., Babcock, B.A., 2010. Impacts of ethanol on planted acreage in market equilibrium. Am. J. Agric. Econ. 92, 789–802.

Ferchaud, F., Vitte, G., Bornet, F., Strullu, L., Mary, B., 2015. Soil water uptake and root distribution of different perennial and annual bioenergy crops. Plant Soil 388, 307–322.

Fernando, A.L., Boleo, S., Barbosa, B., Costa, J., Duarte, M.P., Monti, A., 2015. Perennial grass production opportunities on marginal mediterranean land. BioEnergy Res. 8, 1523–1537.

Fornara, D.A., Tilman, D., 2008. Plant functional composition influences rates of soil carbon and nitrogen accumulation. J. Ecol. 96, 314–322.

Gauder, M., Billen, N., Zikeli, S., Laub, M., Graeff-Honninger, S., Claupein, W., 2016. Soil carbon stocks in different bioenergy cropping systems including subsoil. Soil Tillage Res. 155, 308–317.

Gelfand, I., Zenone, T., Jasrotia, P., Chen, J., Hamilton, S.K., Robertson, G.P., 2011. Carbon debt of Conservation Reserve Program (CRP) grasslands converted to bioenergy production. Proc. Natl. Acad. Sci. 108, 13864–13869.

Gelfand, I., Sahajpal, R., Zhang, X., Izaurralde, R.C., Gross, K.L., Robertson, G.P., 2013. Sustainable bioenergy production from marginal lands in the US Midwest. Nature 493, 514–517.

Gissén, C., Prade, T., Kreuger, E., Nges, I.A., Rosenqvist, H., Svensson, S.-E., et al., 2014. Comparing energy crops for biogas production – yields, energy input and costs in cultivation using digestate and mineral fertilisation. Biomass Bioenergy 64, 199–210.

Gopalakrishnan, G., Negri, M.C., Wang, M., Wu, M., Snyder, S.W., LaFreniere, L., 2009. Biofuels, land, and water: a systems approach to sustainability. Environ. Sci. Technol. 43, 6094–6100.

Gutierrez-Lopez, J., Asbjomsen, H., Helmers, M., Isenhart, T., 2014. Regulation of soil moisture dynamics in agricultural fields using strips of native prairie vegetation. Geoderma 226, 238–249.

Hamilton, S.K., Hussain, M.Z., Bhardwaj, A.K., Basso, B., Robertson, G.P., 2015. Comparative water use by maize, perennial crops, restored prairie, and poplar trees in the US Midwest. Environ. Res. Lett. 10

Harris, Z.M., Spake, R., Taylor, G., 2015. Land use change to bioenergy: a meta-analysis of soil carbon and GHG emissions. Biomass Bioenergy 82, 27–39.

Heaton, E.A., Schulte, L.A., Berti, M., Langeveld, H., Walter, Z.L., Parrish, D., et al., 2013. Managing a second-generation crop portfolio through sustainable intensification: examples from the USA and the EU. Biofuels Bioprod. Biorefin. 7, 702–714.

Helmers, M.J., Eisenhauer, D.E., Dosskey, M.G., Franti, T.G., Brothers, J.M., McCullough, M.C., 2005. Flow pathways and sediment trapping in a field-scale vegetative filter. Trans. ASAE 48, 955–968.
Helmers, M.J., Zhou, X.B., Asbjornsen, H., Kolka, R., Tomer, M.D., Cruse, R.M., 2012. Sediment removal by prairie filter strips in row-cropped ephemeral watersheds. J. Environ. Qual. 41, 1531–1539.
Hernandez-Santana, V., Zhou, X., Helmers, M.J., Asbjornsen, H., Kolka, R., Tomer, M., 2013. Native prairie filter strips reduce runoff from hillslopes under annual row-crop systems in Iowa, USA. J. Hydrol. 477, 94–103.
Herzberger, A.J., Duncan, D.S., Jackson, R.D., 2014. Bouncing back: plant-associated soil microbes respond rapidly to prairie establishment. PLoS ONE, e115775.
Hoagland, K., Ruark, M., Renz, M., Jackson, R., 2013. N rate and harvest timing effects on switchgrass biomass yield, combustion quality, and energy yield. BioEnergy Res. 6, 1012–1021.
Hoffmann, C.C., Kjaergaard, C., Uusi-Kirmppi, J., Christian, H., Hansen, B., Kronvang, B., 2009. Phosphorus retention in riparian buffers: review of their efficiency. J. Environ. Qual. 38, 1942–1955.
Holland, R.A., Eigenbrod, F., Muggeridge, A., Brown, G., Clarke, D., Taylor, G., 2015. A synthesis of the ecosystem services impact of second generation bioenergy crop production. Renew. Sustain. Energy Rev. 46, 30–40.
Hudiburg, T.W., Davis, S.C., Parton, W., Delucia, E.H., 2015. Bioenergy crop greenhouse gas mitigation potential under a range of management practices. Glob. Change Biol. Bioenergy 7, 366–374.
Jach-Smith, L.C., Jackson, R.D., 2015. Nitrogen conservation decreases with fertilizer addition in two perennial grass cropping systems for bioenergy. Agric. Ecosyst. Environ. 204, 62–71.
James, L.K., Swinton, S.M., Thelen, K.D., 2010. Profitability analysis of cellulosic energy crops compared with corn. Agron. J. 102, 675–687.
Jarvie, H.P., Sharpley, A.N., Spears, B., Buda, A.R., May, L., Kleinman, P.J., 2013. Water quality remediation faces unprecedented challenges from "legacy phosphorus". Environ. Sci. Technol. 47, 8997–8998.
Jiang, Y., Swinton, S.M., 2009. Market interactions, farmers' choices, and the sustainability of growing advanced biofuels: a missing perspective? Int. J. Sustain. Dev. World Ecol. 16, 438–450.
Jordan, N., Warner, K.D., 2010. Enhancing the multifunctionality of US agriculture. Bioscience 60, 60–66.
Jordan, N., Boody, G., Broussard, W., Glover, J.D., Keeney, D., McCown, B.H., et al., 2007. Environment - sustainable development of the agricultural bio-economy. Science 316, 1570–1571.
Jordan, N.R., Slotterback, C.S., Cadieux, K.V., Mulla, D.J., Pitt, D.G., Olabisi, L.S., et al., 2011. TMDL implementation in agricultural landscapes: a communicative and systemic approach. Environ. Manage. 48, 1–12.
Jordan, N., Schulte, L.A., Williams, C., Mulla, D.J., Pitt, D.G., Schively Slotterback, C., et al., 2013. Landlabs: an integrated approach to creating agricultural enterprises that meet the triple bottom line. J. High. Educ. Outreach Engagem. 17, 175–200.
Kong, A.Y.Y., Six, J., Bryant, D.C., Denison, R.F., van Kessel, C., 2005. The relationship between carbon input, aggregation, and soil organic carbon stabilization in sustainable cropping systems. Soil Sci. Soc. Am. J. 69, 1078–1085.
Lavelle, P., Decaëns, T., Aubert, M., Barot, S., Blouin, M., Bureau, F., et al., 2006. Soil invertebrates and ecosystem services. Eur. J. Soil Biol. 42, S3–S15.
Liebman, M., Helmers, M.J., Schulte, L.A., Chase, C.A., 2013. Using biodiversity to link agricultural productivity with environmental quality: results from three field experiments in Iowa. Renew. Agric. Food Syst. 28, 115–128.
Liere, H., Kim, T.N., Werling, B.P., Meehan, T.D., Landis, D.A., Gratton, C., 2015. Trophic cascades in agricultural landscapes: indirect effects of landscape composition on crop yield. Ecol. Appl. 25, 652–661.
Meehan, T.D., Gratton, C., Diehl, E., Hunt, N.D., Mooney, D.F., Ventura, S.J., et al., 2013. Ecosystem-service tradeoffs associated with switching from annual to perennial energy crops in riparian zones of the US Midwest. PLoS ONE 8.
Miesel, J.R., Renz, M.J., Doll, J.E., Jackson, R.D., 2012. Effectiveness of weed management methods in establishment of switchgrass and a native species mixture for biofuels in Wisconsin. Biomass Bioenergy 36, 121–131.

Oates, L.G., Duncan, D.S., Gelfand, I., Millar, N., Robertson, G.P., Jackson, R.D., 2015. Nitrous oxide emissions during establishment of eight alternative cellulosic bioenergy cropping systems in the North Central United States. GCB Bioenergy 8, 539–549.

Oates, L.G., Duncan, D.S., Sanford, G.R., Liang, C., Jackson, R.D., 2016. Bioenergy cropping systems that incorporate native grasses stimulate growth of plant-associated soil microbes in the absence of nitrogen fertilization. Agriculture, Ecosystems & Environment 233, 396–403.

Orr, M.-J., Gray, M.B., Applegate, B., Volenec, J.J., Brouder, S.M., Turco, R.F., 2015. Transition to second generation cellulosic biofuel production systems reveals limited negative impacts on the soil microbial community structure. Appl. Soil Ecol. 95, 62–72.

Parrish, D.J., Fike, J.H., 2005. The biology and agronomy of switchgrass for biofuels. CRC Crit. Rev. Plant Sci. 24, 423–459.

Perez-Suarez, M., Castellano, M.J., Kolka, R., Asbjornsen, H., Helmers, M., 2014. Nitrogen and carbon dynamics in prairie vegetation strips across topographical gradients in mixed Central Iowa agroecosystems. Agric. Ecosyst. Environ. 188, 1–11.

Pugesgaard, S., Schelde, K., Larsen, S.U., Laerke, P.E., Jørgensen, U., 2015. Comparing annual and perennial crops for bioenergy production - influence on nitrate leaching and energy balance. GCB Bioenergy 7, 1136–1149.

Qiu, J., Turner, M.G., 2013. Spatial interactions among ecosystem services in an urbanizing agricultural watershed. Proc. Natl. Acad. Sci. 110, 12149–12154.

Rabalais, N.N., Turner, R.E., Wiseman, W.J., 2002. Gulf of Mexico hypoxia, aka "The dead zone". Annu. Rev. Ecol. Syst. 33, 235–263.

Räty, M., Uusi-Kämppä, J., Yli-Halla, M., Rasa, K., Pietola, L., 2009. Phosphorus and nitrogen cycles in the vegetation of differently managed buffer zones. Nutr. Cycl. Agroecosyst. 86, 121–132.

Reed, T., Carpenter, S.R., 2002. Comparisons of P-yield, riparian buffer strips, and land cover insix agricultural watersheds. Ecosystems 5, 568–577.

Roberts, T.L., Johnston, A.E., 2015. Phosphorus use efficiency and management in agriculture. Resour. Conserv. Recycl. 105, 275–281.

Robertson, G.P., Swinton, S.M., 2005. Reconciling agricultural productivity and environmental integrity: a grand challenge for agriculture. Front. Ecol. Environ. 3, 38–46.

Robertson, G.P., Vitousek, P.M., 2009. Nitrogen in agriculture: balancing the cost of an essential resource. Ann. Rev. Environ. Resour. 34, 97–125.

Robertson, G.P., Dale, V.H., Doering, O.C., Hamburg, S.P., Melillo, J.M., Wander, M.M., et al., 2008. Agriculture. Sustainable biofuels redux. Science (New York, NY) 322, 49.

Robertson, G.P., Gross, K.L., Hamilton, S.K., Landis, D.A., Schmidt, T.M., Snapp, S.S., et al., 2014. Farming for ecosystem services: an ecological approach to production agriculture. Bioscience 64, 404–415.

Sanford, G., 2014. Perennial grasslands are essential for long term SOC storage in the mollisols of the North Central USA. In: Hartemink, A.E., McSweeney, K. (Eds.), Soil Carbon. Springer International, Switzerland.

Sanford, G., Oates, L., Hamilton, S., Jasrotia, P., Thelen, K., Robertson, G.P., et al., 2016. Comparative productivity of alternative cellulosic bioenergy cropping systems in the North Central USA. Agric. Ecosyst. Environ. 216, 344–355.

Sanford, G.R., Posner, J.L., Jackson, R.D., Kucharik, C.J., Hedtcke, J.L., Lin, T.-L., 2012. Soil carbon lost from Mollisols of the North Central U.S.A. with 20 years of agricultural best management practices. Agric. Ecosyst. Environ. 162, 68–76.

Sarkar, S., Miller, S.A., 2014. Water quality impacts of converting intensively-managed agricultural lands to switchgrass. Biomass Bioenergy 68, 32–43.

Schmidt, T., Fernando, A.L., Monti, A., Rettenmaier, N., 2015. Life cycle assessment of bioenergy and bio-based products from perennial grasses cultivated on marginal land in the Mediterranean region. BioEnergy Res. 8, 1548–1561.

Schulte, L.A., Liebman, M., Asbjornsen, H., Crow, T.R., 2006. Agroecosystem restoration through strategic integration of perennials. J. Soil Water Conserv. 61, 164A–169A.

Searle, S.Y., Malins, C.J., 2014. Will energy crop yields meet expectations? Biomass Bioenergy 65, 3–12.
Slotterback, C.S., Runck, B., Pitt, D., Kne, L., Jordan, N., Mulla, D., et al., 2015. Collaborative geodesign to advance multifunctional landscapes. Landsc. Urban Plan.
Smith, L.L., Allen, D.J., Barney, J.N., 2015. Yield potential and stand establishment for 20 candidate bioenergy feedstocks. Biomass Bioenergy 73, 145–154.
Smith, R.G., Gross, K.L., Robertson, G.P., 2008. Effects of crop diversity on agroecosystem function: crop yield response. Ecosystems 11, 355–366.
Song, F., Zhao, J., Swinton, S.M., 2011. Switching to perennial energy crops under uncertainty and costly reversibility. Am. J. Agric. Econ. 93, 768–783.
Soranno, P., Hubler, S., Carpenter, S., Lathrop, R., 1996. Phosphorus loads to surface waters: a simple model to account for spatial pattern of land use. Ecol. Appl. 6, 865–878.
Spiertz, J.H.J., Ewert, F., 2009. Crop production and resource use to meet the growing demand for food, feed and fuel: opportunities and constraints. NJAS - Wageningen J. Life Sci. 56, 281–300.
Swinton, S.M., Lupi, F., Robertson, G.P., Hamilton, S.K., 2007. Ecosystem services and agriculture: cultivating agricultural ecosystems for diverse benefits. Ecol. Econ. 64, 245–252.
Tayyebi, A., Meehan, T.D., Dischler, J., Radloff, G., Ferris, M., Gratton, C., 2016. SmartScape™: a web-based decision support system for assessing the tradeoffs among multiple ecosystem services under crop-change scenarios. Comput. Electron. Agric. 121, 108–121.
Thelemann, R., Johnson, G., Sheaffer, C., Banerjee, S., Cai, H.W., Wyse, D., 2010. The effect of landscape position on biomass crop yield. Agron. J. 102, 513–522.
Tilman, D., Hill, J., Lehman, C., 2006. Carbon-negative biofuels from low-input high-diversity grassland biomass. Science 314, 1598–1600.
Tilman, D., Socolow, R., Foley, J.A., Hill, J., Larson, E., Lynd, L., et al., 2009. Beneficial biofuels--the food, energy, and environment trilemma. Science 325, 270–271.
Tisdall, J., Oades, J., 1982. Organic matte and water-stable aggregates in soils. J. Soil Sci. 33, 141–163.
Tufekcioglu, A., Raich, J.W., Isenhart, T.M., Schultz, R.C., 1998. Fine root dynamics, coarse root biomass, root distribution, and soil respiration in a multispecies riparian buffer in Central Iowa, USA. Agrofor. Syst. 44, 163–174.
Tufekcioglu, A., Raich, J.W., Isenhart, T.M., Schultz, R.C., 2003. Biomass, carbon and nitrogen dynamics of multi-species riparian buffers within an agricultural watershed in Iowa, USA. Agrofor. Syst. 57, 187–198.
Turner, R.E., Rabalais, N.N., 2003. Linking landscape and water quality in the Mississippi river basin for 200 years. Bioscience 53, 563–572.
van der Heijden, M.G.A., Bardgett, R.D., van Straalen, N.M., 2008. The unseen majority: soil microbes as drivers of plant diversity and productivity in terrestrial ecosystems. Ecol. Lett. 11, 296–310.
Ventura, S., Hull, S., Jackson, R., Radloff, G., Sample, D., Walling, S., et al., 2012. Guidelines for sustainable planting and harvest of nonforest biomass in Wisconsin. J. Soil Water Conserv. 67, 17A–20A.
Walter, K., Don, A., Flessa, H., 2015. Net N_2O and CH_4 soil fluxes of annual and perennial bioenergy crops in two central German regions. Biomass Bioenergy 81, 556–567.
Webster, C.R., Flaspohler, D.J., Jackson, R.D., Meehan, T.D., Gratton, C., 2010. Diversity, productivity and landscape-level effects in North American grasslands managed for biomass production. Biofuels 1, 451–461.
Werling, B.P., Dickson, T.L., Isaacs, R., Gaines, H., Gratton, C., Gross, K.L., et al., 2014. Perennial grasslands enhance biodiversity and multiple ecosystem services in bioenergy landscapes. PNAS 111, 1652–1657.
Zhou, X., Helmers, M.J., Asbjornsen, H., Kolka, R., Tomer, M.D., Cruse, R.M., 2014. Nutrient removal by prairie filter strips in agricultural landscapes. J. Soil Water Conserv. 69, 54–64.

CHAPTER 16

Biotechnology Impacts on Soil and Environmental Services

Robert J. Kremer
University of Missouri, Columbia, MO, United States

16.1 Introduction

Soil health is defined by Doran (2002) as the capacity of a living soil to function, within natural or managed ecosystem boundaries, to sustain plant and animal productivity, maintain or enhance water and air quality, and promote plant and animal health. Optimal soil health requires a balance between soil functions for productivity, environmental quality, and plant and animal health all of which are greatly affected by management and land-use decisions. Good management practices that consider soil health must consider all functions, rather than focus on single functions, such as crop productivity (Doran, 2002). Therefore, relationships of soil microorganisms formed by intimate associations with plants and animals strongly suggest that they are major contributors to soil health.

There is a current awareness that the quality of human health based on proper nutrition is provided by "healthy foods" grown in soils under sustainable farming practices that minimize the use of synthetic fertilizers and pesticides. The perceived connection of soil condition with the quality of food has led to an interest in soil health because crops grown on biologically rich soils may lead to nutrient-dense healthy foods (Reeve et al., 2016), in contrast to the previous notion of the early 20th century that the role of most microorganisms was plant and animal growth suppression or inhibition and disease causation (Lynch, 1983).

Subsequent to the discovery and realization of the numerous benefits imparted by a majority of environmental microorganisms, many organisms underwent extensive evaluation for applications in agricultural production and resulted in several biologically based practices and products available to farmers. The biotechnological advances were promoted primarily as economic incentives for increased crop yields and improved growth efficiency in livestock by either supplementing or reducing inputs of synthetic chemicals. Consequently, only limited information is available on the direct impacts of biotechnology on soil health and environmental services. The objectives of this chapter are to provide background on impacts

Soil Health and Intensification of Agroecosystems.
DOI: http://dx.doi.org/10.1016/B978-0-12-805317-1.00016-6

of biotechnology on soil health and environmental quality and to present biotechnology-based management practices for maintaining or improving soil health and environmental quality.

16.2 Biotechnology: Agricultural and Environmental Context

Currently, the term "biotechnology" is popularly associated with genetic modification (GM); however, molecular manipulation to achieve genetic alteration of organisms, which is the essence of GM (also frequently referred to as "genetic engineering"), is only one aspect of biotechnology. The encompassing definition of biotechnology is "the integrated use of microbiology, biochemistry and engineering sciences in order to achieve the technological application of the capacities of microorganisms and cultured tissue cells" (Lynch, 1983). The origins of "traditional biotechnology" are thought to have occurred nearly 10,000 years ago, closely paralleling the development of agriculture, and involved simple fermentation of fruits and grains, bread-making, and transformation of milk into cheese and yogurt (Mosier and Ladisch, 2009). Ancient Roman farmers improved soil health by planting leguminous crops to unknowingly increase and maintain soil communities of rhizobia that ensured root nodulation and effective nitrogen (N) fixation by legumes subsequently planted in their fields (Brill, 1981). Biotechnology applications for plant and soil productivity in more recent times were moved forward considerably by Lorenz Hiltner, who developed the first patented rhizobial inoculants and biological agents for seed protection in Germany during the early 1900s (Hartmann et al., 2008). Lynch (1983) advanced the definition of "soil biotechnology" as the study and manipulation of soil microorganisms and their metabolic processes to optimize crop production. "Modern biotechnology" evolved in the latter 20th century with the advancement of molecular biology techniques that allowed intentional manipulation of genes of different organisms to generate changes in the genetic makeup of a target organism using recombinant DNA or direct injection of nucleic acid into cells, resulting in GE or transgenic organisms (Mosier and Ladisch, 2009). Both definitions of biotechnology clearly link biologically based management to soil health and accommodate the possibility of effects due to GE plants or microorganisms introduced into agricultural, grassland, and forest ecosystems on soil health by indirectly manipulating microbial communities and activities.

16.3 Soil Health and Biotechnology

Soil health continues to be an evolving concept, and despite the numerous accepted indicators of chemical and physical properties used in assessment of management impacts on soil health, a complete set of workable biological indicators remains elusive. Although significant advancements in molecular microbiology allow in-depth detection of soil microbial communities to describe biodiversity, the foundation for soil function, microbial diversity, and microbial functional groups are not used as standard soil health indicators in assessment models due to lack of sufficient databases and to difficulty in devising in-field sampling methods that maintain in situ conditions (Lehman et al., 2015). Devising an operational set of biological indicators to provide monitoring is further complicated by the complexity of soil biota and functions (Stone et al., 2016). Several soil health assessment models are available for evaluation of effects of land management on the soil resource (Stott et al., 2013; Morrow et al., 2016). One of the more comprehensive models, the soil management assessment framework (SMAF), uses responses of dynamic soil properties or indicators that affect soil function (Andrews et al., 2004). A set of indicators representing physical, chemical, and biochemical parameters is used to generate a soil quality (health) index to measure the ability of a soil to function. As a dynamic model, SMAF is subject to ongoing revision and is currently comprised of 13 indicators with sufficient databases to provide robust scoring curves necessary for model calculations (Stott et al., 2013). Indicators used in SMAF include macroaggregate stability, water-filled pore space, bulk density, pH, electrical conductivity, salinity, extractable P and K, SOC, microbial biomass C, potentially mineralizable N, and β-glucosidase activity. However, available models including SMAF are considered deficient in indicators representing important measures of soil properties such as soil loss (Morrow et al., 2016), disease suppression (van Bruggen and Semenov, 2000), and microbial diversity and biological/biochemical activity (Stott et al., 2013).

In order to obtain comprehensive information, additional biological indicators or "bioindicators" are often measured to supplement the soil health indexes generated via the various models. The bioindicators may include soil carbon mineralization, active carbon (i.e., permanganate-oxidizable C), water-extractable (soluble) C, nitrification, soil enzymes, soil microbial community structure and biodiversity components, soil fauna (i.e., earthworms), and plant disease criteria (van Bruggen and Semenov, 2000; Killham and Staddon, 2002; Stott et al., 2013; Morrow et al., 2016). Bioindicators that target key microbial groups and processes can be sensitive measurements for overall assessment of soil health. Inclusion of reference soils with similar properties aid in gauging potential changes due to different management systems such as inclusion of biotechnological-based practices (Sparling, 1997). For purposes of describing impacts of products of biotechnology used as management tools on soil health, both "standard" soil health indicators used in soil health models and additional bioindicators will be considered in the following discussion. Also, because the majority of biotechnology practices is focused on improving plant growth and yield, responses involving root growth and rhizosphere microbial function will be related to soil health impacts based on the vital role roots have in driving soil biology including microbial characteristics, building soil organic matter, influencing soil structure, and affecting other soil processes (Garbeva et al., 2004; Marschner, 2012).

16.3.1 Soil Health and Biological Amendments

Crop and soil management using products and practices based on traditional biotechnology emphasizes the role of soil biological processes yet ultimately interacts with chemical and physical properties to synchronize the functional capacity of soil to sustain plant and animal

productivity and minimize environmental disruption (Abbott and Murphy, 2007).
To emphasize the importance of soil biological processes on overall soil health, the term "soil biological fertility" has been proposed to provide more insight on how specific biological properties respond to management practices, including those based on biotechnology applications. Soil biological fertility, considered a component of soil fertility and synonymous with soil biological health, is defined as "the capacity of organisms living in soil to contribute to the nutritional requirements of plants and animals for productivity, reproduction, and quality while maintaining biological processes that contribute positively to soil physical and chemical status" (Abbott and Murphy, 2007). Hatfield and Walthall (2015) further assert that because soil biological fertility is directly related to metabolic activity primarily of the soil heterotrophic microbial community, then management practices promoting soil biological activity, including integration of biotechnological microorganisms and processes, will enhance soil health.

The number of biological products developed through traditional biotechnology have expanded greatly during the last 40 years and are marketed as a means of reducing chemical inputs in agricultural production and for improving or optimizing crop productivity, pest management, soil fertility and plant nutrition, and soil health (Calvo et al., 2014). Discussion of biological products used in management systems and possible effects on soil health is complicated by inconsistent and often contradictory terminology used to classify biotechnology-based inputs. Broad classifications of biological products and some of their components are presented in Table 16.1, however, little standardization of terminology definitions has occurred. Most types of biological products are classified under broad categories as agricultural biologicals (microbial inoculants) or soil additives (Metting, 1990; Schlaeppi and Bulgarelli, 2015); however, descriptions of many preparations are confounding. For example, biofertilizers may be comprised of living organisms able to mobilize and augment nutrients available to plants (Verma, 1993; Sayyed et al., 2012). In contrast, biofertilizers may be considered processed organic materials that rely on the indigenous soil microbial community for nutrient release, exemplified by a stabilized poultry manure product that contains nutrients to accelerate soil biological activity (Hatfield and Walthall, 2015). Biostimulants are preparations that enhance nutrient uptake and crop quality (Calvo et al., 2014), however individual product contents vary widely and may include living microbial cultures only (Vessey, 2003), extracts of microbial cultures only (Sen et al., 2015; Tandon and Dubey, 2015), or microbial extracts supplemented with nutrients (Chen et al., 2002, 2003). Plant probiotics are an emerging group of biological products that refer exclusively to living organisms applied to soil, seed, or the rhizosphere specifically for plant growth enhancement (Islam and Hossain, 2012). Although originally consisting of gut microbiome bacteria such as *Lactobacillus* and *Bifidobacterium* species to benefit gastrointestinal and overall human health (Hume, 2011; Getachew et al., 2016), probiotics developed for nonhuman environmental applications, such as "effective microorganisms" (EM), are composed of

Table 16.1: Terminology used for products developed through biotechnology processes for use in agroecosystems that affect soil health and environmental services

Product or Term	Descriptions	References
Microbial inoculants	Living cultures of selected microbial strains (single or multiple species) designed to mediate specific biological processes. Commercial inoculants currently available for free-living bacteria, rhizobia, fungi, mycorrhizae, and algae	Calvo et al. (2014)
Agricultural biologicals	Collective term for plant beneficial products (microbial inoculants, plant extracts, beneficial insects, etc.) that can be used to promote crop health and productivity	Schlaeppi and Bulgarelli (2015)
Biofertilizers	Preparations of beneficial microorganisms that may include living or latent (inactivated by nondestructive processes such as lyophilization) microbial cells in sufficient numbers with ability to mobilize and augment the availability of plant nutrients that improve plant growth and nutrition; includes groups of P, S, K, Fe solubilizing microorganisms; symbiotic and free-living N-fixing bacteria; mycorrhizae, organic matter decomposers; microbial biomass as organic fertilizer	Bhattacharyya and Jha (2012); Sayyed et al. (2012); Verma (1993); Vessey (2003)
Biostimulants	Substances including microorganisms that are applied to plants, rhizosphere, seed, soil, or other growing media intended to enhance the ability of plants to assimilate applied nutrients, and provide benefits to plant development including tolerance to abiotic stress; do not serve as plant nutrients and facilitate nutrient availability to plants by mechanisms different from fertilizers	Calvo et al. (2014)
Plant probiotics	Living microorganisms that benefit the health of plants when applied in adequate concentrations; primarily consist of PGPR; based on principle of use of live cultures as food supplements to benefit human and animal health	Hume (2011); Islam and Hossain (2012)
Plant prebiotics	Based on mammalian nutrition concept of using nondigestible food ingredients to stimulate growth and activity of beneficial bacteria resident in the colon. Applied to plant and soil health, the basis is stimulation of soil microbial processes and plant growth extracts of metabolites from fermentation cultures of selected microorganisms, which include precursors for plant growth regulator synthesis, various enzymes, chelating agents, amino acids, and organic acids but no living microorganisms	Gibson and Roberfroid (1995)
Plant bioregulators	Organic chemical- and hormone-based substances applied externally to plants to boost plant signaling leading to enhanced growth and crop yield	Srivastava et al. (2016)
Biological soil conditioners	Selected soil microorganisms for improving soil physical properties via production of polysaccharides for formation/stabilization of soil aggregates and soil structure; promote soil fertility and plant growth	Metting (1990); Metting and Rayburn (1983)
Soil additives	A variety of biological products (biofertilizers and biostimulants) applied directly to soil to promote soil fertility and plant growth	Subler et al. (1998)

(*Continued*)

Table 16.1: Terminology used for products developed through biotechnology processes for use in agroecosystems that affect soil health and environmental services (Continued)

Product or Term	Descriptions	References
Soil amendments	Broad category of organic materials for application to soil to improve soil properties, enhance soil microbial activity, promote plant growth; may be composed of processed manure, biosolids, etc., and may or may not include selected microorganisms and added nutrients	Hatfield and Walthall (2015)
Compost tea	The product of compost contained within a porous bag suspended in recirculating water within a container or commercially available "tea brewer" with the intention of maintaining aerobic conditions. The fermented end product contains living microorganisms cultured from the compost, microbial metabolites that inhibit plant pathogens and/or stimulate plant growth, and low nutrient contents	Scheuerell and Mahaffee (2002); Carpenter-Boggs (2005)
Effective microorganisms (EM)	An inoculant consisting of ≈80 species of beneficial microorganisms coexisting in the same culture comprised of lactic acid bacteria, cyanobacteria, yeasts, actinobacteria, fungi, and other heterotrophic bacteria used for enhanced decomposition of organic materials to release nutrients and synthesize plant growth-regulating substances	Higa (1998); Xu et al. (2000)
Plant growth-promoting rhizobacteria (PGPR)	A component of the overall group of plant growth-promoting bacteria, PGPR are free-living bacteria, mainly isolated from rhizoplane and rhizosphere; produce substances that promote plant growth, suppress soilborne phytopathogens; mobilize plant nutrients. Frequently are components in biofertilizers, biostimulants, and probiotics for plants	Calvo et al. (2014); Glick (2015); Islam and Hossain (2012); Sayyed et al. (2012); Vessey (2003)

beneficial environmental microorganisms combined with microorganisms originally isolated from the mammalian gut microbiome (Higa, 1998; Xu et al., 2000; Islam and Hossain, 2012). However, probiotics consisting primarily of gut microbiome bacteria applied to soil increased soil enzyme activity for dehydrogenase, β-glucosidase, and fluorescein diacetate hydrolase and soil fungal community components suggesting a possible practice for improving soil biological health in sustainable crop production systems (Rajper et al., 2016). *Prebiotics*, originally defined as nondigestible food ingredients to stimulate growth and activity of beneficial bacteria to supplement human nutrition (Gibson and Roberfroid, 1995), have been developed for soil and plant health use. Thus, products classified as biostimulants or biofertilizers that consist of only extracts of various microbial metabolites may also be considered as prebiotics based on their intended effects to directly influence plant growth or serve as precursors for the native soil microbial community to synthesize plant growth regulators (Tandon and Dubey, 2015; Sen et al., 2015) or substances for improving soil health indicators such as aggregation or soil organic matter (Sen et al., 2015). Plant bioregulators are products that contain known hormone-based substances formulated to modulate plant signaling to attract indigenous rhizosphere microorganisms for plant growth promotion, pathogen protection, stabilized soil aggregation, and sustained microbial biomass and diversity within a vigorous and dense root system (Srivastava et al., 2016). Microbial soil conditioners, developed for soil aggregation and stabilization, have received limited attention. These bioproducts are based on the ability of certain microorganisms to produce abundant mucilaginous materials including extracellular polysaccharides, glycoproteins, and proteins that bind soil particles into aggregates (Metting, 1990).

The availability of biological products, initially limited to legume inoculants consisting of symbiotic N-fixing rhizobia, expanded in the 1950s when bacterial inoculants consisting of free-living N-fixing *Azotobacter* spp. and phosphorus (P)-mineralizing and solubilizing bacteria (*Bacillus* and *Pseudomonas* species) were introduced in Russia and India, for use as biofertilizers in a wide range of crops to reduce inputs of mineral fertilizers (Brown, 1974). However, the resulting increase in plant growth was found to be due to growth-regulating and antibiotic substances produced by the introduced bacteria in the rhizosphere that stimulated root growth and suppressed root pathogens, respectively (Brown, 1974). Continued evaluation of bacteria specifically for growth-promoting properties led to the categorization of all plant-associated bacteria with beneficial effects, including rhizobia, free-living N fixers, and nutrient-mobilizing bacteria, as plant growth-promoting bacteria (Glick, 2015). Specific characterization of free-living bacteria with the ability to enhance plant growth *through colonization of roots* was first described in the late 1970s (Suslow et al., 1979; Kloepper et al., 1980, 1989). Members of this group of bacteria, collectively known as plant growth-promoting rhizobacteria (PGPR), are included in many of the biological products currently available on the market. PGPR refers to soil (rhizosphere) bacteria able to colonize root systems and promote plant growth and are present either in the rhizosphere, in the root surface, or inhabiting spaces between cortical cells (endo-rhizobacteria). Diverse mechanisms that PGPR use for promoting plant growth, discussed in depth in several reviews (Vessey, 2003; Podile and Kishore, 2006; Rani and Goel, 2012; Glick, 2015), include nonsymbiotic N fixation, P solubilization, increased iron uptake, suppression of plant pathogenic microorganisms, or regulation of different plant hormone levels, which includes developing resistance to drought and salinity stress. With successful demonstration of plant growth enhancement by application of PGPR, inoculants containing single or multiple bacterial isolates, or combined with other organisms including mycorrhizal fungi, have been developed for use on several crops (Kloepper et al., 1989; Podile and Kishore, 2006; Sayyed et al., 2012; Bashan et al., 2014; Glick, 2015) as biofertilizers, biostimulants, or plant probiotics (Table 16.1). For example, a mixture of PGPR strains combined with the arbuscular mycorrhizal fungus *Glomus intraradices* applied to tomato (*Solanum lycopersicum* L.) with 70% of recommended fertilizer rate, produced the same yield as a full fertility application without inoculants, suggest that using inoculants to reduce chemical fertilizer inputs could be a viable component of nutrient management strategies (Adesemoye et al., 2009).

The progress made in biotechnology advances over the past 100 years has firmly established numerous agricultural and environmental microorganisms representing nearly all taxonomic and functional groups as the basis for delivering biological products to agricultural, grassland, and forest ecosystems to enhance plant growth and simultaneously improve or maintain soil health. However, perplexing issues associated with the biotechnology industry include the seeming lack of standards for assuring the quality of commercial inoculants (Bashan et al., 2014) and the reliance on testimonials and anecdotal observations for supporting the performance of products while scientific documentation of the many claims is severely lacking. A limited number of studies describing effects of practices and products developed using biotechnology on soil health indicators are reported here.

One of the most studied microbial inoculants with positive impacts on soil health consists of mycorrhizae. The success in the effectiveness of currently available mycorrhizal inoculants is noteworthy as these obligate symbiotic fungi must be grown in association with a living host plant root system to provide infective propagules for formulation into a suitable and marketable product (Bagyaraj, 1992; Khaliq et al., 2010). Not only are mycorrhizae beneficial in mobilizing nutrients, contributing to available soil P, Zn, and Cu and mineralized N and in building soil microbial biomass, they are very important in aggregate formation and stabilization in soil (Smith and Read, 1997). Tisdall and Oades (1979) point out that the combined density of host plant root including mycorrhizal hyphal length rather than the proportion of roots colonized by mycorrhizae are most important in formation of stable aggregates, suggesting that mycorrhizal effectiveness and type of host plant are critical considerations in evaluating aggregate stability as a soil health indicator. Furthermore, aggregate stabilization mediated by mycorrhizae may conserve soil organic C because recent photosynthetically fixed C retrieved from host plants by the mycorrhizal symbiont is used in synthesizing the polysaccharide and glycoprotein materials needed to aggregate soil particles, while particle adhesion by saprophytic soil fungi require substrates extracted from soil organic matter, which may be in short supply in many soils (Smith and Read, 1997).

Microbial inoculants based on PGPR are primarily aimed at enhancing plant growth but may also aid in improving or maintaining soil health. The major benefits of biofertilizers and biostimulants that contain PGPR are the additions of organic N and available forms of P and K to soil because of PGPR with abilities to fix atmospheric N and solubilize inorganic forms of P and K or mineralize organic forms of P, respectively (Archana et al., 2012; Bhattacharyya and Jha, 2012). Microbial inoculants introduced into the rhizosphere are presumed to modify and enhance soil microbial diversity, an important soil biological health indicator, through effects of plant exudates, which enhance the performance of the introduced microorganism in soils (Ambrosini et al., 2016). The potential effects of microbial inoculants on the soil microbial community structure and diversity have been demonstrated in a limited number of studies with specific plant species, management system, and crop rotation (Schmidt et al., 2015). The paucity of research reported on potential effects on soil processes, functional activities, microbial characterization, and thereby on overall soil health, suggests a critical need for greater understanding into impacts of microbial inoculants in the soil environment (Ambrosini et al., 2016).

Probiotic cultures consisting of EM are used in combination with organic materials that may serve as substrates for proliferation of the EM cultures and production of metabolites during an incubation period prior to application to soil as a biofertilizer or soil additive (Li and Zhang, 2000). After application, microorganisms and their metabolites within the biofertilizer enhance crop growth and improve plant nutrient availability with similar efficiency as legume cover crop green manure or incorporated poultry manure (Yamada and Xu, 2000). A limited number of studies on composts amended with EM show increased soil microbial biomass, soil enzyme activities, and enhanced soil aggregation (Cao et al., 2000; Yamada and Xu, 2000; Means et al., 2005). In contrast some reports indicate that EM added to manure substrates had no effect on microbial diversity or on N mineralization and uptake in plant biomass (van Vliet et al., 2006). While progress has been made in developing optimum organic substrates to support promotion of beneficial EM cultures in enhancing plant growth and health, considerable effort is still required for evaluating effects of EM biofertilizers on soil health in sustainable agricultural systems across a range of different soils and cropping systems.

Prebiotic preparations of extracted metabolites of numerous selected microbial cultures comprise many biostimulants currently marketed as plant growth-promoting amendments that may be applied to seed, in-furrow to soil, or to crop foliage by spraying during one or more recommended plant growth stages (Shen, 2000). Unfortunately, little evidence for the effectiveness of these products appears in the literature. Subler et al. (1998) reported that soil application of a commercially available biostimulant containing plant growth-regulating compounds and other metabolites unidentified because of proprietary protection, stimulated decomposition and mineralization, and altered soil microbial communities and activities. Subsequent studies demonstrated the ability of the biostimulant to increase N mineralization from organic substances in the soil or from added alfalfa and wheat biomass and improved N availability for crop growth (Chen et al., 2001, 2002, 2003). It was concluded that the biostimulants evaluated can significantly affect soil microbial activity and N dynamics in the soil and under the conditions of the studies.

Microbial agents in soil conditioners evaluated for enhancing soil structure include soil microalgae and cyanobacteria, rhizosphere bacteria (rhizobacteria), and mycorrhizal fungi. Cultures of the microalga *Chlamydomonas sojao* applied to the surface of a loamy sand (mixed, mesic xeric Toripsamment) increased soil aggregate stability up to 77% relative to control soils, which was due to increased production of polysaccharides by the microalga (Metting, 1986). Cyanobacterial soil conditioners increased rice production on saline soils by reducing electrical conductivity, and improving fertility status by increasing N, P, and

S content and availability (Hashem, 2001). Multiple strains of cyanobacteria applied to semiarid clay loam alluvium increased organic C accumulation, which contributed to greater water-stable soil aggregates, decreased bulk density, and increased soil microbial activity and nutrient mineralization under pearl millet (*Pennisetum glaucum* L.) and wheat (*Triticum aestivum* L.) crops (Nisha et al., 2007). The most beneficial impacts of cyanobacterial soil conditioners on soil health indicators have been restricted to sandy, poorly structured soils low in organic matter or soils with low pH or high salinity, whereas soils in areas with ample rainfall often sustain native cyanobacterial densities that outcompete and negate efficacy of added soil conditioners (Metting, 1990). In contrast, the rhizobacterium *Pantoea agglomerans*, with high extracellular polysaccharide-producing ability, inoculated onto seeds of wheat densely colonized the developing rhizosphere after planting in a clay soil (Vertisol) leading to improved soil aggregation in the root zone that also mediated nutrient and water uptake through improved soil macroporosity (Amellal et al., 1999). Selected mixed cultures of cyanobacteria combined with amendments consisting of silicate slag and *Azolla* fern biomass increased soil microbial biomass, soil organic C, and total N in rice (*Oryza sativa* L.) fields after two seasons (Ali et al., 2014).

Compost tea (Table 16.1) is a biological product available commercially or produced on-farm that is in widespread use in organic farming and gardening but is gaining popularity in large-scale farming, viticulture, and lawn care enterprises. The microbial population and metabolic products in compost are the most significant factors contributing to the effectiveness of pathogen suppression and plant growth promotion, however, little information is available on the composition and chemistry of these factors, which hinders the full understanding of the mechanisms of the functional capacity of compost tea that is critical in developing a consistent and reliable product (Scheuerell and Mahaffee, 2002; Carpenter-Boggs, 2005). Standards of minimum contents of microbial groups including bacteria, fungi, protozoa, and nematodes required for efficacy are under consideration (Scheuerell and Mahaffee, 2002), which would be helpful in assessing compost tea for potential impact on soil health.

Very few biotechnology applications have been developed to manipulate or enhance soil fauna as a component contributing to soil health. A long-term field study in China in which EM-amended compost was applied to soil at $15\,t\,ha^{-1}$ revealed that earthworm biomass increased in the spring by 30.2% compared with chemical fertilizer, but the effect on earthworm density was not as great (Cao et al., 2000). The added microorganisms (EM) or the soil microbial biomass activity enhanced by EM may stimulate earthworm growth and development, however, the relationship between the two biological groups was complex and in need of further study. An approach to deliver a bioproduct consisting of pellets containing beneficial microorganisms combined with substrates attractive to earthworms that would then disperse beneficial microorganisms within the soil profile has been proposed and is under development (Doube et al., 1994). This "earthworm food-based soil inoculation system" provides multiple purposes including vectoring microorganisms for biocontrol of soilborne pathogens and insects; inoculation of soils with rhizobia and *Frankia* sp. for N fixation; and establishment of PGPR in crop rhizospheres.

16.3.2 Soil Health and Genetically Modified Crops

Modern biotechnology has led to the development of numerous new crop varieties with enhanced disease and insect pest resistance, greater drought and salt tolerance, and resistance to herbicides for simplified weed management (Berg, 2009). However, these efforts have traditionally focused on selection of plant phenotype characteristics while important beneficial plant–microbe interactions that impact soil health have been overlooked during varietal selection (Morrissey et al., 2004; Berg, 2009). Thus, impacts of products of modern biotechnology, primarily transgenic crops, on soil health are in drastic contrast to agricultural biological products of traditional biotechnology.

Genetically modified plants may indirectly affect the structure, function, and diversity of soil and rhizosphere microbial communities. Rhizosphere microorganisms play key roles in the soil–root environment, providing essential ecosystem services including decomposition of crop residues, mediating biogeochemical cycles within the soil food web, and maintenance of environmental quality and productivity. Transgenic crops, primarily corn (*Zea mays* L.) and cotton (*Gossypium hirsutum* L.), are engineered to incorporate the *Cry1Ab* gene from the bacterium *Bacillus thuringiensis* that codes for production of an insecticidal protein (Bt cultivars) that kills Lepidopteran pests but it is also released into soil through root exudation (Saxena et al., 1999). Glyphosate-resistant crop cultivars were developed by insertion of a transgene (*cp4*) from an *Agrobacterium* species that codes for an insensitive version of 5-enolpyruvyl-shikimate-3-phosphatase synthase, a critical enzyme for synthesis of aromatic amino acids that is blocked in glyphosate-sensitive plants (Franz et al., 1997). Herbicide-resistant crops affect rhizosphere microorganisms because glyphosate applied to crops is released through roots into the rhizosphere environment (Kremer et al., 2005; Kremer and Means, 2009). Furthermore, intact transgenic DNA is released through root exudation and may be picked up by soil microorganisms through natural transformation and resulting in intercellular persistence and amplification of foreign DNA sequences within the microbial lateral gene transfer network and may result in unforeseen environmental impacts (Levy-Booth et al., 2007).

Despite a number of reviews that suggest the transgenic cropping systems have no or little significant effect on soil and environmental biological processes (i.e., Cerdeira and Duke, 2010; Duke et al., 2012; Bennett et al., 2013), the literature is replete with reports of adverse effects of transgenic crop management on fundamental soil health indicators. Representative studies describing effects of practices and products developed using modern biotechnology on soil health indicators are reported here.

A recent review of effects of transgenic crops on the soil microbial community found a majority of the effects were due to nonspecific or indirect outcomes of the genetic transformation, or *pleiotropy*, rather than directly attributable to the presence of Bt toxins or glyphosate within the plant (Turrini et al., 2015). Kremer et al. (2005) showed that higher soluble carbohydrate and amino acid exudation from roots of transgenic soybean without glyphosate application altered the composition of rhizosphere microbial communities compared with nontransgenic cultivars. Similarly, the resistance of two Bt cotton lines to the root phytopathogen *Fusarium oxysporum* was inferior compared with the parental lines (Li et al., 2009). A survey of GM crops and trees found several plants that negatively affected the soil fungal community although traits that were expressed were not expected to affect fungi (Stefani and Hamelin, 2010). This is particularly troubling considering the important roles fungi perform in soil structure formation and organic matter processing. Thus, the potential impact on soil microbial communities during development of transgenic crop cultivars was generally overlooked, which emphasizes the need for monitoring key sensitive microbial groups governing important soil health function in future environmental impact assessments of transgenic crops (Turrini et al., 2015).

Transgenic crops modified for the Bt trait are primarily aimed at controlling specific insect pests but cause adverse effects on nontarget soil microbial communities, which may lead to overall deterioration of soil health. A greenhouse study with an "agricultural soil" amended with vegetative residues of Bt corn cultivars and incubated for 4 months showed reduction in soil respiration and mycorrhizal colonization of corn roots and altered rhizosphere bacterial community composition (Castaldini et al., 2005). Other transgenic corn cultivars engineered to produce *Cry* proteins expressing Bt depressed the initial development of mycorrhizal symbiosis under field conditions (Cheeke et al., 2012). Results from several studies have shown alterations in soil microbial community structure, biological processes including soil enzyme activities, and decomposition and nutrient cycling (Singh and Dubey, 2016). Continuous cultivation (≥4 years) of Bt cotton resulted in the persistence of Cry proteins in soil that contributed to significant reductions in soil microbial biomass carbon, microbial activities, and eight of ten soil enzyme activities (Chen et al., 2011). An interesting aspect of this study is the apparent inhibition of soil protease activity by *Cry* proteins may suppress hydrolysis of soil *Cry* proteins, thereby contributing to long-term persistence in soils (Chen et al., 2012), which could adversely affect soil health biological components in soils under continuous Bt cotton. Growth of Bt corn on tropical soils caused significant reductions in abundance of ammonia-oxidizing bacteria (nitrifying bacteria), which were consistently responsive to the extent that this microbial group was proposed as a sensitive bioindicator for assessing impacts of GE crops on soil health (Cotta et al., 2014). Abundance of protist and nematode communities in soils may be reduced in fields cultivated to Bt corn (Griffiths et al., 2005). Because results from some studies also indicate no effect of Bt cultivars on soil microbial community structure and function, it is suggested that the assessments of potential benefits and ecological and environmental risks of GM crops on a case-by-case basis may be the most appropriate approach to ensure a fair monitoring of impacts on soil health status (Singh and Dubey, 2016).

Impacts of GM crops modified for resistance to herbicides on soil health cannot be discussed without consideration of glyphosate because this herbicide is an integral component of the modern biotechnological cropping systems currently in practice. Consequently, glyphosate is the most widely used herbicide in the world with 1.35 million metric tons predicted for use in weed management on a global scale by 2017 (Global Industry Analysts, 2011), which will potentially enter the environment. Thus, impacts on soil health by glyphosate-resistant crops as deployed in the majority of modern crop production are realistically a result of combined contributions of the GM crop and glyphosate necessarily applied during the growing season. Glyphosate interacts with many factors in the environment (Duke, 2011); however, we will focus on the effect on soil biological components in this brief overview since the primary entry into soil is released from roots of both glyphosate-susceptible and -resistant plants where glyphosate may immediately contact microbial communities in the rhizosphere (Kremer et al., 2005). Glyphosate effects on microbial communities, the basis of important ecosystem services including nutrient cycling, are critical to understand because of potential reduction in the functional sustainability of soils (Zabaloy et al., 2012). Despite inconsistencies in reports of glyphosate effects on microbial community function and structure in short-term studies, the changes in microbial metabolism upon exposure to repeated applications merit longer-term field studies (Zabaloy et al., 2012).

Effects of glyphosate on the soil microbial community are mediated by multiple mechanisms of action including disruption of the electron transport system in fungi and accumulation of toxic levels of reactive oxygen species at lower than recommended rates for weed control reducing growth, cellular morphology, and overall metabolic activity of soil fungi, which could impair important functions in the soil ecosystem (Nicolas et al., 2016). The impact of glyphosate on beneficial microorganisms is illustrated by effects on the soybean-*Bradyrhizobium* symbiosis in which reductions in root nodulation and N fixation are common in both transgenic genetically engineered, glyphosate-resistant and nontransgenic cultivars based on data collected under field conditions (Zobiole et al., 2012). Glyphosate also reduced fluorescent pseudomonad bacteria, most species of which are responsible for suppressing fungal growth in the plant rhizosphere, mobilizing nutrients for plant uptake, producing plant growth-promoting substances, and potential biodegradation of synthetic chemicals including herbicides (Kremer and Means, 2009; Zobiole et al., 2012). Similarly, soil bacteria responsible for Mn transformations that have a major impact on plant nutrient availability are affected by glyphosate such that the community involved in Mn reduction (available Mn) is suppressed, whereas the components involved in Mn oxidation (immobilized or unavailable Mn) are enhanced (Kremer and Means, 2009; Zobiole et al., 2012). Glyphosate significantly decreased total microbial biomass in soybean rhizosphere soil not previously exposed to glyphosate, but did not alter the microbial community structure (Lane et al., 2012). Recent research results revealed that repeated applications to soil where glyphosate-resistant soybean and corn were grown reduced bacterial components integral to biogeochemical processes, which suggested significant changes in the nutrient status in crop rhizospheres (Newman et al., 2016).

Glyphosate reduces spore abundance and viability and root colonization of arbuscular mycorrhizal fungi in susceptible plants (Druille et al., 2013, 2015) and with transgenic crops (Powell et al., 2009). Glyphosate reduces mycorrhizal spore germination and growth of hyphal extensions from host plant roots, thereby interfering with water and nutrient uptake and transport to the plant. Zaller et al. (2014) reported that glyphosate affected interactions between essential soil organisms such as earthworms and arbuscular mycorrhizal fungi, whereby glyphosate significantly decreased root mycorrhization, soil mycorrhizae spore biomass, vesicles, and propagules.

Genetically engineered microbial strains as biofertilizers are under development but evaluation has been limited to contained systems (Viebahn et al., 2009). The majority of GM microorganisms are symbiotic and asymbiotic N-fixing bacteria. Several strains of rhizobia genetically engineered for improved N fixation, or nodulation competitiveness were poor competitors with the native soil microbial community and did not persist in soils (Hirsch, 2004). Nontarget effects of GM biofertilizers have been characterized as insignificant relative to the natural variation in indigenous or unaltered strains of similar microbial species. Knowledge on effects of GM microorganisms in the environment, survival in soil, and interactions with the natural microbial community with respect to overall soil health is very limited. Obviously, more information on potential impacts on soil microbial structure and function is necessary before GM microorganisms are developed as an option for agricultural management.

16.4 Environmental Services and Biotechnology

The modern era of agricultural production with emphasis on inputs of chemical fertilizers, synthetic pesticides, and high-yield crop varieties has resulted in soil degradation, decreased soil organic matter, decreased water quality, increased greenhouse gas emissions, depletion of stratospheric ozone, and increased water use (Khan et al., 2007; Motavalli et al., 2008; Rani and Goel, 2012). The continued excessive use of chemicals and tillage under these production conditions will likely lead to deterioration of the physical, chemical, and biological health of soil, potential decreases in crop productivity, as well as environmental degradation. More efforts are needed to optimize soil productivity in an environmentally sustainable approach for nutrient management and ecosystem protection while preserving the capacity of soil to function as a healthy system.

16.4.1 Environmental Services and Biological Amendments

The free-living N-fixing and plant growth-promoting bacteria *Azospirillum* spp. formulated in soil conditioners applied in nonagricultural or disturbed areas prone to soil erosion and dust pollution are under evaluation for soil stabilization in arid regions of Mexico (Bashan et al., 2004). A primary mechanism for improving soil aggregation and structure in these ecosystems is stabilization of soil particles and dust due to *Azospirillum* spp. promotion of root biomass of cactus plants. Biological soil conditioners containing cyanobacteria are effective in sustaining the stability and productivity of soils in desert and semiarid ecosystems by improving soil C and N contents through organic matter inputs, improving soil aggregation and reducing erosion, and improving water-holding capacity and soil aeration (Singh et al., 2011). Similar beneficial effects on fertility status and organic matter buildup by cyanobacterial amendments have been noted for saline and acidic soils where N requirement for rice grown in these areas is reduced by 25–35% (Singh et al., 2011). Mycorrhizal inoculants are extensively used in reforestation and reconstruction or restoration of forest and grassland ecosystems (Smith and Read, 1997), which impart benefits to not only the host plants but also to several soil health parameters. Other microorganisms with ability for degrading xenobiotic or hazardous chemicals are in use as biofertilizers for remediation of degraded ecosystems (Romeh and Hendawi, 2014). Selected PGPR for inducing systemic tolerance to salt and drought in plants are under development for vegetation in stressed environments that will enable increased nutrient uptake by plants, thereby reducing fertilizer use and lessening the effects of water contamination and nutrient run-off from these sites (Yang et al., 2009). Soil-applied biopesticides, which are selected microbial cultures applied in formulation for biological management of specific weeds, indirectly benefit ecosystem quality by reducing inputs of chemical herbicides, which could lead to nontarget effects on other desirable plants and organisms and on water quality (Kremer, 2005).

16.4.2 Environmental Services and Genetically Modified Cropping Systems

Indirect effects of GM cropping systems on ecosystem services include a perceived benefit of reduced soil erosion because the widespread use of glyphosate allows farmers to use no-tillage practices rather than rely on disruptive and intensive tillage, and reduce the total amount of pesticides used (Cerdeira and Duke, 2010). However, according to the Conservation Tillage Information Center (http://www.ctic.purdue.edu/), reduced tillage occurs on only 38% of cropped areas in the United States, indicating a majority of GM cropping systems are under some form of tillage. Furthermore, because of the development of glyphosate-resistant weeds, the actual amount of herbicide active ingredient applied per hectare has increased, e.g., by 25% from 1995 to 2008 in soybean alone (Beckie and Hall, 2014).

Residues of Bt corn often reach streams or bodies of water where the *Cry* toxins can be leached out of plant tissue and contact aquatic organisms. A survey of 217 streams in Indiana, United States found that 85% of the streams contained maize plant tissues and the *Cry1Ab* toxin was detected in water from 23% of the sites (Tank et al., 2009). The toxins were detected 6 months after corn harvest in nearby fields, illustrating the prolonged persistence in aquatic environments. Exposure of the freshwater crustacean *Daphnia magna* to *Cry* toxins caused high mortality, small body size, and low reproductive rates (Bohn et al., 2016). Because *Daphnia magna* is very sensitive to changes in aquatic quality, the adverse response to *Cry* toxins suggests that these compounds may interrupt the aquatic food web with further potential disruption of ecosystem functioning.

Glyphosate movement in runoff from fields planted to GE crops may impact soil and aquatic organisms that are essential components of ecosystem functions. Contardo-Jara et al. (2009) reported increased bioaccumulation of glyphosate by the blackworm *Lumbriculus variegatus* (Muller), a typical sediment-dwelling invertebrate, when feeding on decomposing plant litter and microbial biomass contaminated with glyphosate. The organic surfactants in the herbicide formulation facilitated membrane passage of glyphosate into the body of *L. variegatus*, leading to increased superoxide dismutase (SOD) and glutathione S-transferase activities, which are indicators of environmental stress. Implications of this study are that formulation components of glyphosate herbicides exhibit synergism in effects on organisms in the environment. Activation of biotransformation mechanisms requires energy, e.g., for synthesis of biotransformation enzymes that causes expenditures of energy otherwise used for normal growth and reproduction. Glyphosate interacts synergistically with parasitic organisms to cause diseases in aquatic snails and fish (Kelly et al., 2010). Banks et al. (2014) reported that glyphosate and the surfactant alkylphenol ethoxylate plus alcohol ethoxylate decreased soil bacterial components 7 weeks after application at label rates to a silt loam soil with pH 5.6. However, no effect on the microbial community was detected in a silty clay loam with pH 4.5. Changes in the microbial community due to herbicides or surfactants were minimal in this study of a single application of these chemicals, but could be indicators of potential long-term effects.

16.5 Management Practices for Soil Health and Ecosystem Services

16.5.1 Biotechnology for Restoration of Soil and Environmental Health

Management practices under development relative to exploiting microbial functions to benefit plant growth and health or improve soil health are currently focusing on improved microbial selections or using strategies that promote the desired functions of the native soil microbiome through specific plant interactions (Mitter et al., 2013). For example, using a phytoremediation approach for restoring degraded landscapes might employ fungal inocula developed from soil at the remediation site combined with native plant species to facilitate the establishment of the fungi with desired functions such as degradation of contaminants (Thijs et al., 2016). Alternatively, plants can be selected for improved quality and quantity of root exudates for improved nutrient availability, protect against stresses, and encourage proliferation of beneficial microbes; the selection by plants of beneficial microbial communities including antibiotic-producing microbes for natural suppression of soilborne fungal pathogens can substantially improve sustainability of agricultural systems and ecosystem quality (Ryan et al., 2009; Sessitsch and Mitter, 2015; Thijs et al., 2016). Supplementing biostimulants and biofertilizers with signaling compounds such as the lipo-chitooligosaccharides to attract beneficial microorganisms associated with key plant interactions would advance the effectiveness of agricultural biologicals and better assure positive effects on plant growth and soil health (Tanaka et al., 2015). More deliberate use of organic amendments with agricultural biologicals in crop production would have a greater impact on microbial functions, demonstrated by Hatfield and Walthall (2015), than mineral fertilizers because available carbon substrates are more important than inorganic nutrients for functional soil microbial communities.

16.5.2 Genetically Modified Cropping Systems

Management for soil health and ecosystem sustainability within a GM biotechnology context is nearly completely comprised of strategies devised to lessen detrimental impacts of this cropping system. Indeed, incorporation of traditional practices into these systems is resulting in improved soil and environmental health assessments. Manure applied to soils of various cropping systems, including GM corn-soybean rotations, in five diverse watersheds in the northcentral US significantly increased soil health ratings based on assessments using the SMAF (Karlen et al., 2014). Inclusion of cover crops and organic amendments in an intercropping arrangement with Bt cotton overcame previous negative effects on microbial activity, measured as soil enzymes, and increased bacterial, fungal, and actinobacterial components of the soil microbial community (Singh et al., 2013). Use of biological products to counteract detrimental effects of glyphosate in the field has been successful with application of *Azospirillum* sp. and *Pseudomonas* sp. to corn (Travaglia et al., 2015). This approach not only reduces potential plant damage by glyphosate but also degrades the herbicide to minimize its persistence in the environment. A commercial biostimulant applied to glyphosate-resistant soybean in field trials overcame detrimental effects of glyphosate treatment on soil microorganisms as suggested by increased microbial respiration and soil enzyme activities compared with no biostimulant (Means et al., 2007). Several projects currently underway are developing beneficial microbial consortia specifically to overcome traits of GE crops and/or glyphosate in the rhizosphere that are detrimental to functions necessary for plant growth and for maintaining soil health (Bakker et al., 2012; Schmidt et al., 2015; Wigley et al., 2016). Products containing the newly developed microbial consortia are expected to be available in the near future.

16.6 Conclusions

Modern biotechnology involving use of GM cropping systems has not been directly subjected to evaluations for comparing soil health with more sustainable cropping systems. Many of the studies reviewed for this chapter have examined many soil health indicators and most suggest that GM cropping systems do not promote soil health unless sustainable practices such as cover cropping or organic soil amendments are included in the management plan. The consistent negative effect of continuous planting of Bt and glyphosate-resistant crop varieties on soil fungi and mycorrhizae is particularly alarming due to the critical functions these groups of microorganisms perform relative to soil structure, soil organic matter formation, and nutrient cycling. Application of strategies similar to those for improving agricultural biological functions by manipulating microbial consortia in the rhizosphere of GM crops to either degrade *Cry* proteins and glyphosate or to enhance mechanisms within the root environment to overcome detrimental factors associated with GM crop production will aid in improving soil health under the GM cropping system.

Agricultural biological products developed by traditional biotechnology can be important tools for improving and maintaining soil health. Currently most are targeted at enhancing plant growth and protecting plants from phytopathogens. Because most products have not been rigorously evaluated, it is assumed that beneficial plant production attributes are translated to soil health due to increased root biomass that support higher microbial biomass and production of organic matter, two keys for optimum soil health. With current availability of new microbiological and molecular methods for screening soils and rhizospheres for both composition and function of microorganisms associated with plants, biological composition and accurate modes of action of future biological products should improve considerably in the future (Sessitsch et al., 2012). It is anticipated that as use of biological products becomes more accepted as a component of management, their impact on soil health could be described using a soil health assessment plan, similar to those already in place that can quantify the effects of single practices, such as the use of manure as a soil amendment (Karlen et al., 2014).

References

Abbott, L.K., Murphy, D.V., 2007. What is soil biological fertility?. In: Abbott, L.K., Murphy, D.V. (Eds.), Soil Biological Fertility: A Key to Sustainable Land use in Agriculture. Springer, Dordrecht, The Netherlands, pp. 1–15.

Adesemoye, A.O., Torbert, H.A., Kloepper, J.W., 2009. Plant growth-promoting rhizobacteria allow reduced application of chemical fertilizers. Microbial. Ecol. 58, 921–929.

Ali, M.A., Sattar, M.A., Islam, M.N., Inubushi, K., 2014. Integrated effects of organic, inorganic and biological amendments on methane emission, soil quality and rice productivity in irrigated paddy ecosystem of Bangladesh: field study of two consecutive rice growing seasons. Plant Soil 378, 239–252.

Ambrosini, A., de Souza, R., Passaglia, L.M.P., 2016. Ecological role of bacterial inoculants and their potential impact on soil microbial diversity. Plant Soil 400, 193–207.

Amellal, N., Bartoli, F., Villemin, G., Talouizte, A., Heulin, T., 1999. Effects of inoculation of EPS-producing *Pantoea agglomerans* on wheat rhizosphere aggregation. Plant Soil 211, 93–101.

Andrews, S.S., Karlen, D.L., Cambardella, C.A., 2004. The soil management assessment framework: a quantitative soil quality evaluation method. Soil Sci. Soc. Am. J. 68, 1945–1962.

Archana, G., Buch, A., Kumar, G.N., 2012. Pivotal role of organic acid secretion by rhizobacteria in plant growth promotion. In: Satyanarayana, T., Bhavdish, N.J., Prakash, A. (Eds.), Microorganisms in Sustainable Agriculture and Biotechnology. Springer, Dordrecht, The Netherlands, pp. 35–53.

Bagyaraj, D.J., 1992. Vesicular-arbuscular mycorrhizae: application in agriculture. Methods Microbiol. 24, 359–373.

Bakker, M.G., Manter, D.K., Sheflin, A.M., Weir, T.L., Vivanco, J.M., 2012. Harnessing the rhizosphere microbiome through plant breeding and agricultural management. Plant Soil 360, 1–13.

Banks, M.L., Kennedy, A.C., Kremer, R.J., Eivazi, F., 2014. Soil microbial community response to soil texture, surfactants and herbicides. Appl. Soil Ecol. 74, 12–20.

Bashan, Y., Holgun, G, de-Bashan, L.E., 2004. Azospirill*um*-plant relationships: physiological, molecular, agricultural, and environmental advances (1997–2003). Can. J. Microbiol. 50, 521–577.

Bashan, Y., de-Bashan, L.E., Prabhu, S.R., Hernandez, J.-P., 2014. Advances in plant growth-promoting bacterial inoculant technology: formulations and practical perspectives (1998–2013). Plant Soil 378, 1–33.

Beckie, H.J., Hall, L.M., 2014. Genetically-modified herbicide-resistant (GMHR) crops a two-edged sword? An Americas perspective on development and effect on weed management. Crop Prot. 66, 40–45.

Berg, G., 2009. Plant-microbe interactions promoting plant growth and health: perspectives for controlled use of microorganisms in agriculture. Appl. Microbiol. Biotechnol. 84, 11–18.

Bhattacharyya, P.N., Jha, D.K., 2012. Plant growth-promoting rhizobacteria (PGPR): emergence in agriculture. World J. Microbiol. Biotechnol. 28, 1327–1350.

Bennett, A.B., Chi-Ham, C., Barrows, G., Sexton, S., Zilberman, D., 2013. Agricultural biotechnology: economics, environment, ethics, and the future. Ann. Rev. Environ. Resour. 38, 249–279.

Bohn, T., Rover, C.M., Semenchuk, P.R., 2016. *Daphnia magna* negatively affected by chronic exposure to purified Cry-toxins. Food Chem. Toxicol. 91, 130–140.

Brill, W., 1981. Agricultural microbiology. Sci. Am. 245 (3), 198–215.

Brown, M.E., 1974. Seed and root bacterization. Annu. Rev. Phytopathol. 12, 181–197.

Cao, Z., Li, W., Sun, Q., Ma, Y., Xu, Q., 2000. Effect of microbial inoculation on soil microorganisms and earthworm communities. J. Crop Prod. 3, 275–283.

Calvo, P., Nelson, L., Kloepper, J.W., 2014. Agricultural uses of plant stimulants. Plant Soil 383, 3–41.

Carpenter-Boggs, L., 2005. Diving into compost tea. Biocycle 46, 61–62.

Castaldini, M., Turrini, A., Sbrana, C., Benedetti, A., Marchionni, M., Mocali, S., et al., 2005. Impact of Bt corn on rhizospheric and soil eubacterial communities and on beneficial mycorrhizal symbiosis in experimental microcosms. Appl. Environ. Microbiol. 71, 6719–6729.

Cerdeira, A.L., Duke, S.O., 2010. Effects of glyphosate-resistant crop cultivation on soil and water quality. GM Crops 1, 16–24.

Cheeke, T.E., Rosenstiel, T.N., Cruzan, M.B., 2012. Evidence of reduced arbuscular mycorrhizal fungal colonization in multiple lines of Bt maize. Am. J. Bot. 99, 700–707.

Chen, S.-K., Subler, S., Edwards, C.A., 2002. Effects of agricultural biostimulants on soil microbial activity and nitrogen dynamics. Appl. Soil Ecol. 19, 249–259.

Chen, S.-K., Edwards, C.A., Subler, S., 2003. The influence of two agricultural biostimulants on nitrogen transformations, microbial activity, and plant growth in soil microcosms. Soil Biol. Biochem. 35, 9–19.

Chen, Z.H., Chen, L.J., Zhang, Y.L., Wu, Z.J., 2011. Microbial properties, enzyme activities and the persistence of exogenous proteins in soil under consecutive cultivation of transgenic cottons (*Gossypium hirsutum* L.). Plant Soil Environ. 57, 67–74.

Chen, Z.H., Chen, L.J., Wu, Z.J., 2012. Relationships among persistence of *Bacillus thuringiensis* and cowpea trypsin inhibitor proteins, microbial properties and enzymatic activities in rhizosphere soil after repeated cultivation with transgenic cotton. Appl. Soil Ecol. 53, 23–30.

Contardo-Jara, V., Klingelmann, E., Wiegand, C., 2009. Bioaccumulation of glyphosate and its formulation Roundup Ultra in *Lumbriculus variegatus* and its effects on biotransformation and antioxidant enzymes. Environ. Pollut. 157, 57–63.

Cotta, S.R., Dias, A.C.F., Marriel, I.E., Andreote, F.D., Seldin, L., van Elsas, J.D., 2014. Different effects of transgenic maize and nontransgenic maize on nitrogen-transforming archaea and bacteria in tropical soils. Appl. Environ. Microbiol. 80, 6437–6445.

Doran, J.W., 2002. Soil health and global sustainability: translating science into practice. Agric. Ecosyst. Environ. 88, 119–127.

Doube, B.M., Stephens, P.M., Davoren, C.W., Ryder, M.H., 1994. Earthworms and the introduction and management of soil microorganisms. In: Pankhurst, C.E., Doube, B.M., Gupta, V.V.S.R., Grace, P.R. (Eds.), Soil Biota: Management in Sustainable Farming Systems. CSIRO, East Melbourne, Australia, pp. 32–41.
Druille, M., Cabello, M.N., García Parisi, P.A., Golluscio, R.A., Omacini, M., 2015. Glyphosate vulnerability explains changes in root-symbionts propagules viability in pampean grasslands. Agric. Ecosyst. Environ. 202, 48–55.
Druille, M., Cabello, M.N., Omacini, M., Golluscio, R.A., 2013. Glyphosate reduces spore viability and root colonization of arbuscular mycorrhizal fungi. Appl. Soil Ecol. 64, 99–103.
Duke, S.O., 2011. Glyphosate degradation in glyphosate-resistant and -susceptible crops and weeds. J. Agric. Food Chem. 59, 5835–5841.
Duke, S.O., Lydon, J., Koskinen, W.C., Moorman, T.B., Chaney, R.L., Hammerschmidt, R., 2012. Glyphosate effects on plant mineral nutrition, crop rhizosphere microbiota, and plant disease in glyphosate-resistant crops. J. Agric. Food Chem. 60, 10375–10397.
Franz, J.E., Mao, M.K., Sikorski, J.A., 1997. Glyphosate: a unique global herbicide ACS Monograph 189. American Chemical Society, Washington, DC.
Garbeva, P., van Veen, J.A., van Elsas, J.D., 2004. Microbial diversity in soil: selection of microbial populations by plant and soil type and implications for diseased suppressiveness. Annu. Rev. Phytopathol. 42, 243–270.
Getachew, A., Berhanu, A., Pandian, M., 2016. Improvement of health by probiotics: the roles on human gut and immune modulation. Br. Microbiol. Res. J. 15, 1–14.
Gibson, G.R., Roberfroid, M.B., 1995. Dietary modulation of the human colonic microbiota: introducing the concept of prebiotics. J. Nutr. 125, 1401–1412.
Glick, B.R., 2015. Beneficial Plant-Bacterial Interactions. Springer, New York, NY.
Global Industry Analysts, 2011. Global Glyphosate Market to Reach 1.35 Million Metric Tons by 2017, According to a New Report by Global Industry Analysts, Inc. (10.09.11).
Griffiths, B.S., Caul, S., Thompson, J., Birch, A.N.E., Scrimgeour, C., Anderson, M.N., et al., 2005. A comparison of soil microbial community structure, protozoa and nematodes in field plots of conventional and genetically modified maize expressing the *Bacillus thuringiensis* CryIAb toxin. Plant Soil 275, 135–146.
Hartmann, A., Rothballer, M., Schmid, M., 2008. Lorenz Hiltner, a pioneer in rhizosphere microbial ecology and soil bacteriology research. Plant Soil 312, 7–14.
Hashem, M.A., 2001. Problems and prospects of cyanobacterial biofertilizer for rice cultivation. Aust. J. Plant Physiol. 28, 881–888.
Hatfield, J.L., Walthall, C.L., 2015. Soil biological fertility: foundation for the next revolution in agriculture? Commun. Soil Sci. Plant Anal. 46, 753–762.
Higa, T., 1998. Effective microorganisms for a more sustainable agriculture, environment and society: potential and prospects. In: Parr, J.F., Hornick, S.B. (Eds.), Proceedings of the Fourth International Conference on Kyusei Nature Farming. U.S. Department of Agriculture, Washington, DC, pp. 6–7.
Hirsch, P.R., 2004. Release of transgenic bacterial inoculants – rhizobia as a case study. Plant Soil 266, 1–10.
Hume, M.E., 2011. Historic perspective: prebiotics, probiotics, and other alternatives to antibiotics. Poult. Sci. 90, 2663–2669.
Islam, M.T., Hossain, M.M., 2012. Plant probiotics in phosphorus nutrition to crops with special reference to rice. In: Maheshwari, D.K. (Ed.), Bacteria in Agrobiology: Plant Probiotics. Springer, Heidelberg, pp. 325–363.
Karlen, D.L., Stott, D.E., Cambardella, C.A., Kremer, R.J., King, K.W., McCarty, G.W., 2014. Surface soil quality in five Midwestern cropland Conservation Effects Assessment Project watersheds. J. Soil Water Conserv. 69, 393–401.
Kelly, D.W., Poulin, R., Tompkins, D.M., Townsend, C.R., 2010. Synergistic effects of glyphosate formulation and parasite infection on fish malformations and survival. J. Appl. Ecol. 47, 498–504.
Khaliq, A., Bagyarag, D.J., Alam, M., 2010. Advances in mass production technology of arbuscular mycorrhiza. In: Thangadurai, D., Busso, C.A., Hijri, M. (Eds.), Mycorrhizal Biotechnology. CRC Press, Boca Raton, FL, pp. 43–55.
Khan, S.A., Mulvaney, R.L., Ellsworth, T.R., Boast, C.W., 2007. The myth of nitrogen fertilizer for soil carbon sequestration. J. Environ. Qual. 36, 1821–1832.

Killham, K., Staddon, W.J., 2002. Bioindicators and sensors of soil health and the application of geostatistics. In: Burns, R.G., Dick, R.P. (Eds.), Enzymes in the Environment: Activity, Ecology, and Applications. Marcel Dekker, New York, NY, pp. 391–405.

Kloepper, J.W., Leong, J., Teintze, M., Schroth, M.N., 1980. Enhanced plant growth by siderophores produced by plant growth-promoting rhizobacteria. Nature 286, 885–886.

Kloepper, J.W., Lifshitz, R., Zablotowicz, R.M., 1989. Free-living bacterial inocula for enhancing crop productivity. Trends Biotechnol. 7, 39–44.

Kremer, R.J., 2005. The role of bioherbicides in weed management. Biopestic. Int. 1, 127–141.

Kremer, R.J., Means, N.E., 2009. Glyphosate and glyphosate-resistant crop interactions with rhizosphere microorganisms. Eur. J. Agron. 31, 153–161.

Kremer, R.J., Means, N.E., Kim, S.-J., 2005. Glyphosate affects soybean root exudation and rhizosphere microorganisms. Int. J. Environ. Anal. Chem. 85, 1165–1174.

Lane, M., Lorenz, N., Saxena, J., Ramsier, C., Dick, R.P., 2012. Microbial activity, community structure and potassium dynamics in rhizosphere soil of soybean plants treated with glyphosate. Pedobiologia 55, 53–159.

Lehman, R.M., Acosta-Martinez, V., Buyer, J.S., Cambardella, C.A., Collins, H.P., Ducey, T.F., et al., 2015. Soil biology for resilient, healthy soil. J. Soil Water Conserv. 70, 12A–18A.

Levy-Booth, D.J., Campbell, R.G., Gulden, R.H., Hart, M.M., Powell, J.R., Klironomos, J.N., et al., 2007. Cycling of extracellular DNA in the soil environment. Soil Biol. Biochem. 39, 2977–2991.

Li, X., Liu, B., Heia, S., Liu, D., Han, Z., Zhou, K., et al., 2009. The effect of root exudates from two transgenic insect-resistant cotton lines on the growth of *Fusarium oxysporum*. Transgenic Res. 18, 757–767.

Li, Z., Zhang, H., 2000. Application of microbial fertilizers in sustainable agriculture. J. Crop Prod. 3, 337–347.

Lynch, J.M., 1983. Soil Biotechnology: Microbiological Factors in Crop Productivity. Blackwell Scientific Publications, Oxford, UK.

Marschner, P., 2012. Rhizosphere biology. In: Marschner, P. (Ed.), Mineral Nutrition of Higher Plants. Elsevier, San Diego, CA, pp. 369–388.

Means, N.E., Starbuck, C.J., Kremer, R.J., Jett, L., 2005. Effects of a food-waste-based soil conditioner on soil properties and plant growth. Compost. Sci. Util. 13, 116–121.

Means, N.E., Kremer, R.J., Ramsier, C., 2007. Effects of glyphosate and foliar amendments on activity of microorganisms in the soybean rhizosphere. J. Environ. Sci. Health Part B 42, 125–132.

Metting Jr., F.B., 1986. Population dynamics of *Chlamydomonas sajao* and its influence on soil aggregate stabilization in the field. Appl. Environ. Microbiol. 51, 1161–1164.

Metting Jr., F.B., 1990. Microalgae applications in agriculture. Dev. Ind. Microbiol. 31, 265–270.

Metting Jr., F.B., Rayburn, W.R., 1983. The influence of a microalgal conditioner on selected Washington soils: an empirical study. Soil Sci. Soc. Am. J. 47, 682–685.

Mitter, B., Brader, G., Afzal, M., Compant, S., Naveed, M., Trognitz, F., et al., 2013. Advances in elucidating beneficial interactions between plants, soil, and bacteria. Adv. Agron. 121, 381–445.

Morrissey, J.P., Dow, J.M., Mark, G.L., O'Gara, F., 2004. Are microbes at the root of a solution to world food production? EMBO Rep. 5, 922–926.

Morrow, J.G., Huggins, D.R., Carpenter-Boggs, L.A., Reganold, J.P., 2016. Evaluating measures to assess soil health in long-term agroecosystem trials. Soil Sci. Soc. Am. J. 80, 450–462.

Mosier, N.S., Ladisch, M.R., 2009. Modern Biotechnology: Connecting Innovations in Microbiology and Biochemistry to Engineering Fundamentals. John Wiley, Hoboken, NJ.

Motavalli, P.P., Goyne, K.W., Udawatta, R.P., 2008. Environmental impacts of enhanced-efficiency nitrogen fertilizers. Crop Manage. http://dx.doi.org/10.1094/CM-2008-0730-02-RV.

Newman, M.M., Hoilett, N.O., Lorenz, N., Dick, R.P., Liles, M.R., Ramsier, C., et al., 2016. Glyphosate effects on soil rhizosphere-associated bacterial communities. Sci. Total Environ. 54, 155–160.

Nicolas, V., Oestreicher, N., Vélot, C., 2016. Multiple effects of a commercial Roundup® formulation on the soil filamentous fungus *Aspergillus nidulans* at low doses: evidence of an unexpected impact on energetic metabolism. Environ. Sci. Pollut. Res. 23, 14393–14404.

Nisha, R., Kaushik, A., Kaushik, C.P., 2007. Effect of indigenous cyanobacterial application on structural stability and productivity of an organically poor semi-arid soil. Geoderma 138, 49–56.

Podile, A.R., Kishore, G.K., 2006. Plant growth-promoting rhizobacteria. In: Gnanamanckam, S.S. (Ed.), Plant-Associated Bacteria. Springer, Dordrecht, The Netherlands, pp. 195–230.
Powell, J.R., Levy-Booth, D.J., Gulden, R.H., et al., 2009. Effects of genetically modified, herbicide-tolerant crops and their management on soil food web properties and crop litter decomposition. J. Appl. Ecol. 46, 388–396.
Rajper, A.M., Udawatta, R.P., Kremer, R.J., Lin, C.-H., Jose, S., 2016. Effects of probiotics on soil microbial activity under cover crops in field and greenhouse studies. Agrofor. Syst. 90, 811–827.
Rani, A., Goel, R., 2012. Role of PGPR under different agroclimatic conditions. In: Maheshwari, D.K. (Ed.), Bacteria in Agrobiology: Plant Probiotics. Springer, Heidelberg, pp. 169–184.
Reeve, J.R., Hoagland, L.A., Villalba, J.J., Carr, P.M., Atucha, A., Cambardella, C., et al., 2016. Organic farming, soil health, and food quality: considering possible links. Adv. Agron. 137, 319–367.
Romeh, A.A., Hendawi, M.Y., 2014. Bioremediation of certain organophosphorus pesticides by two biofertilizers, *Paenibacillus (Bacillus) polymyxa* (Prazmowski) and *Azospirillum* lipoferum (Beijerinck). J. Agric. Sci. Technol. 16, 265–276.
Ryan, P.R., Dessaux, Y., Thomashow, L.S., Weller, D.M., 2009. Rhizosphere engineering and management for sustainable agriculture. Plant Soil 321, 363–383.
Sayyed, R.Z., Reddy, M.S., Kumar, K.V., Yellareddygari, S.K.R., Deshmukh, A.M., Patel, P.R., et al., 2012. Potential of plant growth-promoting rhizobacteria for sustainable agriculture. In: Maheshwari, D.K. (Ed.), Bacteria in Agrobiology: Plant Probiotics. Springer, Heidelberg, pp. 287–313.
Saxena, D., Flores, S., Stotzky, G., 1999. Insecticidal toxin in root exudates from Bt corn. Nature 402, 480.
Scheuerell, S., Mahaffee, W., 2002. Compost tea: principles and prospects for plant disease control. Compost. Sci. Util. 10, 313–338.
Schlaeppi, K., Bulgarelli, D., 2015. The plant microbiome at work. Mol. Plant Microbe-Interact. 28, 212–217.
Schmidt, J., Messmer, M., Wilbois, K.-P., 2015. Beneficial microorganisms for soybean [*Glycine max* (L.) Merr.], with a focus on low root-zone temperatures. Plant Soil 397, 411–445.
Sen, A., Srivastava, V.K., Singh, R.K., Singh, A.P., Raha, P., Ghosh, A.K., et al., 2015. Soil and plant responses to the application of *Ascophyllum nodosum* extract to no-till wheat (*Triticum aestivum* L.). Commun. Soil Sci. Plant Anal. 46, 123–136.
Sessitsch, A., Mitter, B., 2015. 21st century agriculture: integration of plant microbiomes for improved crop production and food security. Microbial Biotechnol. 8, 32–33.
Sessitsch, A., Hardoim, P., Döring, J., Weilharter, A., Krause, A., Woyke, T., et al., 2012. Functional characteristics of an endophyte community colonizing rice roots as revealed by metagenomic analysis. Mol. Plant-Microbe Interact. 25, 28–36.
Shen, D., 2000. Beneficial microorganisms and metabolites derived from agricultural wastes in improving plant health and protection. J. Crop Prod. 3, 349–366.
Singh, A.K., Dubey, S.K., 2016. Current trends in Bt crops and their fate on associated microbial community dynamics: a review. Protoplasma 253, 663–681.
Singh, J.S., Pandey, V.C., Singh, D.P., 2011. Efficient soil microorganisms: a new dimension for sustainable agriculture and environmental development. Agric. Ecosyst. Environ. 140, 339–353.
Singh, R.J., Ahlawat, I.P.S., Singh, S., 2013. Effects of transgenic Bt cotton on soil fertility and biology under field conditions in a subtropical inceptisol. Environ. Monit. Assess. 185, 485–495.
Smith, S.E., Read, D.J., 1997. Mycorrhizal Symbiosis. Academic Press, San Diego.
Sparling, G.P., 1997. Soil microbial biomass, activity and nutrient cycling as indicators of soil health. In: Pankhurst, C., Doube, B.M., Gupta, V.V.S.R. (Eds.), Biological Indicators of Soil Health. CABI, Wallingford, UK, pp. 97–120.
Srivastava, A.K., Pasala, R., Minhas, P.S., Suprasanna, P., 2016. Plant bioregulators for sustainable agriculture: integrating redox signaling as a possible unifying mechanism. Adv. Agron. 137, 237–278.
Stefani, F.O.P., Hamelin, R.C., 2010. Current state of genetically modified plant impact on target and non-target fungi. Environ. Rev. 18, 441–475.
Stone, D., Ritz, K., Griffiths, B.G., Orgiazzi, A., Creamer, R.E., 2016. Selection of biological indicators appropriate for European soil monitoring. Appl. Soil Ecol. 97, 12–22.

Stott, D.E., Cambardella, C.A., Karlen, D.L., Harmel, R.D., 2013. A soil quality and metabolic activity assessment after fifty-seven years of agricultural management. Soil Sci. Soc. Am. J. 77, 903–913.

Subler, S., Dominguez, J., Edwards, C.A., 1998. Assessing biological activity of agricultural biostimulants: bioassays for plant growth regulators in three soil additives. Commun. Soil Sci. Plant Anal. 29, 859–866.

Suslow, T.V., Kloepper, J.W., Schroth, M.N., Burr, T.J., 1979. Beneficial bacteria enhance plant growth. California Agriculture 33 (6), 15–17.

Tanaka, K., Cho, S.-H., Lee, H., Pham, A.Q., Batek, J.M., Cui, S., et al., 2015. Effect of lipo-chitooligosaccharide on early growth of C4 grass seedlings. J. Exp. Bot. 66, 5727–5738.

Tandon, S., Dubey, A., 2015. Effects of Biozyme (*Ascophyllum nodusum*) biostimulant on growth and development of soybean [*Glycine max* (L.) Merill]. Commun. Soil Sci. Plant Anal. 46, 845–858.

Tank, J.L., Rosi-Marshall, E.J., Royer, T.V., Whiles, M.R., Griffiths, N.A., Frauendorf, T.C., et al., 2009. Occurrence of maize detritus and a transgenic protein (Cry1Ab) within the stream network of an agricultural landscape. Proc. Natl. Acad. Sci. 107, 17645–17650.

Thijs, S., Sillen, W., Rineau, F., Weyens, N., Vangronsveld, J., 2016. Towards an enhanced understanding of plant-microbe interactions to improve phytoremediation: engineering the metaorganism. Front. Microbiol. 7, 341.

Tisdall, J.M., Oades, J.M., 1979. Stabilization of soil aggregates by the root systems of ryegrass. Aust. J. Soil Res. 17, 429–441.

Travaglia, C., Masciarelli, O., Fortuna, J., Marchetti, G., Cardozo, P., Lucero, M., et al., 2015. Towards sustainable maize production: glyphosate detoxification by *Azospirillum* sp. and *Pseudomonas* sp. Crop Prot. 77, 102–109.

Turrini, A., Sbrana, C., Giovannetti, M., 2015. Belowground environmental effects of transgenic crops: a soil microbial perspective. Res. Microbiol. 166, 121–131.

van Bruggen, A.H.C., Semenov, A.M., 2000. In search of biological indicators for soil health and disease suppression. Appl. Soil Ecol. 15, 13–24.

van Vliet, P.C.J., Bloem, J., de Goede, R.G.M., 2006. Microbial diversity, nitrogen loss and grass production after addition of Effective Micro-organisms (EM) to slurry manure. Appl. Soil Ecol. 32, 188–198.

Verma, L.N., 1993. Biofertilizers in agriculture. In: Thampan, P.K. (Ed.), Organics in Soil Health and Crop Production. Peekay Tree Crops Development Foundation, Cochin, pp. 151–184.

Vessey, J.K., 2003. Plant growth promoting rhizobacteria as biofertilizers. Plant Soil 255, 571–586.

Viebahn, M., Smit, E., Glandorf, D.C.M., Wernars, K., Bakker, P.A.H.M., 2009. Effect of genetically modified bacteria on ecosystem and their potential benefits for bioremediation and biocontrol of plant diseases–a review. In: Lichtfouse, E. (Ed.), Climate Change, Intercropping, Pest Control and Beneficial Microorganisms. Springer, Dordrecht, pp. 45–69.

Wigley, P., George, C., Turner, S., 2016. Accelerated directed evolution of microbial consortia for the development of desirable plant phenotype traits. U.S. Patent 9,260,713 B2. 16.02.16.

Xu, H.-L., Wang, X., Wang, J., 2000. Modeling photosynthesis decline of excised leaves of sweet corn plants grown with organic and chemical fertilizers. J. Crop Prod. 3, 245–253.

Yamada, K., Xu, H., 2000. Properties and applications of an organic fertilizer inoculated with effective microorganisms. J. Crop Prod. 3, 255–268.

Yang, J., Kloepper, J.W., Ryu, C.-M., 2009. Rhizosphere bacteria help plants tolerate abiotic stress. Trends Plant Sci. 14, 1–4.

Zabaloy, M.C., Gómez, E., Garland, J.L., Gómez, M.A., 2012. Assessment of microbial community function and structure in soil microcosms exposed to glyphosate. Appl. Soil Ecol. 61, 333–339.

Zaller, J.G., Heigl, F., Ruess, L., Grabmaier, A., 2014. Glyphosate herbicide affects belowground interactions between earthworms and symbiotic mycorrhizal fungi in a model ecosystem. Sci. Rep. 4, 5634.

Zobiole, L.H., Kremer, R.J., Oliveira, R., Constantin, J., 2012. Glyphosate effects on photosynthesis, nutrient accumulation, and nodulation in glyphosate-resistant soybean. J. Plant Nutr. Soil Sci. 175, 319–330.

Further Reading

Al-Kaisi, M., Hanna, M., 2008. Resource Conservation Practices: Consider the Strip-Tillage Alternative (pdf) PM 1901c. Extension Publication, Iowa State University, Ames, Iowa.

Altieri, M.A., 1995. Agroecology: The Science of Sustainable Agriculture. Westview Press, Boulder.

Altieri, M.A., 2002. Agroecology: the science of natural resource management for poor farmers in marginal environments. Agric. Ecosyst. Environ. 93, 1–24.

ASABE, 2015. Soil and Water Terminology. Standard 526.4 SEP2015. American Society of Agricultural and Biological Engineers, St. Joseph, MI.

Burney, J.A., Davis, S.J., Lobell, D.B., 2010. Greenhouse gas mitigation by agricultural intensification. Proc. Natl. Acad. Sci. 107, 12052–12057.

Carpenter-Boggs, L., 2011. The science behind biodynamics. eOrganic. Available online at <http://articles.extension.org/pages/28756/the-science-behind-biodynamics>.

Carter, M.R. (Ed.), 1993. Conservation Tillage in Temperate Agroecosystems, Lewis Publishers, Boca Raton, FL.

Costanza, R., Daly, H.E., 1992. Natural capital and sustainable development. Conserv. Biol. 6, 37–46.

Costanza, R., d'Arge, R., de Groot, R., Farber, S., Grasso, M., Hannon, B., et al., 1997. The value of the world's ecosystem services and natural capital. Nature 387, 253–260.

DeLong, D.C., 1996. Defining biodiversity. Wildl. Soc. Bull. 24, 738–749.

Ekins, P., Simon, S., Deutsch, L., Folke, C., De Groot, R., 2003. A framework for the practical application of the concepts of critical natural capital and strong sustainability. Ecol. Econ. 44, 165–185.

Giller, K.E., Beare, M.H., Lavelle, P., Izac, A.-M.N., Swift, M.J., 1997. Agricultural intensification, soil biodiversity and agroecosystem function. Appl. Soil Ecol. 6, 3–16. http://dx.doi.org/10.1016/S0929-1393(96)00149-7.

International Institute for Sustainable Development (IISD), 1994. Sustainability of Canada's Agri-Food System—A Prairie Perspective. Faculty of Agricultural and Food Sciences, University of Manitoba, Canada.

Jensen, M.E., Burman, R.D., Allen, R.G. (Eds.), 1990. Evapotranspiration and Irrigation Water Requirements. ASCE Manuals and Reports on Engineering Practices No. 70, pp 3, Published by the American Society of Civil Engineers, New York, NY.

Johnson, J.M.F., Franzluebbers, A.J., Weyers, S.L., Reicosky, D.C., 2007. Agricultural opportunities to mitigate greenhouse gas emissions. Environ. Pollut. 150, 107–124. http://dx.doi.org/10.1016/j.envpol.2007.06.030.

Jury, W.A., Gardner, W.A., Gardner, W.H., 1991. Soil Physics, fifth ed.. John Wiley & Sons, INC, New York, NY.

Lal, R., 1989. Conservation tillage for sustainable agriculture. Adv. Agron. 42, 85–197.

Mosier, A.R., Syers, J.K., Freney, J.R., 2004. Agriculture and the Nitrogen Cycle. Assessing the Impacts of Fertilizer Use on Food Production and the Environment. Scope-65. Island Press, London.

Nelson, A.G., Spaner, D., 2010. Cropping systems management, soil microbial communities, and soil biological fertility. In: Lichtfouse, E. (Ed.), Genetic Engineering, Biofertilization, Soil Quality and Organic Farming, Springer, Dordrecht, pp. 217–242.

Nelson, D.W., Sommers, L.E., 1996. Total carbon, organic carbon, and organic matter. In: Sparks, D.L. (Ed.), Methods of Soil Analysis. Part 3. Chemical Methods. SSSA Book Series No. 5, SSSA and ASA, Madison, WI, pp. 961–1010.

Olson, R.K., Al-Kaisi, M.M., Lal, R., Lowery, B., 2014. Examining the paired comparison method approach for determining soil organic carbon sequestration. Soil Water Cons. J. 69 (6), 193–197. http://dx.doi.org/10.2489/jswc.69.6.193A.

Paull, J., 2011. Attending the first organic agriculture course: Rudolf Steiner's agriculture course at Koberwitz, 1924. J. Soc. Sci. 21, 64–70.

Rittmann, B.E., Hausner, M., Loffler, F., Love, N.G., Muyzer, G., Okabe, S., et al., 2006. A vista for microbial ecology and environmental biotechnology. Environ. Sci. Technol. 40, 1096–1103. http://dx.doi.org/10.1021/es062631k.

Roberts, T.L., 2008. Improving nutrient use efficiency. Turk. J. Agric. For. 32, 177–182.

Schaller, N., 1990. Mainstreaming low-input agriculture. J. Soil Water Conserv. 45, 9–12.

Sherrod, L.A., Dunn, G., Peterson, G.A., Kolberg, R.L., 2002. Inorganic carbon analysis by modified pressure-calcimeter method. Soil Sci. Soc. Am. J. 66, 299–305.

Stevenson, F.J., 1994. Humic Chemistry: Genesis, composition, reactions, second ed.. John Wiley and Sons, New York, NY.

Sylvia, D.M., Fuhrmann, J.J., Hartel, P.G., Zuberer, D.A., 2005. Principles and Applications of Soil Microbiology, second ed.. Pearson/Prentice Hall, Upper Saddle River, NJ.

Viets Jr., F.G., 1962. Fertilizers and the efficient use of water. Adv. Agron. 14, 223–264.

Wagner, S.W., Hanson, J.D., Olness, A., Voorhees, W.B., 1998. A volumetric inorganic carbon analysis system. Soil Sci. Soc. Am. J. 62, 690–693.

Wilhelm, W.W., Johnson, J.M.F., Hatfield, J.L., Voorhees, W.B., Linden, D.R., 2004. Crop and soil productivity response to corn residue removal: a literature review. Agron. J. 96, 1–17.

Wilson, E.O., 1999. The Diversity of Life. W.W. Norton & Company.

Glossary

Aggregate Arrangement of primary soil particles into a secondary group of a definite size, where particles are attached by a combination of clay (resulting from surface charge effects), humus, bacterial secretions, iron, and minerals oxides.

Agricultural Extensification Increasing production of agricultural products by expanding the land base, typically through a change in land use (Angelsen, 2010; Pfaff and Walker, 2010).

Agricultural Intensification A set of patterns of land-use change with the common feature of increased use of the same resources for agricultural production, usually as a result of a switch from intermittent to continuous cultivation of the same area of land (Giller et al., 1997).

Agricultural Sustainable Intensification Increasing production of food, feed, fiber, and fuel (bioenergy) with minimal or no negative environmental and socioeconomic consequences. Implied is balancing/optimizing among multiple competing societal demands including protecting the global environment (Garnet et al., 2013; Lal and Steward, 2015). (See Sustainable Intensification.)

Agroclimatic Indices Combinations of temperature and precipitation used to determine where crops could be grown within an agroecosystem. Current versions utilize soil water dynamics to quantify crop stress.

Agroecology The application of ecological science to the study, design and management of sustainable agroecosystems (Altieri, 2002); the science of understanding the internal functioning of an agricultural production system, viewed as an ecosystem (Altieri, 1995).

Agroecosystem A system where agriculture production, either crops or animals or a combination of both, has been introduced into an ecosystem (USDA-ERS, 2015).

Agroforestry A land use management practice integrating the production of trees and shrubs into an agricultural environment.

Agronomy The science and technology used to produce renewable resources, including food, forage, fiber, and fuel.

Alfalfa A deep-rooted leguminous perennial plant (*Medicago sativa*) widely grown for hay and forage.

Annual Gross Primary Productivity (AGPP) The amount of solar energy fixed by plants during photosynthesis (Woodwell and Whittaker, 1968) and therefore represents the integral rate of photosynthesis on an annual basis (Gilmanov et al., 2003).

Available Water The amount of water held by a soil between water field capacity and permanent wilting point. This is the water that is available for use by plants. (See Water Field Capacity and Permanent Wilting Point.)

Biodiversity Generally refers to the variety of life on Earth (DeLong, 1996). Biodiversity describes complexity and variability at different levels of biological organization, from genes to ecosystems, and the ecological and evolutionary processes that sustain it (Wilson, 1992; Torsvik and Ovreas, 2002; Nelson and Spaner, 2010).

Biodynamic Agriculture A method of farming formulated by Dr. Rudolf Steiner in the 1920s (Paull, 2011) that utilizes a holistic natural and spiritual approach to farming, characterized by the application of "biodynamic preparations," preservation of on-farm biodiversity, limited off-farm inputs and optional use of astrological calendars or other cosmic indicators (Carpenter-Boggs, 2011).

Bioenergy Renewable energy provided or produced by living organisms.

Biofertilizers Preparations of beneficial microorganisms that may include living or latent (inactivated by nondestructive processes such as lyophilization) microbial cells with ability to mobilize and augment the availability of plant nutrients that improve plant growth and nutrition; includes groups of phosphorous (P), selenium (S), potassium (K), iron (Fe) solubilizing microorganisms; symbiotic and free-living nitrogen (N)-fixing bacteria; mycorrhizae, organic matter decomposers; microbial biomass as organic fertilizers (Sayyed et al., 2012).

Biofuels The energy-releasing product (fuel) produced from bioenergy sources.

Biological Soil Conditioners Selected soil microorganisms for improving soil physical properties via production of polysaccharides for formation or stabilization of soil aggregates and soil structure; promote soil fertility and plant growth (Metting, 1990).

Bioproducts Any material (chemicals, fuels, etc.) derived from a biologically based, renewable resource.

Biostimulants Substances, often limited to microbial metabolites, that are applied to plants, rhizosphere, seed, soil, or other growing media intended to enhance the ability of plants to assimilate applied nutrients, and provide benefits to plant development including tolerance to abiotic stress; do not serve as plant nutrients but facilitate nutrient availability to plants by biological mechanisms, different from fertilizers (Calvo et al., 2014).

Biotechnology The integrated use of microbiology, biochemistry, and engineering sciences in order to achieve the technological application of the capacities of microorganisms and cultured tissue cells (Lynch, 1983).

Brassica Any of a large genus (*Brassica*) of Old World temperate-zone herbs of the mustard family with cylindrical pods.

Buffer Strip An area of land maintained in some form of permanent vegetation (grasses, shrubs, trees, etc.) to prevent the loss of soil and nutrients through erosive forces of wind or water, and provide a means to manage soil, water, and air quality.

"Campesino-a-Campesino" Methodology Meaning "Farmer-to-Farmer," is a method of extension whereby one farmer who has a solution or innovates a solution to an agricultural problem promotes or teaches that solution to another farmer, which initiates further exchange of that information to other farmers (Rosset et al., 2011).

Carbon Cycle A process mediated by soil microorganisms begins on land with primary production through photosynthetic plants that takes up inorganic carbon (C) as CO_2 and produces organic compounds (Post et al., 1990). The process of C moving between the atmosphere, organisms including primary producers that fix CO_2 into organic compounds, secondary producers and microorganisms that release CO_2 in respiration, as well as CO_2 that can be trapped in (sea) water or in other solid bicarbonate forms.

Cation or Anion Exchange Capacity The moles of positive or negative charges per unit mass of soil that relates to exchange of cations or anions occurring on the surface of colloidal fractions of inorganic or organic soil particles (primarily clay and humus) (Mengel and Kirkby, 1982; Weil and Brady, 2016).

Cereal Rye A winter-hardy small grain (*Secale cereale*) that is closely related to wheat.

Chemical Fertilizer Synthetically produced nutrients that are applied to the soil or plants to supply one or more plant nutrients essential to growth.

Clover Any of a genus (*Trifolium*) of low leguminous herbs having trifoliate leaves and flowers in dense heads.

Conservation Agriculture System An approach to managing agroecosystems for improved and sustained productivity, increased profits and food security, while preserving and enhancing the resource base and the environment (FAO, 2015; Lal, 2015).

Conservation tillage Any tillage sequence, the object of which is to minimize or reduce loss of soil and water; operationally, a tillage or tillage and planting combination which leaves a 30% or greater cover of crop residue on the soil surface (Lal, 1989; Carter, 1993; SSSA, 1997).

Consumptive Use The total amount of water taken up by vegetation for transpiration or building of plant tissue, plus the unavoidable evaporation of soil moisture, snow, and intercepted precipitation associated with vegetal growth (ASABE, 2015). (See Evapotranspiration.)

Conventional Agricultural Production Systems Any production system that includes synthetic chemicals (i.e., fertilizer, pesticides, hormones, steroids, etc.) in the production of row cropping systems for food and fiber.

Conventional Tillage Primary and secondary tillage operations normally performed in preparing a seedbed and/or cultivating for a given crop grown in a given geographical area, usually resulting in <30% cover of the crop residues remaining on the surface after completion of the tillage sequence (SSSA, 1997; Koller, 2003).

Cover Crop A cover crop is a type of plant grown primarily to suppress weeds, reduce soil erosion, improve soil fertility/quality, scavenge soil nutrients, and control diseases and pests.

Crop Rotation A sequence of different crops, growing one or more crops annually on a given land area within a growing season followed by another crop on the same land area in successive growing seasons.

Cultural Services The nonmaterial benefits that ecosystems provide to humans (e.g., spiritual enrichment or recreational opportunities) (MEA, 2005).

Dryland Agriculture Farming in the absence of irrigation (also known as rainfed agriculture); a term more commonly used in areas where evapotranspiration exceeds precipitation (SSSA, 1997).

Ecological Intensification Analysis, use, and optimization of biological regulation to improve agroecosystem productivity and ecosystem services; use of agroecology theory as a means to improve the performance of agroecosystems to meet societal needs and protect the environment (Doré et al., 2011; Bommarco et al., 2013; Peterson and Snapp, 2015).

Ecosystem All the interacting parts of the physical and biological world (Ricklefs, 1979), typically a term applied to a defined boundary under study, e.g., a mountain ecosystem, a wetland ecosystem, an agroecosystem.

Ecosystem Services The benefits provided to humans by ecosystems; generally partitioned into provisioning, regulating, cultural, and supporting services (MEA, 2005).

Evaporation The physical process by which a liquid or solid is transformed to the gaseous state, which in irrigation usually is restricted to the change of water from liquid to gas (Jensen et al., 1990).

Evapotranspiration The combined process by which water is transformed from the earth surface to the atmosphere; evaporation of liquid or solid water plus transpiration from plants (Jensen et al., 1990). The actual evapotranspiration is the amount lost from the soil and plant and potential evapotranspiration is the amount of atmospheric demands without limitations from the soil or plant. (See Consumptive Use.)

Geological Erosion Erosion of soil or other earth materials caused by natural forces, such as rain, wind, floods, etc.

Gilgai Complexes of circular soil mounds and linear ridges and depressions common to undisturbed soil prone to shrink-swell dynamics (Kishné et al., 2014).

Global Warming Potential An index that estimates the warming effectiveness of unit mass of a given greenhouse gas integrated for specified time interval compared to a unit mass of carbon dioxide. The global warming potential of CO_2 is 1.0 (https://www.ipcc.ch/ipccreports/tar/wg1/518.htm).

Grass Any of a large family (Gramineae syn. Poaceae) of monocotyledonous, mostly herbaceous plants with jointed stems, slender sheathing leaves, and flowers borne in spikelets of bracts.

Grassland Any of a number of ecosystems characterized by the predominance of grass-like species within the Order Poales including Poaceae (syn. Gramineae; grasses), Cyperaceae (sedges), and Juncaceae (rushes).

Gravitational Potential The energy per unit volume of water required to move an infinitesimal amount of pure, free water from the reference elevation to the soil water elevation (Jury et al., 1991).

Greenhouse Gas Atmospheric gases (e.g., water, carbon dioxide, methane, and nitrous oxide) which absorb and emit radiation at specific wavelengths within the spectrum of infrared radiation emitted by the Earth's surface, the atmosphere and clouds (https://www.ipcc.ch/ipccreports/tar/wg1/518.htm).

Greenhouse Gas Intensity Global warming potential per unit of agronomic yield (Mosier et al., 2006).

Hydraulic Conductivity The ability of a soil to transmit water when it is saturated or unsaturated under a hydraulic gradient. A proportionality constant indicates the movement of water through soil pores (also affected by soil structure, size of conducting pores, and total porosity) (Hillel, 1998).

Invasive (species) A biological organism (plant, animal, or pathogen) that is not native to a specific environment that may cause harm due to a tendency to spread or outcompete other organisms in that environment.

Landscape A unit of area within an ecosystem defined by its visible features (e.g., landforms, vegetation cover) and extent.

Legume Any of a large family (Leguminosae syn. Fabaceae) of dicotyledonous herbs, shrubs, and trees having fruits that are legumes or loments, bearing nodules on the roots that contain nitrogen (N)-fixing bacteria.

Low-Input Agriculture Often referred to as "low-input, sustainable agriculture," is an assemblage of management strategies or practices that result in agricultural "systems that rely less on external inputs and more on internal resources" (Ikerd, 1990; Schaller, 1990).

Matrix Potential The energy per unit volume of water required to transfer an infinitesimal quantity of water from a reference pool of soil water at the elevation of the soil to the point of interest in the soil at reference air pressure (Jury et al., 1991).

Microbe (syn. microbiota, microorganism) A microscopic singular or multicellular living organism, typically implying a bacterium, but also including fungi, protozoa, algae, and so forth.

Microbial Biomass See Soil Microbial Biomass.

Microbial Community Structure The phylogenetic makeup (identity and number) of the microorganisms present in a community (Rittmann, 2006).

Mima Mounds Circular soil hillocks found in select natural ecosystems on all continents except Antarctica (Gabet et al., 2003).

Modern Biotechnology Evolved in the latter twentieth century with the advancement of molecular biology techniques that allowed intentional manipulation of genes of different organisms to generate changes in the genetic makeup of a target organism using recombinant DNA or direct injection of nucleic acid into cells, resulting in genetically engineered or transgenic organisms (Mosier and Ladisch, 2009).

Mycorrhiza The symbiotic association between specific fungi with the fine roots of higher plants (Sylvia et al., 2005).

Natural Capital An economic term that refers to a form of stock that possesses the capacity to give rise to the flow of ecosystem goods (e.g., grain, raw materials), and ecosystem services (e.g., erosion control, biodiversity) (Costanza and Daly, 1992; Costanza et al., 1997; Elkins et al., 2003;).

Net Primary Productivity The NPP is the gross primary production (GPP), or the total amount of energy fixed by live plants, minus autotrophic respiration (R_A), and can also be estimated by measuring the dry mass of combined aboveground and belowground components of an agroecosystem during one year of growth (Woodwell and Whittaker, 1968).

Nitrogen Cycle The biological and chemical transformation of different sources of nitrogen through decomposition of organic nitrogen (N) pools that are derived from dead animals, plants, and microbial biomass and N uptake by plants, microbes, and other organisms (Prosser, 2007).

No-tillage Characterized by no soil disturbance except during planting or nutrient application with full residue or mulch cover on the soil surface.

Nutrient Cycling The release of nutrients in the soil environment through mediation of biotic and abiotic activity either from parent materials or decomposition of soil organic matter for plant and soil organism use (Trofymow et al., 1983).

Nutrient Diffusion A process driven by nutrients' concentration gradients, where the nutrient transport occurs from higher to lower concentration by random thermal motion (Brewster and Tinker, 1970).

Nutrient Mass Flow Mass flow occurs when nutrients are transported to the root system through convective flow of water from soil to the plant roots as influenced by water potential gradients and nutrient concentrations (Mengel and Kirkby, 1982).

Nutrient Use Efficiency A concept that describes nutrient uptake by crop, in relation to the amount of nutrient that is applied or available for use, and determined by one of several calculations, e.g., partial factor productivity, apparent recovery efficiency, physiological efficiency, or crop removal efficiency (Mosier et al., 2004; Roberts, 2008).

Organic Agriculture Production Systems Food production systems that avoid the use of most synthetic materials such as pesticides and antibiotics. Organic farmers and ranchers follow a defined set of standards to produce organic food and fiber (USDA).

Organic Matter Consists of various portions of living plant roots, dead forms of organic material (plants and animals) at various stages of decomposition, soil microorganisms and soil animals. (See Soil Organic Matter.)

Osmotic Potential The result of dissolved salts and nutrients (i.e., fertilizers) in the soil water to form the soil solution, which makes soil water less available to plants.

Pasture Land used for feeding of domestic livestock through the grazing of annual or perennial forage crops.

Permanent Wilting Point The amount of water held by soil so tightly that plants cannot extract the water so it is not available for transpiration. This occurs under extreme dry conditions, where the soil micropores are filled mostly with air.

Pit-and-Mound Topography Microtopographic variations in mature forests resulting from fallen trees. Depressions formed as root-mats are removed from the soil, while adjacent mounds develop from the subsequent decomposition of roots and release of formerly held soil (Beatty, 2003).

Plant Growth-Promoting Rhizobacteria (PGPR) A component of the overall group of plant growth-promoting bacteria, PGPR are free-living bacteria, mainly isolated from rhizoplane and rhizosphere; they produce substances that promote plant growth, suppress soilborne phytopathogens; mobilize plant nutrients. Frequently are components in biofertilizers, biostimulants, and probiotics for plants (Vessey, 2003; Calvo et al., 2014; Glick, 2015).

Potential Yield Maximum agricultural yield set by climatic limitations from solar radiation and temperature (Bommarco et al., 2013).

Prairie A specific type of temperate ecosystem characterized by a diverse combination of plants, typically a mixture of perennial grasses, forbs, and shrubs, and animals, and delineated by specific soil characteristics and moisture conditions (rainfall) (IISD, 1994).

Provisioning Services These include the products people acquire from ecosystems, such as food, biofuels, fiber, freshwater, or genetic resources (MEA, 2005).

Radish A member of the *Brassica* genus that has a large taproot.

Realized Yield Actual agricultural yield obtained by producers (Bommarco et al., 2013).

Regulating Services Ecosystem processes that renew basic human needs, such as clean air and water, climate stabilization, water purification, and flood regulation (MEA, 2005).

Resilient Soil Soil that is capable of adapting to imposed management practices or natural constraints/events without showing adverse impacts or reduction in productivity.

Rhizome The underground stem of a plant from which additional roots or shoots can arise.

Rhizosphere Zone of soil under immediate influence of plant roots, usually a few millimeters surrounding the roots. The kinds, numbers, and activities of microorganisms in the rhizosphere differ from that of the bulk soil (Caravaca et al., 2002; Sylvia et al., 2005).

Riparian Areas "The aquatic ecosystem and the portions of the adjacent terrestrial ecosystem that directly affects or is affected by the aquatic environment" (http://www.soil.ncsu.edu/publications/BMPs/buffers.html; Welsch, 1991).

Saturated Soil The condition of the soil when all its pores both large (*macropores*) and small (*micropores*) are filled with water.

Soil Amendments Broad category of organic materials for application to soil to improve soil properties, enhance soil microbial activity, promote plant growth; may be composed of processed manure, biosolids, etc. and may or may not include selected microorganisms and added nutrients (Hatfield and Walthall, 2015).

Soil Biological Fertility The capacity of organisms living in soil to contribute to the nutritional requirements of plants and animals for productivity, reproduction, and quality, while maintaining biological processes that contribute positively to soil physical and chemical status (Abbott and Murphy, 2007).

Soil Bulk Density Dry mass of soil per unit volume of soil (i.e., kg m^{-3}).

Soil Carbon Carbon measured in soil commonly by combustion, can include both inorganic and organic carbon moieties; synonymous with total carbon (Nelson and Sommers, 1996).

Soil Carbon Sequestration The process of transferring CO_2 from the atmosphere into the soil of a land unit through plants, plant residues, and other organic compounds, which are stored or retained in the land unit as part of the soil organic matter (humus) (Stevenson, 1994; Johnson et al., 2007; Olson et al., 2014).

Soil Chemistry The natural chemical composition of a given soil. This composition is characterized by the original parent materials of soil, which define the nutrient supply and holding capacity for organisms and plant use.

Soil Densification and/or Compaction A reduction in soil porosity and increase in bulk density and cone penetration resistance caused by human activities such as agriculture operations that involve machinery use in unsuitable conditions (i.e., wet soil conditions).

Soil Erosion The process of soil material removal mainly by water or wind, which can be accelerated by lack of plant cover on the soil surface.

Soil Health The continued capacity of soil to function as a vital living system, within ecosystem and land-use boundaries, to sustain biological productivity, maintain the quality of air and water environments, and promote plant, animal, and human health (Doran et al., 1996, 1999, 2002). (See Soil Quality.)

Soil Inorganic Carbon Carbon (C) derived from inorganic C compounds, such as calcium carbonate (Nelson and Sommers, 1996; Wagner et al., 1998; Sherrod et al., 2002).

Soil Microbial Biomass Total mass of microorganisms in a given volume or mass of soil (Sylvia et al., 2005).

Soil Organic Carbon (SOC) Carbon (C) derived from organic compounds in the soil after removing visible roots or macroresidues (Nelson and Sommers, 1996; Wilhelm et al., 2004).

Soil Organic Carbon Stock A term used to identify soil organic carbon (SOC) status (gain or loss) based on soil organic carbon concentration and bulk density, typically expressed as mass per unit area for a given depth (e.g., Mg C ha^{-1} in 100cm).

Soil Organic Matter The organic fraction of soil—excluding undecomposed plant or animal residue, synonymous with "humus." Soil organic matter is about 56% carbon (C) (https://www.soils.org/publications/soils-glossary; Stevenson, 1994). (See Organic Matter.)

Soil Organic Matter Mineralization A process mediated by soil microorganisms decomposing and transforming organic materials and organic compounds to mineral forms available for plant growth and soil microbial use. This process is influenced by factors such as oxygen, soil moisture, and temperature.

Soil Parent Material The underlying geological material that consists of bed rock or a superficial or drift deposit exposed to physical, chemical, and biological weathering processes at the surface, which works its way to deepest layers producing soil horizons that have a similar composition to parent materials (Weil and Brady, 2016).

Soil Porosity The void spaces or pores that occupy a fraction of the total soil volume. The size distribution and total porosity in a given soil are determined by soil texture among other things.

Soil Productivity The capacity of a soil to produce food and fiber, and provide an adequate environment for soil organisms.

Soil Quality The capacity of the soil to function (Karlen et al., 1997). (See Soil Health.)

Soil Structure The arrangement of individual soil particles (sand, silt, and clay) into secondary particles or groups called *aggregates* held together by organic matter and mineral oxides (Tisdall and Oades, 1982).

Soil Texture The combination of different portions of sand, silt, and clay particles, which influences soil porosity and other physical and chemical properties of soils. The percentage of these different mineral particles is influenced by parent materials and soil forming processes.

Soil Water Balance Available water in the soil profile, which is determined as the difference between precipitation, runoff, and leaching/drainage.

Soil Water Content The amount of water (on a weight or volume basis) retained in soil. Soil water content is influenced by soil texture.

Strip-Tillage This is characterized by narrow strips tilled normally in the fall as wide as 15–20 cm and 15–20cm deep with 10cm berm, where nutrients such as nitrogen (N), phosphorous (P), and potassium (K) are applied at the same time when tilled zone are prepared (Al-Kaisi and Hanna, 2008).

Supporting Services Ecosystem services needed to maintain provisioning, regulating, and cultural services. Common examples include soil building, nutrient cycling, and pollination (MEA, 2005).

Sustainability From the Latin word *sustinere* (*sus-*, from below and *tenere*, to hold), meaning to keep in existence or maintain, implying long-term support or permanence. As it pertains to agriculture, sustainability is "farming systems that are capable of maintaining their productivity and utility indefinitely. Sustainable systems must be resource-conserving, environmentally compatible, socially supportive, and commercially competitive" (Ikerd, 1990).

Sustainable Intensification Largely considered a goal rather than a defined practice that entails the need to provide food security through increased agricultural production achieved by means of evidence-based social and natural sciences, which are environmentally sound, socially acceptable, and economically viable (Garnett et al., 2013; Peterson and Snapp, 2015).

Traditional Fallow System Leaving the land either uncropped and weed-free or with only volunteer vegetation during at least one period when a crop would traditionally be grown (SSSA, 1997).

Water Field Capacity After a saturation event, field water capacity is the condition of the soil after gravitational water has stopped draining from the soil.

Water Infiltration The movement of water through the soil surface into the soil profile. This influences how much water enters the root zone and how much runs off the soil surface.

Water Potential The potential energy of water per unit volume compared to pure water that allows for determination of the ability of water to move through the soil profile.

Water Use Efficiency In agronomic terms this is the ratio of crop yield (economic yield) to water used to produce the yield, which can be expressed (yield g m^{-2} and water use in mm) in kg m^{-3} on volume basis or g kg^{-1} on mass basis (Viets, 1962).

Weed Any plant occurring in an area or location in which it is undesirable/unwanted.

Yield Gap Difference between potential and realized agricultural yields (Bommarco et al., 2013).

Yield Scaled Emission The expression of greenhouse gas emissions in relation to productivity (Venterea et al., 2011).

Index

Note: Page numbers followed by "*f*" and "*t*" refer to figures and tables, respectively.

M

N

O

P

R